Amar

D0907194

Places and Regions in Global Context

Human Geography

Places and Regions in Global Context

Human Geography

Paul L. Knox

Virginia Polytechnic Institute and State University

Sallie A. Marston

University of Arizona

PRENTICE HALL
Upper Saddle River, New Jersey 07458

Library of Congress Cataloging-in-Publication Data

Knox, Paul L.
    Places and regions in global context: human geography / Paul L. Knox,
       Sallie A. Marston
           p. cm.
    Includes index.
    ISBN 0-13-141491-7
    1. Human geography.   I. Marston, Sallie A.   II. Title
  GF41.K56 1998   97-10607
  304.2—dc21   CIP

Editor: *Daniel Kaveney*
Editor-in-Chief: *Paul F. Corey*
Assistant Vice President of Production and Manufacturing: *David W. Riccardi*
Executive Managing Editor: *Kathleen Schiaparelli*
Assistant Managing Editor: *Shari Toron*
Marketing Manager: *Leslie Cavaliere*
Marketing Assistant: *David Stack*
Manufacturing Manager: *Trudy Pisciotti*
Creative Director: *Paula Maylahn*
Art Manager: *Gus Vibal*
Art Director: *Joseph Sengotta*
Cover and Interior Design: *Amy Rosen*
Front Cover Illustration: *"Figures with Books, World," José Ortéga, Inc.,*
   Stock Illustration Source
Editorial Assistants: *Betsy Williams* and *Nancy Gross*
Associate Editor: *Wendy Rivers*
Copyediting, Text Composition, and Electronic Page Makeup: *Thompson Steele Production Services*

 © 1998 by Prentice-Hall, Inc.
Simon and Schuster/A Viacom Company
Upper Saddle River, New Jersey 07458

Printed in the United States of America

10 9 8 7 6 5 4 3 2 1

ISBN 0-13-141491-7

Prentice-Hall International (UK) Limited, *London*
Prentice-Hall of Australia Pty. Limited, *Sydney*
Prentice-Hall Canada, Inc., *Toronto*
Prentice-Hall Hispanoamericana, S.A., *Mexico*
Prentice-Hall of India Private Limited, *New Delhi*
Prentice-Hall of Japan, Inc., *Tokyo*
Simon & Schuster Asia Pte. Ltd., *Singapore*
Editora Prentice-Hall do Brasil, Ltda., *Rio de Janeiro*

Contents

Chapter 2 The Changing Global Context 51

Chapter 3 Geographies of Population 97

C h a p t e r 4   Nature, Society, and Technology   147

C h a p t e r 5 Mapping Cultural Identities   189

C h a p t e r   6   Interpreting Places and Landscapes   231

Chapter 7 The Geography of Economic Development 269

Chapter 8 Agriculture and Food Production 317

Chapter 9 The Politics of Territory and Space 357

Chapter 10 Urbanization   407

Chapter 11 City Spaces: Urban Structure   441

Chapter 12   Future Geographies   493

Preface

*A highly embroiled quarter, a network of streets that I had avoided
for years, was disentangled at a single stroke when one day a person
dear to me moved there. It was as if a searchlight set up at this
person's window dissected the area with pencils of light*

Walter Benjamin, *One-Way Street and Other Writings.*
London: New Left Books, p. 85.

Most people have an understanding of what their own lives are like and some
knowledge of their own areas. Yet, even as the countries and regions of the world
become more interconnected, most of us still know very little about the lives of
people in other societies or about the ways in which the lives of those people connect to our own.

The quotation above reminds us that in order to understand places, they
must first be made meaningful to us. This book provides an introduction to
human geography that will make places and regions meaningful. To study human
geography, to put it simply, is to study the dynamic and complex relationships
between peoples and the worlds they inhabit. Our book gives students the basic
geographical tools and concepts needed to understand the complexity of places
and regions and to appreciate the interconnections between their own lives and
those of people in different parts of the world.

Objective and Approach

The objective of the book is to introduce the study of human geography by providing not only a body of knowledge about the creation of places and regions but
also an understanding of the interdependence of places and regions in a globalizing world. The approach is aimed at establishing an intellectual foundation that
will enable a life-long and life-sustaining geographical imagination.

The book takes a fresh approach to human geography, reflecting the major
changes that have recently been impressed on global, regional, and local landscapes. These changes include the globalization of industry, the breakup of the
Soviet empire, the upwelling of ethnic regionalisms on the heels of decolonization
and new state formation, the physical restructuring of cities, the transformation
of traditional agricultural practices throughout much of the world, and the
emerging trend toward transnational political and economic organizations. The
approach used in *Places and Regions in Global Context* provides access not only
to the new ideas, concepts, and theories that address these changes but also to the
fundamentals of human geography: the principles, concepts, theoretical frameworks, and basic knowledge that are necessary to more specialized studies.

The most distinctive feature of this approach is that it employs the concept of
geographical scale and emphasizes the interdependence of both places and
processes at different scales. In overall terms, this approach is designed to provide
an understanding of global/local relationships and their outcomes. It follows that
one of the chief organizing principles is how globalization frames the social and
cultural construction of particular places and regions at various scales.

This approach has several advantages.

- It captures aspects of human geography that are among the most compelling
  in the contemporary world—the geographical bases of cultural diversity and
  their impacts on everyday life, for example.
- It encompasses the salient aspects of new emphases in academic human
  geography—the new geopolitics and its role in the social construction of
  spaces and places, for example.

- It makes for an easier marriage between topical and regional material by emphasizing how processes link them—technological innovation and the varying ways technology is adopted and modified by people in particular places, for example.

- It facilitates meaningful comparisons between places in different parts of the world—how the core-generated industrialization of agriculture shapes gender relations in households both in the core and the periphery, for example.

In short, the textbook is designed to focus on geographical processes and to provide an understanding of the interdependence among places and regions without losing sight of their individuality and uniqueness.

Several important themes are woven into each chapter, making them integral to the overall approach:

- the relationship between global processes and their local manifestations,

- the interdependence of people and places, especially the interactive relationships between core regions and peripheral regions,

- the continuing transformation of the political economy of the world system, and of nations, regions, cities, and localities,

- the social and cultural differences that are embedded in human geographies (especially the differences that relate to race, ethnicity, gender, age, and class).

Chapter Organization

The organization of the book is innovative in several ways. First, the chapters are organized in such a way that the conceptual framework—why geography matters in a globalizing world—is laid out in Chapters 1 and 2 and then deployed in thematic chapters (Chapters 3 through 11). The concluding chapter, Chapter 12, provides a coherent summary of the main points of the text by showing how future geographies may unfold, given what is known about present geographical processes and trends. Second, the conceptual framework of the book requires the inclusion of two introductory chapters rather than the usual one. The first describes the basics of a geographic perspective; the second explains the value of the globalization approach.

Third, the distinctive chapter ordering within the book follows the logic of moving from less complex to more complex systems of human social and economic organization, always highlighting the interaction between people and the world around them. The first thematic chapter (Chapter 3) focuses on human population. Its placement reflects the central importance of people in understanding geography. Chapter 4 deals with the relationship between people and the environment as it is mediated by technology. This chapter capitalizes on the growing interest in environmental problems and establishes a central theme: that all human geographical issues are about how people negotiate their environment—whether the natural or the built environment. No other introductory human geography textbook includes such a chapter.

The chapter on nature, society and technology is followed by Chapter 5 on cultural geography. The intention in positioning the cultural chapter here is to signal that culture is the primary medium through which people operate and understand their place in the world. In Chapter 6, the impact of cultural processes on the landscape is explored, together with the ways in which landscape shapes cultural processes.

In Chapter 7 the book begins the move toward more complex concepts and systems of human organization by concentrating on economic development. The focus of Chapter 8 is agriculture. The placement of agriculture after economic development reflects the overall emphasis on globalization. This chapter shows how processes of globalization and economic development have led to the industrialization of agriculture at the expense of more traditional agricultural systems and practices.

The final three thematic chapters cover political geography (Chapter 9), urbanization (Chapter 10), and city structure (Chapter 11). The division of urban geography into two chapters, rather than a more conventional single chapter, is an important indication of how globalization increasingly leads to urbanization of the world's people and places. The final chapter, on future geographies (Chapter 12), gives a sense of how a geographic perspective might be applied to the problems and opportunities to be faced in the twenty-first century.

Features

To signal the freshness of the approach, the pedagogy of the book employs a unique cartography program, two different boxed features, "Visualizing Geography" and "Geography Matters," as well as more familiar pedagogical devices such as chapter overviews and end-of-chapter exercises.

 **Cartography:** The signature projection is Buckminster Fuller's Dymaxion™ projection, which centers the globe on the Arctic Circle and arrays the continents around it. This projection helps illustrate the global theme of the book because no one continent commands a central position over and above any other. (*Source:* The word Dymaxion and the Fuller Projection Dymaxion™ Map design are trademarks of the Buckminster Fuller Institute, Santa Barbara, California, ©1938, 1967 & 1992. All rights reserved.)

 Geography Matters: Geography Matters boxes examine one of the key concepts of the chapter, providing an extended example of its meaning and implications through both visual illustration and text. The Geography Matters features demonstrate to students that the focus of human geography is upon real world problems.

 Visualizing Geography: Visualizing Geography boxes treat key concepts of the chapter by using a photographic essay. This feature helps students recognize that visual landscape contains readily accessible evidence about the impact of globalization on people and places.

Pedagogical Structure Within Chapters: Each chapter opens with a brief vignette that introduces the theme of the chapter and illustrates why a geographical approach is important. Throughout each chapter, key terms are printed in boldface as they are introduced, with capsule definitions of the term in the margin of the same page. These key terms are listed alphabetically, together with their location in the text, at the end of the chapter. Figures with extensive captions are provided to integrate illustration with text.

Each chapter ends with a Conclusion, a list of the Main Points covered by the chapter, and the list of Key Terms.

Finally, to facilitate the exploration of the resources available on the World Wide Web, Internet exercises are provided. These exercises require students to put into practice several of the key concepts of a chapter. Also included are more traditional types of exercises for those students and instructors who do not have access to electronic resources.

Supplements

The book includes a complete supplements program for both students and teachers.

For the Student

Study Guide (0-13-888710-1): The study guide includes additional learning objectives, a complete chapter outline, critical thinking exercises, problems and short essay work, and a self-test with an answer key in the back of the book.

Internet Support: The *Places and Regions in Global Context* web site gives students an opportunity to further explore topics presented in the book using the Internet. Additionally, it provides support for the Internet-based activities at the end of each chapter, links to other interesting web sites, and on-line review exercises.

Earth on the Internet: A Field Guide for Geoscience Students (0-13-779828-8) by Andrew T. Stull and Duane Griffin is a student's guide to the Internet and World Wide Web specific to geography. It is available free as a shrink-wrap with the text.

For the Instructor

Instructor's Resource Manual (0-13-236506-5): The Intructor's Resource Manual, intended as a resource for both new and experienced teachers, includes a variety of lecture outlines, additional source materials, teaching tips, advice on how to integrate visual supplements, and various other ideas for the classroom.

Test Bank (0-13-236530-8): An extensive array of test questions accompanies the book. These questions are available in hard copy and also on Windows (0-13-236548-0) and Macintosh (0-13-236563-4) disks.

Overhead Transparencies (0-13-236522-7): The transparencies feature 125 illustrations from the text, all enlarged for excellent classroom visibility. The illustrations are also available as a slide set (0-13-236514-6).

Prentice Hall-*New York Times* **Themes of the Times Supplements for Geography:** Prentice Hall–*New York Times* Themes of the Times Supplements for Geography reprints significant articles from the *New York Times*. The Geography *New York Times* article is available to students free of charge. Please contact your local Prentice Hall representative for details.

Conclusion

The idea for this book evolved from conversations between the authors and colleagues about how to teach human geography in colleges and universities. Our intent was to find a way not only to capture the exciting changes that are rewriting the world's landscapes and reorganizing the spatial relationships between people, but also demonstrate convincingly why the study of geography matters. Our aim was to show why a geographical imagination is important, how it can lead to an understanding the world and its constituent places and regions, and how it has practical relevance in many spheres of life.

The book was written at the culmination of a significant period of reform in geographic education. One important outcome of this reform was the inclusion of geography as a core subject in *Goals 2000: Educate America Act* (Public Law 103-227). Another was the publication of a set of national geography standards for K–12 education (*Geography for Life,* published by National Geographic Research and Education for the American Geographical Society, the Association of American Geographers, the National Council for Geographic Education, and the National Geographic Society, 1994). This book builds on these reforms, offering a fresh and compelling approach to college-level geography.

Acknowledgments

We are indebted to many people for their assistance, advice, and constructive criticism in the course of preparing this book. Among those who provided comments on various drafts of the book are the following professors:

Christopher A. Airriess (*Ball State University*)
Stuart Aitken (*University of California at San Diego*)
Brian J. L. Berry (*University of Texas at Dallas*)
George O. Brown, Jr. (*Boston College*)
Henry W. Bullamore (*Frostburg State University*)
Edmunds V. Bunske (*University of Delaware*)
Craig Campbell (*Youngstown State University*)
Fiona M. Davidson (*University of Arkansas*)
Vernon Domingo (*Bridgewater State College*)
Nancy Ettlinger (*The Ohio State University*)
Paul B. Frederic (*University of Maine*)
Gary Fuller (*University of Hawaii at Manoa*)
Wilbert Gesler (*University of North Carolina*)
Melissa Gilbert (*Temple University*)
Douglas Heffington (*Middle Tennessee State University*)
Peter Hugill (*Texas A&M University*)
David Icenogle (*Auburn University*)
Mary Jacob (*Mount Holyoke College*)
Douglas L. Johnson (*Clark University*)
James Kus (*California State University, Fresno*)
John C. Lowe (*George Washington University*)
Byron Miller (*University of Cincinnati*)
Roger Miller (*University of Minnesota*)
John Milbauer (*Northeastern State University*)
Don Mitchell (*Syracuse University*)
Woodrow W. Nichols, Jr. (*North Carolina Central University*)
Richard Pillsbury (*Georgia State University*)
James Proctor (*University of California, Santa Barbara*)
Jeffrey Richetto (*University of Alabama*)
Andrew Schoolmaster (*University of North Texas*)
Debra Straussfogel (*University of New Hampshire*)

Gerald R. Webster (*University of Alabama*)
Joseph S. Wood (*George Mason University*)
Wilbur Zelinsky (*Penn State University*)

In addition, Michael Wishart (*World Bank Photo Library*), Earthaline Harried (*Photography Division, U.S Department of Agriculture*), and Gray Tappan (*U.S. Geological Survey, Earth Resources Observation Systems Data Center*) were especially helpful in our photo research. Special thanks go to our development editor, Barbara Muller; to our editors at Prentice Hall, Ray Henderson and Dan Kaveney; to our production editor, Andrea Fincke; to our indexer, Penny Waterstone; and to our research assistants: Karen Barton, Barbara Morehouse, Mark Patterson, and (especially) Lynn Slosberg at the University of Arizona; Malathi Ananthakrishnan, Stephanie Brown, Jayashree Narayana, Robert Humphries, Robert Warwick, and Ian Watson at Virginia Tech.

Finally, a number of colleagues gave generously of their time and expertise in guiding our thoughts, making valuable suggestions, and providing materials: Thabit Abu-Rass (*University of Arizona*), Michael Bonine (*University of Arizona*), Martin Cadwallader (*University of Wisconsin*), Judith Carney (*University of California, Los Angeles*), Efiong Etuk (*Virginia Tech*), Antonio Luna Garcia (*Universitas de Pompeu Fabra*), Jane Goldman, George Henderson (*University of Arizona*), Miranda Joseph (*University of Arizona*), Cindi Katz (*City University of New York*), Diana Liverman (*University of Arizona*), Beth Mitchneck (*University of Arizona*), Mark Nichter (*University of Arizona*), Mimi Nichter (*University of Arizona*), Gearóid Ó Tuathail (*Virginia Tech*), Leland Pederson (*University of Arizona*), Ralph Saunders (*University of Arizona*), Neil Smith (*Rutgers University*), Dick Walker (*University of California, Berkeley*), Marv Waterstone (*University of Arizona*), Michael Watts (*University of California, Berkeley*), and Emily Young (*University of Arizona*).

About the Authors

Paul L. Knox

Paul Knox recieved his Ph.D. in Geography from the University of Sheffield, England. In 1985, after teaching in the United Kingdom for several years, he moved to the United States to take up a position as professor of urban affairs and planning at Virginia Tech. His teaching centers on urban and regional development, with an emphasis on comparative study. In 1989 he received a university award for teaching excellence. He has written several books on aspects of economic geography, social geography, and urbanization. He is co-editor of an international journal, *Environment & Planning,* serves on the editorial board of several other scientific journals, and is co-editor of a series of books on World Cities. In 1996 he was appointed to the position of University Distinguished Professor at Virginia Tech, where he currently serves as interim Dean of the College of Architecture and Urban Studies.

Sallie A. Marston

Sallie Marston received her Ph.D. in Geography from the University of Colorado, Boulder. She has been a faculty member at the University of Arizona since 1986. Her teaching focuses on the historical, social, and cultural aspects of American urbanization with particular emphasis on race, class, gender, and ethnicity issues. She received the College of Social and Behavioral Sciences Outstanding Teaching Award in 1989. She is the author of numerous journal articles and book chapters, serves on the editorial board of several scientific journals, and is the co-editor of the *Sage Urban Affairs Annual Review* series. In 1994–1995 she served as Interim Director of Women's Studies and the Southwest Institute for Research on Women and is currently an associate professor in the Department of Geography and Regional Development.

Chapter 1

Geography Matters

Part of a sixteenth-century map of the world.

A few years ago, an international Gallup survey ranked Americans aged 18 and 24 last in geographic knowledge, compared with their counterparts in other developed countries. Fewer than half of all American college graduates were able accurately to locate New York on a map of the United States, Zürich on a map of Europe, or Cairo on a map of Africa. Fewer still were able to say much about Zürich as a place, or about the ways in which events in New York, Zürich, and Cairo affect one another. In today's world, where places are increasingly interdependent, it is important to know something about human geography and to understand how places affect, and are affected by, one another.

Consider, for example, some of the prominent news stories of 1996. At first glance, they were a mixture of achievements, disputes, and disasters that might seem to have little to do with geography, apart from the international flavor of the coverage. There was civil war in Liberia; ethnic strife in the Congo and Sri Lanka; the deployment of NATO troops in Croatia and Bosnia; IRA terrorist bombs in London; Hamas terrorist bombs in Jerusalem; the explosion of French nuclear bombs in the Pacific; the near-bankruptcy of Mexico; Chinese military exercises just offshore from Taiwan; U.S. air strikes on Iranian military installations; and, in the United States, major workforce reductions by several large corporations. Off the front pages, we read of the continuing debate over global warming; the continuing diffusion of the AIDS epidemic; and the dismantling of the welfare system.

Most of these stories did, in fact, reflect important geographical dimensions. The diffusion of AIDS, for example, is a geographical as well as a social and cultural phenomenon. Thus, in the United States as in other countries, the AIDS epidemic has diffused through the country in a very distinctive geographical pattern. The dismantling of the welfare state also involves some important spatial patterns. The withdrawal of welfare policies by federal and state governments has had a much greater impact on some communities than others because of differences from place to place in social composition and economic opportunity. Behind some of the major news stories, geographical processes played a more central role. Stories about local economic development (such as workforce reductions), local territorial disputes (as between Taiwan and the People's Republic of China), the globalization of both the economy and popular culture, and global warming, for example, all involve a strong geographical element.

Human geography is about recognizing and understanding the interdependence among places and regions, without losing sight of the individuality and uniqueness of specific places. Basic tools and fundamental concepts enable geographers to study the world in this way. Geographers learn about the world by finding out where things are and why they are there. Maps and mapping, of course, play a key role in how geographers analyze and portray the world. They are also key in introducing to others geographers' ideas about the way that places and regions are made and altered.

Why Places Matter

An appreciation of the diversity and variety of peoples and places is a theme that runs through the entire span of **human geography**, the study of the spatial organization of human activity and of people's relationships with their environments. This theme is inherently interesting to nearly all of us. *National Geographic* magazine (Figure 1.1) has become a national institution by drawing on the wonder and fascination of the seemingly endless variety of landscapes and communities around the world. Almost 8 million households, representing about 30 million regular readers, subscribe to this one magazine for its intriguing descriptions and striking photographs. Millions more read it occasionally in offices, lobbies, or waiting rooms.

Yet many Americans often seem content to confine their interest in geography to the pages of glossy magazines, to television documentaries, or to vacations. It has become part of the conventional wisdom that many Americans have little real appreciation or understanding of people and places beyond their daily

Human geography: the study of the spatial organization of human activity and of people's relationships with their environments.

Figure 1.1 Geography's popularity
The *National Geographic* magazine has the third-largest subscription rate in the United States (after *TV Guide* and *Reader's Digest*). Its popularity reflects people's interest in the variety of landscapes and communities around the world (though more than half of its subscribers look only at the photographs and their captions). Part of its popularity is almost certainly a result of the care that the editors have taken to present the world as inoffensively as possible, with upbeat coverage and a simplified view of the world as a collection of separate places, each eventually to be visited through photojournalism.

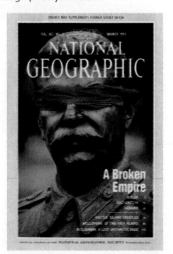

routines. This is perhaps putting it too mildly. Surveys have revealed that there is widespread ignorance among a high proportion of Americans not only of the fundamentals of the world's geography but also of the diversity and variety within the United States itself. This ignorance and apathy have been lampooned by Garry Trudeau (Figure 1.2), whose Doonesbury cartoons so often serve to capture our shortcomings.

So, although most people are fascinated by different places, relatively few have a systematic knowledge of them. Fewer still have an understanding of how different places came to be the way they are, or of why places matter in the broader scheme of things. This lack of understanding is an important issue because geographic knowledge can take us far beyond simply glimpsing the inherently interesting variety of peoples and places.

The Uniqueness of Places

Places are dynamic, with changing properties and fluid boundaries that are the product of the interplay of a wide variety of environmental and human factors. This dynamism and complexity are what make places so fascinating for readers of *National Geographic*. They are also what make places so important in shaping people's lives and in influencing the pace and direction of change. Places provide the settings for people's daily lives. It is in these settings that people learn who and what they are, and how they should think and behave.

A young, white female growing up in an affluent suburb of a small, Midwestern city, for example, learns (from her family, friends, and neighbors,

Doonesbury   BY GARRY TRUDEAU

Figure 1.2 Garry Trudeau lampoons American geographic literacy In a 1994 survey commissioned by the National Geographic Society, over 50 percent of Americans could not locate Britain or France on a map of Europe, and only 55 percent could locate the state of New York on a map of the United States. Over 20 percent could not identify the Pacific Ocean on a map of the world; 19 percent could not identify Mexico, and 14 percent could not even find the United States on a world map. (*Source:* Doonesbury © 1988 G. B. Trudeau. Reprinted with permission of UNIVERSAL PRESS SYNDICATE. All rights reserved.)

and from her own experiences of her physical environment) that she is a unique individual; that she has to respect other people's ideas and—especially—their property; that her standing as an individual will be measured by her educational achievements, sports performance, and material possessions; that her destiny will be a function of her own efforts; and so on. A young female growing up in a different setting—let's say a peripheral neighborhood in an Iranian city—learns a different set of beliefs and behaviors. Her experience will center on her home and immediate neighborhood and teach her that she is an integral part of her family and community; that she has to respect other people's roles within this social environment; and that her own standing will be measured by her comportment within a relatively close circle of kin and friends.

Not only do these differences of emphasis result in rather different values, attitudes and behaviors, they also make it difficult for people raised in different settings to understand and appreciate one another. Americans are often shocked at what they perceive as the subjugated lifestyles of Iranian women. Conversely, Iranians are often shocked at what they perceive as the amoral and materialistic lifestyles of American women.

Places also exert a strong influence, for better or worse, on people's physical well-being, their opportunities, and their lifestyle choices. Living in a small town dominated by petrochemical industries, for example, means a higher probability than elsewhere of being exposed to toxic air and water pollution, a higher probability of having only a limited range of job opportunities, and a higher probability of having a relatively narrow range of lifestyle options because of a lack of amenities. Living in the central neighborhoods of a large metropolitan area, on the other hand, usually means having a wider range of job opportunities, and having a greater choice of lifestyle options because of the variety of amenities that is accessible within a short distance; but it also means living with a relatively high exposure to crime.

Places also contribute to people's collective memory and become powerful emotional and cultural symbols. Think of the evocative power of places like Times Square in New York, the Mall in Washington, DC, Hollywood Boulevard in Los Angeles, and Graceland in Memphis, to most Americans (Figure 1.3). And for many people there are also ordinary places with special meaning: a childhood neighborhood, a college campus, a baseball stadium, or a family vacation spot.

Finally, places are the sites of innovation and change, of resistance and conflict. The unique characteristics of specific places can provide the preconditions for new agricultural practices (for example, the development of seed agriculture and the use of plow and draft animals that sparked the first agricultural revolution in the Middle East in prehistoric times); new modes of economic organization (for example, the Industrial Revolution that began in the English Midlands in the late eighteenth century); new cultural practices (the punk movement that began in disadvantaged British housing projects, for example); and new lifestyles (for example, the "hippie" lifestyle that began in San Francisco in the late 1960s). It is in specific locales that important events happen; and it is from them that significant changes spread.

Nevertheless, the influence of places is by no means limited to the occasional innovative change. Because of their distinctive characteristics, places always modify and sometimes resist the imprint of even the broadest economic, cultural, and political trends. Consider, for example, the way that a global cultural trend—rock 'n' roll—was modified in Jamaica to produce reggae, while in Iran and North Korea rock 'n' roll has been resisted by the authorities, with the result that it has acquired an altogether different kind of value and meaning for the citizens of those countries. Or, to take a very different illustration, think of the way that some communities have declared themselves to be "nuclear free" zones: places where nuclear weapons and nuclear reactors are unwelcome or even banned by local laws. By establishing such zones, individual communities are seeking to challenge national trends toward the use of nuclear energy and toward national policies of

Times Square, New York

Hollywood Boulevard, Los Angeles

The Mall, Washington, DC

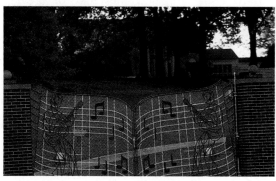
Graceland, Memphis

Figure 1.3 The power of place
Some places acquire a strong symbolic value because of the buildings, events, people, or histories with which they are associated. These examples are all places with a strong symbolic value for most Americans and, indeed, for many residents of other countries.

maintaining nuclear arms. They are, to borrow a phrase, thinking globally and acting locally. In doing so, they may influence thinking in other communities, so that, eventually, their challenge could result in a reversal of established trends.

The Interdependence of Places

Places, then, have an importance of their own. Yet at the same time most places are *interdependent,* each filling specialized roles in complex and ever-changing geographies. Consider, for example, the way that Manhattan, New York, operates as a specialized global center of corporate management, business, and financial services while relying on thousands of other places to satisfy its needs. For labor it draws on analysts and managers from the nation's business schools; blue- and pink-collar workers from neighboring boroughs; and skilled professional immigrants from around the world. For food it draws on fruits and vegetables from Florida, dairy produce from upstate New York, and specialty foods from Europe, the Caribbean, and Asia. For energy it draws on coal from southwest Virginia to fuel its power stations. And for consumer goods it draws on specialized manufacturing settings all over the world.

This interdependence means that individual places are tied in to wider processes of change that are reflected in broader geographical patterns. New York's attraction for business-school graduates, for example, is reflected in the overall pattern of migration flows that, cumulatively, affect the size and composition of labor markets around the country: New York's gain is somewhere else's loss. An important issue for human geographers—and a central theme of this book—is to recognize these wider processes and broad geographical patterns without losing sight of the individuality and uniqueness of specific places. This means that we have to recognize another kind of interdependence: the interdependence that exists *between different geographic scales*.

The Interdependence of Geographic Scales

In today's world, some of the most important aspects of the interdependence between geographic scales are provided by the relationships between the *global* and the *local* scales. The study of human geography not only shows how global trends influence local outcomes but also how events in particular localities can come to influence patterns and trends elsewhere.

New York again can illustrate both ends of this relationship. In New York's Stock Exchange and financial markets, brokers and clients must, in their own interest, take a global view. Their collective decisions influence stock prices, currency rates, and interest rates around the world—decisions that often have very direct outcomes at the local level. Factories in certain localities may be closed and workers laid off because changed currency rates make their products too expensive to export successfully; elsewhere, new jobs may be created because the same change in currency rates puts a different local economy at an advantage within the global marketplace. On the other hand, local events can reverberate through New York's Stock Exchange and financial markets with global effects. Political instability in a region that produces a key commodity, for instance, can result in changes in global pricing. A striking example of this was provided in August 1990. Within 24 hours of Iraq's invasion of Kuwait, a major oil-producing country, gasoline prices in Europe and North America had risen by 10 percent. By the time United Nations forces interceded in January 1991, gasoline prices had increased by 36 percent, and the stock prices of many companies—especially those dependent on high inputs of oil or gasoline—had fallen significantly.

Interdependence As a Two-Way Process

One of the most important tenets of human geography is that places are not just distinctive outcomes of geographical processes; they are part of the processes themselves. Think of any city neighborhood, its distinctive mix of buildings and its people. This mix will be the product of a combination of processes, including real estate development; the dynamics of the city's housing market; the successive occupancy of residential and commercial buildings by particular groups who move in and then out of the neighborhood; the services and upkeep provided by the city; and so on. Over time, these processes result in a distinctive physical environment with an equally distinctive population profile, social atmosphere, image, and reputation. Yet these neighborhood characteristics exert a strong influence, in turn, on the continuing processes of real estate redevelopment, housing market dynamics, and migration in and out of the neighborhood.

Places, then, are dynamic phenomena. They are created by people responding to the opportunities and constraints presented by their environments. As people live and work in places, they gradually impose themselves on their environment, modifying and adjusting it to suit their needs and express their values. At the same time, people gradually accommodate both to their physical environment and to the people around them. There is thus a *continuous two-way process* in which people create and modify places while at the same time being influenced by the settings in which they live and work (Figure 1.4).

Place making is always incomplete and ongoing, and it occurs simultaneously at different scales. Processes of geographic change are constantly modifying and reshaping places, and places are constantly coping with change. It is important for geographers to be sensitive to this kind of interdependence without falling into the trap of overgeneralization, or losing sight of the diversity and variety that constitute the heart of human geography. It is equally important not to fall into the trap

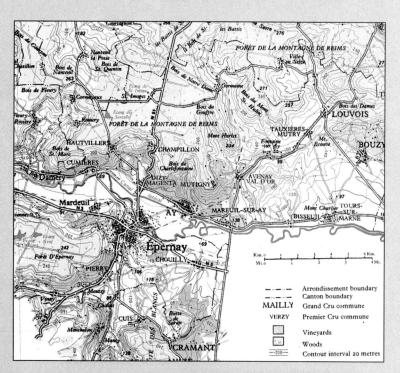

> It is man who reveals a country's individuality by moulding it to his own use. He establishes a connection between unrelated features, substituting for the random effects of local circumstances a systematic cooperation of forces. Only then does a country acquire a specific character, differentiating it from others, till at length it becomes, as it were, a medal struck in the likeness of a people.
>
> Paul Vidal de la Blache
> *Tableau de la Geographie de la France,* 1903, p. viii

Figure 1.4 Placemaking People develop patterns of living that are attuned to the opportunities and constraints of the local physical environment. When this happens, distinctive regional landscapes are produced. These photographs and the map show part of Champagne, in northeastern France. In Champagne there has been an agricultural society from the Middle Ages, its chalky soils and temperate climate being particularly suited to viticulture, or vine-growing. The result has been a distinctive landscape of small winemaking settlements surrounded by rolling slopes of carefully tended vineyards. (*Source:* Map, H. Johnson, *World Atlas of Wine.* London: Mitchell Beazley, 1971, p. 107.)

of singularity, or treating places and regions as separate entities, as the focus of study in and of themselves.

Why Geography Matters

As a subject of scientific observation and study, geography has made important contributions both to the understanding of the world and to its development. As we move further into the Information Age, geography continues to contribute to the understanding of a world that is more complex and fast-changing than ever before. With such an understanding, it is possible not only to appreciate the diversity and variety of the world's peoples and places, but also to be aware of their relationships to one another and to be able to make positive contributions to local, national, and global development.

The Greeks were the first to develop and take advantage of geographical knowledge, and their contributions were, in turn, built upon by Roman scholars and leaders. After the fall of the Roman Empire, during the Middle Ages, the most significant advances in geographic knowledge came from Chinese and Middle

Eastern scholars. Then, in the fifteenth century, the focus of geographical scholarship shifted back to Europe (see Feature 1.1, "Geography Matters—The Development of Geographic Thought"). The timing is no coincidence: The foundations of modern geography were established in conjunction with the beginnings, in the fifteenth century, of a worldwide system of exploration and trade.

Geography and Exploration

The Portuguese, under the sponsorship of Dom Henrique (known as "Prince Henry the Navigator"), established a school of navigation and cartography, and began to explore the Atlantic and the coast of Africa. **Cartography** is the name given to the system of practical and theoretical knowledge about making distinctive visual representations of Earth's surface in the form of maps. Figure 1.5 shows the key voyages of discovery. Portuguese explorer Bartholomeu Dias

Cartography: the body of practical and theoretical knowledge about making distinctive visual representations of Earth's surface in the form of maps.

Figure 1.5   The European Age of Discovery   The European voyages of discovery can be traced to Portugal's Prince Henry the Navigator (1394–1460), who set up a school of navigation and financed numerous expeditions with the objective of circumnavigating Africa in order to establish a profitable sea route for spices from India. The knowledge of winds, ocean currents, natural harbors, and watering places built up by Henry's captains was an essential foundation for the subsequent voyages of Columbus, da Gama, de Magalhães (Magellan), and others. The end of the European Age of Discovery was marked by Captain James Cook's voyages to the Pacific. (*Source:* Map projection, Buckminster Fuller Institute and Dymaxion Map Design, Santa Barbara, CA. The word Dymaxion and the Fuller Projection Dymaxion™ Map design are trademarks of the Buckminster Fuller Institute, Santa Barbara, California, ©1938, 1967 & 1992. All rights reserved.)

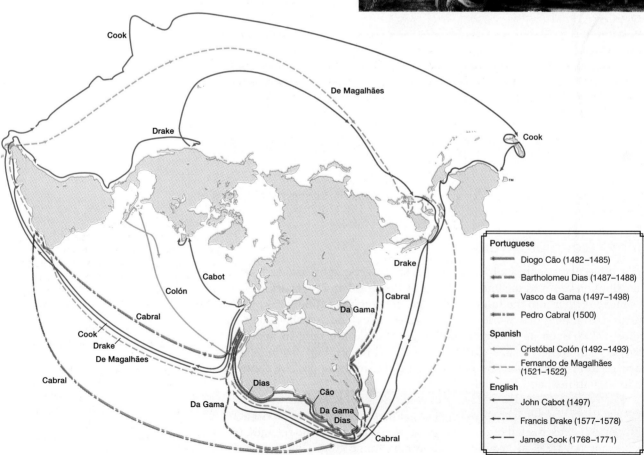

Portuguese	
⬤═══	Diogo Cão (1482–1485)
⬤▬ ▬	Bartholomeu Dias (1487–1488)
⬤▬▬ ▬▬	Vasco da Gama (1497–1498)
⬤▬▬▬	Pedro Cabral (1500)
Spanish	
———	Cristóbal Colón (1492–1493)
– – – –	Fernando de Magalhães (1521–1522)
English	
———	John Cabot (1497)
—·—·—	Francis Drake (1577–1578)
◀———	James Cook (1768–1771)

reached the Cape of Good Hope (the southern tip of Africa) in 1488. In 1492, Cristóbal Colón (Christopher Columbus), from Genoa, Italy, sailed to Hispaniola (the island that is now Haiti and the Dominican Republic) under the sponsorship of the Castillian (Spanish) monarchy. Six years later, Vasco da Gama reached India; two years after that, Pedro Cabral crossed the Atlantic from Portugal to Brazil. A small fleet of Portuguese ships reached China in 1513, and the first circumnavigation of the globe was completed by Portuguese navigator Fernando de Magalhães (better known as Magellan) in 1519. Portuguese successes inspired other countries to attempt their own voyages of discovery, all of them in pursuit of commercial advantage and economic gain. Between them, these explorations led to an invaluable body of knowledge about ocean currents, wind patterns, coastlines, peoples, and resources.

Geographical knowledge acquired during this so-called Age of Discovery was crucial to the expansion of European political and economic power in the sixteenth century. In societies that were becoming more and more commercially oriented and profit-conscious, geographical knowledge became a valuable commodity in itself. Information about regions and places was a first step to control and influence over them, and this in turn was an important step to wealth and power. At the same time, every region began to be opened up to the influence of other regions because of the economic and political competition that was unleashed by geographical discovery. Not only was the New World affected by European colonists, missionaries, and adventurers; but also the countries of the Old World found themselves pitched into competition with one another for overseas resources. New crops like corn and potatoes, introduced to Europe from the New World, meanwhile had a profound impact on local economies and ways of life.

The growth of a commercial world economy meant that objectivity in cartography and geographical writing became essential. Navigation, political boundaries, property rights, and rights of movement all depended on accuracy and impartiality. Success in commerce depended on clarity and reliability in describing the opportunities and dangers presented by one region or another. International rivalries required sophisticated understandings of the relationships among nations, regions, and places. Geography became a key area of knowledge. The historical period in Europe known as the Renaissance (from the mid-fourteenth to the mid-seventeenth centuries) saw an explosion of systematic map-making (Figure 1.6) and geographical description (Figure 1.7, page 14).

Figure 1.6 Renaissance cartography
Accurate mapping was important to Europeans' ability to open up the world for commerce. Gerardus Mercator's world map was specially devised, in 1569, as a navigational chart of the world on which mariners could plot the exact compass distance between any two points as a straight line. The projection that he devised became very popular as a general-purpose map of the world, and was widely in use in atlases and textbooks until the mid-twentieth century. This example is from an atlas created by Willem Blau, published in 1635. The use of the term "atlas" for a collection of maps stems from Mercator himself, who used an illustration of the mythological figure Atlas holding the world on his shoulders to decorate one of his books of maps.

1.1 Geography Matters
The Development of Geographic Thought

Although elementary geographical knowledge had existed since prehistoric times, the ancient Greeks were the first to demonstrate the intellectual importance and utility of geographic knowledge. They showed that places embody fundamental relationships between people and the natural environment, and that geography provides the best way of addressing the *interdependencies* between places, and between people and nature. The Greeks were also among the first to appreciate the practical importance and utility of geographic knowledge, not least in politics, business, and trade. The word "geography" is in fact derived from the Greek language, the literal translation meaning "Earth-writing" or "Earth describing." As the Greek empire developed, descriptive geographical writing came to be an essential tool for recording information about sea and land routes and for preparing colonists and merchants for the challenges and opportunities they would encounter in far-away places.

As the Greek civilization blossomed, other aspects of geography also came to be valued. One important strand of scholarship, drawing on mathematics and astronomy, was concerned with the measurement of Earth and its accurate representation. Serious attempts at cartography, or map-making, included the first map of the known world (See Figure 1.1.1), attributed by the ancient Greeks themselves to Anaximander of Miletus (611–547 B.C.).

Another important strand of Greek scholarship was philosophical. It was from this philosophical strand that the foundations of human geography were established.

Strabo's 17-volume *Geography*, written between 8 B.C. and 18 A.D., was particularly influential. Strabo systematically described places in order to address what he saw as the distinctive local relationships between nature and society. Today, this sort of descriptive approach to geographical differentiation is known as *chorology* (or *chorography*), and is often referred to as the *regional approach* to geography. The problem with chorology is that, in focusing on the distinctiveness of individual parts of Earth's surface (that is, distinctive places or homogeneous regions), it can lose sight not only of the whole but also of the relationships between places and regions. This problem was recognized by another Greek scholar, Ptolemy (c. 90–168 A.D.), whose eight-volume *Guide to Geography* was concerned with developing a comprehensive view of the world (See Figure 1.1.2).

As Greeks, both Strabo and Ptolemy were writing at the height of the Roman Empire, and their work was part of the mainstream of classical scholarship. The Romans, however, were less interested than the Greeks in the scholastic and philosophical aspects of geography, though they did appreciate geographical knowledge as an aid to conquest, colonization, and political control. With the decline and fall of the Roman Empire, however, geography and geographical knowledge were neglected. In medieval Europe, from around 500 A.D. until after 1400 A.D., there was little use or encouragement of science or philosophy of any kind. Meanwhile, though, the base of geographic knowledge was preserved and expanded by Chinese and Islamic scholars (see Figure 1.1.3). Chinese

Figure 1.1.1 A Reconstruction of one of the earliest maps of the world, by Anaximander of Miletus   Almost no details remain of Anaximander's map, but from the written accounts of his contemporaries, we do know roughly what it looked like. This version is attributed to the Greek geographer Hecataeus, and dates from around 500 B.C. At about the same time as Anaximander drafted his map, Pythagoras was theorizing about the curvature of Earth's surface, speculating that Earth was spherical rather than flat. Some 300 years later, Eratosthenes (273–192 B.C.), who is supposed to have coined the term "geography," was the first person to measure accurately the circumference of Earth. He also developed a system of latitude and longitude, which allowed the exact location of places to be plotted on the world's spherical surface. (*Source:* J. B. Harley and D. Woodward (Eds.), *The History of Cartography,* Vol. 1. Chicago: University of Chicago Press, Fig. 8.5, p. 135.)

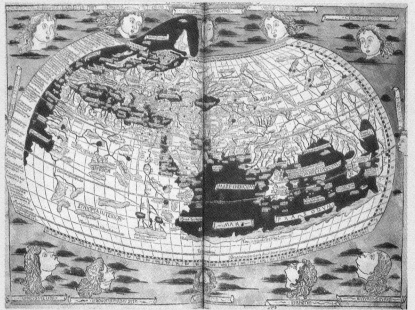

Figure 1.1.2 Ptolemy's map of the world
Ptolemy began his *Guide to Geography* with an explanation of how to construct a globe, together with its parallels and meridians, and then showed how to project the world onto a plane surface. His map of the world stood for centuries as the basis for cartography. This example, published in 1482 by Leinhart Holle in Ulm, Germany, was typical of the basic map of the world in use at the time that Columbus was considering the feasibility of going to China by sailing west. Unfortunately, Ptolemy had disregarded Eratosthenes' earlier calculations on the circumference of Earth. As a result, Ptolemy's painstaking work, apparently so precise, contained fundamental inaccuracies that led generations of philosophers, monarchs, and explorers to under-estimate the size of the globe. When Columbus did sail west from Spain in search of Asia, it was partly as a result of reading Ptolemy, whose miscalculations had led to an exaggerated conception of the size of Asia.

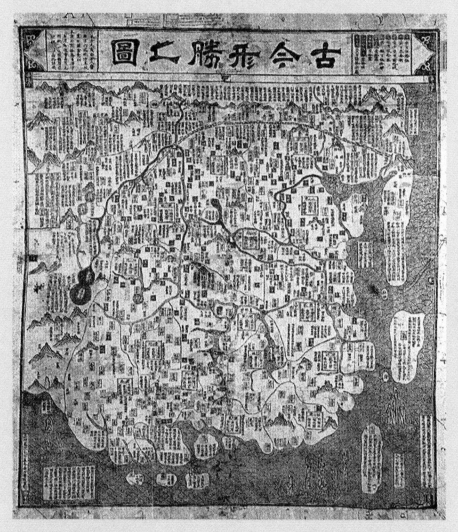

Figure 1.1.3 Early Chinese contributions to geography   Chinese geographical writing dates back to the fifth century B.C., when Chinese writers began to compile travelers' guides and descriptions of the folk customs of unfamiliar regions: the first attempts at cultural geographies. By the third and fourth centuries A.D., detailed descriptions of places were being assembled into geographical encyclopedias. For a thousand years, between roughly 300 A.D. and 1300 A.D., cartographers slowly but steadily added to the body of knowledge about China and adjacent parts of Asia. This knowledge was summarized by Chu Ssu-Pên, whose map of China, prepared between 1311 and 1320, came to be a standard work of reference for over two hundred years. This map, drawn in 1555, represents territory from Samarkand in Central Asia to Japan, and from present-day Mongolia to Java and Sumatra in Southeast Asia. (*Source:* Archivo General de Indias, Seville, Spain. Plate 1, facing p. 324, in J. B. Harley and D. Woodward (Eds.) *The History of Cartography,* Vol. 2, Book 2. Chicago: University of Chicago Press, 1994.)

11

maps of the world from the same period were more accurate than those of European cartographers because they were able to draw on information brought back by Imperial China's eunuch admirals, who successfully navigated much of the Pacific and the Indian Ocean. They recognized, for example, that Africa was a southward-pointing triangle, whereas on European and Arabic maps of the time it was always represented as pointing eastward.

With the rise of Islamic power in the Middle East and the Mediterranean in the seventh and eighth centuries A.D., there resulted the emergence of centers of scholarship throughout these regions, including Baghdad, Damascus, Cairo, and Granada, Spain. Here, surviving Greek and Roman texts were translated into Arabic. Islamic scholars were also able to draw on Chinese geographical writing and cartography, brought back by traders. The requirement that the Islamic religious faithful should undertake at least one pilgrimage to Mecca created a demand for travel guides. It also brought scholars from all over the Arab world into contact with one another, stimulating considerable debate over different philosophical views of the world, and of people's relationship with nature.

Moving forward in time to Western Europe, we can trace the foundations of geography as a formal academic discipline to the writing of a few key scholars in the eighteenth and nineteenth centuries. These included Immanuel Kant (Figure 1.1.4), Alexander von Humboldt (Figure 1.1.5), Carl Ritter (Figure 1.1.6), and Friedrich Ratzel (Figure 1.1.7). Their contribution, in general terms, was to move geography away from straightforward descriptions of parts of Earth's surface toward explanations and generalizations about the relationships of different phenomena within (and among) particular places.

Figure 1.1.4 Immanuel Kant Writing in the latter part of the eighteenth century, Immanuel Kant (1724–1804) distinguished between specific fields of knowledge represented by disciplines such as physics and biology, and two general fields of knowledge: geography, which classified things according to space (that is, regions), and history, which classified things according to time (that is, periods, or epochs). Kant was a very influential philosopher, and his belief in the intellectual importance of geography was an important element in establishing the subject as a formal discipline. His interpretation of the field of geography was also influential. He recognized the importance of commercial geography, theological geography, moral geography (concerning people's customs and ways of life), mathematical geography (concerning Earth's shape and motion), and political geography—and saw them all as being heavily influenced by the underlying physical geography.

Figure 1.1.5 Alexander von Humboldt German geographer Alexander von Humboldt's statue stands outside the Berlin university that is named after him. Von Humboldt (1769–1859) set about the task of collecting and analyzing data about the relationships between the spatial distribution of rocks, plants, and animals. He traveled for five years (1799–1804) in South America, collecting data in order to be able to identify relationships and make generalizations. Von Humboldt emphasized the mutual causality among species and their physical environment: the *interdependence* of people, plants, and animals with one another, within specific physical settings. In this way, von Humboldt showed how people, like other species, have to adapt to their environment, and how their behaviors also affect the environment around them.

Figure 1.1.6 Carl Ritter German geographer Carl Ritter (1779–1859) was the founder of the tradition of regional geography. Ritter's 20-volume Erdkunde, published between 1817 and 1859, was a monumental work of comparison and classification. Like von Humboldt, Ritter saw geography as an integrative science, able to reveal the interdependencies between people and nature. Ritter's approach was to establish a framework for scientific comparisons and generalizations by dividing the continents into broad physical units and then subdividing these into coherent regions with distinctive attributes.

A decisive development in modern geographic thought came in 1925 with a paper written by Carl Sauer, a professor of geography at the University of California, Berkeley. Sauer argued that landscapes should provide the focus for the scientific study of geography because they reflect the outcome, over time, of the interdependence of physical and human factors in the creation of distinctive places and regions. Sauer stressed that, while everything in a particular landscape is interrelated, the physical elements do not necessarily determine the nature of the human elements. Sometimes they do, sometimes it is the other way around, while sometimes it is a bit of both. Sorting these relationships out, and explaining how they constitute distinctive landscapes, were the objectives of Sauer's geography.

From this point onward, academic geography became much more sophisticated. An explosion of scholarship provided geographers with an extensive array of analytical methods and theoretical concepts that they could deploy in their examinations of places, regions, and spatial relationships. By the 1960s a "quantitative revolution" within the social sciences had led to a much greater ability to develop and test "laws" and generalizations about complex relationships within and between places. This sort of approach, using direct measurements of observable phenomena and an established scientific method in order to verify hypotheses and construct universal laws and theories, is known as *positivism*. Positivism is an approach to knowledge that was originally developed in order to distinguish science from metaphysics and religion. It is based on scientific statements that are derived only from verifiable observations. It has provided geographers with insights into spatial relationships and principles of spatial organization in every sphere of human activity.

Today, most geographers accept that such principles never actually work themselves out in pure form. There are always specific histories, unusual circumstances, and special local factors. The key issue thus becomes maintaining a perspective on general principles and broad outcomes without losing sight of the attributes of specific places. Recent developments in geographic thought have focused on understanding *spatial interdependence*. By studying spatial interdependence, geographers are able to address diversity within the framework of broader relationships; to see the uniqueness of individual places and regions within the context of other places and regions (and, indeed, the whole globe); and to see how general relationships play out within particular local settings.

Figure 1.1.7 Friedrich Ratzel and Ellen Churchill Semple One of the most influential of the pioneer academic geographers was Friedrich Ratzel (1844–1904). Ratzel, a German, was strongly influenced by Charles Darwin's theories about species' adaptation to environmental conditions and competition for living space. (Darwin's *On the Origin of Species* was published in 1859.) Ratzel's own ideas were carried to the United States by Ellen Churchill Semple, a former student of his who went on to teach at Clark University. One of the best-known U.S. proponents of Ratzel's ideas was Ellsworth Huntington, who taught geography at Yale University between 1907 and 1917. Though it may seem implausible now, one of his assertions was that civilization and successful economic development are largely the result of invigorating climates, which he defined as temperate climates with marked seasonal variations and varied weather but without prolonged extremes of heat, humidity, or cold. Huntington used this assertion to provide a rationale for the domination of northwestern Europe and the northeastern United States within the world economy.

Ratzel

Semple

Figure 1.7 Vermeer's geographer   Dutch master Johannes Vermeer's painting of *The Geographer* (1668–1669). In Renaissance Europe, the study of geography not only contributed to the growth of scientific knowledge but also helped to support European overseas expansion. Vermeer's geographer is surrounded with accurately rendered cartographic objects, including a wall chart of the sea coasts of Europe, published by Willem Blau in 1658, and a globe made by Jodocus Hondius in 1618.

Ethnocentrism: the attitude that one's own race and culture are superior to others'.

Imperialism: the extension of the power of a nation through direct or indirect control of the economic and political life of other territories.

Masculinism: the assumption that the world is, and should be, shaped mainly by men, for men.

Environmental determinism: a doctrine that holds that human activities are controlled by the environment.

Throughout the seventeenth and eighteenth centuries, the body of geographic knowledge increased steadily as more and more of the world was explored, using increasingly sophisticated techniques of survey and measurement. By the mid-nineteenth century, thriving geographical societies had been established in a number of cities, including Berlin, London, Frankfurt, Moscow, New York, and Paris. By 1899, there were 62 geographical societies worldwide, and university chairs of geography had been created in many of the most prestigious universities around the world. It has to be said, however, that geography was seen at first in narrow terms, as the discipline of exploration. Because the importance of geography was linked so clearly to European commercial and political ambitions, ways of geographic thinking also changed. Places and regions tended to be portrayed from a distinctly European point of view, and from the perspective of particular national, commercial, and religious interests. Geography mattered, but only as an instrument of colonialism.

The result was that geography began to develop a disciplinary tradition that was strongly influenced by ethnocentrism, imperialism, and masculinism. **Ethnocentrism** is the attitude that one's own race and culture are superior to those of others. **Imperialism** is the extension of the power of a nation through direct or indirect control of the economic and political life of other territories. **Masculinism** is the assumption that the world is, and should be, shaped mainly by men, for men. These trends became more and more explicit as European dominance increased, reaching a peak in the late nineteenth century, at the height of European geopolitical influence.

We should also note that most of the geographic writing in the nineteenth century was strongly influenced by environmental determinism. **Environmental determinism** is a doctrine that holds that human activities are controlled by the environment. It rests on a belief that the physical attributes of geographical settings are the root not only of people's physical differences (skin color, stature, facial features, for example), but also of differences from place to place in people's economic vitality, cultural activities, and social structures. Environmental determinists thus tend to think in terms of the influence of the physical environment on people, rather than the other way around. The idea that people's social and economic development and their behavior are fundamentally shaped by their physical environment is one that lasted well into the twentieth century, though geographers now regard it as simplistic.

Geography in a Globalizing World

Today, in a world that is experiencing rapid changes in economic, cultural, and political life, geographic knowledge is especially important and useful. In a fast-changing world, when our fortunes and our ideas are increasingly bound up with those of other peoples in other places, the study of geography provides an understanding of the crucial interdependencies that underpin everyone's lives. One of the central themes throughout this book will be the *interdependence* of people and places.

Another central theme of this book will be globalization. Globalization involves the increasing interconnectedness of different parts of the world through common processes of economic, environmental, political, and cultural change. A world economy has been in existence for several centuries, and with it there has developed a comprehensive framework of sovereign nation-states and an international system of production and exchange. This system has been reorganized several times. Each time it has been reorganized, however, major changes have resulted, not only in world geography but also in the character and fortunes of individual places.

Globalization: the increasing interconnectedness of different parts of the world through common processes of economic, environmental, political, and cultural change.

The most recent round of reorganization has created a highly interdependent world. The World Bank, on page one of the 1995 edition of its *World Development Report,* noted that "These are revolutionary times in the global economy." The report shows how globalization now affects the lives of four very different people in very different places: a Vietnamese peasant, a Vietnamese city dweller, a Vietnamese immigrant to France, and a French garment worker.

> Duong is a Vietnamese peasant farmer who struggles to feed his family. He earns the equivalent of $10 a week for 38 hours of work in the rice fields, but he works full-time only six months of the year—during the off-season he can earn very little. His wife and four children work with him in the fields, but the family can afford to send only the two youngest to school. Duong's eleven-year-old daughter stays at home to help with housework, while his thirteen-year-old son works as a street trader in town. By any standard Duong's family is living in poverty. Workers like Duong, laboring in family farms in low- and middle-income countries, account for about 40 percent of the world's labor force.
>
> Hoa is a young Vietnamese city dweller experiencing relative affluence for the first time. In Ho Chi Minh City she earns the equivalent of $30 a week working 48 hours in a garment factory—a joint venture with a French firm. She works hard for her living and spends many hours looking after her three children as well; her husband works as a janitor. But Hoa's family has several times the standard of living of Duong's and, by Vietnamese standards, is relatively well off. There is every expectation that both she and her children will continue to have a vastly better standard of living than her parents had. Wage employees like Hoa, working in the formal sector in low- and middle-income countries, make up about 20 percent of the global labor force.
>
> Françoise is an immigrant in France of Vietnamese origin who works long hours as a waitress to make ends meet. She takes home the equivalent of $220 a week, after taxes and including tips, for 50 hours work. By French standards she is poor. Legally, Françoise is a casual worker and has no job security, but she is much better off in France than she would have been in Viet Nam. Her wage is almost eight times that earned by Hoa in Ho Chi Minh City. Françoise and other service

workers in high-income countries account for about 9 percent of the global workforce.

Jean-Paul is a fifty-year-old Frenchman whose employment prospects look bleak. For ten years he has worked in a garment factory in Toulouse, taking home the equivalent of $400 a week—twelve times the average in Viet Nam's garment industry. But next month he will lose his job when the factory closes. Unemployment benefits will partly shield him from the shock, but his chances of matching his old salary in a new job are slim. Frenchmen of Jean-Paul's age who lose their jobs are likely to stay unemployed for more than a year, and Jean-Paul is encouraging his son to work hard in school so he can go to college and study computer programming. Workers in industry in high-income countries, like Jean-Paul, make up just 4 percent of the world's labor force.

These four families—two living in Viet Nam, two in France—have vastly different standards of living and expectations for the future. Employment and wage prospects in Toulouse and Ho Chi Minh City are worlds apart, even when incomes are adjusted for differences in the cost of living. Françoise's poverty wage would clearly buy Hoa a vastly more affluent lifestyle. And much of the world's workforce, like Duong, works outside the wage sector, on family farms and in casual jobs, generally earning even lower incomes. The lives of all workers in different parts of the world, however, are increasingly intertwined. French consumers buy the product of Hoa's labor, and Jean-Paul believes it is Hoa's low wages that are taking his job; while immigrant workers like Françoise feel the brunt of Jean-Paul's anger. Meanwhile, Duong struggles to save so that his children can be educated and leave the countryside for the city, where foreign companies advertise new jobs at better wages.

Recently there has been a pronounced change in both the pace and the nature of globalization. New telecommunication technologies, new corporate strategies, and new institutional frameworks have all combined to create a dynamic new framework for real-world geographies. New information technologies have helped create a frenetic international financial system, while transnational corporations are now able to transfer their production activities from one part of the world to another in response to changing market conditions (Figure 1.8). This locational flexibility has meant that a high degree of functional integration now exists between economic activities that are increasingly dispersed, so that products, markets, and organizations are both spread and linked across the globe. Governments, in their attempts to adjust to this new situation, have sought new ways of dealing with the consequences of globalization, including new international political and economic alliances.

The sheer scale and capacity of the world economy means that humans are now capable of altering the environment at the global scale. In addition to the specter of global warming (a result of emissions of gaseous materials into the atmosphere), we also face the reality of serious global environmental degradation through deforestation, desertification, acid rain, loss of genetic diversity, smog, soil erosion, groundwater decline, and the pollution of rivers, lakes, and oceans (Figure 1.9, page 18).

All this adds up to an intensification of global connectedness and the beginnings of the world as one place. Or, to be more precise, this is how it adds up for the 800 million or so of the world's population who are directly tied to global systems of production and consumption, and who have access to global networks of communication and knowledge. All of us in this globalizing world are in the middle of a major reorganization of the world economy and a radical change in our relationships to other people and other places.

At first glance, it might seem that globalization will render geography obsolete—especially in the more developed parts of the world. High-tech communica-

Figure 1.8 The globalization of manufacturing industry   The globalization of
the world economy has been made possible by the emergence of commercial corporations that are transnational in scope, and which can take advantage of modern
production technologies and computer-based information systems in order to keep
track of materials and parts, inventories of finished products, and consumer demand.

tions and the global marketing of standardized products seem as if they might
soon wash away the distinctiveness of people and places, permanently diminishing the importance of differences between places. Far from it. The new mobility
of money, labor, products, and ideas actually increases the significance of place in
some very real and important ways.

- The more universal the diffusion of material culture and lifestyles, the more
  valuable regional and ethnic identities become. One example of this is the
  way that the French government has actively resisted the Americanization of
  French language and culture by banning the use of English words and
  phrases, and by subsidizing the French domestic movie industry.

- The faster the information highway takes people into cyberspace, the more
  they feel the need for a subjective setting—a specific place or community—
  that they can call their own. Examples of this can be found in the private,
  master-planned, residential developments that have sprung up around every
  U.S. metropolitan area since the mid-1980s. Unlike most previous suburban
  developments, each of these master-planned projects has been carefully
  designed to create a sense of community and identity for their residents.

- The greater the reach of transnational corporations, the more easily they are
  able to respond to place-to-place variations in labor markets and consumer
  markets, and the more often and more radically that economic geography has
  to be reorganized. Athletic shoe manufacturers like Nike, for example, frequently switch production from one less-developed country to another in response to the changing international geography of wage rates and currencies.

- The greater the integration of transnational governments and institutions,
  the more sensitive people have become to localized cleavages of race, ethnicity, and religion. An example is the resurgence of nationalism and regionalism, as in the near-secession of Quebec from Canada in 1995 and the
  electoral success of the Forza Italia party in Italy in the early 1990s.

Industrial air pollutants include sulfur and nitrogen oxides, which form dilute acids in moist airstreams, leading to acid rain, which damages forests and pollutes lakes. Four thousand Swedish lakes are now devoid of fish because of acid rain.

Firewood gatherers, such as these in Timbuctu, Mali, must go increasingly farther from home to find wood for fuel because they have already harvested nearer sources.

Like this area on the southern shore, some 24,000 square kilometers (11,000 square miles) of former seabed in the Aral Sea have become a desert of sand and salt.

Deforestation in Russia

Figure 1.9 Environmental degradation   Many of the important issues facing modern society are the consequences—intended and unintended—of human modifications of the physical environment. Humans have altered the balance of nature in ways that have brought economic prosperity to some areas and created environmental dilemmas and crises in others. Clearing land for settlement, mining, and agriculture provides livelihoods and homes for some but alters physical systems and transforms human populations, wildlife, and vegetation. The inevitable byproducts—garbage, air and water pollution, hazardous waste, and so forth—place enormous demands on the capacity of physical systems to absorb and accommodate them.

All in all, the reality is that globalization is variously embraced, resisted, subverted, and exploited as it makes contact with specific cultures and settings. In the process, places are modified or reconstructed, rather than being destroyed or homogenized.

Making a Difference: The Power of Geography

The study of geography has become an essential basis for understanding a world that is more complex and faster-changing than ever before. Through an appreciation of the diversity and variety of the world's peoples and places, geography provides real opportunities not only for contributing to local, national, and global development but also for understanding and promoting multicultural, international, and feminist perspectives on the world.

In the United States, a decade of debate about geography education has resulted in a widespread acceptance that being literate in geography is essential in equipping citizens to earn a decent living, enjoy the richness of life, and participate responsibly in local, national, and international affairs. In response to the inclusion of geography as a core subject in the *Goals 2000: Educate America Act* (Public Law 103–227), a major report on the goals of geographic education was produced jointly by the American Geographical Society, the Association of American Geographers, the National Council for Geographic Education, and the National Geographic Society.[1] Published in 1994, the report emphasizes the importance of being geographically informed; understanding that geography is the study of people, places, and environments from a spatial perspective; and appreciating the interdependent worlds in which we live:

> The power and beauty of geography allow us to see, understand, and appreciate the web of relationships between people, places, and environments.
>
> At the everyday level, for example, a geographically informed person can appreciate the locational dynamics of street vendors and pedestrian traffic or fast-food outlets and automobile traffic; the routing strategies of school buses in urban areas and of backpackers in wilderness areas; the land-use strategies of farmers and of real estate developers.
>
> At a more expanded spatial scale, that same person can appreciate the dynamic links between severe storms and property damage or between summer thunderstorms and flash floods; the use of irrigation systems to compensate for lack of precipitation . . . ; the seasonal movement of migrant laborers in search of work and of vacationers in search of sunshine and warmth.
>
> At a global level, the geographically informed person can appreciate the connections between cyclical drought and human starvation in the Sahel or between the Chernobyl nuclear disaster and the long-term consequences to human health and economic activities throughout eastern and northwestern Europe; the restructuring of human migration and trade patterns as the European Union becomes increasingly integrated or as the Pacific rim nations develop a commonality of economic and political interests; and the uncertainties associated with the possible effects of global warming on human society or the destruction of tropical rain forests on global climate.[2]

The report cited four different reasons for being geographically informed:

- *The Existential Reason.* In 1977, the U.S. spacecraft Voyager 1 set out on its epic journey to the outer solar system and beyond. When it had passed the most distant planet, its camera was turned back to photograph the solar system. Purely by chance, the camera recorded a pale blue dot in the vastness of space. Every human who has ever lived has lived on that blue dot—Earth. Humans want to understand the intrinsic nature of their home. Geography enables them to understand where they are, literally and figuratively.
- *The Ethical Reason.* Earth will continue to whirl through space for untold millennia, but it is not certain that it will exist in a condition in which humans can thrive or even live. . . . Geography provides knowledge of

---

[1]Geography Education Standards Project, *Geography for Life. National Geography Standards 1994.*Washington, D.C., National Geographic Research and Exploration, 1994.
[2]*Geography for Life*, p. 29.

Earth's physical and human systems and of the interdependency of living things and physical environments. That knowledge, in turn, provides a basis for humans to cooperate in the best interests of our planet.

- *The Intellectual Reason.* Geography captures the imagination. It stimulates curiosity about the world and the world's diverse inhabitants and places, as well as about local, regional, and global issues. By understanding our place in the world, humans can overcome parochialism and ethnocentrism. Geography focuses attention on exciting and interesting things, on fascinating people and places, on things worth knowing because they are absorbing and because knowing them lets humans make better-informed and, therefore, wiser decisions.

- *The Practical Reason.* Geography has utilitarian value in the modern world. As the interconnectedness of the world accelerates, the practical need for geographic knowledge becomes more critical. . . . Imagine a doctor who treats diseases without understanding the environment in which the diseases thrive and spread, or a manufacturer who is ignorant of world markets and resources, or a postal worker who cannot distinguish Guinea from Guyana. With a strong grasp of geography, people are better equipped to solve issues not only at the local level but also at the global level.[3]

Geographers at Work

Geography, then, is very much an applied discipline as well as a means of understanding the world. Geographers employed in business, industry, and government are able to use geographic theories and techniques to understand and solve a wide variety of specific problems. In addition, a great deal of the research undertaken by geography professors has an applied focus. As a result, geography is able to make a direct and significant contribution to society. Because of the broad nature of the field, these contributions cover every aspect of human activity, and every scale from the local to the global. A number of examples reflect this.

Remote sensing: the collection of information about parts of Earth's surface by means of aerial photography or satellite imagery designed to record data on visible, infrared, and microwave sensor systems.

Geographic information systems (GIS): integrated computer tools for the handling, processing, and analyzing of geographical data.

- *Cartography and Remote Sensing.* Advanced techniques in cartography and remote sensing are being applied by private industry and by governments throughout the world for a wide variety of purposes: monitoring crop production, measuring deforestation, surveying endangered species, and preparing military maps. **Geographic information systems (GIS)**—integrated computer tools for the handling, processing, and analyzing of geographical data—are being used in many different spheres. Marketing companies, for example, use them in order to target advertising campaigns, ecologists to track deforestation, and planners to develop land-use policies (see Feature 1.2, "Geography Matters—Geographic Information Systems").

- *Location of Public Facilities.* Geographers use specialized techniques to analyze the location patterns of particular population groups, to analyze transportation networks, and to analyze patterns of geographic accessibility to alternative sites. Such analysis enables geographers to determine the most effective locations for new public facilities such as clinics, emergency rooms, social centers for the elderly, and shelters for the homeless.

- *Marketing and the Location of Industry.* Similar techniques are used in order to determine the most efficient, or most profitable, location for new factories, stores, and offices. Geographical research is also used to analyze the changing geography of supply and demand, allowing industry to determine whether, and where, to relocate.

---

[3]*Geography for Life,* pp. 23–24.

- *Geography and the Law.* Geographical analysis is increasingly required in order to help resolve complex social and environmental issues. One important example is the issue of racial segregation and the implications of residential segregation for policies such as school busing. Another example is the issue of property development and the implications of environmental hazards such as flooding, coastal erosion, toxic waste dumps, and earthquake fault zones for policies, codes, and regulations affecting development. A third is the issue of drawing up geographical boundaries for political units in ways that ensure equal representation by population size and racial and political composition—even as populations are changing between one election and another.

- *Disease Ecology.* By analyzing social and environmental aspects of human diseases, geographers are able to shed light on the causes of disease, to predict the spread of particular outbreaks, and to suggest ways in which the incidence of disease might be controlled.

- *Urban and Regional Planning.* Urban and regional planning adopts a systematic, creative approach to address and resolve physical, social, and economic problems of neighborhoods, cities, suburbs, metropolitan areas, and larger regions. Planners work directly on the preservation and enhancement of the quality of life in communities, on the protection of the environment, the promotion of equitable economic opportunity, and the management of growth and change of all kinds. Planning has roots in engineering, law, architecture, social welfare, and government, but it is geography, because of its focus on the interdependence of peoples and places, that provides the best preparation for specialized professional training in urban and regional planning.

- *Economic Development.* Geographers' ability to understand the interdependence of places and to analyze the unique economic, environmental, cultural, and political attributes of specific regions means that they are able to contribute effectively to strategies and policies aimed at economic development. Geographers are involved in applied research and policy formulation concerning economic development all over the world, addressing the problems not only of individual places and regions but also of the entire world economy.

These are just a few examples that illustrate how geography is critical in today's world. The career choices for geography majors are diverse, challenging, and exciting. The single most popular choice for geography majors is, in fact, a career in marketing for retailing or industrial companies. Another popular choice is an administrative, managerial, or analytical post in local, state, or federal government. Most geography graduates are able to find careers in which they have the opportunity to make a positive contribution to the world through their skills in understanding and analyzing it. These careers include cartography, GIS, laboratory analysis, private consulting, urban and regional planning, international development, teaching, and management in private industry.

Studying Human Geography

The study of geography involves the study of Earth as created by natural forces and as modified by human action. This, of course, covers an enormous amount of subject matter. There are two main branches of geography, physical geography and human geography. *Physical geography* deals with Earth's natural processes and their outcomes. It is concerned, for example, with climate, weather patterns, landforms, soil formation, and plant and animal ecology. *Human geography*

1.2 Geography Matters

Geographic Information Systems (GIS)

Geographic Information Systems (GIS) have rapidly grown to become one of the most important methods of geographic analysis, particularly in the military and the commercial world. In the United States, employment in GIS is now one of the 10 fastest-growing technical fields in the private sector. A *Geographic Information System* is an organized collection of computer hardware, software, and geographic data that is designed to capture, store, update, manipulate, and display geographically referenced information. The software in GIS incorporates programs to store and access spatial data, to manipulate those data, and to draw maps.

GIS technology can render visible many aspects of geography that were previously unseen. GIS can, for example, produce incredibly detailed maps based on millions of pieces of information—maps that could never have been drawn by human hands. One example of such a map is the satellite image reconstruction of the vegetation cover of the United States shown in Figure 1.2.1. At the other extreme of spatial scale, GIS can put places under the microscope, creating detailed new insights using huge databases and effortlessly browsable media. An example is the TargetView™ marketing package (Figure 1.2.2), which enables users to analyze and target markets within the United States.

Many of the advances in GIS have come from military applications. In the military, GIS allow infantry commanders to calculate line-of-sight from tanks and defensive emplacements; allows cruise missiles to fly below enemy radar; and provides a comprehensive basis for military intelligence. Beyond the military, GIS technology allow an enormous range of problems to be addressed. GIS can be used, for example, to decide how best to route emergency vehicles to accidents; to monitor the spread of infectious diseases; to identify the location of potential business customers; to identify the location of potential criminals; and to provide a basis for urban and regional planning. The digital media used by GIS make these applications very flexible. Using GIS it is possible to zoom in and out, evaluating spatial relationships at different spatial scales. Similarly, it is possible to vary the appearance and presentation of maps, using different colors and rendering techniques.

The most important aspect of GIS, from an analytical point of view, is that they allow data from several different sources, on different topics, and at different scales, to be merged together. This allows analysts to emphasize the spatial relationships among the objects being mapped. A geographic information system makes it possible to link, or integrate, information that is difficult to associate through any other means. For example, using GIS technology and water company billing information, it is possible to simulate the discharge of materials in the septic

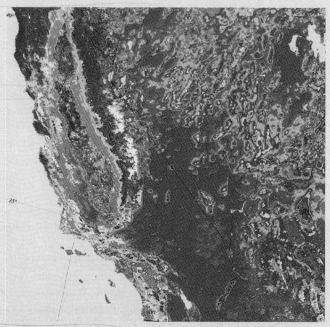

Figure 1.2.1 Map of land cover This extract is from a map of land cover in the United States that was compiled from several data sets using GIS technology. These data sets included 1-kilometer resolution, Advanced Very High Resolution Radiometer (AVHRR) satellite imagery and digital data sets on elevation, climate, water bodies, and political boundaries. Each of the 159 different colors on the U.S. map represents a different vegetation region. The purples and blues on this extract represent various subregions of western coniferous forests; the yellows are grasslands, and the reds are shrublands. The gray-brown region is the barren area of the Mojave Desert. (*Source:* United States Geological Survey, *Map of Seasonal Land Cover Regions*, 1993; see T. Loveland, J. W. Merchant, J. F. Brown, D. O. Ohlen, B. C. Reed, P. Olson and J. Hutchinson, Seasonal Land Cover Regions of the United States, *Annals, Association of American Geographers*, 85 (1995), 339–355.)

systems in a neighborhood upstream from a wetland. The bills show how much water is used at each address. Because the amount of water a customer uses will roughly predict the amount of material that will be discharged into the septic systems, areas of heavy septic discharge can be located using a GIS.

The primary requirement for data to be used in GIS is that the locations for the variables are known. Location may be annotated by x, y, and z coordinates of longitude, latitude, and elevation, or by such systems as ZIP codes or highway mile markers. Any variable that can be located spatially can be fed into a GIS.

Data capture—putting the information into the system—is the time-consuming component of GIS work. Identities of the objects on the map must be specified, as well as their location. Different sources of data, using different systems of measurement, different scales, and different systems of representation, must be integrated with one another. Changes must be tracked and updated.

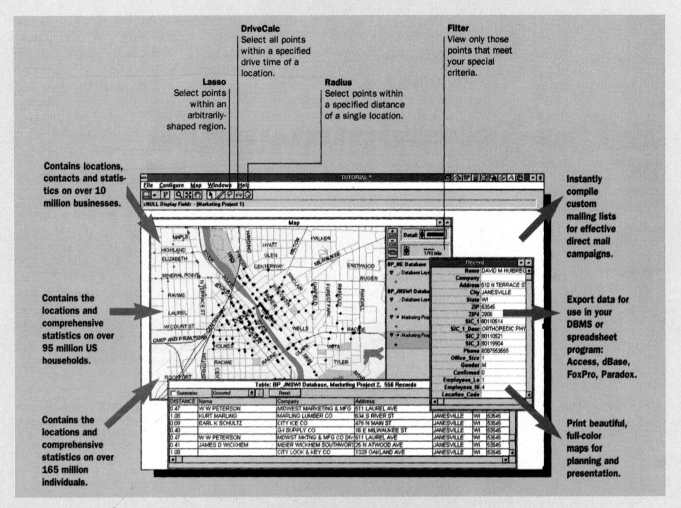

DriveCalc
Select all points within a specified drive time of a location.

Filter
View only those points that meet your special criteria.

Lasso
Select points within an arbitrarily-shaped region.

Radius
Select points within a specified distance of a single location.

Contains locations, contacts and statistics on over 10 million businesses.

Instantly compile custom mailing lists for effective direct mail campaigns.

Contains the locations and comprehensive statistics on over 95 million US households.

Export data for use in your DBMS or spreadsheet program: Access, dBase, FoxPro, Paradox.

Contains the locations and comprehensive statistics on over 165 million individuals.

Print beautiful, full-color maps for planning and presentation.

Figure 1.2.2 GIS software This extract from an advertisement for a marketing software package shows very clearly the scope and flexibility of commercial applications of geographical information systems. (*Source:* Copyright Etak, Inc. All rights reserved.)

Editing of information that is automatically captured can also be difficult. Electronic scanners record blemishes on a map just as faithfully as they record the map features. For example, a fleck of dirt might connect two lines that should not be connected. All this means that, although the costs of GIS hardware and software have fallen significantly (and will probably continue to do so), the costs of acquiring and updating reliable data continue to rise. Currently, for every dollar spent on GIS hardware, 10 dollars must be spent on software and training, and 100 dollars on acquiring and updating data. As a result, it is only the more developed nations and the larger and more prosperous organizations that can take full advantage of GIS technologies.

Within the past decade, GIS have resulted in the creation of more maps than were created in all previous human history. One result is that, as maps have become more commonplace, more people, and more businesses, have become more spatially aware. Nevertheless, some critics have argued that GIS does not represent any real advance in geographers' understanding of places and regions. The results of GIS, they argue, may be useful but are essentially mundane. This is to miss the point that however routine their subject may be, all maps constitute powerful and influential ways of representing the world. A more telling critique, perhaps, is that the real impact of GIS has been to increase the level of surveillance of the population by those who already possess power and control. The fear is that GIS may be helping to create a world in which people are not treated and judged by who they are and what they do, but more by where they live. People's credit ratings, their ability to buy insurance, and their ability to secure a mortgage, for example, are all routinely judged, in part, by GIS-based analyses that take into account the attributes and characteristics of their neighbors.

Source: This feature draws from the GIS page of the United States Geological Service (http://www.usgs.gov/research/gis/title.html) and from Chapter 7 in D. Dorling and D. Fairbairn, *Mapping: Ways of Representing the World,* London: Addison Wesley Longman, 1997.

deals with the spatial organization of human activity and with people's relationships with their environments. This focus necessarily involves looking at natural, physical environments insofar as they influence, and are influenced by, human activity. This means that the study of human geography must cover a wide variety of phenomena. These phenomena include, for example, agricultural production and food security, population change, the ecology of human diseases, resource management, environmental pollution, regional planning, and the symbolism of places and landscapes. Regional geography combines elements of both physical and human geography. **Regional geography** is concerned with the way that unique combinations of environmental and human factors produce territories with distinctive landscapes and cultural attributes. The concept of **region** is used by geographers to apply to larger-sized territories that encompass many places, all or most of which share similar attributes in comparison with the attributes of places elsewhere.

What is distinctive about the study of human geography is not so much the phenomena that are studied as the *way* they are approached. The contribution of human geography is to reveal, in relation to a wide spectrum of natural, social, economic, political, and cultural phenomena, *how and why geographical relationships are important.* Thus, for example, human geographers are not interested only in patterns of agricultural production but also in the geographical relationships and interdependencies that are both cause and effect of such patterns. To put it in concrete terms, geographers are not interested only in what specialized agricultural subregions such as the dairy farming area of Jutland, Denmark, are like (just what and how the region produces its agricultural output, what makes its landscapes and culture distinctive, and so on), but also in the role of subregions such as Jutland in national and international agro-food systems (their interdependence with producers, distributors, and consumers in other places and regions—see Chapter 8).

Regional geography: the study of the ways in which unique combinations of environmental and human factors produce territories with distinctive landscapes and cultural attributes.

Region: larger-sized territory that encompasses many places, all or most of which share similar attributes in comparison with the attributes of places elsewhere.

Basic Tools

In general terms, the basic tools employed by geographers are similar to those in other disciplines. Like other social scientists, human geographers usually begin with *observation.* Information must be collected, and data recorded. This can involve many different methods and tools. Field work (surveying, asking questions, using scientific instruments to measure and record things), laboratory experiments, and archival searches are all used by human geographers to gather information about geographical relationships. In addition, remote sensing (see Figure 1.10) is used to obtain information about parts of Earth's surface by means of aerial photography or satellite imagery designed to collect data on visible, infrared, and microwave sensor systems.

Once data have been obtained through some form of observation, the next important step is to portray and describe them through *visualization* or *representation.* This can involve a variety of tools, including written descriptions, charts, diagrams, tables, mathematical formulae, and maps. Visualization and representation are important activities because they allow large amounts of information to be explored, summarized, and presented to others. They are nearly always a first step in the analysis of geographical relationships, and they are important in conveying the findings and conclusions of geographic research.

At the heart of geographic research, as with other kinds of research, is the *analysis* of data. The objective of analysis, whether of quantitative or qualitative data, is to discover patterns and establish relationships so that hypotheses can be established and models can be built. *Models,* in this sense, are abstractions of reality that help explain the real world. Such models require tools that allow us to generalize about things. Once again we find that geographers are like other social

Landsat satellite images are digital images captured from spectral bands both visible and invisible to the human eye. Different kinds of vegetation cover, soils, and built environments are reflected by different colors in the processed image. This Landsat image is of part of southern Florida (Miami Beach is to the right of the image).

Aerial photographs can be helpful in explaining problems and processes that would otherwise require expensive surveys and detailed cartography. They are especially useful in working with multidisciplinary teams. This example shows residential development encroaching on wetlands near the edge of the city of Oakland in eastern Contra Costa County, California. The water in the middleground is Big Break, a flooded tract; in the background is the San Joaquin River.

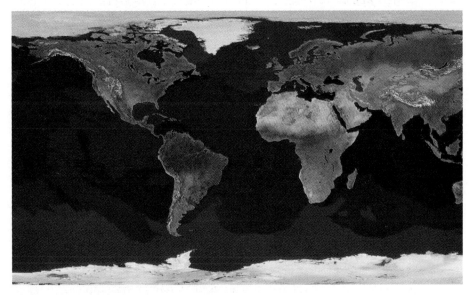

This GeoSphere image, although apparently a natural and objective "snapshot" of Earth, is derived from multispectral scanning that generated over 35 million pixels, which then had to be filtered, corrected for distortions, mathematically transformed to conform to conventional map projections, and hand tinted in order to give it appeal.

Figure 1.10 Remotely sensed images Remotely sensed images can provide new ways of seeing the world, as well as unique sources of data on all sorts of environmental conditions. They can be used to identify change in the environment, and to monitor and analyze the rate of change. Examples of such applications have included studies of the deforestation of the Amazon rainforest, of urban encroachment onto farmland, of water pollution, and of bottlenecks in highway systems.

scientists in that they utilize a wide range of analytical tools, including conceptual and linguistic devices, maps, charts, and mathematical equations.

In many ways, therefore, the tools and methods of human geographers are parallel to those used in other sciences, especially the social sciences. In addition, geographers increasingly use some of the tools and methods of the humanities—interpretive analysis and inductive reasoning, for example, together with ethnographic research and textual analysis. The most distinctive tool in the geographer's kit bag is, of course, the map (see Feature 1.3, "Geography Matters—Understanding Maps"). As we have seen, maps can be used not only to describe data, but also to serve as important sources of data and tools for analysis. Because of their central importance to geographers, they can also be objects of study in their own right.

1.3 Geography Matters

Understanding Maps

Maps are representations of the world. They are usually two-dimensional, graphic representations that use lines and symbols to convey information or ideas about spatial relationships. Maps that are designed to represent the *form* of Earth's surface and to show permanent (or, at least, long-standing) features such as buildings, highways, field boundaries, and political boundaries, are called *topographic maps* (see, for example, Figure 1.3.1). The usual device for representing the form of Earth's surface is the *contour,* a line that connects points of equal vertical distance above or below a zero data point, usually sea level.

Maps that are designed to represent the spatial dimensions of particular conditions, of processes, or of events are called *thematic maps.* These can be based on any one of a number of devices that allow cartographers or map makers to portray spatial variations or spatial relationships. One of these is the isoline, a line (similar to a contour) that connects places of equal data value (for example, air pollution levels, as in Figure 1.3.2). Maps based on isolines are known as *isopleth maps.* Another common device used in thematic maps is the *proportional symbol.* Thus, for example, circles, squares, spheres, cubes, or some other shape can be drawn in proportion to the frequency of occurrence of some particular phenomenon or event at a given location. Symbols such as arrows or lines can also be drawn proportionally, in order to portray flows of things between particular places. Figure 1.3.3 shows two examples of proportional symbols: flow lines and proportional circles. Simple distributions can be effectively portrayed through *dot maps,* in which a single dot or other symbol represents a speci-

Figure 1.3.1 Topographic maps Topographic maps are maps that represent the form of Earth's surface in both horizontal and vertical dimensions. This extract is from a British Ordnance Survey map of the area just to the south of Edinburgh, Scotland. The height of landforms is represented by contours (lines that connect points of equal vertical distance above sea level), which on this map are drawn every 50 feet, with contour values shown to the nearest meter. Features such as roads, powerlines, built-up areas, etc. are shown by stylized symbols. Note how the closely spaced contours of the hill slopes are able to represent the shape and form of the land. (*Source:* Extract from Sheet 66, 1:5,000 Series, Ordnance Survey of the United Kingdom. Copyright © Crown copyright (87375M).)

Figure 1.3.2 Isoline maps Isoline maps portray spatial information by connecting points of equal data value. Contours on topographic maps (see Figure 1.3.1) are isolines. This map shows air pollution in the eastern United States. (*Source:* Reprinted with permission of Prentice Hall, from J. M. Rubenstein, *The Cultural Landscape: An Introduction to Human Geography,* © 1996, p. 584. Adapted from William K. Stevens, "Study of Acid Raid Uncovers Threat to Far Wider Area," New York Times, January 16, 1990, p. 21, map.)

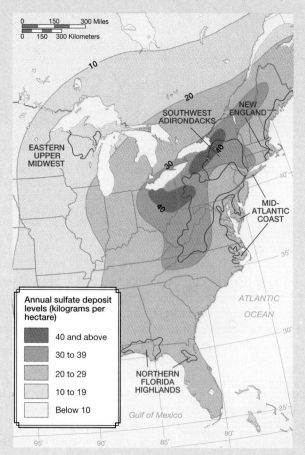

fied number of occurrences of some particular phenomenon or event (Figure 1.3.4, page 28). Yet another device is the *choropleth map,* in which tonal shadings are graduated to reflect area variations in numbers, in frequencies, or in densities (Figure 1.3.5, page 28). Finally, thematic maps can be based on *located charts,* in which graphs or charts are located by place or by region. In this way, a tremendous amount of information can be conveyed in one single map (Figure 1.3.6, page 29).

Map Scales

A *map scale* is simply the ratio between linear distance on a map and linear distance on Earth's surface. It is usually expressed in terms of corresponding lengths, as in "one centimeter equals one kilometer," or as a *representative fraction* (in this case, 1/100,000) or ratio (1 : 100,000). *Small-scale* maps are maps based on small representative fractions (for example, 1/1,000,000 or 1/10,000,000). Small-scale maps cover a large part of Earth's surface on the printed page. A map drawn on this page to the scale of 1 : 10,000,000 would cover about half of the United States; a map drawn to the scale

of 1 : 16,000,000 would easily cover the whole of Europe. *Large-scale* maps are maps based on larger representative fractions (e.g. 1/25,000 or 1/10,000). A map drawn on this page to the scale 1 : 10,000 would cover a typical suburban subdivision; a map drawn to the scale of 1 : 1,000 would cover just a block or two of it.

Map Projections

Because Earth's surface is curved, it is impossible to project onto small-scale maps without some considerable distortion of distance, direction, shape, and area. Cartographers, therefore, have devised a number of different techniques of map projection, each with advantages and disadvantages. The Mercator projection (top left, Figure 1.3.7), for example, preserves directional relationships between places, but distorts area more and more toward the poles—so much so that the poles cannot be shown as single points. The Mollweide projection (top, Figure 1.3.7, page 30) preserves area but distorts directional relationships. The Robinson projection (middle, Figure 1.3.7) is a compromise projection that distorts both area and directional relationships.

Figure 1.3.3 Two examples of proportional symbols in thematic mapping (*Source:* Reprinted with permission of Prentice Hall, from J. M. Rubenstein, *The Cultural Landscape: An Introduction to Human Geography,* © 1996, p. 107 and p. 540.)

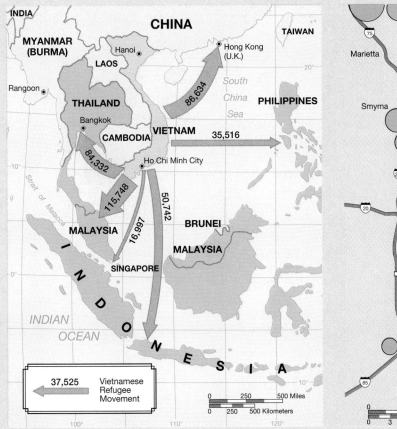

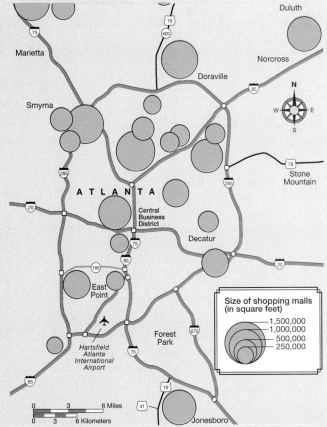

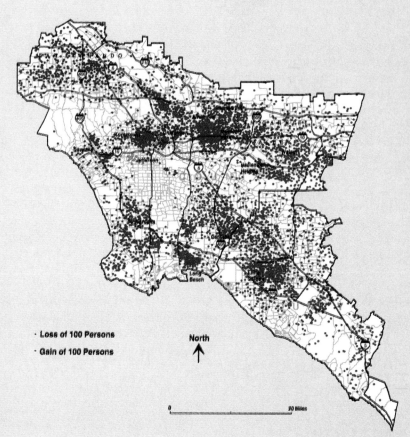

· **Loss of 100 Persons**

· **Gain of 100 Persons**

North

↑

0   20 Miles

Figure 1.3.4 Dot maps Dot maps show the spatial distribution of particular phenomena by means of simple, located symbols (usually dots or small circles, but they can be any shape). This map shows the changing distribution of the Asian population in Metropolitan Los Angeles and Orange Counties between 1980 and 1990. Each blue dot symbolizes the loss of 100 Asian inhabitants from a census tract (a small areal unit used in collecting and reporting data from the U. S. Census of Population and Housing); each red dot shows an equivalent gain. (*Source:* E. Turner and J. P. Allen, *An Atlas of Population Patterns in Metropolitan Los Angeles and Orange Counties, 1990.* Northridge: Department of Geography, California State University, 1991, p. 6.)

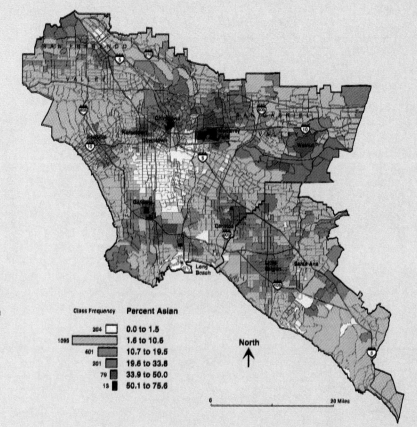

Figure 1.3.5 Choropleth maps Choropleth maps use tonal shadings to represent data based on areal units such as census tracts, counties, or regions. An important feature of such maps is that the the tonal shading is proportional to the values of the data (that is, the higher the value, the darker the shading, or vice versa). This map shows the percentage of the population in each census tract that was Asian for Metropolitan Los Angeles and Orange counties in 1990. On this map, the darker the tone of the shaded area, the higher the percentage of the area's population that was Asian. (*Source:* E. Turner and J. P. Allen, *An Atlas of Population Patterns in Metropolitan Los Angeles and Orange Counties, 1990.* Northridge: Department of Geography, California State University, 1991, p. 7.)

Class Frequency		Percent Asian
204	☐	0.0 to 1.5
1096		1.6 to 10.6
401		10.7 to 19.5
201		19.6 to 33.8
79		33.9 to 50.0
13	■	50.1 to 75.6

North

↑

0   20 Miles

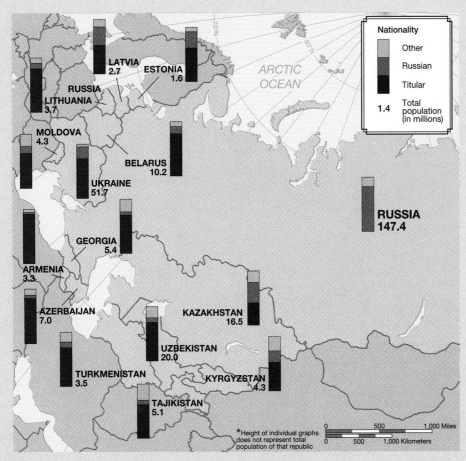

Figure 1.3.6 Located charts By combining graphs or charts with base maps, a great deal of information can be conveyed in a single figure. (*Source:* Reprinted with permission of Prentice Hall, from E. F. Bergman, *Human Geography: Cultures, Connections, and Landscapes,* © 1995, p. 474.)

The choice of map projection depends largely on the purpose of the map. Thus, for example, the Mercator projection (bottom, Figure 1.3.7) is useful for navigation because the exact compass distance between any two points can be plotted as a straight line. On the other hand, equal-area projections such as the Mollweide projection are more useful for thematic maps showing economic, demographic, or cultural data. There are also political and aesthetic considerations. Countries may appear larger and so more "important" on one projection rather than another. The Peters projection, for example (Figure 1.3.8, page 31), is a deliberate attempt to give prominence to the underdeveloped countries of the equatorial regions and the southern hemisphere. As such, it was officially adopted by the World Council of Churches and by numerous agencies of the United Nations and other international institutions. Its unusual shapes give it a shock value that gets people's attention. For some, however, its unusual shapes are ugly: It has been likened to wash hung out to dry. In a thinly veiled attack on the Peters projection, seven North American professional

organizations in geography and cartography signed a resolution urging publishers, the media, and government agencies not to use rectangular-bordered maps of the world for general purposes or artistic displays.

In this book, we shall often use another striking projection, the Dymaxion projection devised by Buckminster Fuller (Figure 1.3.9, page 31). Fuller was a prominent modernist architect and industrial designer who wanted to produce a map of the world with no significant distortion to any of the major land masses. The Dymaxion projection does this, though it produces a world that, at first, may seem disorienting. This is not necessarily such a bad thing, for it can force us to take a fresh look at the world, and at the relationships between places. Because Europe, North America, and Japan are all located toward the center of this map projection, it is particularly useful for illustrating two of the central themes of this book: the relationships among these prosperous regions, and the relationships between this prosperous core group and the less prosperous, peripheral countries of the world. On Fuller's projection, the economically peripheral countries

Figure 1.3.7 Comparison of the Mollweide, Robinson and Mercator, projections Different map projections distort the sizes and shapes of land masses in different ways. On the Mollweide projection (top), relative sizes are true but shapes are distorted. On the Robinson projection (center), both size and shape are partially distorted: It is designed purely for appearance. On the Mercator projection (bottom), the shapes of land masses are true but their relative size is distorted. (*Source:* Center and bottom, reprinted with permission of Prentice Hall, from E. F. Bergman, *Human Geography: Cultures, Connections, and Landscapes,* © 1995, pp. 12–13.)

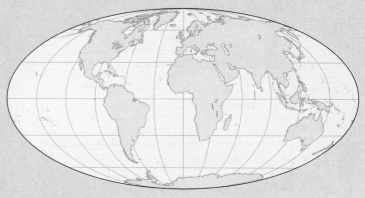

Mollweide projection

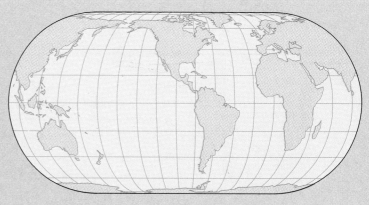

Robinson projection

Mercator projection

Figure 1.3.8 The Peters projection This equal-area projection was an attempt to offer an alternative to traditional projections which, Arno Peters argued, exaggerated the size and apparent importance of the higher latitudes—that is, the world's core regions—and so promoted the "Europeanization" of Earth. While it has been adopted by the World Council of Churches, the Lutheran Church of America, and various agencies of the United Nations and other international institutions, it has been criticized by cartographers in the United States on the grounds of esthetics: One of the consequences of equal-area projections is that they distort the shape of land masses. (*Source:* Reprinted with permission of Prentice Hall, from E. F. Bergman, *Human Geography: Cultures, Connections, and Landscapes,* © 1995, p. 13.)

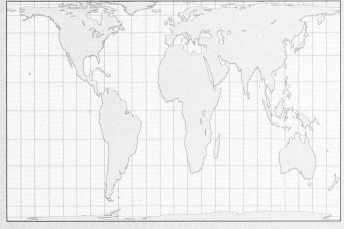

Figure 1.3.9 Fuller's Dymaxion projection   This striking map projection was designed by Buckminster Fuller (1895–1983). As this figure shows, he achieved his objective of creating a map with the minimum of distortion to the shape of the world's major land masses by dividing the globe into triangular areas. Those areas not encompassing major land masses were cut away, allowing the remainder of the globe to be "unfolded" into a flat projection. (*Source:* Map projection, Buckminster Fuller Institute and Dymaxion Map Design, Santa Barbara, CA. The word Dymaxion and the Fuller Projection Dymaxion™ Map design are trademarks of the Buckminster Fuller Institute, Santa Barbara, California, ©1938, 1967 & 1992. All rights reserved.)

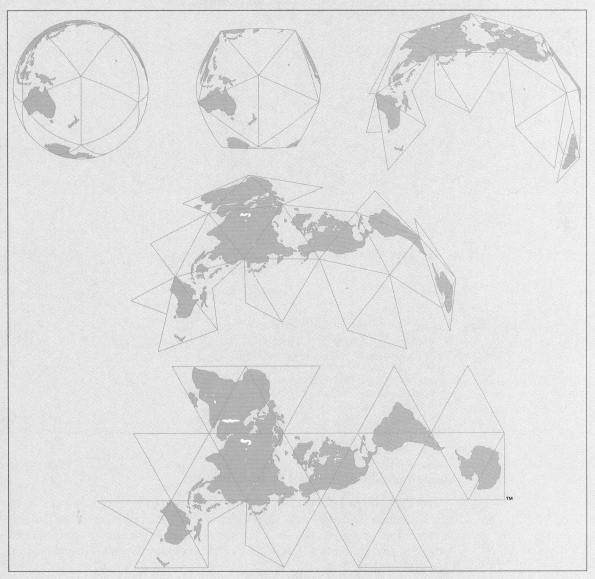

of the world are shown as being cartographically peripheral, too.

One particular kind of map projection that is sometimes used in small-scale thematic maps is the *cartogram*. In this kind of projection, space is transformed according to statistical factors, with the largest mapping units representing the greatest statistical values. Figure 1.3.10 shows two different cartograms of Britain: one in which regional administrative boundaries are represented as proportional to their population, the other in which electoral constituencies are shown by facelike symbols, which represent five different pieces of information about each constituency.

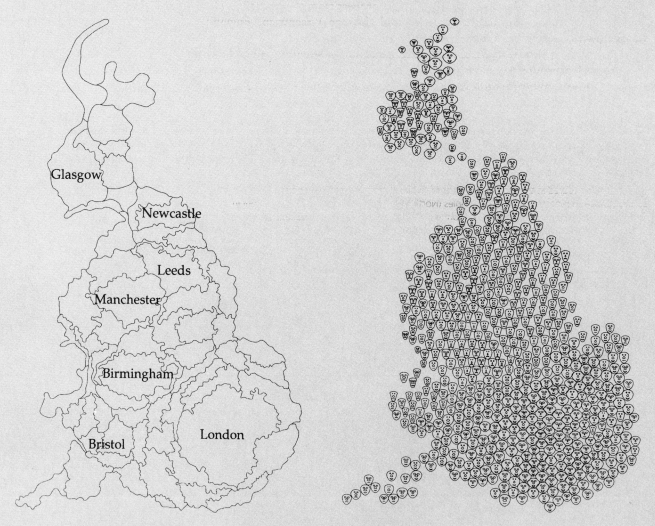

Figure 1.3.10 Examples of cartograms (a) counties (in England and Wales) and regions (in Scotland) shown proportional to their populations and (b) the socioeconomic characteristics of electoral constituencies in Britain, shown by Chernoff faces, a system showing five different attributes in one symbol—In these Chernoff faces, the shape of the mouth is determined by unemployment levels, the cheek by wealth, the position of the eyes by age of local industries, the size of the nose by electoral turnout, and the size of the face by population. (*Source:* D. Dorling, *Environment & Planning A,* 17, 1995, pp. 357 and 359.)

Fundamental Concepts of Spatial Analysis

The study of human geography is easily distinguished by its fundamental concepts. *Location, distance, space, accessibility,* and *spatial interaction* are five concepts that geographers use in analyzing spatial patterns and distributions. Although these concepts may be familiar from everyday language, they do require some elaboration.

Location Often, location is _nominal,_ or expressed solely in terms of the names of regions and places. We speak, for example, of Washington, DC, or Georgetown, a location within Washington, DC. Location can also be used as an _absolute_ concept, whereby locations are fixed mathematically through coordinates of latitude and longitude (Figure 1.11). **Latitude** refers to the angular distance of a point on Earth's surface, measured in degrees, minutes, and seconds north or south from the equator. **Longitude** refers to the angular distance of a point on Earth's surface, measured in degrees, minutes, and seconds east or west from the _prime meridian_ (the line that passes through both poles and through Greenwich, England). Georgetown's coordinates are precisely 38° 55′ N, 77° 00′ E.

Location can also be _relative_, fixed in terms of site or situation. **Site** refers to the physical attributes of a location: its terrain, soil, vegetation, and water sources, for example. **Situation** refers to the location of a place relative to other

Latitude: the angular distance of a point on Earth's surface, _measured north or south_ from the equator, which is 0°.

Longitude: the angular distance of a point on Earth's surface, _measured east or west_ from the prime meridian (the line that passes through both poles and through Greenwich, England, and which has the value of 0°).

Site: the physical attributes of a location: its terrain, soil, vegetation, water sources, for example.

Situation: the location of a place relative to other places and human activities.

Figure 1.11 Latitude and longitude   Lines of latitude and longitude provide a grid that covers Earth, allowing any point on Earth's surface to be accurately referenced. Latitude is measured in terms of angular distance (i.e., degrees and minutes) north or south of the equator, as shown in (a). Longitude is measured in the same way, but east and west from the prime meridian, a line around Earth's surface that passes through both poles (North and South) and the Royal Observatory in Greenwich, just to the east of central London, in England, as shown in (b). Locations are always stated with latitudinal measurements first. The location of Paris, France, for example, is 48° 51′ N and 2° 20′ E; Washington, DC, is 38° 50′ N and 77° 00′ W; Sydney, Australia, is 33° 55′ S and 151° 17′ E, as shown in (c). (_Sources:_ (a) and (c), reprinted with permission of Prentice Hall, from R. W. Christopherson, _Geosystems: An Introduction of Physical Geography,_ 2nd Ed, © 1994, pp. 13 and 15. (b), reprinted with permission of Prentice Hall, from E. F. Bergman, _Human Geography: Cultures, Connections, and Landscapes,_ © 1995, Figs. 1-10 and 1-13.)

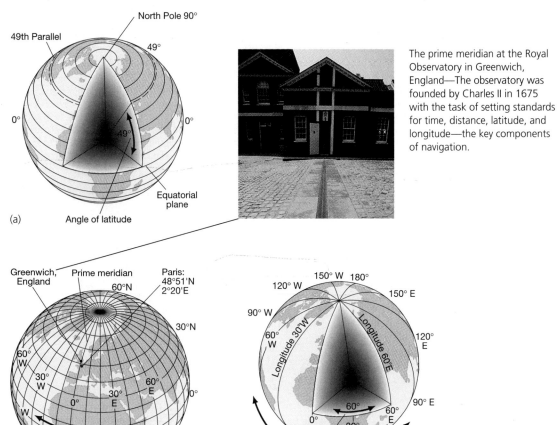

The prime meridian at the Royal Observatory in Greenwich, England—The observatory was founded by Charles II in 1675 with the task of setting standards for time, distance, latitude, and longitude—the key components of navigation.

places and human activities: its accessibility to routeways, for example, or its nearness to population centers (Figure 1.12). Washington, DC, has a low-lying, riverbank site, and is situated at the head of navigation of the Potomac River, on the Eastern seaboard of the United States.

Finally, location also has a *cognitive* dimension, in that people have cognitive images of places and regions, compiled from their own knowledge, experience, and impressions. **Cognitive images** (sometimes referred to as **mental maps**) are psychological representations of locations that spring from people's individual ideas and impressions of these locations. These representations can be based on people's direct experiences, on written or visual representations of actual locations, on hearsay, on people's imaginations, or on a combination of these sources. Location in these cognitive images is fluid, depending on people's changing information and perceptions of the principal landmarks in their environment.

Some things, indeed, may not be located in a person's cognitive image at all! Figure 1.13 shows one person's cognitive image of Washington, DC. Georgetown is given a location within this mental map, even though it is some distance from the residence of the person who sketched her image of the city. Less well-known and less distinctive places do not appear at all on this particular image.

Distance Distance is also useful as an *absolute* physical measure, whose units we may count in kilometers or miles, and as a *relative* measure, expressed in terms of time, effort, or cost. It can take more (or less) time, for example, to travel 10 kilo-

Cognitive images (mental maps): psychological representations of locations that are made up from people's individual ideas and impressions of these locations.

Figure 1.12 The importance of site and situation   The location of telecommunications activities in Denver, Colorado, provides a good example of the significance of the geographic concepts of site (the physical attributes of a location) and situation (the location of a place relative to other places and human activities). Denver has become a major center for cable television, with the headquarters of giant cable companies such as Tele-Communications and DirecTV, an industrywide research lab, and a cluster of specialized support companies that, together, employ over 3,000 people. Denver's *site,* one mile above sea level, is important because it gives commercial transmitters and receivers a better "view" of communications satellites. Its *situation,* on the 105th meridian and equidistant between the telecommunications satellites that are in geosynchronous orbit over the Pacific and Atlantic oceans, allows it to send cable programming directly not just to the whole of the Americas but also to Europe, the Middle East, India, Japan, and Australia—to every continent, in fact, except Antarctica. This is important because it avoids "double-hop" transmission (in which a signal goes up to a satellite, then down, then up and down again), which increases costs and decreases picture quality. Places east or west of the 105th meridian would have to double-hop some of their transmissions because satellite dishes would not have a clear "view" of both the Pacific and Atlantic telecommunications satellites.

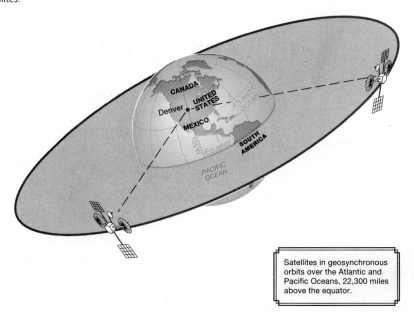

Satellites in geosynchronous orbits over the Atlantic and Pacific Oceans, 22,300 miles above the equator.

Figure 1.13 One person's cognitive image of Washington, DC   This sketch was drawn by Rasheda DuPree, an Urban Affairs major at Virginia Tech, as part of a class exercise in recalling locations within students' hometowns. Rasheda has included many of the District's most prominent landmarks and some of its distinctive districts, including Georgetown. In contrast, there are no recorded locations in the city's southeastern quarter (marked by a skull-and-crossbones in Rasheda's sketch) or in the eastern outskirts (marked as "the burbs").

meters from point A to point B than it does to travel 10 kilometers from point A to point C. Similarly, it can cost more (or less). Geographers also have to recognize that distance can sometimes be in the eye of the beholder. It can seem longer (or shorter), more (or less) pleasant going from A to B as compared to going from A to C. This is **cognitive distance,** the distance that people perceive to exist in a given situation. Cognitive distance is based on people's personal judgments about the degree of spatial separation between points.

The importance of distance as a fundamental factor in determining real-world relationships is a central theme in geography. It was once described as the "first law" of geography: "Everything is related to everything else, but near things are more related than distant things." Waldo Tobler, the geographer from the University of California, Santa Barbara, who put it this way, is one of many who have investigated the **friction of distance,** the deterrent or inhibiting effect of distance on human activity. The friction of distance is a reflection of the time and cost of overcoming distance.

What these geographers have established is that these effects are not uniform, that is, not directly proportional to distance itself. This is true whether distance is

Cognitive distance: the distance that people perceive to exist in a given situation.

Friction of distance: the deterrent or inhibiting effect of distance on human activity.

measured in absolute terms (kilometers, for example) or in relative terms (time- or cost-based measures, for example). What happens is that the deterrent effects of extra distance tend to lessen as greater distances are involved. Thus, for example, while there is a big deterrent effect in having to travel two kilometers rather than one to get to a grocery store, the deterrent effect of the same extra distance (one kilometer) after already traveling 10 kilometers is relatively small.

Distance-decay function: the rate at which a particular activity or process diminishes with increasing distance.

This sort of relationship creates what geographers call a distance-decay function. A **distance-decay function** describes the rate at which a particular activity or phenomenon diminishes with increasing distance (Figure 1.14). The rate of AIDS-infected population, for example, declines with distance from Washington, DC: sharply for the first 15 to 30 kilometers (9.3 to 18.6 miles) or so between the downtown and outlying suburban centers such as Falls Church and Alexandria, after which rates level off among the more distant towns in the region. The explanation for this particular distance-decay function is that, as in other U.S. metropolitan regions, the farther away from the inner city, the lower the proportion of high-risk subgroups tends to be.

Utility: the usefulness of a specific place or location to a particular person or group.

Distance-decay functions reflect people's behavioral response to opportunities and constraints in time and space. As such, they are a reflection of the utility of particular locations to people. The **utility** of a specific place or location refers to its usefulness to a particular person or group. In practice, utility will be thought of in different ways by different people in different situations. The behavior of private firms and their agents or employees, for example, will most often be guided by a bottom-line notion of utility that relates to dollar costs or profits. The same individuals will probably use a different notion of utility when it comes to their own, private lifestyle and the decisions they make in pursuing it. Prestige, convenience, or feelings of personal safety, communality, or happiness may well modify or override financial costs as the measure of place utility. The business manager of a supermarket chain, for example, will almost certainly decide on the utility of potential locations for a new store by weighing criteria based on the projected costs and revenues for each potential site. In deciding on the utility of potential locations in which to retire, however, that same manager will almost certainly decide on the utility of potential locations by weighing criteria based not only on costs but also on a wide range of quality-of-life aspects of potential retirement places.

The common, unifying theme here is that, however place utility is thought of, people will, in most circumstances, tend to *seek to maximize the net utility of location*. The supermarket chain's business manager, for example, will seek to find the location for the chain's new store that is most likely to yield the greatest profit. On retirement, he or she will choose to live in the place that represents the best trade-off among housing costs, the cost of living, and the quality of life. Seeking to maximize the net utility of location means that a great deal of human activity is influenced by what University of Washington geographer Richard Morrill once called the "nearness principle." According to this principle—a more explicit version of Tobler's "first law"—people will seek to:

- maximize the overall utility of places at minimum effort,
- maximize connections between places at minimum cost, and
- locate related activities as close together as possible.

The result is that patterns of behavior, locational decisions, and interrelations between people and places come to take on fairly predictable, organized patterns.

Space Like distance, space can be measured in absolute, relative, and cognitive terms. Table 1.1 (see page 38) lists the concepts that are used in talking about space in these various terms. Absolute space is a mathematical space, described through points, lines, areas, planes, and configurations whose relationships can be fixed precisely through mathematical reasoning. Several ways of analyzing

Figure 1.14 The friction of distance
The effects of distance on people's behavior can be charted on graphs like this. The farther people have to travel, the less likely they are to do so. In this example, we can see clearly the deterrent effects of distance on people's attendance at a free health clinic.

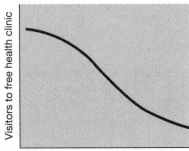

Visitors to free health clinic

Distance from residence to clinic

space mathematically are of use to geographers. The conventional way is to view space as a container, defined by rectangular coordinates and measured in absolute units of distance (kilometers or miles, for example). Other mathematical conceptions of space that geographers sometimes find useful also exist, however. One example is **topological space,** defined as the connections between, or connectivity of, particular points in space (see Figure 1.15). Topological space is measured not in terms of conventional measures of distance, but rather in terms of the nature and degree of connectivity between locations.

Topological space: space that is defined and measured in terms of the nature and degree of connectivity between locations.

Relative measurements of space can also take the form of socioeconomic space or of experiential or cultural space (see Table 1.1, page 38). Socioeconomic space can be described in terms of sites and situations, routes, regions, and distribution patterns. In these terms, spatial relationships have to be fixed through measures of time, cost, profit, or production, as well as through physical distance. Experiential or cultural space is the space of groups of people with common ties, and it is described through the places, territories, and settings whose attributes carry special meaning for these particular groups. Finally, **cognitive space** is defined and measured in terms of people's values, feelings, beliefs, and perceptions about locations, districts, and regions. Cognitive space can be described, therefore, in terms of behavioral space—landmarks, paths, environments, and spatial layouts.

Cognitive space: space defined and measured in terms of people's values, feelings, beliefs, and perceptions about locations, districts, and regions.

Figure 1.15 Topological space   There are some dimensions of space and aspects of spatial organization that do not lend themselves to description simply in terms of distance. In other words, it is often the connectivity of people and places that is important: whether they are linked together, how they are linked, and so on. These attributes of connectivity define a special kind of space, known by mathematicians as topological space. This map of the London Underground is a topological map, showing how specific points are joined within a particular network. The most important aspects of networks of any kind, from the geographer's point of view, are their attributes in terms of connectivity. These attributes determine the flows of people and things (goods, information) and the centrality of places. As most Londoners know, the Underground system provides Paddington with a very high degree of connectivity because trains on the District, Circle, Bakerloo, and Hammersmith & City lines all stop there. Paddington is, therefore, relatively central within the "space of flows" of passenger traffic in central London. Edgeware Road, nearby in absolute terms, is much less central, however, and much less the focus of passenger flows. (*Source:* London Regional Transport.)

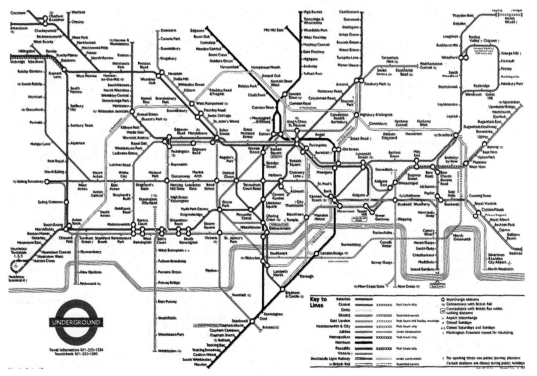

TABLE 1.1 Different Kinds of Spaces Analyzed by Human Geographers

Absolute Space: Mathematical Space	Relative Space: Socioeconomic Space	Relative Space: Experiential/Cultural Space	Cognitive Space: Behavioral Space
Points	Sites	Places	Landmarks
Lines	Situations	Ways	Paths
Areas	Routes	Territories	Districts
Planes	Regions	Domains	Environments
Configurations	Distributions	Worlds	Spatial Layouts

Source: Based on H. Couclelis, "Location, Place, Region and Space" in R. Abler et al., *Geography's Inner Worlds.* New Brunswick, NJ: Rutgers University Press, 1992, Table 10.1, p. 231.

Accessibility: the opportunity for contact or for interaction from a given point or location, in relation to other locations.

Accessibility Given that people tend to pursue the nearness principle, the concept of accessibility is very important. **Accessibility** is generally defined by geographers in terms of relative location: the opportunity for contact or for interaction from a given point or location, in relation to other locations. It implies proximity, or nearness, to something. Because it is a fundamental influence on the utility of locations, distance is an important influence on people's behavior. Distance is one aspect of accessibility, but it is by no means the only important aspect.

Connectivity is also an important aspect of accessibility, because contact and interaction are dependent on channels of communication and transportation: streets, highways, telephone lines, and wavebands, for example. Effective accessibility is, therefore, a function not only of distance but also of the configuration of networks of communication and transportation. Commercial airline networks provide many striking examples of this. Cities that operate as airline hubs are much more accessible than cities that are served by fewer flights and fewer airlines. Charlotte, NC, for example (a US Airways hub), is more accessible from Albany, NY, than Richmond, Virginia, even though Richmond is 400 kilometers (248 miles) closer to Albany than Charlotte. To get to Richmond from Albany, travelers have to fly to Charlotte or another hub and change—a journey that takes longer and often costs more.

Accessibility is often a function of economic, cultural, and social factors. In other words, relative concepts and measures of distance are often as important as absolute distance. A nearby facility, such as a health-care clinic, is only accessible to us if we can actually afford the cost of getting there; if it seems close according to our own standards of distance; if we can afford to use the facility; if we feel that it is socially and culturally acceptable for us to use it; and so on. To take another example, a day-care center may be located just a few blocks from a single-parent family, but the center is not truly accessible if it only opens after the parent has to be at work; or if the parent feels that the staff, children, or other parents at the center are from an incompatible social or cultural group.

Spatial Interaction Interdependence between places and regions can only be sustained through movement and flows. Geographers use the term *spatial interaction* as shorthand for all kinds of movement and flows that involve human activity. Freight shipments, commuting, shopping trips, telecommunications, electronic cash transfers, migration, and vacation travel are all examples of spatial interaction. The fundamental principles of spatial interaction can be reduced to four basic concepts: complementarity, transferability, intervening opportunities, and diffusion.

- *Complementarity.* A precondition for interdependence between places is complementarity. In order for any kind of spatial interaction to take place between two places, there must be a demand in one place and a supply that matches, or *complements* it, in the other. This complementarity can occur as a result of several factors. One important factor is the variation in physical environments and resource endowments from place to place. For example, a heavy flow of vacation travel from Scottish cities to Mediterranean resorts is largely a function of climatic complementarity. To take another example, the flow of crude oil from Saudi Arabia (with vast oil reserves) to Japan (with none) is a function of complementarity in natural resource endowments (Figure 1.16).

A second factor that contributes to complementarity is the international division of labor that derives from the evolution of the world's economic systems. The more-developed countries of the world have sought to establish overseas suppliers for their food, raw materials, and exotic produce, allowing the more-developed countries to specialize in more profitable manufacturing and knowledge-based industries (see Chapter 2). Through a combination of colonialism, imperialism, and sheer economic dominance on the part of these more developed countries, less powerful countries have found themselves with economies that directly complement the needs of the more developed countries. Among the many flows resulting from this complementarity are shipments of sugar from Barbados to the United Kingdom, bananas from Costa Rica and Honduras to the United States, palm oil from Cameroon to France, automobiles from France to Algeria, school textbooks from the United Kingdom to Kenya, and investment capital from the United States to most less-developed countries.

A third contributory factor to complementarity is the operation of principles of specialization and economies of scale. Places, regions, and countries can derive economic advantages from the efficiencies created through specialization, which allows for larger-scale operations. **Economies of scale** are

Economies of scale: cost advantages to manufacturers that accrue from high-volume production, since the average cost of production falls with increasing output.

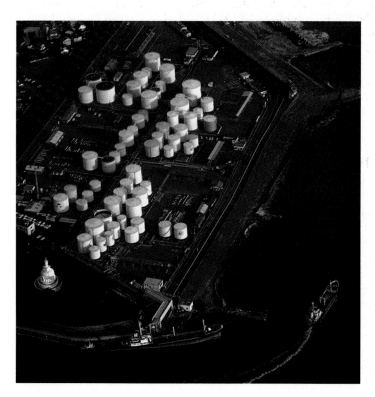

Figure 1.16 Trade as a result of complementarity Japan's needs are complementary to those of Saudi Arabia. Japan's industrial economy requires vast amounts of crude oil, but Japan itself has no oil supplies. Saudi Arabia, on the other hand, is one of the world's biggest suppliers of crude oil but has to import most high-tech products and consumer goods. The result is a high degree of spatial interaction between the two countries. In 1995, Japan imported 1.7 billion barrels of crude oil, 79 percent of it from the Middle East.

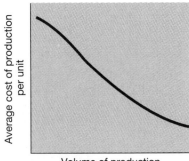

Figure 1.17 Economies of scale In many manufacturing enterprises, the higher the volume of production, the lower the average cost of producing each unit. This is partly because high-volume production allows for specialization and division of labor, which can increase efficiency and hence lower costs. It is also partly because most manufacturing activities have significant fixed costs (such as product design and the cost of renting or buying factory space) that must be paid for irrespective of the volume of production, so that the larger the output, the lower the fixed cost per unit. These savings are known as economies of scale.

Time–space convergence: the rate at which places move closer together in travel or communication time or costs.

cost advantages to manufacturers that accrue from high-volume production, since the average cost of production falls with increasing output (Figure 1.17). Among other things, fixed costs (for example, the cost of renting or buying factory space, which will be the same—fixed—whatever the level of output from the factory) can be spread over higher levels of output, so that the average cost of production falls. Economic specialization results in complementarities, which in turn contribute to patterns of spatial interaction. One example is the specialization of Israeli farmers in high-value fruit and vegetable crops for export to the European Union, which in return exports grains and root crops to Israel.

- *Transferability.* In addition to complementarity, another precondition for interdependence between places is *transferability,* which depends on the frictional or deterrent effects of distance. Transferability is a function of two things: the costs of moving a particular item, measured in real money and/or time, and the ability of the item to bear these costs. If, for example, the costs of moving a product from one place to another make it too expensive to sell successfully at its destination, then that product does not have transferability between those places.

Transferability varies between places, between kinds of items, and between modes of transportation and communication. The transferability of coal, for example, is much greater between places that are connected by rail or by navigable waterways than between places connected only by highways. This is because it is much cheaper to move heavy, bulky materials by rail, barge, or ship. The transferability of fruit and salad crops, on the other hand, depends more on the speed of transportation and the availability of specialized, refrigerated vehicles so the fruits and vegetables stay fresh. While the transferability of money capital is much greater by telecommunications than it is by surface transportation, it is also higher between places where banks are equipped to deal routinely with electronic transfers. Computer microchips have high transferability because they are easy to handle and transport costs are a small proportion of their value. Computer monitors, on the other hand, have lower transferability because of their fragility and because of their relatively lower value by weight and volume.

Transferability also varies over time, as successive innovations in transport and communications technologies and successive waves of infrastructure development (canals, railways, harbor installations, roads, bridges, and so on) alter the geography of transport costs. New technologies and new or extended infrastructures have the effect of altering the transferability of particular things between particular places. *As a result, the spatial organization of many different activities is continually changing and readjusting.* The consequent tendency toward a shrinking world gives rise to the concept of **time–space convergence,** the rate at which places move closer together in travel or communication time or costs. Time–space convergence results from a decrease in the friction of distance as new technologies and infrastructure improvements successively reduce travel and communications time between places. Such space-adjusting technologies have, in general, brought places "closer" together over time. Overland travel between New York and Boston, for example, has been reduced from 3.5 days (in 1800) to 5 hours (in the 1990s) as the railroad displaced stagecoaches and was in turn displaced by interstate automobile travel. Other important space-adjusting innovations include air travel and air cargo; telegraphic, telephonic, and satellite communications systems; national postal services, package delivery services, and facsimile (fax) machines; and modems, fiber-optic networks, and electronic-mail software.

What is most significant about the latest developments in transport and communication is that they are not only global in scope but also able to penetrate to local scales. As this is happening, some places that are distant in kilometers are becoming close together, while some that are close in terms of absolute space are becoming more distant in terms of their ability to reach one another electronically (Figure 1.18). Much depends on the mode of communication—the extent to which people in different places are "plugged in" to new technologies. Older wire cable can carry only small amounts of information; microwave channels are good for person-to-person communication, but depend on line-of-sight; telecommunications satellites are excellent for reaching remote areas but involve significant capital costs for users, while fiber-optic cable is excellent for areas of high population density but not feasible for remoter, rural areas.

Figure 1.18 A town's electronic address The town hall in Blacksburg, Virginia, announces its address in both traditional and cyberspace forms. Blacksburg, together with Virginia Tech, was a pioneer in promoting the development and use of digital communications throughout the community, through Blacksburg Electronic Village.

- *Intervening Opportunity.* While complementarity and transferability are preconditions for spatial interaction, intervening opportunities are more important in determining the *volume* and *pattern* of movements and flows. Intervening opportunities are simply alternative origins and/or destinations. Such opportunities are not necessarily situated directly between two points, or even along a route between them. Thus, to take one of our previous examples, for Scottish families considering a Mediterranean vacation in Greece, resorts in Spain, southern France, and Italy are all intervening opportunities because they can be reached more quickly and more cheaply than resorts in Greece.

  The size and relative importance of alternative destinations are also important aspects of the concept of intervening opportunity. For our Scottish families, Spanish resorts probably offer the greatest intervening opportunity because they contain the largest aggregate number of hotel rooms and holiday apartments. We can, therefore, state the principle of intervening opportunity as follows: Spatial interaction between an origin and a destination will be proportional to the number of opportunities at that destination, and inversely proportional to the number of opportunities at alternative destinations.

- *Spatial Diffusion.* Disease outbreaks, technological innovations, political movements, and new musical fads all originate in specific places and subsequently spread to other places and regions. The way that things spread through space and over time—**spatial diffusion**—is one of the most important aspects of spatial interaction and is crucial to an understanding of geographic change.

  Diffusion seldom occurs in an apparently random way, jumping unpredictably all over the map. Rather, it occurs as a function of statistical probability, which is often based on fundamental geographic principles of distance and movement. The diffusion of a contagious disease, for example, is a function of the probability of physical contact, modified by variations in individual resistance to the disease. The diffusion of an agricultural innovation, such as hybrid wheat, is a function of the probability of information flowing between members of the farming community (itself partly a function of distance), modified by variations in individual farmers' receptivity to innovative change. The result is typically a "wave" of diffusion that describes an S-curve, with a slow build-up, rapid spread, and final leveling off (Figure 1.19, page 42).

Spatial diffusion: the way that things spread through space and over time.

It is possible to recognize several different spatial tendencies in patterns of diffusion. In *expansion diffusion* (Figure 1.20a, page 42), a phenomenon spreads

Figure 1.19 Spatial diffusion The spatial diffusion of many phenomena tends to follow an S-curve of slow build-up, rapid spread, and leveling off. In the case of the diffusion of an innovation, for example, it usually takes a while for enough potential adopters to get to know about the innovation, and even longer for a critical mass of them to adopt it. After that, the innovation spreads quite rapidly, until most of the potential adopters have been exposed to it. (*Source:* D. J. Walmsley and G. J. Lewis. *Human Geography: Behavioural Approaches.* London: Longman, 1984, Fig. 5.3, p. 52. Reprinted by permission of Addison Wesley Longman Ltd.)

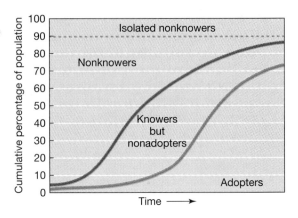

Figure 1.20 Patterns of spatial diffusion (a) expansion diffusion (for example, the spread of a contagious disease, such as cholera, across a city or the spread of an innovative agricultural practice, such as the use of hybrid seed-stock, across a rural area); (b) relocation diffusion (for example, the diffusion of a traditional architectural style through human migration); and (c) hierarchical diffusion (for example, the spread of a fashion trend from large metropolitan areas to smaller cities and towns). (*Source:* L. A. Brown and E. G. Moore, "Diffusion research: A perspective." In C. Board, R. J. Chorley, P. Haggett, and D. R. Stoddart (Eds.), *Progress in Geography.* London: Edward Arnold, 1969, Fig. 3. Reprinted by permission of Edward Arnold/Hodder & Stoughton Educational.)

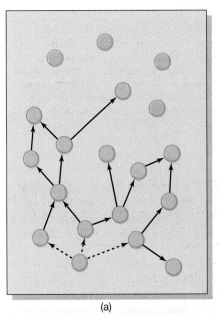

(a)

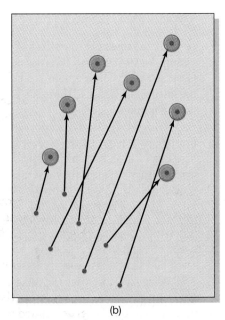

(b)

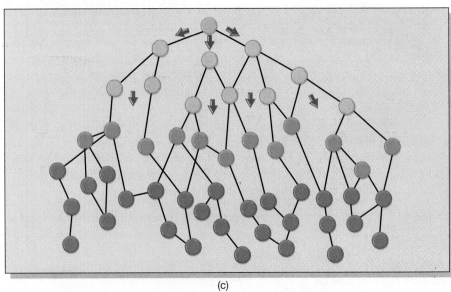

(c)

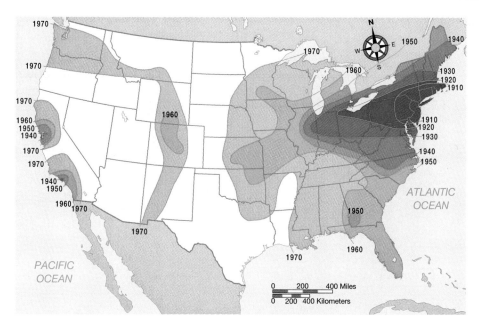

Figure 1.21 The spread of soccer as an intercollegiate sport Soccer spread from one campus to another in a process of expansion diffusion. This process depends on the proximity of fixed "carriers," in this case, nearby campuses in regional sports conferences. Soccer first spread from the Northeast into the urban industrial centers of the Ohio Valley, Illinois, and Indiana. Eastward spread from Los Angeles and San Francisco was markedly slower, largely because of the lower density of colleges in the West. (*Source:* J. Bale, *Sports Geography.* London: E. & F. Spon, 1989, Fig. 3.9, p. 62.)

because of the proximity of carriers, or agents of change, who are fixed in their location. An example is the spread of soccer as a college sport: westward from campuses in New York, Connecticut, Pennsylvania, and New Jersey, and eastward from California (Figure 1.21). With *relocation diffusion,* a phenomenon is diffused as an initial carrier or group of carriers moves from one location to another, taking the phenomenon with it as it travels or migrates (Figure 1.20b). An example would be the spread of a particular kind of building design favored by migrants from one region to another, as happened with the different styles of barns and farm buildings associated with the different streams of European immigrants that settled the American Midwest in the nineteenth century. Last, with *hierarchical diffusion,* a phenomenon can be diffused from one location to another without necessarily spreading to places in between (Figure 1.20c). An example would be the spread of a fashion trend from large metropolitan areas to successively smaller cities, towns, and rural settlements.

Actual patterns and processes of diffusion can reflect several of these tendencies, as different aspects of human interaction come in to play in different geographic settings. The spread of AIDS cases in the United States, for example (Figure 1.22, page 44), reflects a combination of two kinds of diffusion. The epidemic spread through a process of hierarchical diffusion from large metropolitan centers like New York and Los Angeles to smaller ones like Atlanta and Phoenix, and then to even smaller ones like Bismarck, North Dakota, and Bangor, Maine. At the same time, the epidemic spread through expansion diffusion as it moved outward locally from each of these centers. Another medical example—the diffusion of cholera outbreaks in the United States in the nineteenth century—suggests a combination of hierarchical and relocation diffusion (Figure 1.23, page 45). In addition, it reflects the way that changing networks of transportation resulted in different patterns and sequences of contagion.

Figure 1.22 The changing "AIDS surface" over the continental United States, 1982–1990 After initial outbreaks in major metropolitan areas, AIDS cases began to spread hierarchically, to smaller metropolitan areas; and by expansion, along major transport corridors and outward from major metropolitan cores. (*Source:* Reprinted with permission of Prentice Hall, from J. M. Rubenstein, *The Cultural Landscape: An Introduction to Human Geography,* © 1996, p. 33.)

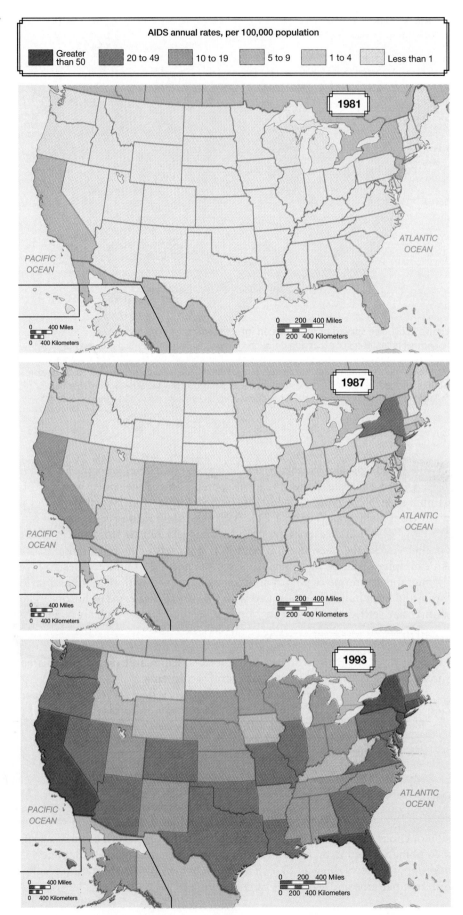

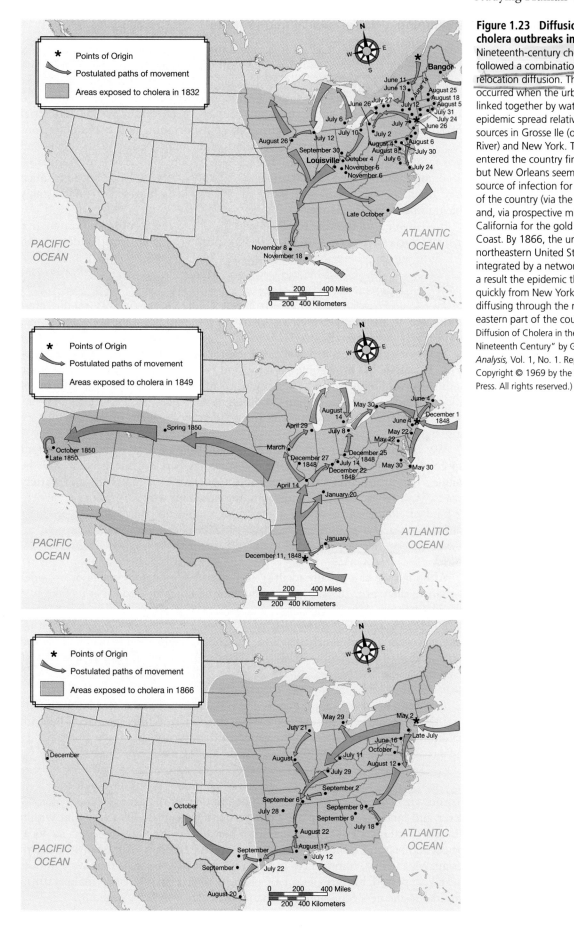

Figure 1.23 Diffusion patterns of cholera outbreaks in the United States
Nineteenth-century cholera epidemics followed a combination of hierarchical and relocation diffusion. The 1832 epidemic occurred when the urban system was loosely linked together by water transport, and the epidemic spread relatively slowly from its sources in Grosse Ile (on the St. Lawrence River) and New York. The 1849 epidemic entered the country first through New York, but New Orleans seems to have been the source of infection for most of the interior of the country (via the Mississippi system) and, via prospective miners traveling to California for the gold rush, for the West Coast. By 1866, the urban system of the northeastern United States had been closely integrated by a network of railroads, and as a result the epidemic that year spread quickly from New York to Detroit before diffusing through the remainder of the eastern part of the country. (*Source:* "The Diffusion of Cholera in the United States in the Nineteenth Century" by G. F. Pyle, *Geographical Analysis,* Vol. 1, No. 1. Reprinted by permission. Copyright © 1969 by the Ohio State University Press. All rights reserved.)

Developing a Geographical Imagination

A **geographical imagination** gives us the capacity to understand changing patterns, changing processes, and changing relationships among people, places, and regions. Developing this capacity is increasingly important as the pace of change around the world increases to unprecedented levels. Whereas much of the world had remained virtually unchanged for decades, even centuries, the Industrial Revolution and long-distance, high-speed transportation and communications brought a rapid series of rearrangements to the countryside and to towns and cities in many parts of the world. Today, with a globalized economy and global telecommunications and transportation networks, places have become much more interdependent, and still more of the world is exposed to increasingly urgent imperatives to change.

It is often useful to think of places and regions as representing the cumulative legacy of successive periods of change. For example, the present-day downtown of Edinburgh, Scotland, embodies elements of medieval, Georgian, Victorian, and modern urban fabric, while downtown Manchester, England, is mainly the product of nineteenth-century development, with a veneer of more recent additions and modifications. The riverfront of Köln, Germany, was rebuilt after World War II, using reproductions of its prewar buildings, most of which had been entirely destroyed (Figure 1.24).

Following this approach, we can look for superimposed layers of development. We can show how some patterns and relationships last, while others are modified or obliterated. We can show how different places bear the imprint of different kinds of change, perhaps in different sequences, and with different outcomes. In order to do so, we must be able to identify the kinds of changes that are most significant.

We can prepare our geographical imagination to deal with an important aspect of spatial change by making a distinction between the general and the unique. This distinction helps account for geographical diversity and variety because it provides a way of understanding how and why it is that one particular kind of change can result in a variety of spatial outcomes: It is because the general effects of a particular change always involve some degree of modification as they are played out in different environments, giving rise to unique outcomes.

Although we can usually identify some general outcomes of major episodes of change, there are almost always some unique outcomes, too. Let us take two related examples. The Industrial Revolution of nineteenth-century Europe provides a good example of a major period of change. A few of the general spatial outcomes were increased urbanization, regional specialization in production, and increased inter-regional and international trade. At one level, places could be said to have become increasingly alike: generic coalfield regions, industrial towns, ports, downtowns, worker housing, and suburbs.

It is clear, however, that these general outcomes were mediated by the different physical, economic, cultural, and social attributes of different places. Beneath the dramatic overall changes in the geography of Europe, new layers of diversity and variety also existed. Industrial towns developed their own distinctive character as a result of their manufacturing specialties, their politics, the personalities and objectives of their leaders, and the reactions and responses of their residents. Downtowns were differentiated from one another as the general forces of commerce and land economics played out across different physical sites and within different patterns of land ownership. They also were differentiated as local socioeconomic and political factors gave rise to different expressions of urban design. Meanwhile, some places came to be distinctive because they were almost entirely bypassed by this period of change, their characteristics making them unsuited to the new economic and spatial order (Figure 1.25, page 48).

(a) Royal Mile, in Edinburgh's old town

(b) Edinburgh Castle

(c) Manchester, England

(d) Köln, Germany

Figure 1.24 Places as the cumulative legacy of change
(a) and (b) These photographs of the downtown area of Edinburgh, Scotland, reflect the legacy of fragments of different layers of development, from the medieval period onward. To understand Edinburgh as a place, we have to see it as the cumulative legacy, not just of the buildings from different periods in the past, but also of the laws, institutions, customs, and so on, that developed in each of these periods; (c) Manchester, England, was a center of industrial growth in the nineteenth century. Its downtown is mainly the product of that era, patched over with more recent additions and modifications; (d) Köln (Cologne), Germany, suffered heavy damage during World War II (though the thirteenth-century cathedral survived). It has been rebuilt, but in such a way as to deliberately reinstate the legacy of past periods of development.

The second example of general and unique outcomes of change is the introduction of the railroad, one of the specific changes involved in the Industrial Revolution. In general terms, the railroad contributed to time–space convergence, to the reorganization of industry into larger market areas, to an increase in inter-regional and international trade, and to the interconnectedness of urban systems. Other unique outcomes, however, have also contributed to distinctive regional geographies. In Britain, the railroad was introduced to an environment that was partially industrialized and densely settled, and one of the main outcomes was that the increased efficiencies provided by the railroad helped to turn Britain's economy into a highly integrated and intensively urbanized national economy. In Spain, however, the railroad was introduced to an environment that was less urbanized and industrialized, and less able to afford the costs of railroad construction. The result was that the relatively few Spanish towns connected by the railroads gained a massive comparative advantage. This situation laid the foundation for a modern space-economy that was much less integrated than Britain's, with an urban system dominated by just a few towns and cities.

Figure 1.25 Bernkastel, Germany   Bernkastel, a small town in Germany, grew up as a market center because of its strategic situation: on the River Mosel, in the heart of an area of high-quality wine production. It suddenly stopped growing in the nineteenth century, however, when it was bypassed by new railway routes that provided other towns in the region with a locational advantage. It is now enjoying a new phase of prosperity as an attractive stop for tourists.

Conclusion

Human geography is the systematic study of the location of peoples and human activities across Earth's surface, and of their relationships to one another. An understanding of human geography is important, both from an intellectual point of view (that is, understanding the world around us) and from practical points of view (for example, contributing to environmental quality, human rights, social justice, business efficiency, political analysis, and government policymaking).

While modern ideas about the study of human geography developed from intellectual roots going back to the classical scholarship of ancient Greece, as the world itself has changed, our ways of thinking about it have also changed. What is distinctive about the study of human geography today is not so much the phenomena that are studied as the way they are approached. The contribution of human geography is to reveal, in relation to economic, social, cultural, and political phenomena, how and why geographical relationships matter, in terms of cause and effect.

Geography matters because it is in specific places that people learn who and what they are, and how they should think and behave. Places are also a strong influence, for better or worse, on people's physical well-being, their opportunities, and their lifestyle choices. Places also contribute to people's collective memory and become powerful emotional and cultural symbols. Places are the sites of innovation and change, of resistance and conflict.

To investigate specific places, however, we must be able to frame our studies of them within the compass of the entire globe. This is important for two reasons. First, the world consists of a complex mosaic of places and regions that are interrelated and interdependent in many ways. Second, place-making forces—especially economic, cultural, and political forces that influence the distribution of human activities and the character of places—are increasingly operating at global and international scales. The interdependence of places and regions means that individual places are tied in to wider processes of change that are reflected in broader geographical patterns. An important issue for human geographers is to recognize these wider processes and broad geographical patterns without losing sight of the individuality and uniqueness of specific places.

This global perspective leads to the following principles.

- Each place, each region, is largely the product of forces that are both local and global in origin.
- Each is, ultimately, linked to many other places and regions through these same forces.
- The individual character of places and regions cannot be accounted for by general processes alone. Some local outcomes are the product of unusual circumstances or special local factors.

Main Points

- Geography matters because it is specific places that provide the settings for people's daily lives. It is in these settings that important events happen; and it is from them that significant changes spread and diffuse.
- Places and regions are highly *interdependent,* each filling specialized roles in complex and ever-changing networks of interaction and change.
- In today's world, some of the most important aspects of the interdependence between geographic scales are provided by the relationships between the *global* and the *local.*
- Human geography provides ways of understanding places, regions, and spatial relationships as the products of a series of interrelated forces that stem from nature, culture, and individual human action.

- The first law of geography is that "everything is related to everything else, but near things are more related than distant things."
- Distance is one aspect of this law, but connectivity is also important, because contact and interaction are dependent on channels of communication and transportation: streets, highways, and telephone lines, for example.

Key Terms

accessibility (p. 38)
cartography (p. 8)
cognitive distance (p. 35)
cognitive images (mental maps (p. 34)
cognitive space (p. 37)
distance-decay function (p. 36)
economies of scale (p. 39)
environmental determinism (p. 14)

ethnocentrism (p. 14)
friction of distance (p. 35)
geographical imagination (p. 46)
geographic information systems (GIS) (p. 20)
globalization (p. 15)
human geography (p. 2)
imperialism (p. 14)

latitude (p. 33)
longitude (p. 33)
masculinism (p. 14)
region (p. 24)
regional geography (p. 24)
remote sensing (p. 20)
site (p. 33)
situation (p. 33)
spatial diffusion (p. 41)

time–space convergence (p. 40)
topological space (p. 37)
utility (p. 36)

Exercises

At the end of each chapter, you will find exercises and activities based on using the Internet, along with some that do not require access to the Internet. This book has its own "home page" on the Internet, where you will find additional resources—maps, photographs, data—as well as exercises and activities that relate to each chapter. You will also find an evaluation check sheet and suggestion form that you can mail to the authors electronically.

 On the Internet

The Internet exercises for this chapter explore the interdependence of places. Drawing on the many resources that the Internet offers for studying places, these exercises will provide you with an understanding of why geography matters.

Unplugged

1. Consider geographical interdependence from the point of view of your own life.

   (a) Take an inventory of your clothes, noting, where possible, where each garment was manufactured. Where did you buy the garments? Was the store part of a regional, national, or international chain? What can you find out about the materials used in the garments? Where were they made?

   (b) Try to establish the "commodity chains" for the ingredients of your next meal: Where were they produced? Where did they go to be processed, warehoused, distributed, and so on?

2. Describe, as exactly and concisely as possible, the *site* (see page 34) of your campus. Then describe its *situation* (see page 34). Can you think of any reasons why the campus is sited and situated just where it is? Would there be a better location; and, if so, why?

3. Choose a local landscape, one with which you are familiar. Make notes on what you think may have contributed to this landscape. Is there any evidence that physical, environmental conditions have shaped any of the human elements in the landscape? Is there any evidence of people having modified the physical landscape?

Chapter 2

The Changing Global Context

Downtown Tokyo

One cliché about the Information Age is that instantaneous global telecommunications, satellite television, and the Internet will soon overthrow all but the last vestiges of geographical differentiation in human affairs. Companies, according to this view, will need no headquarters—they will be able to locate their activities almost anywhere in the world. Employees will work as effectively from home, car, or beach as they could in the offices that need no longer exist. Events halfway across the world will be seen, heard, and felt with the same immediacy as events across town. National differences and regional cultures will dissolve, the cliché has it, as a global marketplace brings a uniform dispersion of people, tastes, and ideas.

Such developments are, in fact, highly unlikely. Even in the Information Age, geography will still matter, and may well become more important than ever. Places and regions will undoubtedly change as a result of the new global context of the Information Age. But *geography will still matter* because of several factors: transport costs, differences in resource endowments, fundamental principles of spatial organization, people's territorial impulses, the resilience of local cultures, and the legacy of the past. An editorial in *The Economist* magazine, debunking the cliché of a spaceless information economy, and pointing out that place and space *do* matter, explained, "The main reason is that history counts: where you are depends very much on where you started from."[1]

The Changing World

The essential foundation for an informed human geography is an ability to understand places and regions as components of a constantly changing global system. In this sense, all geography is historical geography. Built into every place and each region is the legacy of a sequence of major changes in world geography. One of the key features of this aspect of human geography, however, is that the sequence of changes has not been the same everywhere. We can best understand these changes and their consequences for different places and regions by thinking in terms of the world as an evolving, competitive, political-economic system.

The modern world-system was first established over a long period that began in the late fifteenth century and lasted until the mid-seventeenth century. A **world-system** is an *interdependent* system of countries linked by political and economic competition. The hyphenation in the term "world-system," which was coined by historian Immanuel Wallerstein, is meant to emphasize the interdependence of places and regions around the world.

In parts of fifteenth-century Europe, exploration beyond European shores began to be seen as an important way of opening up new opportunities for trade and economic expansion. By the sixteenth century, new techniques of shipbuilding and navigation had begun to bind more and more places and regions together through trade and political competition. As a result, more and more peoples around the world became exposed to one another's technologies and ideas. Their different resources, social structures, and cultural systems resulted in quite different pathways of development, however. Some societies were incorporated into the new, European-based international economic system faster than others; some resisted incorporation; and some sought alternative systems of economic and political organization.

Since the seventeenth century, the world-system has been consolidated with stronger economic ties between countries. It has also been extended, with all the

World-system: an *interdependent* system of countries linked by economic and political competition.

---

[1]"Does It Matter Where You Are?" *The Economist*, July 30, 1994, pp. 13–14.

world's countries eventually becoming involved, to some extent, in the interdependence of the capitalist system. There have, however, been many instances of resistance and adaptation, with some countries (Tanzania, for example) attempting to become self-sufficient and others (China and Cuba, for example) seeking to opt out of the system altogether in order to pursue a different path to development—communism. Today, the overall result is that a highly structured relationship between places and regions has emerged. This relationship is organized around three tiers: *core, semiperipheral,* and *peripheral regions.* These broad geographic divisions have been created through a combination of processes of private economic competition and competition among States. **States are independent political units with territorial boundaries that are internationally recognized by other States** (see the discussion of States, nations, and nation-states in Chapter 9).

The **core regions** of the world-system are those that dominate trade, control the most advanced technologies, and have high levels of productivity within diversified economies. As a result, they enjoy relatively high *per capita* incomes. The first core regions of the world-system were the trading hubs of Holland and England. Later, these were joined by manufacturing and exporting regions in other parts of Western Europe and in North America; later still, by Japan.

The success of these core regions depends on their dominance and exploitation of other regions. This dominance, in turn, depends on the participation of these other regions within the world-system. Initially, such participation was achieved by military enforcement, then by European colonialism, and finally by the sheer economic and political influence of the core regions. **Colonialism** involves the establishment and maintenance of political and legal domination by a State over a separate and alien society. This domination usually involves some colonization (that is, the physical settlement of people from the colonizing State) and always results in economic exploitation by the colonizing State.

Regions that have remained economically and politically unsuccessful throughout this process of incorporation into the world-system are peripheral. **Peripheral regions** are characterized by dependent and disadvantageous trading relationships, by primitive or obsolescent technologies, and by undeveloped or narrowly specialized economies with low levels of productivity.

In between core regions and peripheral regions are semiperipheral regions. **Semiperipheral regions** are able to exploit peripheral regions but are themselves exploited and dominated by the core regions. They consist mostly of countries that were once peripheral. This semiperipheral category underlines the fact that neither peripheral status nor core status is necessarily permanent. The United States and Japan both achieved core status after having been peripheral; Spain and Portugal, part of the original core in the sixteenth century, became semiperipheral in the nineteenth century but are now once more part of the core. Quite a few countries, including Brazil, India, Mexico, South Korea, and Taiwan, have become semiperipheral after first having been incorporated into the periphery of the world-system and then developing a successful manufacturing sector that moved them into semiperipheral status.

An important determinant of these changes in status is the effectiveness of States in ensuring the international competitiveness of their domestic producers. They can do this in several ways: by manipulating markets (protecting domestic manufacturers by charging taxes on imports, for example); by regulating their economies (enacting laws that help to establish stable labor markets, for example); and by creating physical and social infrastructures (spending public funds on road systems, ports, educational systems, and so on). Because some States are more successful than others in pursuing these strategies, the hierarchy of three geographical tiers is not rigid. Rather, it is fluid, providing a continually changing framework for geographical transformation within individual places and regions (Figure 2.1, page 54).

States: independent political units with territorial boundaries that are internationally recognized by other States.

Core regions: regions that dominate trade, control the most advanced technologies, and have high levels of productivity within diversified economies.

Colonialism: the establishment and maintenance of political and legal domination by a State over a separate and alien society.

Peripheral regions: regions with undeveloped or narrowly specialized economies with low levels of productivity.

Semiperipheral regions: regions that are able to exploit peripheral regions but are themselves exploited and dominated by core regions.

Figure 2.1 The world-system core, semiperiphery, and periphery (a) in 1795, (b) in 1895, and (c) in 1995—Note how the Dymaxion projection used in these maps (see Chapter 1) emphasizes the relative proximity of core regions and accentuates the geographic isolation of the economically peripheral regions. While Europe and Africa are oriented conventionally in the Dymaxion projection (with North toward the top of the page), the Americas, Asia, and Australia are shown at right angles to their conventional North-South orientation. (*Source:* Map projection, Buckminster Fuller Institute and Dymaxion Map Design, Santa Barbara, CA. The word Dymaxion and the Fuller Projection Dymaxion™ Map design are trademarks of the Buckminster Fuller Institute, Santa Barbara, California, ©1938, 1967 & 1992. All rights reserved.)

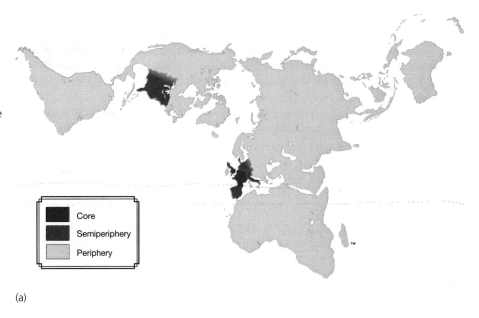

(a)

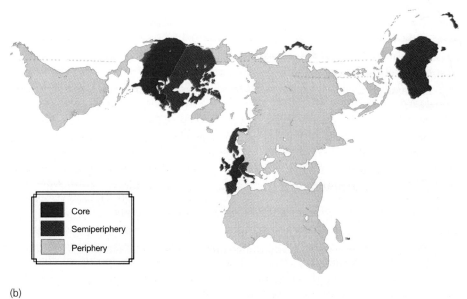

(b)

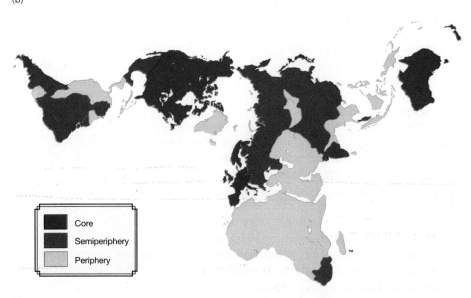

(c)

Geographic Expansion, Integration, and Change

The world-system evolved in successive stages of geographic expansion and integration. This evolution has affected the roles of individual places and the nature of the interdependence among places. It also explains why places and regions have come to be distinctive and how this distinctiveness has formed the basis of geographic variability.

Hearth Areas

Systematically differentiated human geographies began with minisystems. A **minisystem** is a society with a single cultural base and a *reciprocal* social economy. That is, each individual specializes in particular tasks (tending animals, cooking, or making pottery, for example), freely giving any excess product to others, who reciprocate by giving up the surplus product of their own specialization. Such societies are found only in subsistence-based economies (such as hunting, gathering, and livestock raising). Because they do not have the ability (or the need) to sustain an extensive physical infrastructure, they are limited in geographic scale. Before the first agricultural revolution, in prehistoric times, minisystems had been based on hunting and gathering societies that were finely tuned to local physical environments. They were all very small in geographical extent, and very vulnerable to environmental change. After this first agricultural revolution, minisystems were both more extensive and more stable. These qualities eventually contributed to new forms of spatial organization, including urbanization and long-distance trading.

The first agricultural revolution was a transition from hunter-gatherer minisystems to agricultural-based minisystems that began in the Proto-Neolithic, or early Stone Age, period between 9000 and 7000 B.C. The transition was based on a series of technological preconditions: the use of fire to process food, the use of grindstones to mill grains, and the development of improved tools to prepare and store food. One key breakthrough was the evolution and diffusion of a system of slash-and-burn agriculture that is now known as "swidden" cultivation (see Chapter 8). **Slash-and-burn** is a system of cultivation in which plants are cropped close to the ground, left to dry for a period, and then ignited. In Neolithic times this system involved planting familiar species of wild cereals or tubers on scorched land. Slash-and-burn did not require special tools, weeding, or fertilization, provided that cultivation was switched to a new plot after a few crops had been taken from the old one. Another key breakthrough was the domestication of cattle and sheep, a technique that had become established in a few regions by Neolithic, or Stone Age, times (7000 to 5500 B.C.).

As cultural geographer Carl Sauer pointed out in his book *Agricultural Origins and Dispersals* (1952), these agricultural breakthroughs, which triggered the first agricultural revolution, could only take place in certain geographic settings: where natural food supplies were plentiful; where the terrain was diversified (thus offering a variety of habitats and a variety of species); where soils were rich and relatively easy to till; and where there was no need for large-scale irrigation or drainage. Archeological evidence suggests that the breakthroughs took place independently in several different agricultural hearth areas, and that agricultural practices diffused slowly outward from each (Figure 2.2, page 56). **Hearth areas** are geographic settings where new practices have developed, and from which they have subsequently spread. The main hearth areas were situated in four broad regions:

- In the Middle East: in the so-called Fertile Crescent around the foothills of the Zagros Mountains (parts of present-day Iran and Iraq), around the Dead Sea Valley (Jordan and Israel), and on the Anatolian Plateau (Turkey)
- In South Asia: along the floodplains of the Ganges, Brahmaputra, Indus, and Irawaddy Rivers (Assam, Bangladesh, Burma, and northern India)

Minisystem: a society with a single cultural base and a reciprocal social economy.

Slash-and-burn: system of cultivation in which plants are cropped close to the ground, left to dry for a period, and then ignited.

Hearth areas: geographic settings where new practices have developed, and from which they have subsequently spread.

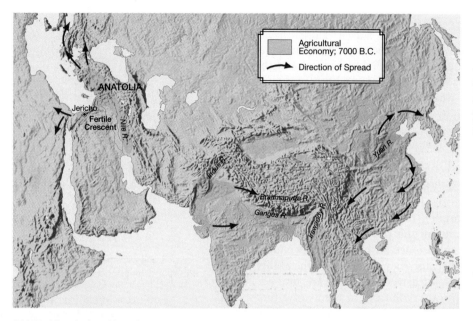

Old World agricultural hearths

Archeological evidence from this site in Jericho has shown that a prosperous agricultural economy based on cultivated wheat and barley was in place as long ago as 7000 B.C. The community depended on domestic sheep and goats as the dominant sources of meat.

Uxmal, on the Yucatán peninsula, was a Mayan city that flourished between 600 and 900 A.D. and was finally abandoned in the mid-fifteenth century.

Figure 2.2 Hearth areas of the first agricultural revolution The first Agricultural Revolution took place independently in several hearth areas, where hunter-gatherer communities began to experiment with locally available plants and animals in ways that eventually led to their domestication. From these hearth areas, improved strains of crops, domesticated animals, and new farming techniques diffused slowly outward. Farming supported larger populations than were possible with hunting and gathering, and the extra labor allowed the development of other specializations, such as pottery making and jewelry.

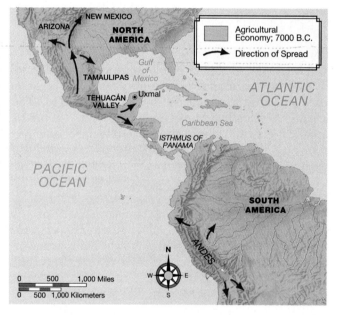

New World agricultural hearths

- In China: along the floodplain of the Yuan River
- In the Americas: in Mesoamerica (the middle belt of the Americas that extends north to the North American Southwest and south to the Isthmus of Panama) around Tamaulipas and the Tehuacán Valley (Mexico) and in Arizona and New Mexico, and along the western slopes of the Andes in South America

This transition to food-producing minisystems had several important implications for the long-term evolution of the world's geographies. First, it allowed much higher population densities and encouraged the proliferation of settled villages. Second, it brought about a change in social organization, from loose communal systems to systems that were more highly organized on the basis of kinship. Kin groups provided a natural way of assigning rights over land and resources and of organizing patterns of land use. Third, it allowed some specialization in nonagricultural crafts, such as pottery, woven textiles, jewelry, and weaponry. This specialization led to a fourth development: the beginnings of barter and trade between communities, sometimes over substantial distances.

Most minisystems vanished a long time ago, although some remnants of minisystems have survived long enough to provide material for *Discovery Channel* "life-in-a-time-warp" specials (Figure 2.3). Examples of these residual and fast-disappearing minisystems are the bushmen of the Kalahari, the hill tribes of Papua New Guinea, and the tribes of the Amazonian rain forest. They contribute powerfully to regional differentiation and sense of place in a few enclaves around the world, but their most important contribution to contemporary human geographies is in the way that they provide a stark counterpoint to the landscapes and practices of the contemporary world-system.

The Growth of Early Empires

The higher population densities, changes in social organization, craft production, and trade brought about by the first agricultural revolution provided the preconditions for the emergence of several world-empires. A **world-empire** is a group of minisystems that have been absorbed into a common political system while retaining their fundamental cultural differences. The social economy of world-empires can be characterized as redistributive-tributary. That is, wealth is appropriated from producer classes by an elite class in the form of taxes or tribute. This redistribution of wealth is most often achieved through military coercion or religious persuasion, or a combination of the two. The best-known world-empires are the largest and longest-lasting of the ancient civilizations—Egypt, Greece, China, Byzantium, and Rome. Figure 2.4 (see page 58) shows the geographic extent of some of the later world-empires: the Ottoman empire, the Chinese empire of the Ch'ing dynasty, the Inca empire, and Mughal India. All these world-empires brought two important new elements to the evolution of the world's geographies. The first was the emergence of *urbanization* (see Chapter 10). Towns and cities became essential as centers of administration, as military garrisons, and as theological centers for the ruling classes who were able to use a combination of military and theological authority to hold their empires together. As long as these early world-empires were successful, they not only gave rise to monumental capital cities but also to a whole series of secondary settlements, which acted as intermediate centers in the flow of tribute and taxes from colonized territories.

The most successful world-empires, such as the Greek and Roman empires, established quite extensive urban systems. In general, the settlements in these urban systems were not very large—typically ranging from a few thousand inhabitants to about 20,000. The seats of empire grew quite large, however. The Mesopotamian city of Ur, in present-day Iraq, for example, has been estimated to have reached a population of around 200,000 by 2100 B.C., while Thebes, the capital of Egypt, is thought to have had more than 200,000 inhabitants in 1600 B.C. Athens and Corinth, the largest of the cities of ancient Greece, had populations between 50,000 and 100,000 by 400 B.C.. Rome, at the height of the Roman Empire (around 200 A.D.), may have had as many as a million inhabitants. The most impressive thing about these cities, though, was not so much their size as their degree of sophistication: elaborately laid out, with paved streets, piped water, sewage systems, massive monuments, grand public buildings, and impressive city walls.

The second important contribution of world-empires to evolving world geographies was *colonization*. In part, this was an indirect consequence of the operation of the **law of diminishing returns.** This law refers to the tendency for productivity to decline, after a certain point, with the continued addition of

Figure 2.3 A remnant minisystem: the head man of a seminomadic clan in Swaziland Minisystems were rooted in subsistence-based social economies that were organized around reciprocity, each person specializing in certain tasks and freely giving any surplus to others, who in turn passed on the surplus from their own work. Today, few of the world's remnant minisystems are "pure," unaffected by contact with the rest of the world. Their economies and societies have become more complex, using money, for example, as the basis for exchange. Some of the most striking illustrations of the imprint of globalization on local communities are those in which mass-produced consumer goods have found their way into remnant minisystems.

World-empire: minisystems that have been absorbed into a common political system while retaining their fundamental cultural differences.

Law of diminishing returns: the tendency for productivity to decline, after a certain point, with the continued application of capital and/or labor to a given resource base.

Figure 2.4 Selected world empires (a) the Ottoman empire in 1350, 1480 and 1590; (b) China under the Ch'ing dynasty (1775); (c) Mughal India (1707); and (d) the Inca empire in 1300 and 1500.

(a) The Ottoman empire in 1350, 1480, and 1590

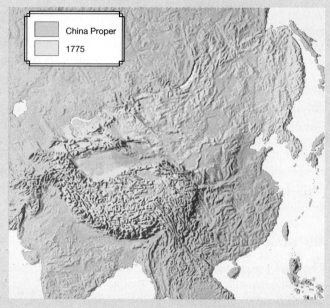

(b) China under the Ch'ing dynasty (1775)

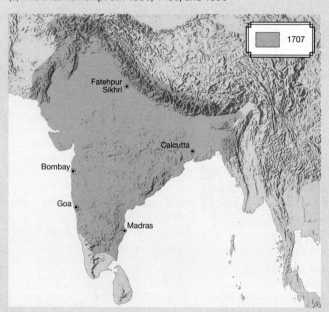

(c) Mughal India

(d) The Inca empire in 1300 and 1500

Fatehpur Sikri, a city in northern India that was founded by the Mughal emperor Akbar in 1569

capital and/or labor to a given resource base. Because of the law of diminishing returns, world-empires could only support growing populations if overall levels of productivity could be increased. While some productivity gains could be achieved through better agricultural practices, harder work, and improvements in farm technology, a fixed resource base meant that, as populations grew, overall levels of productivity fell. For each additional person working the land, the gain in production per worker was less. The usual response was to enlarge the resource base by colonizing nearby land. This colonization had immediate spatial consequences in terms of establishing dominant/subordinate spatial relationships between hearths and colonies. It was also important in establishing hierarchies of settlements and creating improved transportation networks. The military underpinnings of colonization also meant that new towns and cities now came to be carefully sited for strategic and defensive reasons.

The legacy of these important changes is still apparent in today's landscapes. The clearest examples are in Europe, where the Roman world-empire colonized an extensive territory that was controlled through a highly developed system of towns and connecting roads (see Feature 2.1, "Visualizing Geography—Legacies of the Roman Empire"). Most Western European cities that are important today had their origin as Roman settlements. In quite a few it is possible to trace their original street layouts. In some it is possible to glimpse remnants of defensive city walls and of paved streets, aqueducts, viaducts, arenas, sewage systems, baths, and public buildings. In the modern countryside, meanwhile, we can still read the legacy of the Roman world-empire in arrow-straight roads, built by their engineers and maintained and improved by successive generations.

Some world-empires were exceptional in that they were based on a particularly strong central State, with totalitarian rulers who were able to organize large-scale, communal land improvement schemes using forced labor. These world-empires were found in China, India, the Middle East, Central America, and the Andean region of South America. Their dependency on large-scale land improvement schemes (particularly irrigation and drainage schemes) as the basis for agricultural productivity has led some scholars to characterize them as *hydraulic societies*. Today, their legacy can be seen in the landscapes of terraced fields that have been maintained for generations in places like Sikkim, India, and East Java, Indonesia (Figure 2.5).

Figure 2.5 Terraced landscapes These landscapes are the legacy of hydraulic societies—world-empires in which despotic rulers once organized labor-intensive irrigation schemes that allowed for significant increases in agricultural productivity. Irrigation schemes such as these were widely copied and, having been maintained for generations, have become the basis for local economies and ways of life in a number of subregions in Asia and Latin America.

Rice cultivation in Sikkim, formerly a kingdom but now an Indian state

Terrace maintenance, East Java

The Roman Empire in the second century A.D.

Remains of the Temple of Vespasian in the Forum in Rome.

The Roman theater at Mérida, Spain, which dates from around 25 B.C. Mérida was the capital of the Roman province of Lusitania (which encompassed modern Portugal) and became one of the most important towns on the Iberian peninsula, large enough to contain a garrison of 90,000.

This spectacular aqueduct was built in the time of Augustus to supply the city of Nîmes, in southern France, with water from Uzès, 50 kilometers (31 miles) away. The whole project is a testament to the surveying and engineering skills of the Romans. The water dropped only 17 meters (56 feet) over the whole distance.

Roman road building was an important aspect of empire building, and many of the roads that were built by Romans became established as major routeways throughout Europe. Wherever they could, Roman surveyors laid out roads in straight lines. This example is the Appian Way, outside Rome.

The Forum was Rome's political center. Here, in courts and assemblies, consuls and senators took decisions on the internal state of the republic and the fate of the empire. Successive emperors vied with one another to embellish the triumphal architecture of the area.

The Colosseum in Rome was opened in A.D. 80. The arena seated about 50,000 spectators and was covered by a huge awning. After the fall of the Roman Empire, the structure was neglected. Its marble and brick were pillaged for the construction of churches and palaces, and it was converted for a while into a fortress.

Ercolano (Herculaneum), like Pompeii, was smothered with mud and cinders from the eruption of Mount Vesuvius in A.D. 79. It was discovered by well-diggers in the eighteenth century, and is still being excavated. Ercolano was a town of wealthy citizens, with streets of curbed sidewalks and two-storied houses with balconies and overhanging roofs for shade.

Part of Trajan's Column in Rome, a 35-meter monument covered with a spiral frieze that depicts the emperor Trajan's wars with the Dacians in A.D. 101–102 and 105–106.

Water fountain, Herculaneum

Hadrian's Wall, a Roman defensive barrier guarding the northern frontier of the province of Britain. Completed in A.D. 136, it extended almost 110 kilometers (75 miles), with an original height of 6 meters (20 feet), a thickness of 2.5 meters (8–9 feet), and a flat-bottomed ditch that ran from coast to coast in northern Britain. The wall was protected by a series of small forts, but it was evacuated in A.D. 383 after several incursions by northern tribes.

The Geography of the Pre-Modern World

Figure 2.6 shows the generalized framework of human geographies in the Old World as they existed around 1400 A.D. The following characteristics of this period are important. First, harsher environments in continental interiors were still peopled by isolated, subsistence-level, kin-ordered hunting and gathering minisystems. Second, the dry belt of steppes and desert margins stretching across the Old World from the western Sahara to Mongolia was a continuous zone of kin-ordered pastoral minisystems. Third, the hearths of sedentary agricultural production extended in a discontinuous arc from Morocco to China, with two main outliers: in the central Andes and in Mesoamerica. The dominant centers of global civilization were China, northern India (both of them hydraulic variants of world-empires), and the Ottoman Empire of the eastern Mediterranean. Other important world-empires were based in Southeast Asia, in Muslim city-states of coastal North Africa, in the grasslands of West Africa, around the gold and copper mines of East Africa, and in the feudal kingdoms and merchant towns of Europe.

These more-developed realms were interconnected through trade, which meant that several emerging centers of capitalism existed. Port cities were particularly important, and among the leading centers were the city-state of Venice; the

Figure 2.6 The precapitalist Old World, circa 1400 Principal areas of sedentary agricultural production are shaded. Some long-distance trade took place from one region to another, but for the most part it was limited to a series of overlapping regional circuits of trade. (*Source:* After R. Peet, *Global Capitalism: Theories of Societal Development.* New York: Routledge, 1991; J. Abu-Lughod, *Before European Hegemony: The World-System A.D. 1200–1350.* New York: Oxford University Press, 1989; and E. R. Wolf. *Europe and the People Without History.* Berkeley: University of California Press, 1983.)

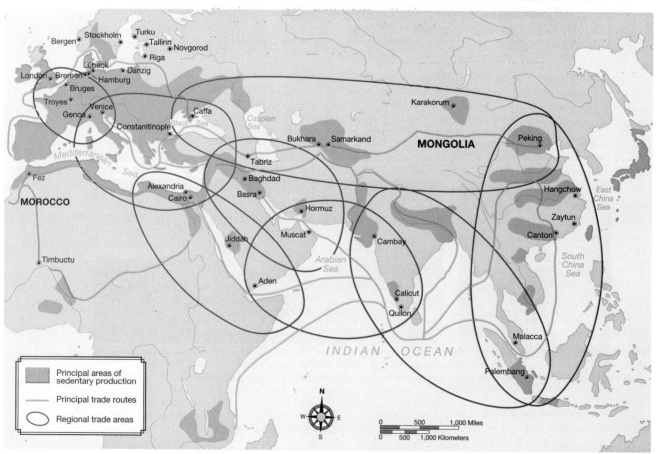

Hanseatic League of independent city-states in northwestern Europe (including Bergen, Bremen, Danzig, Hamburg, Lübeck, Riga, Stockholm, and Tallin; and affiliated trading outposts in other cities, including Antwerp, Bruges, London, Turku, and Novgorod); Cairo, Calicut (now Kozhikode), Canton, Malacca, and Sofala. Traders in these port cities began to organize the production of agricultural specialties, textiles, and craft products in their respective hinterlands. The hinterland of a town or city is its sphere of economic influence—the tributary area from which it collects products to be exported and through which it distributes imports. By the fifteenth century several regions of budding capitalism existed: northern Italy, Flanders, southern England, the Baltic, the Nile Valley, Malabar, Coromandel, Bengal, northern Java, and southeast coastal China (Figure 2.6).

Hinterland: the sphere of economic influence of a town or city.

Mapping a New World Geography

With the emergence of the modern world-system at the beginning of the sixteenth century, a whole new geography began to emerge. Although several regions of budding capitalist production existed, and although imperial China could boast of sophisticated achievements in science, technology, and navigation, it was European merchant capitalism that reshaped the world. Several factors motivated European overseas expansion. A relatively high-density population and a limited amount of cultivable land meant that a continuous struggle occurred to provide enough food. Meanwhile, the desire for overseas expansion was intensified by both competition among a large number of small monarchies and by inheritance laws that produced large numbers of impoverished aristocrats with little or no land of their own. Many of these were eager to set out for adventure and profit.

Added to these motivating factors were the enabling factors of innovations in shipbuilding, navigation, and gunnery. In the mid-1400s, for example, the Portuguese developed a cannon-armed ship—the caravel—that could sail anywhere, defend itself against pirates, pose a threat to those who were initially unwilling to trade, *and* carry enough goods to be profitable. The quadrant (1450) and the astrolabe (1480) enabled accurate navigation and mapping of ocean currents, prevailing winds, and trade routes. Naval power enabled the Portuguese and the Spanish to enrich their economies with capital from gold and silver plundered from the Americas.

Europeans were able not only to send adventurers for gold and silver, but also to take land, decide on its use, and exploit coerced labor to produce high-value crops (such as sugar, cocoa, and indigo) on plantations. A **plantation** is a large landholding that usually specializes in the production of one particular crop for market. Some regions, whose populations were resistant to European disease, and which also had high population densities, a good resource base, and strong States, were able to keep Europeans at arm's length. For the most part, these regions were in South and East Asia. Their dealings with Europeans were conducted through a series of coastal trading stations.

Plantation: a large landholding that usually specializes in the production of one particular crop for market.

Within Europe, innovations in business and finance (banking, loan systems, credit transfers, commercial insurance, and courier services, for example) helped to increase savings, investment, and commercial activity. European merchants and manufacturers also became adept at **import substitution**—copying and making goods previously available only by trading. The result was the emergence of Western Europe as the core region of a world-system that had penetrated and incorporated significant portions of the rest of the world (Figure 2.7, page 64).

Import substitution: the process by which domestic producers provide goods or services that formerly were bought from foreign producers.

For Europe, this overseas expansion stimulated further improvements in technology. These included new developments in nautical map making, naval artillery, shipbuilding, and the use of sail. The whole experience of overseas expansion also

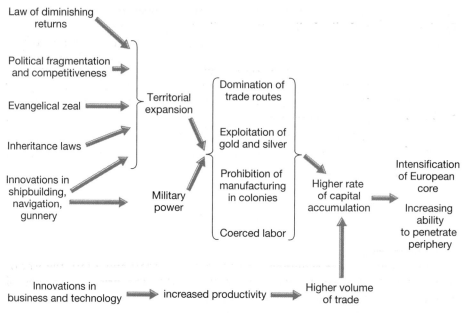

Figure 2.7 Factors in the emergence of a European-based world-system The single most important factor driving European overseas expansion was probably the law of diminishing returns. European growth could be sustained only as long as productivity could be improved, and after a point this could only be obtained by the conquest—peaceful or otherwise—of new territories. (*Source:* P. Knox and J. Agnew, *The Geography of the World Economy,* 2nd Ed., London: Edward Arnold, 1994, Fig. 5.3. Reprinted by permission of Edward Arnold/Hodder & Stoughton Educational.)

provided a great practical school for entrepreneurship and investment. In this way, the self-propelling growth of merchant capitalism was intensified and consolidated.

For the periphery, European overseas expansion meant dependency (as it has ever since for many of the world's peripheral regions). At worst, territory was forcibly occupied and labor systematically exploited. At best, local traders were displaced by Europeans, who imposed their own terms of economic exchange. Europeans soon destroyed most of the Muslim shipping trade in the Indian Ocean, for example (Figure 2.8), and went on to capture a large share of the

Figure 2.8 The Dutch navy attacking the Spanish Canary Islands in 1599 The expansion of European trade and the protection of trade routes required strong navies and a willingness to use them. England's Sir Walter Raleigh expressed the sentiment succinctly in 1608: "Whoso commands the sea commands the trade of the world; whoso commands the trade of the world commands the riches of the world."

oceangoing trade within Asia, selling Japanese copper to China and India, Persian carpets to India, Indian cotton textiles to Japan, and so on.

As revolutionary as these changes were, however, they were constrained by a technology that rested on wind and water power, on wooden ships, wooden structures, and on wood for fuel. Grain mills, for example, were built of wood and powered by water or wind. They could generate only modest amounts of power, and only at sites determined by physical geography, not human choice. Within the relatively small European landmass, wood for structural use and for fuel competed for acreage with food and textile fibers.

More important, however, the size and strength of timber imposed structural limits on the size of buildings, the diameter of waterwheels, the span of bridges, and so on. In particular, it imposed limits on the size and design of ships, which in turn imposed limits on the volume and velocity of world trade. The unsuitability of such technology for overland transportation also meant that, for a long time, the European world-system could only penetrate into continental interiors along major rivers.

After 300 years of evolution, roughly between 1450 and 1750, the world-system had incorporated only parts of the world. The principal spheres of European influence were Mediterranean North Africa; Portuguese and Spanish colonies in the Americas; Indian ports and trading colonies; the East Indies; African and Chinese ports; the Greater Caribbean; and North America. The rest of the world functioned more or less as before, with slow-changing geographies based around modified minisystems and world-empires that were only partially and intermittently penetrated by market trading.

Industrialization and Geographic Change

With the new production and transportation technologies of the Industrial Revolution (from the late 1700s), capitalism truly became a global system that reached into virtually every part of the world and into virtually every aspect of people's lives. Human geographies were recast again, this time with a more interdependent dynamic. New production technologies, based on more efficient energy sources, helped raise levels of productivity and create new and better products that stimulated demand, increased profits, and created a pool of capital for further investment. New transportation technologies triggered successive phases of geographic expansion, allowing for internal development as well as for external colonization and imperialism, the deliberate exercise of military power and economic influence by core States in order to advance and secure their national interests (see Chapter 7).

The colonization and imperialism that accompanied the expansion of the world-system was closely tied to the evolution of world leadership cycles. **Leadership cycles** are periods of international power established by individual States through economic, political, and military competition. In the long term, success in the world-system depends on economic strength and competitiveness, which brings political influence and pays for military strength. With a combination of economic, political, and military power, individual States can dominate the world-system, setting the terms for many economic and cultural practices and imposing their particular ideology by virtue of their pre-eminence. This kind of dominance is known as hegemony. **Hegemony** refers to domination over the world economy, exercised—through a combination of economic, military, financial, and cultural means—by one national State in a particular historical epoch. Over the long run, the costs of maintaining this kind of power and influence tend to weaken the hegemon. This phase of the cycle is known as *imperial overstretch*. It is followed by another period of competitive struggle, which brings the possibility of a new dominant world power (see Feature 2.2, "Geography Matters— World Leadership Cycles").

Leadership cycles: periods of international power established by individual States through economic, political, and military competition.

Hegemony: domination over the world economy, exercised by one national State in a particular historical epoch through a combination of economic, military, financial, and cultural means.

The modern world-system has so far experienced five full leadership cycles. In much simplified terms, and omitting for the moment any details of the consequent geographic changes, they have involved dominance by Portugal, the Netherlands, Great Britain, and the United States.

Portuguese dominance was established through initial advantages derived from Atlantic exploration, trade, and plunder. The Treaty of Tordesillas, arbitrated by Pope Alexander VI in 1494, consolidated these advantages by limiting direct competition with Portugal's chief rival, Spain. The treaty allowed Portugal to lay claim to any territory to the east of a line drawn north–south, 370 leagues (1850 kilometers, or 1110 miles) west of Cape Verde (that is, between 48° and 49° West), which included the eastern coast of Brazil (Figure 2.2.1). Spain was allowed to claim the less accessible lands to the west of this line. Subsequently allied together through royal marriage, Spain and Portugal sought to head off competition from England and Holland through imposing a global world-empire based on plunder rather than trade. By 1550, Portugal had established a number of important outposts around the world (Figure 2.2.2). The Portuguese attempt to establish a global world empire was thwarted by English naval superiority, but it was the Dutch whose superior merchant economy gave them the resources to become the next hegemon.

Dutch dominance began with the failure of the Portuguese-backed Spanish Armada (against England) in 1588. The foundation of Dutch economic prosperity lay in the fishing and shipping industries. Dutch ports and Dutch shipping dominated European trade, and the Dutch government coordinated Dutch merchants' long-distance trade through the Dutch East India Company and the Dutch West India Company (see Figure 2.2.3).

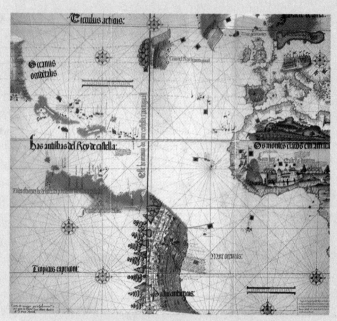

Figure 2.2.2 Portugal's claims to the New World The Treaty of Tordesillas (1494) gave Portugal control over what were the more accessible parts of the New World. This map, drawn in 1502 by Alberto Contino, shows Portuguese territories marked by flags with a blue interior and a red border. Spanish territories, farther west, are marked by the red, gold, and black flag of Spain.

The hegemony of the United Provinces of the Netherlands (the so-called Dutch Republic) continued until the 1660s, when both the English and the French were able to mount a serious challenge. A series of three Anglo-Dutch wars (1652–1654, 1664–1667, and 1672–1678) helped to settle the issue in favor of England, which consolidated its power soon afterwards through the Act of Union (1707) with Scotland.

British dominance was sustained, in spite of a relatively poor domestic resource base, by overseas trade and

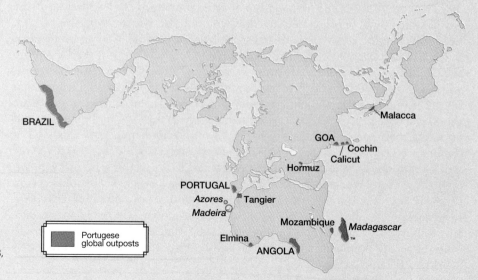

Figure 2.2.1 Portuguese global outposts, circa 1550 (*Source:* Map projection, Buckminster Fuller Institute and Dymaxion Map Design, Santa Barbara, CA. The word Dymaxion and the Fuller Projection Dymaxion™ Map design are trademarks of the Buckminster Fuller Institute, Santa Barbara, California, ©1938, 1967 & 1992. All rights reserved.)

Portuguese global outposts

BRAZIL
Malacca
GOA
Cochin
Calicut
Hormuz
PORTUGAL
Azores
Tangier
Madeira
Mozambique
Madagascar
Elmina
ANGOLA

colonization, backed up by a powerful navy. In the 1770s, this dominance was shaken by an alliance between France and the American revolutionaries, and there followed a period of struggle between Britain and France that came to involve much of Europe. A series of Napoleonic Wars marked the competitive phase of the leadership cycle, which was brought to a close by decisive victories by Admiral Lord Nelson at Trafalgar (1805; see Figure 2.2.4) and Lord Wellington at Waterloo (1815). By then, the Industrial Revolution had given Britain an economic and military edge that proved decisive.

The second cycle of British dominance was based on the economic advantages of early industrialization, which allowed for an unprecedented degree of incorporation of the world under British imperial and economic hegemony. But from the 1860s onward, imperial overstretch and increasing economic competition from the United States and Germany put an end to hegemony and marked the beginning of a period of struggle. The German challenge culminated in the Great War of 1914–1918 (World War I), which left Germany defeated, Britain weakened, and the United States strengthened.

The **United States** was economically dominant within the world-system by 1920, but did not achieve hegemonic power because of a failure of political will. It took a second World War, prompted by the resurgence of Germany and the sudden rise of Japan, to force the United States out of its isolationism. After World War II, the United States was unquestionably the hegemonic power. Its economic and cultural dominance within the world-system has not yet been seriously challenged, though its

Figure 2.2.4   The Battle of Trafalgar   On the afternoon of October 21, 1805, a few miles offshore from the Spanish port of Cádiz in the Bay of Trafalgar, 27 line-of-battle sailing ships commanded by Vice-Admiral Lord Horatio Nelson annihilated a combined French and Spanish fleet of 33 ships-of-the-line under French Admiral Pierre Villeneuve. Nelson's victory that afternoon established a naval supremacy that enabled Britain to build and maintain a global empire, and to enjoy a leadership cycle of such unchallenged dominance that the following 100 years came to be known as the *Pax Brittanica*.

political and military superiority was in question for several decades as the Soviet Union sought a noncapitalist path to modernization and power. This challenge disappeared in 1989, but by then the economic foundation for U.S. hegemony had come under serious threat because of the resurgence of Japanese and European industry and the globalization of economic activity through transnational corporations (see Figure 2.2.5).

Figure 2.2.5   The Gulf War   All hegemonic powers must protect the economic foundations of their power. They must also resist challenges to their political, cultural, and ideological dominance. Occasionally, this calls for military intervention. The Gulf War, fought in 1991, was an example of this sort of intervention. It was prompted by the invasion of Kuwait in 1990 by Iraq, where president Saddam Hussein had consolidated popularity through an aggressive stance toward American culture and business interests. The invasion of Kuwait not only threatened to disrupt the supply of crude oil to the world economy but also posed a threat to the U.S.-sponsored regional balance of power in the Middle East. On January 15, 1991, the United States, with 500,000 troops and massive air power, led a 100-hour ground war against Iraq, driving Iraqi troops out of Kuwait and destroying Iraq's military infrastructure.

Figure 2.2.3   Merchant vessels of the Dutch and English East India Companies   The English East India Company was established in 1600 as a monopolistic trading company and agent of British imperialism in India. In response, the Dutch East India Company was founded in 1602 by the government of the United Provinces of the Netherlands in order to protect its trading interests in the Far East. The company was almost a branch of the government. It was granted a trade monopoly in the waters of the Indian and Pacific Oceans, together with the rights to conclude treaties with local rulers, to build forts and maintain armed forces, and to carry on administrative functions through officials who were required to take an oath of loyalty to the Dutch government.

Industrialization, meanwhile, resulted not only in the complete reorganization of the human geography of the original European core of the world-system, but also in an extension of the world-system core to the United States and Japan.

Europe In Europe, three distinctive waves of industrialization occurred. The first, between about 1790 and 1850, was based on the initial cluster of industrial technologies (steam engines, cotton textiles, and ironworking) and was very localized. It was limited to a few regions in Britain where industrial entrepreneurs and workforces had first exploited key innovations and the availability of key resources (coal, iron ore, water). These included north Cornwall, eastern Shropshire (where Abraham Darby II built the world's first iron bridge—at Ironbridge, on the River Severn), south Staffordshire, south Lancashire, southwestern Yorkshire, Tyneside, and Clydeside (Figure 2.9). Although these regions shared the common impetus of certain key innovations, each of them retained its own technological traditions and industrial style. From the start, then, industrialization was a regional-scale phenomenon.

The second wave, between about 1850 and 1870, involved the diffusion of industrialization to most of the rest of Britain and to parts of northwest Europe, particularly the coalfield areas of northern France, Belgium, and Germany (Figure 2.10). This second wave also brought a certain amount of reorganization to the first-wave industrial regions as a new cluster of technologies (steel, machine tools, railroads, steamships) brought new opportunities, new locational requirements, new business structures, and new forms of societal organization. New opportunities were created as railroads and steamships made more places

Figure 2.9 The first wave of industrialization in Britain The Industrial Revolution was very much a regional phenomenon, taking place simultaneously in several locales because of the industry's need to be near mineral resources and because of the importance of local canal systems.

The bridge at Ironbridge, Shropshire, was erected in 1779. The first cast-iron bridge, it served as an advertisement for Abraham Darby's Coalbrookdale ironworks, and as an inspiration for the first wave of industrialization in Britain.

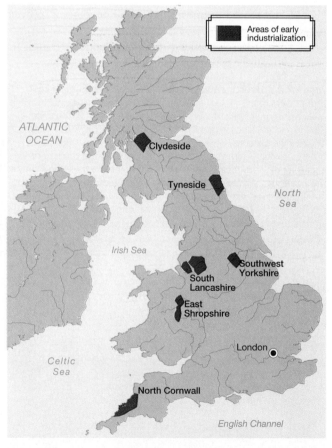

Figure 2.10 The spread of industrialization in Europe   European industrialization began with the emergence of small industrial regions in several different parts of Britain, where early industrialization drew on local mineral resources, water power, and early industrial technologies. As new rounds of industrial and transportation technologies emerged, industrialization spread to other regions with the right locational attributes: access to raw materials and energy sources, good communications, and large labor markets.

Manchester was in the vanguard of the Industrial Revolution. On the basis of a cotton textile industry, Manchester grew from less than 70,000 inhabitants in 1800 to 750,000 in 1880 when it was earning half of all Britain's export revenues.

The Ruhr coalfield was developed in the 1860s by the Krupp and Thyssen steelmaking firms; it came into its own as the heartland of German industry after Germany had to return coal- and iron-ore–rich Alsace-Lorraine to France at the end of World War I.

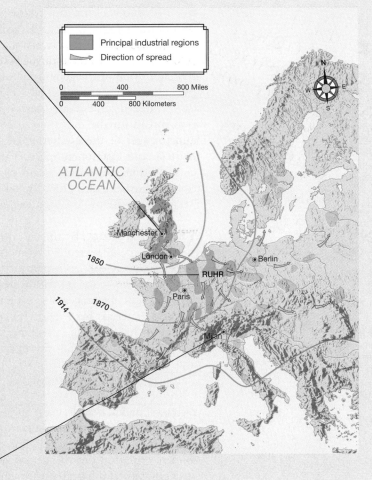

Milan, with only 7 percent of Italy's population, accounts for 28 percent of its national income. Its industrial base (automobiles, appliances, engineering, textiles), developed early in the twentieth century, is now complemented by finance and business services.

accessible, bringing their resources and their markets into the sphere of industrialization. New materials and new technologies (steel, machine tools) created opportunities to manufacture and market new products. These new activities brought some significant changes in the logic of industrial location.

The importance of railway networks, for example, attracted industry away from smaller towns on the canal systems toward larger towns with good rail connections. The importance of steamships for coastal and international trade attracted industry to larger ports. At the same time, the importance of steel produced concentrations of heavy industry in places with nearby supplies of coal, iron ore, and limestone. The scale of industry increased as new technologies and improved transportation made larger markets accessible to firms. Local, family firms became small companies that were regional in scope. Small companies grew to become powerful firms serving national markets, and specialized business, legal, and financial services emerged in larger cities to assist them. The growth of new occupations transformed the structure of social classes, and this transformation in turn came to be reflected in the politics and landscapes of industrial regions.

The third wave of industrialization, between 1870 and 1914, saw a further reorganization of the geography of Europe as yet another cluster of technologies (including electricity, electrical engineering, and telecommunications) imposed different needs and created new opportunities. During this period, industrialization spread for the first time to remoter parts of Britain, France, and Germany and to most of the Netherlands, southern Scandinavia, northern Italy, eastern Austria, and Catalonia, Spain. The overall result was to create a core-within-a-core. Within the world-system core, processes of industrialization, modernization, and urbanization had forged a core of prosperity centered on the "Golden Triangle" stretching between London, Paris, and Berlin.

The United States By the end of the nineteenth century, the core of the world-system had itself extended to include the United States and Japan. The United States, politically independent just before the onset of the Industrial Revolution, was able to make the transition from the periphery to the core because of several favorable circumstances. Its vast natural resources of land and minerals provided the raw materials for a wide range of industries that could grow and organize without being hemmed in and fragmented by political boundaries. Its population, growing quickly through immigration, provided a large and expanding market and a cheap and industrious labor force. Its cultural and trading links with Europe provided business contacts, technological know-how, and access to capital (especially British capital) for investment in a basic infrastructure of canals, railways, docks, warehouses, and factories.

As in Europe, industrialization developed around pre-existing centers of industrialization and population, and was shaped by the resource needs and market opportunities of successive clusters of technology. America's industrial strength was established at the beginning of the twentieth century with the development of a new cluster of technologies that included the internal combustion engine, oil and plastics, electrical engineering, and radio and telecommunications (see Chapter 7). The outcome was a distinctive manufacturing region, known as the Manufacturing Belt, that by 1920 stretched from Boston and Baltimore in the East to Milwaukee and St. Louis in the West (Figure 2.11)—another core-within-a-core.

Japan Japan remained a feudal world-empire for a long time before vaulting into the core of the world-system. In 1868, a revolution deposed the feudal Tokugawa regime and restored an old imperial dynasty—the Meiji—whose backers deliberately set out to modernize the country as quickly as possible, whatever the cost.

Figure 2.11   The manufacturing belt of the United States   The cities of this region, already thriving industrial centers that were well connected through the early railroad system, were ideally placed to take advantage of a series of crucial shifts between 1880 and 1920: a general upsurge in demand for consumer goods; the increased efficiency of the telegraph system and postal services; advances in manufacturing technologies; and the opening-up of the national market through the extension of the railroad system. Individual cities began to specialize in their industrial products (e.g., grain milling in Minneapolis, brewing in Milwaukee and St. Louis, coachbuilding and furniture in Cincinnati, agricultural machinery in Springfield, Illinois), producing in volume for the national market rather than local ones. This specialization in turn required an increase in commodity flows between the cities of the Manufacturing Belt, which bound the region more closely together.

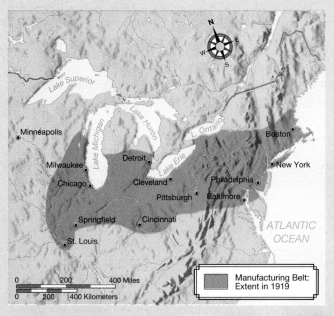

Chicago grew from 500,000 to 2,500,000 in the 40 years between 1880 and 1920. Specializing in meatpacking, furniture manufacture, and printing and publishing, it was also a key transportation hub. This photograph of the downtown area was taken in 1929.

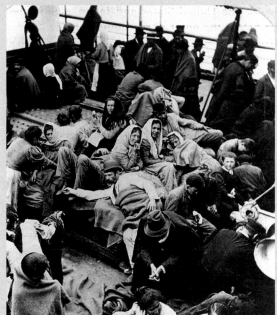

The growth of the Manufacturing Belt was fueled by immigration. Between 1860 and 1920, more than 25 million immigrants arrived in East Coast ports, with New York receiving the bulk of them.

The Manufacturing Belt as seen through a mosiac of nighttime satellite images. Shores and boundaries have been added to the satellite image to enhance identification. (*Source:* Department of the Interior, U.S. Geological Survey, EROS Data Center.)

71

Under the slogan "National Wealth and Military Strength" the government took an unprecedented role, strong and deliberate, in creating capitalist enterprises and establishing a modern infrastructure. The costs were borne chiefly by a domestic workforce (whose rewards for hard work were meager) and by the victims of Japanese military aggression in Korea, Formosa (now Taiwan), and Manchuria.

Just as Japanese industry was becoming established, with a base in textiles and shipbuilding, the First World War provided a timely opportunity to expand productive capacity. With much European and American industry diverted to supply war materials, Japanese textile manufacturers were able to expand into Asian and Latin American markets. Meanwhile, with a great deal of merchant shipping destroyed by the hostilities, Japanese shipping industries took the opportunity to expand their merchant fleet. The profits from this commercial activity paid for the rebirth of the Japanese navy, which by 1918 had a dozen battleships and battle cruisers, with 16 more under construction. The United States at the time had only 14 battleships, with three under construction. Within 50 years of the Meiji revolution, Japan had joined the core of the world-system.

Internal Development of the Core Regions

Within the world's core regions, the transformation of regional geographies hinged on successive innovations in transport technology. These innovations opened up agrarian interiors and intensified interregional trading networks. Farmers were able to mechanize their equipment, while manufacturing companies were able to take over more resources and more markets.

Canals and the Growth of Industrial Regions   The first phase of this internal geographic expansion and regional integration was in fact based on an old technology: the canal (Figure 2.12). Merchant trade and the beginnings of industrialization in both Britain and France were underpinned by extensive navigation systems that joined one river system to another. By 1790, France had just over 1,000 kilometers (620 miles) of canals and canalized rivers; Britain had nearly 3,600 kilometers (2,230 miles). The Industrial Revolution provided both the need and the capital for a spate of additional canal building that began to integrate and extend emerging industrial regions.

In Britain, 2,000 more kilometers (1,240 miles) of canals were built between 1790 and 1810. In France, which did not industrialize as early as Britain, 1,600 additional kilometers (990 miles) were built between 1830 and 1850. In the United States, the landmark was the opening of the Erie Canal in 1825. This breakthrough enabled New York, a colonial gateway port, to reorient itself toward the nation's growing interior. The Erie Canal was so profitable that it set off a "canal fever" that resulted in the construction of some 2,000 kilometers (1,240 miles) of navigable waterways in the next 25 years. This canal system helped to bind the emergent manufacturing belt together (Figure 2.13, page 74).

Steamboats, Railroads, and Internal Development   The scale of the United States was such that a network of canals was only a viable proposition in more densely settled areas. The effective colonization of the interior could not take place until the development of steam-powered transportation—first riverboats, and then railroads. The first steamboats were developed in the early 1800s, offering the possibility of opening up the vast interior by way of the Mississippi and its tributaries. By 1830, the technology and design of steamboats had been perfected, and navigable channels had been established. The heyday of the river steamboat was between 1830 and 1850. During this period, vast acreages of the U.S. interior were opened up to commercial, industrialized agriculture—especially cotton production for export to British textile manufacturers. At the same

Figure 2.12 European canal systems The canal systems that opened up the interiors of Europe and North America in the eighteenth century were initially dependent on horse power. Early examples, like the restored Rochdale Canal in northern England (top), carried horse-drawn barges. Later, barges were able to utilize steam- and oil-powered engines, giving a new lease on life to larger canals and navigable rivers. Barges are still an important mode of transport for bulky goods. This example (bottom) shows coal being moved up the River Rhine from coalfields in the Ruhr to power stations in southern Germany. The map shows how the Rhine-Mosel canal system links with west European rivers to provide an extensive waterway system.

(a) The Rhine-Mosel canal system

(b) Rochdale Canal

(c) Coal barge headed upstream on the Rhine

time, river ports such as New Orleans, St. Louis, Cincinnati, and Louisville grew rapidly, extending the frontier of industrialization and modernization.

By 1860, the railroads had taken over the task of internal development, further extending the frontier of settlement and industrialization and intensifying the use of previously developed regions (see Feature 2.3, "Visualizing Geography—Railroads and Geographic Change"). The railroad originated in Britain, where George Stephenson engineered the world's first commercial railroad, a 20-kilometer (12.4-mile) line between Stockton and Darlington that was opened in 1825. The famous *Rocket*, the first-ever locomotive for commercial passenger trains, was designed mainly by Stephenson's son Robert for the Liverpool & Manchester line that opened four years later. The economic success of this line sparked two railroad-building booms that eventually created a highly efficient transportation network for Britain's manufacturing industry. In other

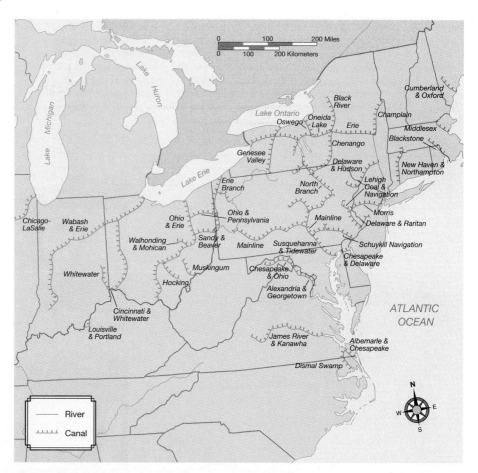

Figure 2.13 The canal network of the northeastern United States in 1860 The only economically profitable canal, the Erie, was designed to compensate for the geographic problems of the colonial city of New York when it needed to turn to inland trade after the Revolutionary War cut off Britain as a trading partner. (*Source:* After J. E. Vance, Jr., *Capturing the Horizon: The Historical Geography of Transportation Since the Transportation Revolution of the Sixteenth Century.* © 1990 The Johns Hopkins University Press.)

core countries, where sufficient capital existed to license (or copy) the locomotive technology and install the track, railroad systems led to the first full stage of economic and political integration.

While the railroads integrated the economies of entire countries and allowed vast territories to be colonized, they also brought some important regional and local restructuring and differentiation. In the United States, for example, the railroads led to the consolidation of the Manufacturing Belt. They also contributed to the mushrooming of Chicago as the focal point for railroads that extended the Manufacturing Belt's dominance over the West and South. This reorientation of the nation's transportation system effectively ended the role of the cotton regions of the South as outliers of the British trading system. Instead, they became outliers of the U.S. Manufacturing Belt, supplying factories in New England and the Mid-Atlantic Piedmont. This left New Orleans, which had thrived on cotton exports, to cope with an abrupt end to its phenomenal growth.

Tractors, Trucks, Road Building, and Spatial Reorganization In the twentieth century the internal combustion engine has powered further rounds of internal development, integration, and intensification (Figure 2.14). The replacement of horse power by lightweight, internal combustion–engined tractors, beginning

Figure 2.14 The geographical impacts of the internal combustion engine The internal combustion engine revolutionized the geography of the more affluent and developed parts of the world between 1920 and 1970. The production of trucks in the United States increased from 74,000 in 1915 to 750,000 in 1940 and 1.75 million in 1965. Automobile production increased from 896,000 in 1915 to 3.7 million in 1940 and 9.3 million in 1965. Meanwhile, road building opened up interior regions and opened out metropolitan regions. Between 1945 and 1965, federal highways increased from a total of 456,936 kilometers (308,741 miles) to 1,344,908 kilometers (908,722 miles). Later, limited-access highways (autobahns in Germany, autostrada in Italy, interstate highways in the United States, and motorways in Britain) provided such an efficient and flexible means of long-distance passenger and freight movement that railways were eclipsed as the principal framework for the transportation geography of industrial countries. With this eclipse, the spatial organization of industries and land uses was radically reorganized, and a new round of geographical change took place.

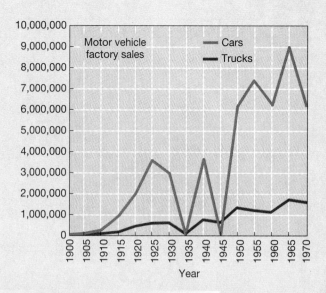

Model T assembly line

Henry Ford (1863–1947) had an unusual combination of skills as an engineer and an entrepreneur. He left his original company (which subsequently reorganized as the Cadillac Motor Car Company) in order to concentrate on perfecting a vehicle that could be the ordinary family's utility rather than the rich family's luxury. His clashes with big business and the Eastern establishment made him a popular figure. When the Model T, his "motor car for the great multitude," was finally ready in 1908, it was a huge success. In the 19 years of its production, over 16 million were sold—half of the auto output of the whole world for the period. The Model T helped to revolutionize the urban geography of America, just as Ford's trucks and tractors helped to change the geography of rural regions. His system of assembly-line mass production, coupled with a carefully organized division of labor and a vertically integrated manufacturing empire (the ownership and control of raw materials and distribution systems as well as manufacturing and assembly plant), was widely copied throughout the industrial world and had a profound effect on the spatial organization of manufacturing industries.

Railroad service in 1880. Areas shown in black are no more than 15 miles (24 kilometers) from a railroad line. (*Source:* P. Hugill, *World Trade Since 1431*. Baltimore: Johns Hopkins University Press, 1993, p. 180.)

Railroad service in 1860. Areas shown in black are no more than 15 miles (24 kilometers) from a railroad line. (*Source:* P. Hugill, *World Trade Since 1431*. Baltimore: Johns Hopkins University Press, 1993, p. 179.)

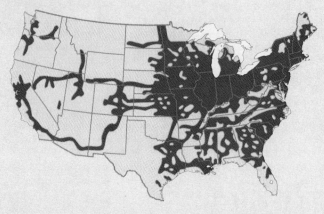

Railroad construction was labor-intensive, but required little more than common farm equipment applied to a graded surface. Most of the labor force consisted of immigrants: European immigrants on railroads driven westward from the East Coast and Chinese immigrants on those driven eastward from the West Coast.

Chicago's railyards at their peak, in 1942

The first passenger train services averaged little more than 20 to 35 kilometers per hour (15–20 m.p.h.), but locomotive technology changed very rapidly, making it easier, faster, and cheaper to conquer the vast interior distances of the United States. Between 1830 and 1845, the United States created the world's largest rail system: some 5,458 kilometers (3,688 miles), compared to Britain's 3,083 kilometers (2,069 miles), Germany's 2,956 kilometers (1,997 miles), France's 817 kilometers (552 miles), and Belgium's 508 kilometers (343 miles).

Grand Central Station, New York City, with the Hotel Commodore, behind

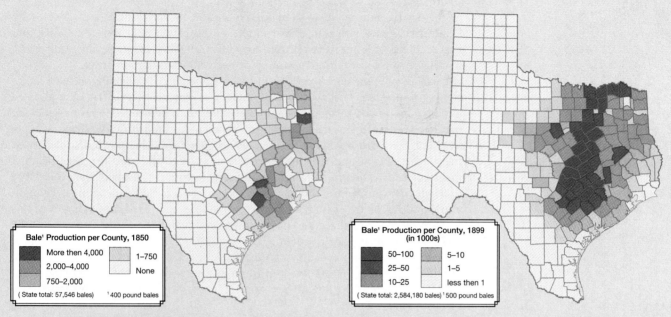

Bale¹ Production per County, 1850

More then 4,000	1–750
2,000–4,000	None
750–2,000	

(State total: 57,546 bales) ¹ 400 pound bales

Bale¹ Production per County, 1899
(in 1000s)

50–100	5–10
25–50	1–5
10–25	less then 1

(State total: 2,584,180 bales) ¹ 500 pound bales

The railroads both reorganized the geography of America's cities and opened up its interior regions. Railway stations reordered land use patterns in central business districts, where new hotels and department stores vied for sites adjacent to the station's entrance. Farther away, the railroad lines cut swathes through the urban fabric, separating neighborhoods from one another and establishing the basis for a radial framework for the physical development of the city.

Before the railroads reached Texas, cotton production was concentrated in a few areas along short sections of navigable rivers (left). The railroads, which reached the interior of Texas after 1850, allowed cotton production to extend inland, up onto the high plains (right). (*Source:* P. Hugill, "The Macro-Landscape of the Wallerstein World Economy," *Geoscience and Man,* 25, 1988, Figs. 1 and 5, pp. 78 and 80, Department of Geography and Anthropology, Louisiana State University.)

The Illinois Central Railroad issued a promotional poster in the 1860s (left), celebrating the role it envisaged for its trunk line from Chicago to New Orleans might play in the settling of the American interior. Note the deliberate juxtaposition of the high technology of the railroad locomotive and the telegraph line, contrasting with the obsolescent technologies of the stagecoach (inset top right) and barge (inset bottom right). Even the ocean-going ship is shown with masts and sails, in order to diminish its status in comparison to the railroad locomotive. The poster issued 50 years later (right), on the other hand, shows the railroad locomotive in the company of several important new technologies: aircraft, automobiles and trucks, and electric street lighting.

in the 1910s, amounted to a major revolution in agriculture (see Chapter 8). Productivity was increased, the frontiers of cultivable land were extended, and vast amounts of labor were released for industrial work in cities. The result was a parallel revolution in the geographies of both rural and urban areas.

The development of trucks in the 1910s and 1920s suddenly released factories from locations tied to railroads, canals, and waterfronts. Trucking allowed goods to be moved farther, faster, and cheaper than before. As a result, trucking made it feasible to locate factories on inexpensive land on city fringes, and in smaller towns and peripheral regions where labor was cheaper. It also increased the market area of individual factories, and reduced the need for large product inventories. This decentralization of industry, in conjunction with the availability of buses, private automobiles, and massive road-building programs, brought about another phase of spatial reorganization (Figure 2.14). The outcomes of this phase were the specialized and highly integrated regions and urban systems of the modern core of the world-system. This integration was not simply a question of their being interconnected through highway systems. It also involved close economic linkages among manufacturers, suppliers, and distributors—linkages that enabled places and regions to specialize and develop economic advantages (see Chapter 7).

Organizing the Periphery

Parallel with the internal development of core regions were changes in the geographies of the periphery of the world-system. Indeed, the growth and internal development of the core regions simply could not have taken place without the foodstuffs, raw materials, and markets provided by the colonization of the periphery and the incorporation of more and more territory into the sphere of industrial capitalism.

As soon as the Industrial Revolution had gathered momentum in the early nineteenth century, the industrial core nations embarked on the inland penetration of the world's midcontinental grassland zones in order to exploit them for grain or stock production. This led to the settlement, through the emigration of European peoples, of the temperate prairies and pampas of the Americas; the veld in southern Africa; the Murray-Darling Plain in Australia; and the Canterbury Plain in New Zealand. At the same time, as the demand for tropical plantation products increased, most of the tropical world came under the political and economic control—direct or indirect—of one or another of the industrial core nations. In the second half of the nineteenth century, and especially after 1870, there was a vast increase in the number of colonies and the number of people under colonial rule.

The International Division of Labor

The fundamental logic behind all this colonization was economic: the need for an extended arena for trade, an arena that could supply foodstuffs and raw materials in return for the industrial goods of the core. The outcome was an international division of labor, driven by the needs of the core, and imposed through its economic and military strength. This **division of labor** involved the specialization of different people, regions, and countries in certain kinds of economic activities. In particular, colonies began to specialize in the production of foodstuffs and raw materials meeting certain criteria:

- Where an established demand existed in the industrial core (for foodstuffs and industrial raw materials, for example);

Division of labor: the specialization of different people, regions, or countries in particular kinds of economic activities.

- Where colonies held a **comparative advantage** in that their productivity was higher than for other possible specializations; and

- Which did not duplicate or compete with domestic suppliers within core countries (tropical agricultural products like cocoa and rubber, for example, simply could not be grown in core countries).

The result was that colonial economies were founded on narrow specializations that were oriented to, and dependent upon, the needs of core countries. Examples of these specializations are many: bananas in Central America; cotton in India; coffee in Brazil, Java, and Kenya; copper in Chile; cocoa in Ghana; jute in East Pakistan (now Bangladesh); palm oil in west Africa; rubber in Malaya (now Malaysia) and Sumatra; sugar in the Caribbean islands; tea in Ceylon (now Sri Lanka); tin in Bolivia; and bauxite in Guyana and Surinam. Most of these specializations have continued through to the present. Thus, for example, 48 of the 55 countries in sub-Saharan Africa still depend on just three products—tea, cocoa, and coffee—for more than half of their export earnings.

This new world economic geography took some time to establish, and the details of its pattern and timing were heavily influenced by technological innovations. The incorporation of the temperate grasslands into the commercial orbit of the core countries, for example, involved successive changes in regional landscapes as critical innovations such as barbed wire, the railroad, and refrigeration were introduced. The single most important innovation behind the international division of labor, however, was the development of metal-hulled, oceangoing steamships (Figure 2.15). This development was, in fact, cumulative, with

Comparative advantage: principle whereby places and regions specialize in activities for which they have the greatest advantage in productivity relative to other regions—or for which they have the least disadvantage.

Figure 2.15 Isambard Kingdom Brunel   Brunel (1806–1859) was one of Victorian Britain's great engineers. He began his career as resident engineer on the Thames Tunnel before going on to design a suspension bridge over the gorge of the River Avon. As Chief Engineer to the Great Western Railway, he introduced a broad-gauge track that allowed much higher speeds than previous railway systems, contributing directly to the integration of the economic geography of the United Kingdom. He also worked on railways in Austria, Italy, and India. His most spectacular successes, though, were in marine engineering, where his designs created ships with technological breakthroughs that helped to change the economic geography of the world. The *Great Western* (completed in 1837), a wooden paddle vessel, was the first steamship to provide a transatlantic service. The *Great Britain* (1843) was an iron-hulled vessel, the first of any size to be driven by a screw propeller. The *Great Eastern* (1858) was propelled by both paddles and screw and was the first ship with a double iron hull. After service as a passenger ship, it laid the first successful transatlantic cable.

The *Great Eastern*

Isambard Kingdom Brunel

improvements in engines, boilers, transmission systems, fuel systems, and construction materials adding up to produce dramatic improvements in carrying capacity, speed, range, and reliability. The construction of the Suez Canal (opened in 1869) and the Panama Canal (opened in 1914) was also critical, providing shorter and less hazardous routes between core countries and colonial ports of call. By the eve of the First World War the world economy was effectively integrated by a system of regularly scheduled steamship trading routes (Figure 2.16). This integration in turn was supported by the second most important innovation behind the international division of labor: a network of telegraph communications (Figure 2.17, page 82) that enabled businesses to monitor and coordinate supply and demand across vast distances on an hourly basis.

The international division of labor brought about a substantial increase in trade and a huge surge in the overall size of the capitalist world economy. The peripheral regions of the world contributed a great deal to this growth. By 1913 Africa and Asia provided more *exports* to the world economy than either North America or the British Isles. Asia alone was *importing* almost as much, by value, as North America. The industrializing countries of the core bought increasing amounts of foodstuffs and raw materials from the periphery, financed by profits from the export of machinery and of manufactured goods. Britain, the hegemonic power of the period, drew on a trading empire that was truly global (Figure 2.18, page 83).

Patterns of international trade and interdependence became increasingly complex. Britain used its capital to invest not just in peripheral regions but also in profitable industries in other core countries, especially the United States. At the same time, these other core countries were able to export cheap manufactures to Britain. Britain financed the purchase of these goods, together with imports of food from its dominion States (Canada, South Africa, Australia, and New Zealand) and colonies, through the export of its own manufactured goods to peripheral countries. India and China, with large domestic markets, were especially important. *Thus there developed a widening circle of exchange and dependence, with constantly switching patterns of trade and investment.*

Imperialism: Imposing New Geographies on the World

The incorporation of the periphery was by no means entirely motivated by this basic logic of free trade and investment. Although Britain was the hegemonic power in the late nineteenth century, several other European countries (notably Germany, France, and the Netherlands), together with the United States and, later, Japan, were competing for global influence. This competition developed into a scramble for territorial and commercial domination. The core countries engaged in pre-emptive geographic expansionism in order to protect their established interests and to limit the opportunities of others. They also wanted to secure as much of the world as possible—through a combination of military oversight, administrative control, and economic regulations—in order to ensure stable and profitable environments for their traders and investors. This combination of circumstances defined a new era of imperialism.

In the final quarter of the nineteenth century, a second wave of imperialism brought a competitive form of colonialism that resulted in a scramble for territory. Africa, more than any other peripheral region, was given an entirely new geography. It was carved up into a patchwork of European colonies and protectorates in just 34 years, between 1880 and 1914, with little regard for either physical geography or the preexisting human geographies of minisystems and world-empires. Whereas European interest had previously focused on coastal trading stations and garrison ports, it now extended to the entire continent. The Berlin Conference of

Figure 2.16   Major steamship routes in 1920   The shipping routes reflect (1) the transatlantic trade between the bipolar core of the world-system at the time; and (2) the colonial and imperial relations between the world's core economies and the periphery. Transoceanic shipping boomed with the development of steam-turbine engines for merchant vessels and with the construction of shipping canals, such as the Kiel Canal, the Suez canal, the St. Lawrence Seaway, and the Panama Canal. When the 51 miles of the Panama Canal opened in 1914, shipping could move between the Atlantic and the Pacific without having to go around South America, saving thousands of miles of steaming. (*Source:* Map projection, Buckminster Fuller Institute and Dymaxion Map Design, Santa Barbara, CA. The word Dymaxion and the Fuller Projection Dymaxion™ Map design are trademarks of the Buckminster Fuller Institute, Santa Barbara, California, ©1938, 1967 & 1992. All rights reserved.)

▶ Manchester, accessible to transoceanic shipping via the 36-mile Manchester Ship Canal (completed in 1894), imported cotton from the American South and exported cotton textiles to markets throughout the world.

▶ Established by the British as a trading port in 1898, Hong Kong was intended to provide a point of exchange between European and Eastern goods. The lease for the small colony (898 square kilometers, or 410 sq. mi.) ran for 99 years.

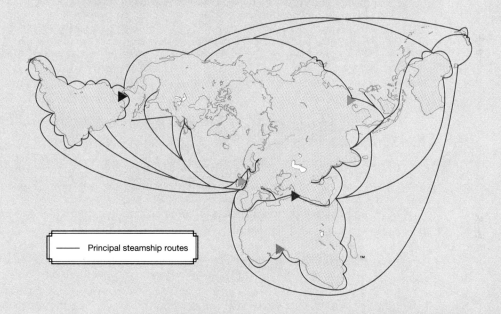

— Principal steamship routes

▶ After several years of construction, work on the Panama Canal was halted in 1889 because of political instability in Panama. After intervention by the United States in 1903, the canal was completed, opening for transoceanic trade in 1914.

▶ Port Harcourt, 41 miles upstream from the Gulf of Guinea on the Bonny River in Nigeria, was one of scores of trading ports established under colonial rule. Today, it handles exports of coal, oil, timber, peanuts, and tin.

▶ Built by the French, the 105-mile Suez Canal was completed in 1869, providing the shortest maritime route between Europe and the Indian and western Pacific Oceans.

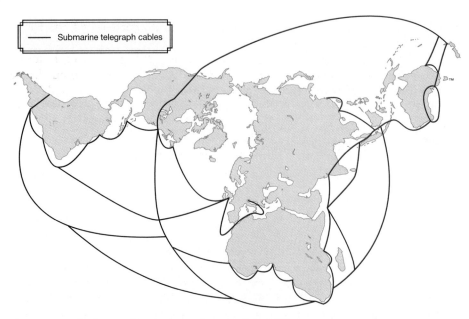

Figure 2.17   The international telegraph network in 1900   For Britain, submarine telegraph cables were the nervous system of its empire. Of the global network of 153,000 miles of submarine cable, Britain had laid 105,000 miles. (*Source:* Map projection, Buckminster Fuller Institute and Dymaxion Map Design, Santa Barbara, CA. The word Dymaxion and the Fuller Projection Dymaxion™ Map design are trademarks of the Buckminster Fuller Institute, Santa Barbara, California, ©1938, 1967 & 1992. All rights reserved.)

1885–1886, organized by European powers in order to establish some ground rules for imperialist expansion in Africa, made three decisions:

1. Notice of intent to lay claim to sovereignty over particular parts of Africa was to be given in advance of any military occupation of those parts

2. Effective occupation of a territory was to be regarded as grounds for a legitimate claim to sovereignty

3. Disputes between European powers were to be settled by arbitration, in order to avoid direct conflict between European powers

Within just a few years, the whole of Africa became incorporated into the modern world-system, with a geography that consisted of a hierarchy of three kinds of spaces. One consisted of regions and localities organized by European colonial administrators and European investors to produce commodities for the world market. A second consisted of zones of production for local markets, where peasant farmers produced food for consumption by laborers engaged in commercial mining and agriculture. The third consisted of widespread regions of subsistence agriculture whose connection with the world-system was as a source of labor for the commercial regions.

Meanwhile, the major powers jostled and squabbled over small Pacific islands that had suddenly become valuable as strategic coaling stations for their navies and merchant fleets. Resistance from indigenous peoples was quickly brushed aside by imperial navies with iron steamers, high-explosive guns, and troops with rifles and cannon. European weaponry was so superior that Otto von Bismarck, the founder and first Chancellor (1871–1890) of the German empire, referred to these conflicts as "sporting wars." Between 1870 and 1900, European countries added almost 22 million square kilometers (10 million square miles) and 150 million people to their spheres of control—20 percent of Earth's land surface and 10 percent of its population.

The imprint of imperialism and colonization on the geographies of the newly incorporated peripheries of the world-system was immediate and profound. The periphery was brought to be almost entirely dependent on European and North American capital, shipping, managerial expertise, financial services, and news and communications. As a consequence of this, it also came to be dependent on European cultural products: language, education, science, religion, architecture, and planning. All of this came to be etched into the landscapes of the periphery in a variety of ways as new places were created, old places were remade, and regions were reorganized.

One of the most striking changes in the periphery was the establishment and growth of externally oriented port cities through which commodity exports and manufactured imports were channeled. Often, these major ports were also colonial administrative and political capitals, so that they became overwhelmingly important, growing rapidly to sizes far in excess of other settlements. Good examples include Georgetown (Guyana), Lagos (Nigeria), Luanda (Angola), Karachi (Pakistan), and Rangoon (Burma). Meanwhile, the interior geography of peripheral countries was restructured as smaller settlements were given the new functions of colonial administration and commercial marketing. As with the interior development of the core countries, transport networks were vital to this process. Railroads provided the principal means of spatial reorganization, and in the colonies of Africa, Central America, and South and Southeast Asia they evolved into linear patterns with simple feeder routes and limited interconnections that focused almost exclusively on major port cities (Figure 2.19, page 84).

Figure 2.18 The British Empire in the late 1800s   Protected by the all-powerful Royal Navy, the British merchant navy established a web of commerce that collected food for British industrial workers and raw materials for its industries, much of it from colonies and dependencies appropriated by imperial might and developed by British capital. So successful was the trading empire that Britain also became the hub of trade for other States. (*Sources:* After P. Hugill, *World Trade Since 1431,* p. 136. © 1993 The Johns Hopkins University Press. Map projection, Buckminster Fuller Institute and Dymaxion Map Design, Santa Barbara, CA. The word Dymaxion and the Fuller Projection Dymaxion™ Map design are trademarks of the Buckminster Fuller Institute, Santa Barbara, California, ©1938, 1967 & 1992. All rights reserved.)

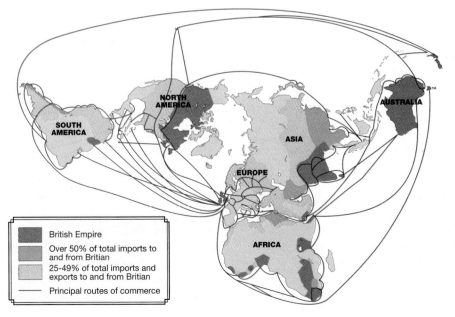

Figure 2.19 Railway penetration of peripheral interiors

Colonial railroad networks were usually established in order to connect port cities with interior regions that produced foodstuffs and raw materials for export. The simple rail networks that resulted have formed the basis for subsequent urban development in ex-colonial countries; it has been very difficult for towns not served by the railroads to compete with those that are. Meanwhile, the port cities that were the focus of the colonial rail systems have grown disproportionately large, dominating their national urban systems.

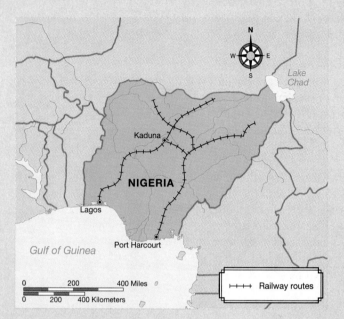

Nigeria's colonial rail system

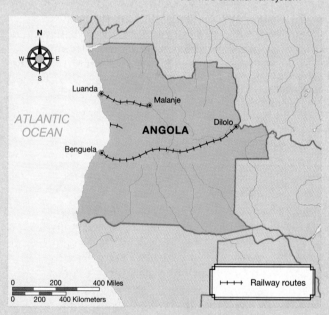

Burma's colonial rail system

Angola's colonial rail system

Globalization

The imperial world order began to disintegrate shortly after World War II, however. The United States emerged as the new hegemonic power, the dominant State within the world-system core. This core came to be called the "First World." The Soviet Union and China, opting for alternative paths of development for themselves and their satellite countries, were seen as a "Second World," withdrawn from the capitalist world economy. Their pursuit of alternative political economies was based on radically different values.

By the 1950s, many of the old European colonies began to seek political independence. Some of the early independence struggles were very bloody, because the colonial powers were initially reluctant to withdraw from colonies where strategic resources or large numbers of European settlers were involved. In Kenya, for example, a militant nationalist movement known as the Mau Mau launched a campaign of terrorism, sabotage, and assassination against British colonists in the early 1950s. Their actions killed over 2,000 white settlers between 1952 and 1956; in return, 11,000 Mau Mau rebels were killed by the colonial army and 20,000 put into detention camps by the colonial administration. By the early 1960s, however, the process of decolonization had become relatively smooth. (In Kenya, Jomo Kenyatta, who had been jailed as a Mau Mau leader in 1953, became prime minister of the newly independent country in 1962.) The periphery of the world-system now consisted of a "Third World" of politically independent States, some of which adopted a policy of nonalignment vis-à-vis the geopolitics of the First and Second Worlds. They were, nevertheless, still highly dependent, in economic terms, on the world's core countries.

As newly independent peripheral States struggled to be free of their economic dependence through industrialization, modernization, and trade from the 1960s onward, so the capitalist world-system became increasingly integrated and interdependent. The old imperial patterns of international trade broke down and were replaced by more complex patterns. Nevertheless, the newly independent States were still influenced by many of the old colonial links and legacies that remained intact. The result was a neo-colonial pattern of international development. **Neo-colonialism** refers to economic and political strategies by which powerful States in core economies indirectly maintain or extend their influence over other areas or people. Instead of formal, direct rule (colonialism), controls are exerted through such strategies as international financial regulations, commercial relations, and covert intelligence operations. Because of this neo-colonialism, the human geographies of peripheral countries continued to be heavily shaped by the linguistic, cultural, political, and institutional influence of the former colonial powers, and by the investment and trading activities of their firms.

Neo-colonialism: economic and political strategies by which powerful States in core economies indirectly maintain or extend their influence over other areas or people.

At about the same time, a new form of imperialism was emerging. This was the commercial imperialism of giant corporations. These corporations had grown within the core countries through the elimination of smaller firms by mergers and takeovers. By the 1960s, quite a few of them had become so big that they were *transnational* in scope, having established overseas subsidiaries, taken over foreign competitors, or simply bought into profitable foreign businesses.

These **transnational corporations** have investments and activities that span international boundaries, with subsidiary companies, factories, offices, or facilities in several countries. By the mid-1990s nearly 40,000 transnational corporations were operating, 90 percent of which were headquartered in the core States. Between them, these corporations control about 180,000 foreign subsidiaries and account for over $6 trillion in worldwide sales. Transnational corporations have been portrayed as imperialist by some geographers because of their ability and willingness to exercise their considerable power in ways that adversely affect peripheral States. They have certainly been central to a major new phase of geographical restructuring that has been under way for the last 25 years or so.

Transnational corporation: company with investments and activities that span international boundaries and with subsidiary companies, factories, offices, or facilities in several countries.

This phase has been distinctive because an unprecedented amount of economic, political, social, and cultural activity has spilled beyond the geographic and institutional boundaries of States. It is a phase of *globalization,* a much fuller integration of the economies of the worldwide system of States and a much greater interdependence of individual places and regions from every part of the world-system.

Globalization has, of course, been under way since the inception of the modern world-system in the sixteenth century. The basic framework for globalization has been in place since the nineteenth century, when the competitive system of States fostered the emergence of international agencies and institutions; global networks of communication; a standardized system of global time; international competitions and prizes; international law; and internationally shared notions of citizenship and human rights.

The distinctive feature of globalization over the past 25 years or so is that a decisive increase has occurred in the proportion of the world's economic and cultural activities that are international in scope. This increase is linked to a significant shift in the nature of international economic activity. Flows of goods, capital, and information that take place within and between transnational corporations are becoming more important than imports and exports between countries. (Between 1990 and 1995, U.S. overseas investment in manufacturing grew at twice the rate of exports of U.S.-manufactured goods.) At the same time, all these flows and activity have helped to spread new values around the world. These new values range from consumer lifestyle preferences to altruistic concerns with global resources, global environmental change, and famine relief.

Commodity chain: network of labor and production processes beginning with the extraction or production of raw materials and ending with the delivery of a finished commodity.

The contemporary world economy is constituted through the myriad commodity chains that criss-cross global space. **Commodity chains** are networks of labor and production processes whose origin is in the extraction or production of raw materials and whose end result is the delivery and consumption of a finished commodity. These networks often span countries and continents, linking into vast global assembly lines the production and supply of raw materials; the processing of raw materials; the production of components; the assembly of finished products; and the distribution of finished products. As we shall see in Chapter 7, these global assembly lines are increasingly important in shaping places and regions.

This globalization of the contemporary world—its causes and effects in terms of specific aspects of human geographies at different spatial scales—will be a recurring theme through the rest of this book. For the moment, we need only note in broad outline its principal causes and outcomes.

A New International Division of Labor

The globalization of the past quarter-century has been caused by four important and interrelated factors: a new international division of labor, an internationalization of finance, a new technology system, and a homogenization of international consumer markets.

The new international division of labor has involved three main changes. First, the United States has declined as an industrial producer, relative to the spectacular growth of Japan and the resurgence of Europe as industrial producers. The second result of the new international division of labor is that manufacturing production has been decentralized from all of these core regions to some semiperipheral and peripheral countries. In 1995, U.S.-based companies employed about 5.5 million workers overseas, 80 percent of whom were in manufacturing jobs. An important reason for this trend has been the prospect of keeping production costs low by exploiting the huge differential in wage rates around the world (Figure 2.20). Example: A Taiwanese transnational company invested $80

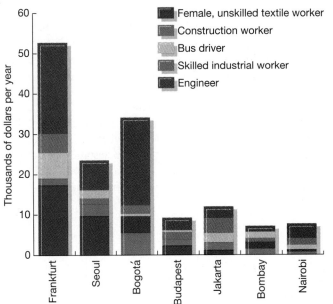

Figure 2.20 International differences in wage rates Wages vary substantially across countries and regions. Adjusted for differences in their currencies' purchasing power, the earnings of engineers in Frankfurt, Germany, are seven times those of engineers in Mumbai (formerly Bombay), India. International differences in the pay of unskilled workers are even greater—unskilled female textile workers in Frankfurt are paid 18 times as much as their counterparts in Nairobi, Kenya. (*Source:* World Bank, *World Development Report 1995.* New York: Oxford University Press, 1996, Fig. 1.2, p. 11.)

million in 1993 to build a huge factory in Pouchen, Indonesia, in order to make sports shoes for companies like Nike, Converse, and LA Gear. The factory now makes Cortez running shoes for Nike, whom it charges $12.50 a pair; Nike sells them to stores for $27.50, and stores sell them to the public for $50.00. These margins are made possible to a large extent by the $4-a-day wages paid to the factory's 9,000 assembly-line workers (compared to, say, $80 a day plus benefits that Nike would have to pay to factory workers in the United States). A third result of the new international division of labor is that new specializations have emerged within the core regions of the world-system: high-tech manufacturing and **producer services** (that is, services such as information services, insurance, and market research that enhance the productivity or efficiency of other firms' activities or that enable them to maintain specialized roles). One significant reflection of this new international division of labor is that global trade has grown much more rapidly over the past 25 years than global production—a clear indication of the increased economic integration of the world-system.

The second factor contributing to today's globalization is the internationalization of finance: the emergence of global banking and globally integrated financial markets. These changes are, of course, tied in to the new international division of labor. In particular, they are a consequence of massive increases in levels of international direct investment. Between 1988 and 1996, the flow of investment capital from core to semiperipheral and peripheral countries increased twentyfold. These increases include transnational investments by individuals and businesses as well as cross-border investments undertaken within the internal structures of transnational corporations. In addition, the capacity of computers and information systems to deal very quickly with changing international conditions has added a speculative component to the internationalization of finance. All in all, about $100 billion worth of currencies are traded every day. The volume of international investment and financial trading has created a need for banks and financial institutions that could handle investments on a large scale, across great distances, quickly and efficiently. The nerve-centers of the new system are located in just a few places—London, Frankfurt, New York, and Tokyo, in particular. Their activities are interconnected around the clock (Figure 2.21, page 88), and their networks penetrate into every corner of the globe.

Producer services: services that enhance the productivity or efficiency of other firms' activities or that enable them to maintain specialized roles.

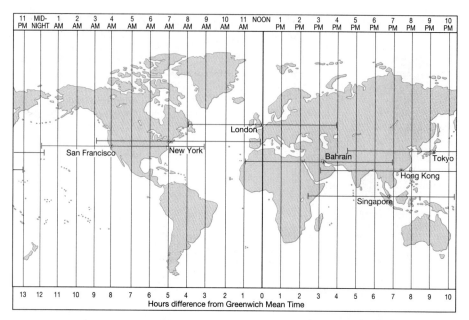

Figure 2.21 Twenty-four-hour trading between major financial markets   Office hours in the most important financial centers—New York, London, and Tokyo—overlap one another because the three cities are situated in broadly separated time zones. This means that, between them, they span the globe with 24-hour trading in currencies, stocks, and other financial instruments.

The third factor contributing to globalization is a new technology system based on a combination of innovations, including solar energy, robotics, microelectronics, biotechnology, digital telecommunications, and computerized information systems. This new technology system has required the geographical reorganization of the core economies. It has also extended the global reach of finance and industry and permitted a more flexible approach to investment and trade. Especially important in this regard have been new and improved technologies in transport and communications—the integration of shipping, railroad, and highway systems through containerization (Figure 2.22); the introduction of wide-bodied cargo jets (Figure 2.23, page 90); and the development of fax machines, fiber-optic networks, communications satellites, and electronic mail and information-retrieval systems (Figure 2.24, page 90). Finally, many of these telecommunications technologies have also introduced a wider geographical scope and faster pace to many aspects of political, social, and cultural change, as we shall see in subsequent chapters.

A fourth factor in globalization has been the growth of consumer markets. Among the more affluent populations of the world, similar trends in consumer taste have been created by similar social processes. A new and materialistic international culture has taken root, in which people save less, borrow more, defer parenthood, and indulge in affordable luxuries that are marketed as symbols of style and distinctiveness. This culture is easily transmitted through the new telecommunications media, and it has been an important basis for transnational corporations' global marketing of "world products" (German luxury automobiles, Swiss watches, British raincoats, French wines, American soft drinks, Italian shoes and designer clothes, and Japanese consumer electronics, for example). It is also a culture that has been easily reinforced through other aspects of globalization, including the internationalization of television, especially CNN, MTV, Star Television, and the syndication of TV movies and light entertainment series (Figure 2.25, page 91).

Figure 2.22 The impact of containerization on world trade   Containerization revolutionized long-distance transport because it did away with the slow, expensive, and unreliable business of loading and unloading ships with manual labor. Before containerization, ships spent one day in port for every one day at sea; after, they spent a day in port for every 10 days at sea. By 1965, an international standard for containers had been adopted, making it possible to transfer goods directly from ship to rail to road, and allowing for a highly integrated global transport infrastructure. Containerization requires a heavy investment in both vessels and dockside handling equipment, however. As a result, container traffic has quickly become concentrated in a few ports that handle high-volume transatlantic and transpacific trade. (*Source:* Map projection, Buckminster Fuller Institute and Dymaxion Map Design, Santa Barbara, CA. The word Dymaxion and the Fuller Projection Dymaxion™ Map design are trademarks of the Buckminster Fuller Institute, Santa Barbara, California, ©1938, 1967 & 1992. All rights reserved.)

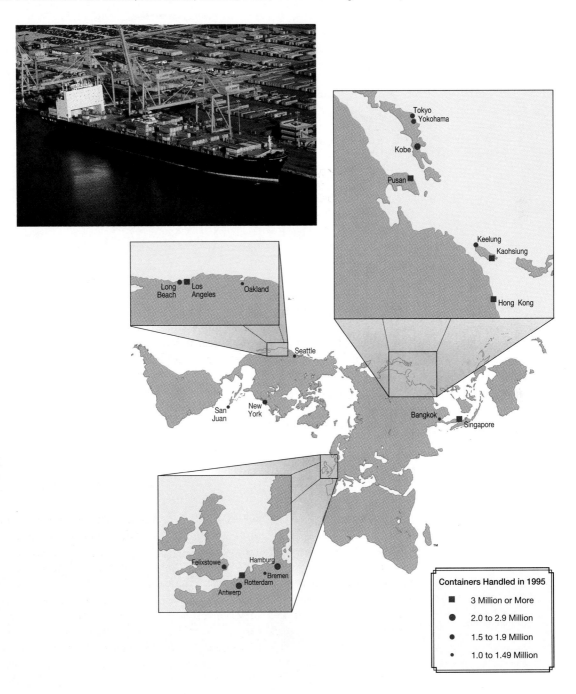

Figure 2.23 Global air cargo traffic in 1990 The introduction of wide-bodied cargo jets (like the Boeing 747) in the 1970s was an important factor in contributing to the globalization of the world economy. Within a few years, specialized parcel services had established regular routes handling a high volume of documents and freight with a high value-to-weight ratio. This Dymaxion projection shows how the pattern of air freight reflects the three-cornered structure of the contemporary world economy, with the highest-volume flows going between Western Europe, North America, and Japan. (*Source:* Map projection, Buckminster Fuller Institute and Dymaxion Map Design, Santa Barbara, CA. The word Dymaxion and the Fuller Projection Dymaxion™ Map design are trademarks of the Buckminster Fuller Institute, Santa Barbara, California, ©1938, 1967 & 1992. All rights reserved.)

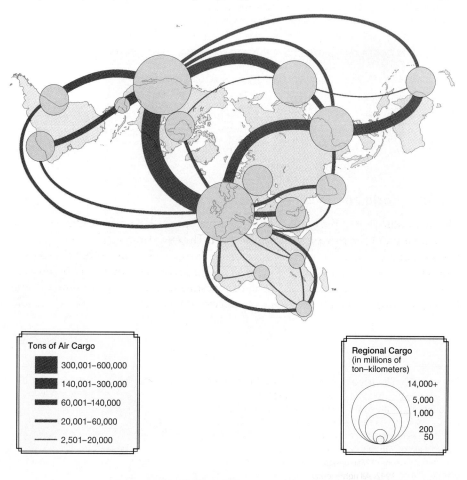

Tons of Air Cargo

300,001–600,000

140,001–300,000

60,001–140,000

20,001–60,000

2,501–20,000

Regional Cargo (in millions of ton–kilometers)

14,000+

5,000

1,000

200
50

Figure 2.24 Teleport Teleports are office parks that are equipped with satellite Earth stations and linked to local fiber-optics lines. Within the context of an expanding and evermore integrated global communications network, teleports offer an important local competitive advantage. Just as major ports handle the transshipment of cargo, teleports serve as vital information transmission facilities in the age of global capital. The world's first teleport was built on Staten Island, New York, in 1981. By 1995 there were over 60 teleports around the world.

Figure 2.25   Global marketing of television programming   The globalization of culture has been facilitated more than anything else by television broadcasting via satellite and by the sales of popular television programs to markets around the world. This photograph shows a television fair held in Miami Beach, Florida.

The Fast World and the Slow World

The main outcome of the globalization that has resulted from these changes is the consolidation of the core of the world-system. The core is now a close-knit triad of the geographic centers of North America, the European Union of Western Europe, and Japan (see Figure 2.26). These three geographic centers are connected through three main circuits, or flows, of investment, trade, and communication: between Europe and North America, between Europe and the Far East, and

Figure 2.26   The tripolar core of the world economy   In general terms, the world economy is now structured around a "core" with three centers: the United States, Japan, and the European Union. Most of the flows of goods, capital, and information are within and between these three centers. Among them, they dominate the world's periphery, with each center having particular influence in its own regional expansion zone: its nearest peripheral region. (*Source:* Map projection, Buckminster Fuller Institute and Dymaxion Map Design, Santa Barbara, CA. The word Dymaxion and the Fuller Projection Dymaxion™ Map design are trademarks of the Buckminster Fuller Institute, Santa Barbara, California, ©1938, 1967 & 1992. All rights reserved.)

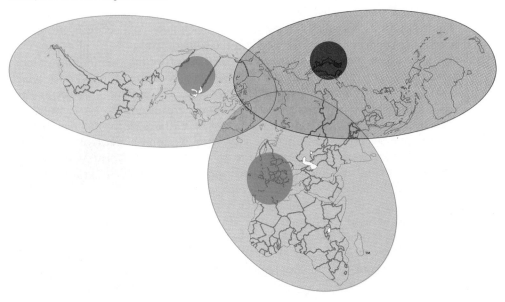

Fast world: people, places, and regions directly involved, as producers and consumers, in transnational industry, modern telecommunications, materialistic consumption, and international news and entertainment.

Slow world: people, places, and regions whose participation in transnational industry, modern telecommunications, materialistic consumption, and international news and entertainment is limited.

among the regions of the Pacific Rim. Within this core of centers and flows, a new hierarchy of regional specialization has been imposed by the locational strategies of transnational corporations and international financial institutions. As we shall see in Chapter 7, this geographic specialization is having some profound effects on the geographies of the world's core regions.

Globalization, although incorporating more of the world, more completely, into the capitalist world-system, has intensified the differences between the core and the periphery. According to the United Nations Development Program, the gap between the poorest 20 percent of the world's population and the wealthiest 20 percent increased threefold between 1960 and 1990. Some parts of the periphery have almost slid off the economic map. In sub-Saharan Africa, economic output fell by one-third during the 1980s, and now people's standard of living there is, on average, lower than it was in the early 1960s. Meanwhile, differences between the core and the periphery are now less easily captured in terms of the framework of States. Along with the broad geographic division of countries into core, semiperiphery, and periphery, an increasing division now exists between the "fast world" and the "slow world." The **fast world** consists of people, places, and regions directly involved, as producers and consumers, in transnational industry, modern telecommunications, materialistic consumption, and international news and entertainment. The **slow world,** which accounts for about 85 percent of the world's population, consists of people, places, and regions whose participation in transnational industry, modern telecommunications, materialistic consumption, and international news and entertainment is limited. The slow world consists chiefly of the impoverished periphery, but it also includes many rural backwaters, declining manufacturing regions, and disadvantaged slums in core countries, all of them bypassed by this latest phase in the evolution of the modern world-system.

The center of gravity of the fast world is the triadic core of the world-system. The United States, for example, with less than five percent of the world's population, accounts for more than 40 percent of the world's telephone stock. Similarly, the fast world also extends throughout the world to the more affluent regions, neighborhoods, and households that are "plugged in" to the contemporary world economy, whether as producers or consumers of its products and culture. The leading edge of the fast world is the Internet, the global web of computer networks that began in the United States in the 1970s as a decentralized communication system sponsored by the U.S. Department of Defense. Until the mid-1970s, there were less than 50 nodes (servers) in the whole system. Then, in the early 1980s, the original network (ARPANET) was linked with two important new networks: CSnet (funded by the National Science Foundation) and BITNET (funded by IBM). In July 1988 a high-speed backbone (NSFnet) was established in order to connect regional networks in the United States.

Today, these early networks have become absorbed into the Internet, a loose confederation of thousands of small, locally run computer networks for which there is no clear center of control or authority. The Internet has become the world's single most important mechanism for the transmission of scientific and academic knowledge. Roughly 50 percent of its traffic is electronic mail; the rest consists of scientific documents, data, bibliographies, electronic journals, bulletin boards, and a user interface to the Internet, the World Wide Web. In 1995 more than 2 million nodes existed in more than 130 countries; about 70 million people had access to the Internet; and somewhere between 25 and 30 million people worldwide had Internet E-mail addresses. The Internet has been doubling in networks and users every year since 1990, but most Internet users are still in the world's core regions: About 70 percent are in North America, and another 20 percent are in Europe (Figure 2.27). The rest are in Japan, Australia, and New Zealand, and in the fragmentary outposts of the fast world that are

Figure 2.27 Global Internet connectivity   Like all previous revolutions in transportation and communications, the Internet is effectively reorganizing space. Although often spoken of as "shrinking" the world and "eliminating" geography, the Internet is highly uneven in its availability and use. About 90 percent of all traffic on the Internet originates from, or is addressed to, North America. In contrast, some peripheral countries have no Internet connectivity, while in others the costs are prohibitive. Even in Europe, Japan, and North America, Internet connectivity is decidedly uneven in socioeconomic terms. It is the medium of the "fast world" of big business and affluent consumers. (*Source:* Map projection, Buckminster Fuller Institute and Dymaxion Map Design, Santa Barbara, CA. The word Dymaxion and the Fuller Projection Dymaxion™ Map design are trademarks of the Buckminster Fuller Institute, Santa Barbara, California, ©1938, 1967 & 1992. All rights reserved.)

▶ In 1996, the United States accounted for 60 percent of the world's 9.5 million Internet hosts. It also accounted for 80 percent of the commercial hosts (".com") and 80 percent of the educational hosts (".edu").

▶ Japan lags behind other developed countries in its use of the Internet. In 1996 there were 100,000 Internet hosts in Japan, compared with more than 250,000 in Britain and 220,000 in Germany. Fewer than 10 percent of Japanese offices are computerized, compared with 42 percent in the United States. Similarly, only 9 percent of Japanese personal computers are hooked in to a network of some sort, compared to 52 percent in the United States.

▷ China is connected to the Internet, but through lines with only a very small capacity. In early 1996 this amounted to less than half a dozen lines, each with 64-kilobit-per-second capacity—much less than the capacity of a small U.S. campus. Meanwhile, the Chinese government has announced plans to censor Internet access, as soon as it becomes feasible to do so.

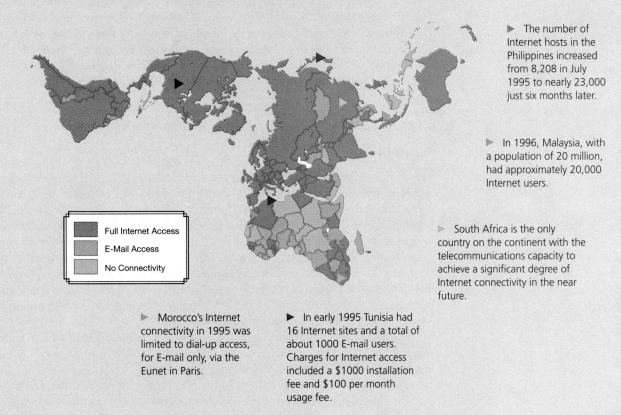

▶ The number of Internet hosts in the Philippines increased from 8,208 in July 1995 to nearly 23,000 just six months later.

▶ In 1996, Malaysia, with a population of 20 million, had approximately 20,000 Internet users.

▷ South Africa is the only country on the continent with the telecommunications capacity to achieve a significant degree of Internet connectivity in the near future.

Full Internet Access
E-Mail Access
No Connectivity

▷ Morocco's Internet connectivity in 1995 was limited to dial-up access, for E-mail only, via the Eunet in Paris.

▶ In early 1995 Tunisia had 16 Internet sites and a total of about 1000 E-mail users. Charges for Internet access included a $1000 installation fee and $100 per month usage fee.

embedded within the larger metropolitan areas of the periphery and semiperiphery. Overall, 90 percent of all Internet traffic originates in, or is destined for, North America.

This division between fast and slow worlds is, of course, something of a caricature. In fact, the fast world encompasses almost every*where* but not every*body*. As a result, human geography now has to contend with the apparent paradox of people whose everyday lives are lived part in one world, part in another. Consider, for example, the shantytown residents of Mexico City. With extremely low incomes, only makeshift housing, and little or no formal education, they somehow are knowledgeable about international soccer, music, film, and fashion, and are even able to copy fast-world consumption through cast-offs and knock-offs. Much the same could be said about the impoverished residents of rural Appalachia (substitute NASCAR racing for international soccer) and, indeed, about most regions of the slow world. Very few regions remain largely untouched by globalization.

This distinction between the fast world and the slow world brings us back to the themes of place, scale, and change that will recur throughout the rest of this book. At first glance, the emergence of the fast world, with its transnational architectural styles, dress codes, retail chains, and popular culture, and its ubiquitous immigrants, business visitors, and tourists, seems as if it might have brought a sense of placelessness and dislocation—a loss of territorial identity and an erosion of the distinctive sense of place associated with certain localities. Yet the common experiences associated with globalization are still modified by local geographies. The structures and flows of the fast world are variously embraced, resisted, subverted, and exploited as they make contact with specific places and specific communities. *In the process, places and regions are reconstructed rather than effaced.* Often, this involves deliberate attempts by the residents of a particular area to create or re-create territorial identity and a sense of place. Inhabitants of the fast world, in other words, still feel the need for enclaves of familiarity, centeredness, and identity. Human geographies change, but they don't disappear.

Conclusion

Places and regions everywhere carry the legacy of a sequence of major changes in world geography. The evolution of world geography can be traced from the prehistoric hearths of agricultural development and human settlement, through the trading systems of the pre-capitalist, pre-industrial world to the foundations of the geography of modern world. These foundations were cast through industrialization, the colonization of the world, and the spread of an international market economy. As *The Economist* magazine pointed out, "History counts: where you are depends very much on where you started from." Today, these foundations can be seen in the geography of the Information Age, a geography that now provides a new, global context for places and regions.

Today's world is highly integrated. Places and regions have become increasingly interdependent, linked together through complex and rapidly changing commodity chains that are orchestrated by transnational corporations. Using new technology systems that allow for instantaneous global telecommunications and flexible patterns of investment and production, these corporations span the fast world of the core and the slow worlds of the periphery. This integration does tend to blur some national and regional differences as the global marketplace brings a dispersion of people, tastes, and ideas. The overall result, though, has been an intensification of the differences between the core and the periphery.

Within this new global context, local differences in resource endowments remain, and people's territorial impulses endure. Many local cultures continue to be resilient or adaptive. Fundamental principles of spatial organization also continue to operate. All this ensures that, even as the world-system becomes more and more integrated, places and regions continue to be made and re-made. The new global context is filled with local variety that is constantly changing, just as the global context itself is constantly responding to local developments.

Main Points

- Places and regions fit together within a world-system that has been created as a result of processes of private economic competition and political competition between States.

- Each individual place and individual region carries out its own particular role within the competitive world-system.

- Because of these different roles, places and regions are dependent on one another. The development of each place affects, and is affected by, the development of many other places.

- The evolution of the world-system has exhibited distinctive stages, each of which has left its legacy in different ways on particular places, depending on their changing role within the world-system.

- The new technologies of the Industrial Revolution brought about the emergence of a global economic system that reached into almost every part of the world and into virtually every aspect of people's lives.

- Within the world's core regions, the new technologies of the Industrial Revolution began to transform regional geographies.

- The growth and internal colonization of the core regions could only take place with the foodstuffs, raw materials, and markets provided by the colonization of the periphery.

- Today, the world-system is highly structured, and is characterized by three tiers: *core regions, semiperipheral regions,* and *peripheral regions.*

- Globalization has intensified the differences between the core and the periphery and has contributed to the emergence of an increasing division between a fast world (about 15 percent of the world's population) and a slow world (about 85 percent of the world's population) with contrasting lifestyles and levels of living.

Key Terms

colonialism (p. 53)
commodity chain (p. 86)
comparative advantage (p. 79)
core regions (p. 53)
division of labor (p. 78)
fast world (p. 92)
hearth areas (p. 55)

hegemony (p. 65)
hinterland (p. 63)
import substitution (p. 63)
law of diminishing returns (p. 57)
leadership cycles (p. 65)
minisystem (p. 55)
neo-colonialism (p. 85)

peripheral regions (p. 53)
plantation (p. 63)
producer services (p. 87)
semiperipheral regions (p. 53)
slash-and-burn (p. 55)
slow world (p. 92)
States (p. 53)

transnational corporation (p. 85)
world-empire (p. 57)
world-system (p. 52)

Exercises

 On the Internet

1. The Internet offers many opportunities to explore various aspects of globalization. The Internet exercises for this chapter explore the idea of globalization and the changing international division of labor, contrasting the experiences of the fast world and the slow world. After completing the exercises for this chapter you should have a better understanding of current trends in the evolution of the world-system.

Unplugged

1. Transport and communications technologies have been crucial in the evolution of the world's human geographies. With reference to your own region, compile a list of examples of changes that have taken place as a consequence of changing transport and/or communications technologies.

2. We have seen that the historical development of the world order was driven successively by Portuguese, Spanish, Dutch, and British colonial expansion. Find out how many countries around the world now have Portuguese, Spanish, Dutch, and English as their official language (or as one of their official languages). How many Spanish- and English-speaking people are there in the world today? You will find the data you need in a good statistical yearbook, such as the Encyclopaedia Brittanica's *Yearbook*.

3. On page 52, we quoted *The Economist* magazine as saying that places and regions are important because "where you are depends very much on where you started from." Illustrate this point with reference to any region with which you are familiar.

4. Study the diagram of global air cargo in Figure 2.24. What are the five largest flows of air cargo? What reasons can you give for the dominance of these flows?

Chapter 3

Geographies of Population

Old Town Square, Prague

News reports of the international AIDS epidemic indicate that by the year 2010 the disease will shorten average life expectancy in Zambia by 50 percent and take 30 years off the life of the average resident of Thailand. In Zambia, if AIDS did not exist, the life expectancy would be about 66 in the year 2010. Yet because the epidemic has become so geographically widespread there, the average life expectancy is just 33 years. Life expectancies for other sub-Saharan African residents will also be significantly shortened as the AIDS epidemic reaches its peak in the first decade of the twenty-first century. The variations in the impact of AIDS on different populations in different places prompt geographically oriented questions. Why, for example, is the AIDS epidemic so much more threatening to African nations than to countries like Finland? What impact will AIDS have on population growth rates in various places?

Mortality is one of the chief concepts used by researchers who study population. For population geographers, knowing the mortality, or death, figures is not enough: They also want to know where mortality is occurring, why it is occurring, and what the consequences of it are for the remaining population. In this chapter we examine population distribution and structure as well as the dynamics of population growth and change, with a special focus on spatial variations and implications. In short, we want to know the locations of population clusters, the numbers of men and women and old and young and the ways these combine to create overall change, either as growth or decline. We also look at population movements and the models and concepts that population experts have developed to understand better the potential and problems posed by human populations.

The Demographer's Toolbox

Demography: the study of the characteristics of human populations.

Demography, or the study of the characteristics of human populations, is an interdisciplinary undertaking. Geographers study population to understand the areal distribution of Earth's population. They are also interested in the reasons for, and the consequences of, the distribution of population from the international to the local level. While historians study the evolution of demographic patterns and sociologists the social dynamics of human populations, it is geographers who focus special attention on the spatial patterns of human populations, the implications of such patterns, and the reasons for them. Using many of the same tools and methods of analysis as other population experts, geographers think of population in terms of the places that populations inhabit. They also consider populations in terms of the way that places are shaped by populations and in turn shape the populations that occupy them.

Censuses and Vital Records

Population experts rely on a wide array of instruments and institutions to carry out their work. Government entities, schools, and hospitals collect information on births, deaths, marriages, immigration, and other aspects of population change. The most widely known instrument for assessing the state of the population is the census, a survey that originated with the Romans and intended to provide information for tax collection.

Census: count of the number of people in a country, region, or city.

At a simple level, a **census** is a straightforward count of the number of people in a country, region, or city. Censuses, however, are not usually so simple. Most are also directed at gathering other information about the population, such as previous residence, marital status, income, and other personal data.

Many of the developed countries of the western hemisphere comprehensively assess the characteristics of their national populations every 10 years. In the United States, for example, the Bureau of the Census has conducted a survey of the population every 10 years since 1790. The information gathered is used to apportion seats in the U.S. House of Representatives, as well as to distribute federal tax funds and other revenues back to states, counties, and cities. Figure 3.1 (see page 100) is an abbreviated copy of the short form of the 1990 U.S. Census that was distributed to households throughout the country. As the form indicates, the questionnaire includes items about age, gender, and ethnicity, among other things.

In addition to the census, population experts also employ other data sources to assess population characteristics. One such source is **vital records,** which report births, deaths, marriages, divorces, and the incidence of certain infectious diseases. These data are collected and records of them kept by local, county, state, and other levels of government. Schools, hospitals, police departments, prisons, and other public agencies such as the Immigration and Naturalization Service in the United States, and international organizations like the World Health Organization, also collect demographic statistics that are useful to population experts.

Vital records: information about births, deaths, marriages, divorces, and the incidence of certain infectious diseases.

Limitations of the Census

Census enumerations are an extremely expensive and labor-intensive undertaking for any governmental jurisdiction and, as a result, they occur rather infrequently—usually no more than every five or 10 years. Historically, the United States has undertaken a population census every 10 years, but in 1985 it introduced a quinquennial (every five years) census to augment the decennial (every 10 years) one. Most prominent among the reasons for initiating a quinquennial census was the fact that data collected every 10 years quickly became obsolete. In addition, in many peripheral and semiperipheral countries, governments are not always able to finance a decennial census such as the kinds of comprehensive surveys undertaken in the more developed countries like France or Germany.

The last census undertaken in the former Soviet Union occurred in 1989. Unlike the U.S. census, enumeration did not occur at the individual level but at a district level, and the questions asked made it impossible to compare it with the U.S. census or even with previous censuses in the former USSR. Furthermore, some countries, such as Anguilla, have never conducted complete censuses. The Democratic Republic of Korea (North Korea) conducted its last census in 1944 (we think).

Additional problems with censuses include different enumeration dates (in terms of years, months, and days) for different countries around the world. The incompatibility of enumeration dates makes comparisons among, between, and within countries quite difficult. For example, the United States conducted its most recent decennial census on April 1, 1990. Poland conducted its decennial census on December 6, 1988, a difference in collection times between the two of a year and a half. Even a few months difference can be significant; between April 1, 1990, and April 1, 1992, for example, it is estimated that the United States may have added nearly 5.5 million people to its total population.

Self-identification of ethnicity or race by indicators such as chief language spoken at home or skin color mask any number of other, often more meaningful, cultural attachments. Perhaps the most problematic ethnicity category in the 1980 U.S. census was "Hispanic," an umbrella term used to refer to residents whose cultural heritage derives from a Spanish-speaking country. What Hispanics have in common is that they have ancestral ties to Latin America or Spain. Yet this apparent commonality masks a great deal of cultural differentiation because a self-identified "Hispanic" can be Mexican-American, Puerto Rican, Cuban-American, Spaniard, or someone from any of the South or Central American countries.

In short, no census is entirely comprehensive. All tend to underrepresent nonmainstream kinds of households as well as homeless individuals. Many

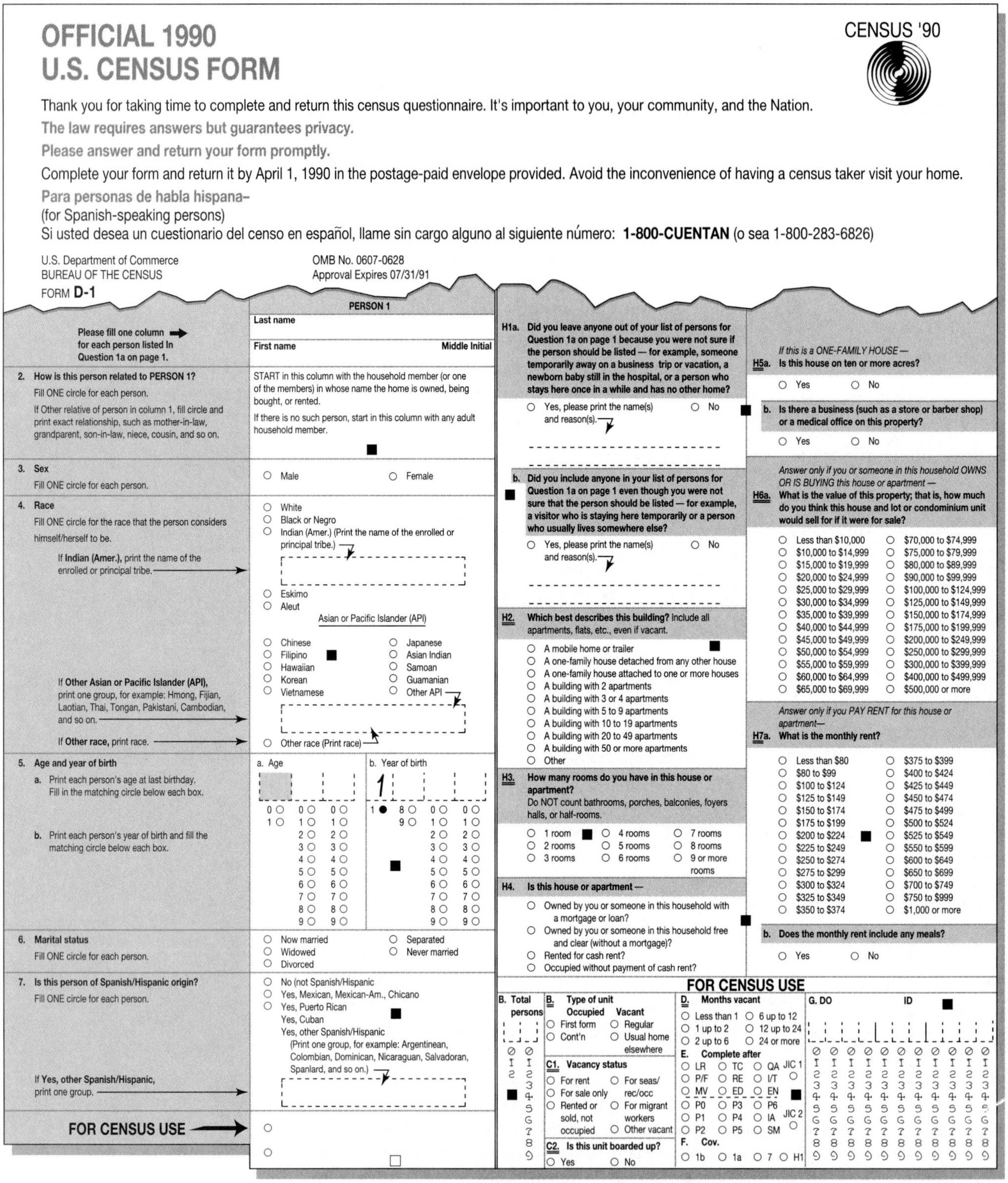

OFFICIAL 1990
U.S. CENSUS FORM

CENSUS '90

Thank you for taking time to complete and return this census questionnaire. It's important to you, your community, and the Nation.

The law requires answers but guarantees privacy.

Please answer and return your form promptly.

Complete your form and return it by April 1, 1990 in the postage-paid envelope provided. Avoid the inconvenience of having a census taker visit your home.

Para personas de habla hispana–
(for Spanish-speaking persons)
Si usted desea un cuestionario del censo en español, llame sin cargo alguno al siguiente número: **1-800-CUENTAN** (o sea 1-800-283-6826)

U.S. Department of Commerce
BUREAU OF THE CENSUS
FORM **D-1**

OMB No. 0607-0628
Approval Expires 07/31/91

PERSON 1

Please fill one column ➡ for each person listed in Question 1a on page 1.

Last name

First name — Middle Initial

2. How is this person related to PERSON 1?
Fill ONE circle for each person.
If Other relative of person in column 1, fill circle and print exact relationship, such as mother-in-law, grandparent, son-in-law, niece, cousin, and so on.

START in this column with the household member (or one of the members) in whose name the home is owned, being bought, or rented.
If there is no such person, start in this column with any adult household member.

3. Sex
Fill ONE circle for each person.
○ Male ○ Female

4. Race
Fill ONE circle for the race that the person considers himself/herself to be.
If **Indian (Amer.)**, print the name of the enrolled or principal tribe. ——➡
If **Other Asian or Pacific Islander (API)**, print one group, for example: Hmong, Fijian, Laotian, Thai, Tongan, Pakistani, Cambodian, and so on. ——➡
If **Other race**, print race. ——➡

○ White
○ Black or Negro
○ Indian (Amer.) (Print the name of the enrolled or principal tribe.)
○ Eskimo
○ Aleut

Asian or Pacific Islander (API)
○ Chinese   ○ Japanese
○ Filipino   ○ Asian Indian
○ Hawaiian   ○ Samoan
○ Korean   ○ Guamanian
○ Vietnamese   ○ Other API
○ Other race (Print race)

5. Age and year of birth
a. Print each person's age at last birthday. Fill in the matching circle below each box.
b. Print each person's year of birth and fill the matching circle below each box.

a. Age   b. Year of birth

6. Marital status
Fill ONE circle for each person.
○ Now married   ○ Separated
○ Widowed   ○ Never married
○ Divorced

7. Is this person of Spanish/Hispanic origin?
Fill ONE circle for each person.
○ No (not Spanish/Hispanic)
○ Yes, Mexican, Mexican-Am., Chicano
○ Yes, Puerto Rican
○ Yes, Cuban
○ Yes, other Spanish/Hispanic
(Print one group, for example: Argentinean, Colombian, Dominican, Nicaraguan, Salvadoran, Spaniard, and so on.)
If **Yes, other Spanish/Hispanic,** print one group. ——➡

FOR CENSUS USE ——➡

H1a. Did you leave anyone out of your list of persons for Question 1a on page 1 because you were not sure if the person should be listed — for example, someone temporarily away on a business trip or vacation, a newborn baby still in the hospital, or a person who stays here once in a while and has no other home?
○ Yes, please print the name(s) and reason(s).
○ No

b. Did you include anyone in your list of persons for Question 1a on page 1 even though you were not sure that the person should be listed — for example, a visitor who is staying here temporarily or a person who usually lives somewhere else?
○ Yes, please print the name(s) and reason(s).
○ No

H2. Which best describes this building? Include all apartments, flats, etc., even if vacant.
○ A mobile home or trailer
○ A one-family house detached from any other house
○ A one-family house attached to one or more houses
○ A building with 2 apartments
○ A building with 3 or 4 apartments
○ A building with 5 to 9 apartments
○ A building with 10 to 19 apartments
○ A building with 20 to 49 apartments
○ A building with 50 or more apartments
○ Other

H3. How many rooms do you have in this house or apartment?
Do NOT count bathrooms, porches, balconies, foyers halls, or half-rooms.
○ 1 room   ○ 4 rooms   ○ 7 rooms
○ 2 rooms   ○ 5 rooms   ○ 8 rooms
○ 3 rooms   ○ 6 rooms   ○ 9 or more rooms

H4. Is this house or apartment —
○ Owned by you or someone in this household with a mortgage or loan?
○ Owned by you or someone in this household free and clear (without a mortgage)?
○ Rented for cash rent?
○ Occupied without payment of cash rent?

If this is a ONE-FAMILY HOUSE —
H5a. Is this house on ten or more acres?
○ Yes   ○ No

b. Is there a business (such as a store or barber shop) or a medical office on this property?
○ Yes   ○ No

Answer only if you or someone in this household OWNS OR IS BUYING this house or apartment —
H6a. What is the value of this property; that is, how much do you think this house and lot or condominium unit would sell for if it were for sale?
○ Less than $10,000   ○ $70,000 to $74,999
○ $10,000 to $14,999   ○ $75,000 to $79,999
○ $15,000 to $19,999   ○ $80,000 to $89,999
○ $20,000 to $24,999   ○ $90,000 to $99,999
○ $25,000 to $29,999   ○ $100,000 to $124,999
○ $30,000 to $34,999   ○ $125,000 to $149,999
○ $35,000 to $39,999   ○ $150,000 to $174,999
○ $40,000 to $44,999   ○ $175,000 to $199,999
○ $45,000 to $49,999   ○ $200,000 to $249,999
○ $50,000 to $54,999   ○ $250,000 to $299,999
○ $55,000 to $59,999   ○ $300,000 to $399,999
○ $60,000 to $64,999   ○ $400,000 to $499,999
○ $65,000 to $69,999   ○ $500,000 or more

Answer only if you PAY RENT for this house or apartment—
H7a. What is the monthly rent?
○ Less than $80   ○ $375 to $399
○ $80 to $99   ○ $400 to $424
○ $100 to $124   ○ $425 to $449
○ $125 to $149   ○ $450 to $474
○ $150 to $174   ○ $475 to $499
○ $175 to $199   ○ $500 to $524
○ $200 to $224   ○ $525 to $549
○ $225 to $249   ○ $550 to $599
○ $250 to $274   ○ $600 to $649
○ $275 to $299   ○ $650 to $699
○ $300 to $324   ○ $700 to $749
○ $325 to $349   ○ $750 to $999
○ $350 to $374   ○ $1,000 or more

b. Does the monthly rent include any meals?
○ Yes   ○ No

FOR CENSUS USE

B. Total persons

B. Type of unit
Occupied / Vacant
○ First form   ○ Regular
○ Cont'n   ○ Usual home elsewhere

C1. Vacancy status
○ For rent   ○ For seas/rec/occ
○ For sale only
○ Rented or sold, not occupied   ○ For migrant workers
○ Other vacant

C2. Is this unit boarded up?
○ Yes   ○ No

D. Months vacant
○ Less than 1   ○ 6 up to 12
○ 1 up to 2   ○ 12 up to 24
○ 2 up to 6   ○ 24 or more

E. Complete after
○ LR   ○ TC   ○ QA
○ P/F   ○ RE   ○ I/T
○ MV   ○ ED   ○ EN
○ P0   ○ P3   ○ P6
○ P1   ○ P4   ○ IA
○ P2   ○ P5   ○ SM

F. Cov.
○ 1b   ○ 1a   ○ 7   ○ H1

G. DO   **ID**

Figure 3.1 Abbreviated U.S. Census short form, 1990 Census forms are among the most important tools a nation possesses for gathering information about the characteristics of its citizens. Far from just counting the number of people in the country, the U.S. census compiles a wealth of information on the makeup of households, people's racial and ethnic associations, age, marital status, occupation, respondent's type of residential structure, and many other details. These data provide the basis for an almost unlimited array of analyses ranging from mapping the geographical distribution of any of the attributes surveyed, for example, to studying the implications of those attributes for public policy or private marketing strategies.

municipalities have complained that the U.S. Bureau of the Census undercounted their city's population greatly. For example, the city of Inglewood in metropolitan Los Angeles estimates that 13,400 individuals were not found by the enumerators, leading to an undercount of 10.9 percent, the largest of any city in the country. Because federal revenue-sharing formulas as well as the apportionment of U.S. House seats are tied to population numbers, urban political officials from many of America's large and medium-sized cities have complained extensively about the numbers generated from the 1990 census.

Population Distribution and Structure

Because human geographers explore the interrelationships and interdependencies between people and places, they are interested in demography, or the systematic analysis of the numbers and distribution of human populations. Population geographers bring to demography a special perspective—the spatial perspective—which emphasizes description and explanation of the "where" of population distribution, patterns, and processes. For instance, the seemingly simple fact that as of 1993, the world was inhabited by 5.685 billion people is one that geographers like to think about in a more complex way. While this number is undeniably phenomenal and, furthermore, increasing daily, for geographers the most important aspect of this number is its uneven spatial expression from region to region and place to place. Equally important are the implications and impacts of these differences. Looking at population numbers, geographers ask themselves two questions: Where are these populations concentrated, and what are the causes and consequences of such a population distribution?

Population Distribution

At a basic level many geographic reasons exist for the distribution of populations throughout the globe. As the world population density map demonstrates (Figure 3.2, page 102), some areas of the world are very heavily inhabited, while others only sparsely. Other areas contain no people whatsoever, as Figure 3.3 (see page 103) so elegantly illustrates. Bangladesh and the Netherlands, for example, have overall high population densities. India, however, displays a different pattern, with especially high population concentrations along the coasts and rivers, but a relatively low population density in its western sector. Egypt presents an interesting pattern, with virtually all its population concentrated on three percent of its total land area: along the Nile River, the country's main source of water. The sparsest inland population concentrations in the world occur in the densely forested frontier regions of the Amazon. Thus, environmental and physical factors are important influences on population distributions and concentrations.

Degree of accessibility, topography, soil fertility, climate and weather, water availability and quality, and type and availability of other natural resources are some of the factors that shape important population distribution. Other factors are also crucial—first and foremost is a country's political and economic experiences and characteristics. For example, the high population concentrations along Brazil's Atlantic coast date back to the trade patterns set up during the years of Portuguese colonial control. Another important factor is culture as expressed in religion, tradition, or historical experience. Cities like Medina and Mecca, in the Middle East, comprise important population concentrations in no small part because they are Islamic sacred sites.

Table 3.1 (see page 103) shows population estimates in terms of traditional continental distributions. From the table it is clear that Asia is far and away the most populous continent, including 56 percent of the world's inhabitants in more than 40 countries. In Asia, the two countries of China and India constitute

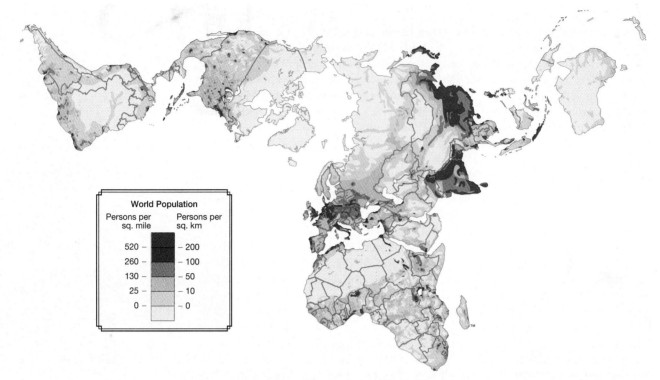

Figure 3.2 World population density, 1995 As this map shows, the world's population is not uniformly distributed across the globe. Such maps are useful in understanding the relationships between population distribution and the national contexts within which they occur. Note how China and India are the largest countries in the world with respect to population whereas the core countries have much smaller populations. (*Source:* Map projection, Buckminster Fuller Institute and Dymaxion Map Design, Santa Barbara, CA. The word Dymaxion and the Fuller Projection Dymaxion™ Map design are trademarks of the Buckminster Fuller Institute, Santa Barbara, California, ©1938, 1967 & 1992. All rights reserved.)

World Population

Persons per sq. mile | Persons per sq. km

520 – – 200
260 – – 100
130 – – 50
25 – – 10
0 – – 0

37 percent of the continental total and 21 percent of the world total. Running a distant second and third are Africa with 14 percent of the world's population, and South America with 10 percent.

The population clusters that take shape have a number of physical similarities. Almost all of the world's inhabitants live on 10 percent of the land. Most live near the edges of land masses, near the oceans, seas, or along rivers with easy access to a navigable waterway. Approximately 90 percent live north of the equator, where the largest proportion of the total land area (63 percent) is located. Finally, most of the world's population live in temperate, low-lying areas with fertile soils.

Population numbers are significant not only at the global scale but also at other levels. Population concentrations within countries, regions, and even metropolitan areas are also important for showing us where people are (Figures 3.4 and 3.5, page 104). For example, much of the population of Australia, which totaled 16,806,730 in 1989, is distributed along the coastal areas where most of the large cities are. In Mexico, with a total population of 70,562,202 in 1990, the coastal areas are relatively sparsely populated, while the heavily populated cities, such as Mexico City, are located in the interior of the country (Figure 3.6, page 105).

Population Density and Composition

Another way to explore population is in terms of density, a numerical measure of the relationship between the number of people and some other unit of interest expressed as a ratio. For example, crude density (sometimes referred to as arith-

Figure 3.3 World illumination map, 1993 This photograph of the globe at night illustrates population distribution by way of electrical illumination. It not only provides insight into population density, but also demonstrates, in a very simple way, differences in levels of development. Notice in particular the contrast between Europe and Africa, core and periphery. (*Sources: Hammond Atlas of the World.* 1993. New York: Oxford University Press, p. 7.)

metic density) is probably the most common measurement of population density. **Crude density,** also called **arithmetic density,** is the total number of people divided by the total land area. Metropolitan Phoenix, for example, with a population of roughly one million and a land area of 842 square kilometers (325 square miles), has a population density of a little over 1,188 persons per square

Crude density (arithmetic density): total number of people divided by the total land area.

TABLE 3.1 World Population Estimates by Continents		
Continent	**Number of Inhabitants**	**% of Total Population**
Africa	795,900,000	14
Asia	3,183,600,000	56
Australia	56,850,000	1
Europe	511,650,000	9
Former U.S.S.R.	284,250,000	5
North America	284,250,000	5
South America	568,500,000	10

Figure 3.4 Population distribution of Australia, 1990 As is the case in many other countries and regions of the globe, Australia's population is largely clustered along the coasts, particularly the east and west coasts. Much of this population is descended from Europeans, while scattered population in the interior of the continent is largely aboriginal. The most important cities politically and economically are also located along the coasts. (*Source: Atlas of Australian Resources: Population,* 3d Series, Vol. 2. Canberra: Division of Natural Mapping, 1991, p. 4.)

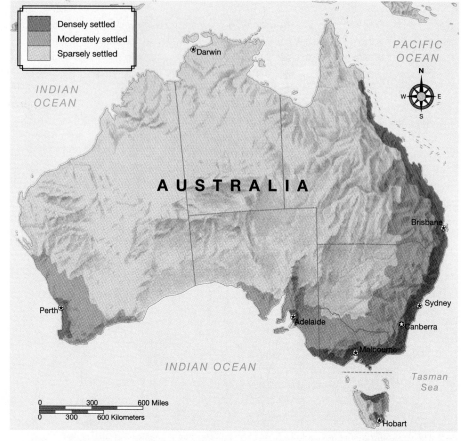

Figure 3.5 Mexico's population distribution, 1993 Mexico's population distribution is very much the opposite of Australia's with most of its population located in the interior of the country away from the mountains and in areas where soil fertility is highest. Mexico City is an inland city and is one of the largest—in terms of population—in the world. (Source: James B. Pick and Edgar W. Butler, *The Mexico Handbook*. Boulder: Westview Press, 1994. With permission of the authors.)

kilometer (3,000 persons per square mile) (Figure 3.7). By comparison, metropolitan New York, with a population of over 7 million and a land area of 803 square kilometers (310 square miles), has a population density of nearly 8,717 persons per square kilometer (24,000 persons per square mile) (Figure 3.8, page 106).

The limitation of the crude density ratio—and hence the reason for its "crudeness"—is that it is one-dimensional. It tells us very little about the opportunities and obstacles that the relationship between people and land contains. For that, we need other tools for exploring population density such as nutritional density or agricultural density. **Nutritional density** is the ratio between the total population and the amount of land under cultivation in a given unit of area. The **agricultural density** ratio is the relationship between the number of agriculturists—people earning their living or subsistence from working the land—per unit of farmable land in a specific area. Figure 3.9 (see page 107) is a map of the ratio of population to the number of physicians, an indicator of the quality of a country's health care.

Nutritional density: ratio between the total population and the amount of land under cultivation in a given unit of area.

Agricultural density: ratio between the number of agriculturists per unit of arable land in a specific area.

Figure 3.6 Crowding in Mexico City Although attempts are being made to redistribute Mexico City's population by encouraging development around its edges and beyond, the city is still very densely occupied. Mexico City experiences many of the problems that accompany dense urbanization in core countries such as traffic congestion and air pollution.

Figure 3.7 Population density, Phoenix The urban landscape of Phoenix is low-profile and widely spread. Unlike Manhattan, New York City, which is constrained by the edges of the island, few if any physical boundaries exist to inhibit growth in Phoenix. More important, however, is that land is relatively cheap in Phoenix—in part because it is more abundant, but also because it is not as important in the world-system. Phoenix is more regionally important, whereas New York City's importance in the world economic system is national and international.

Figure 3.8 Population density, New York City Urban form in New York City is the result of many factors. Most important in terms of its population size is its importance as a central—if not the central—node in the world-system of cities. Such a prominent role is built upon both a varied and extensive population base. The urban economy is complex, requiring a wide range of labor skills from professional and managerial to low-skilled, service-oriented workers.

In addition to exploring patterns of distribution and density, population geographers also examine population in terms of its composition, that is, in terms of the subgroups that constitute it. Understanding population composition enables geographers to gather important information about population dynamics. For example, knowing the composition of a population in terms of the total number of males and females, number and proportion of old people and children, and number and proportion of people active in the work force provides valuable insights into the ways in which the population behaves.

Facing unique challenges are countries with a population that contains a high proportion of old people—a situation many core countries will soon be facing as their "baby boom" population ages. The **baby boom** population includes those individuals born between the years 1946 and 1964. A considerable amount of that country's resources and energies is necessary to meeting the needs of a large number of people who may no longer be contributing in any significant fashion to the creation of the wealth necessary for their maintenance. There might also be a need to import workers to supplement the small working-age population. Knowing the number of women of childbearing age in a population, along with other information about their status and opportunities, can provide valuable information about the future growth potential of that population.

For example, populations in core countries, like Denmark, with a small number of women of childbearing age relative to the total population size, but who have high levels of education, socioeconomic security, and wide opportunities for work outside the home, will generally grow very slowly, if at all. Populations in peripheral countries like Kenya, on the other hand—where great numbers exist of women of childbearing age who have low levels of education, socioeconomic security, and relatively few employment opportunities—will continue to experience very high rates of population growth, barring unforeseen changes. The variety that exists within a country's population very much shapes the opportunities and challenges it must confront nationally, regionally, and locally.

Understanding population composition, then, not only can tell us much about the future demographics of regions but it is also quite useful in the present. For example, businesses use population composition data to make marketing decisions and to decide where to locate their businesses. For many years, businesses used laborious computer models to help target their markets. With the recent development of geographic information systems (GIS), however, this process has

Baby boom: population of individuals born between the years 1946 and 1964.

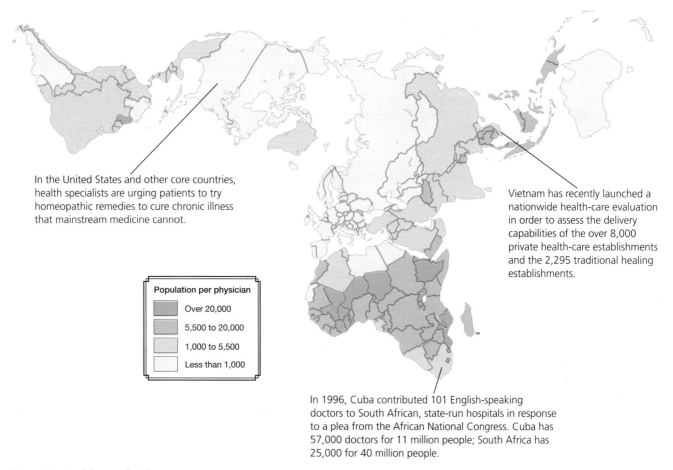

In the United States and other core countries, health specialists are urging patients to try homeopathic remedies to cure chronic illness that mainstream medicine cannot.

Vietnam has recently launched a nationwide health-care evaluation in order to assess the delivery capabilities of the over 8,000 private health-care establishments and the 2,295 traditional healing establishments.

Population per physician

Over 20,000

5,500 to 20,000

1,000 to 5,500

Less than 1,000

In 1996, Cuba contributed 101 English-speaking doctors to South African, state-run hospitals in response to a plea from the African National Congress. Cuba has 57,000 doctors for 11 million people; South Africa has 25,000 for 40 million people.

Figure 3.9 Health-care density   Another measure of population density is reflected in this map, which shows the number of doctors per 1,000 people in the population. The different shades of green reflect doctors per 1,000 population going from light to dark. The core countries as well as China have the highest ratio of doctors per population. Most of the continent of Africa, excepting South Africa and the Maghreb of North Africa, has the lowest ratio, reflecting another dimension of core–periphery inequality. (*Source:* Map projection, Buckminster Fuller Institute and Dymaxion Map Design, Santa Barbara, CA. The word Dymaxion and the Fuller Projection Dymaxion™ Map design are trademarks of the Buckminster Fuller Institute, Santa Barbara, California, ©1938, 1967 & 1992. All rights reserved.)

been greatly simplified. Some examples of assessing the location and composition of particular populations, known as **geodemographic analysis,** are provided in Feature 3.1, "Geography Matters—GIS Marketing Applications."

Geodemographic analysis: practice of assessing the location and composition of particular populations.

Age–Sex Pyramids

Areal distributions are not the only way that demographers portray population distributions. In order to display the variation within particular subgroups of a population or with respect to certain descriptive aspects of a population such as births or deaths, demographers also use a graphical device called a histogram. A **histogram** is a vertical bar graph. The height of the bar represents frequency of occurrence of a given phenomenon; the width depicts the various frequency categories. Figure 3.10 (see page 110) is an example of three very differently shaped histograms.

The most common way for demographers to represent graphically the composition of the population is to construct an **age–sex pyramid,** which is a

Histogram: vertical bar graph in which bar height represents frequency and in which width depicts the various frequency categories.

Age–sex pyramid: a representation of the population based on its composition according to age and sex.

3.1 Geography Matters
GIS Marketing Applications

Geographic information systems, or GIS, possess the powerful ability to reveal spatial patterns in demographic data. Recognizing the potential applications for marketing, marketing firms have readily incorporated GIS into their daily operations. Prior to using GIS for marketing, businesses made crude estimates of the location and behavior of their clientele. By using a GIS to analyze demographic data, marketing companies are now able to provide businesses with maps identifying areas of potential customers. With this information, businesses can engage in a promotional "sales blitz," targeting specific customers who have been identified from GIS analysis.

The key to using GIS effectively in marketing is the ability to link demographic data to particular locations. This is known as *georeferencing*. Demographic data are linked to zip code areas and telephone prefix zones. Then analyses are performed on the data to determine whether meaningful spatial relationships exist among specified areal units, based on given demographic data. For example, the residences of people with a particular level of education may cluster around a few contiguous zip code regions (see Figure 3.1.1).

Demographic data such as income level, education level, home value, age, and number of household residents provide marketing firms with information that can be linked to certain consumer goods. For example,

people who have high income and high education levels tend to own luxury cars. Luxury car dealerships, for example, would want to target such households. A GIS could produce a map that would show the general location of such households, and advertisements could then be distributed throughout this area.

An example of a business using demographic data with a GIS to target its clients is a company that distributes medical supplies to health-care professionals, nursing homes, and hospitals. Entered into a GIS were demographic data such as the age–sex pyramid; number of physicians per county; population density; percent of local funding spent on health care; and the number of nursing homes per 1,000 people. A statistical analysis was conducted on the demographic data, and a map was created to illustrate predicted sales volume. This predicted sales volume map was compared to a map showing actual sales. The residuals—what was left out on both maps—were then used to create a third map that displayed potential untapped markets. As a result of using GIS to identify underrepresented and unrepresented markets, the company rearranged its sales team to target these areas.

Another area in which GIS is used in marketing is in locational analysis—especially determining where to locate a business. The success of businesses such as gasoline stations and fast-food restaurants depends on locational factors such as traffic density. GIS is aptly suited for modeling traffic patterns. Traffic patterns are frequently determined by using demographic data that indicate how far, how often, and in which direction people travel to work. Many GIS systems have models that can be used to predict traffic patterns. By identifying areas with high traffic volumes, these businesses can be strategically located. Other conditions such as zoning restrictions and proximity to competition can also be factored into a GIS to reduce the number of potential retail sites.

The spending patterns of consumers are another source of information that demographic databases often contain. These data are typically collected by marketing firms and are used to assist clients in determining business sites. US WEST provided such a service in Denver, Phoenix, and Omaha. By distributing Your Value Cards, a plastic credit cardlike card that entitled users to special promotions, US WEST was able to track the spending patterns of consumers. Whenever the card was used, US WEST compiled point-of-sale data of the household using the card. By integrating this demographic database with a GIS, US WEST was able to produce maps identifying not only where consumers shopped, but also the area where they lived. As of December 1995, approximately 2,000 businesses were participating in the US WEST pro-

Figure 3.1.1 Your Value Card, Denver This graphic shows how zip code can be used to help marketers differentiate their market based on place of residence. In order to make this generalization, marketers must assume that the populations who live in these zip codes are relatively homogeneous. (*Source: Business Geographics,* November/December 1995, p. 70. Map courtesy of US WEST.)

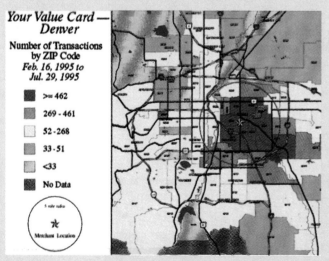

Figure 3.1.2   Gourmet Coffee Buyers day and night markets (*Source:* Claritas, 1995.)

gram. During 1995, over 1.4 million Your Value Card transactions had been recorded.

One well-known family restaurant chain and participating Your Value Card merchant had two restaurants within five kilometers (three miles) of one another. The corporate headquarters wanted to relocate one restaurant to help realize both restaurants' full potential. When a bank turned down the application for a loan to relocate, the restaurant approached US WEST for assistance. US WEST provided the restaurant chain with a GIS-based market analysis, which the restaurant presented to the bank. This time the loan was approved. The restaurant chain has benefited greatly from relocating, with Your Value Card transactions accounting for $90,000/month in revenues between the two restaurants.

The marriage of demographic data and GIS creates a powerful tool for the marketing industry. By using GIS to analyze selected demographic data, marketing firms are able to uncover spatial patterns such as residence, traffic flows, and consumer spending. These patterns can be used by businesses to help implement more effective marketing campaigns or to locate a new outlet. These examples illustrate how GIS and demographic data may be applied to overcome problems and suggest solutions in a commercial setting.

It should also be pointed out, however, that the potential for invasion of privacy through the coupling of

different sorts of data sets is very real. For instance, when you shop in U.S. grocery stores with electronic inventory systems and pay for your groceries by credit card or check, a record of your purchases can now be joined with your name. Such a coupling allows marketers to do more than target you more specifically by providing you with information on sales and manufacturers' coupons. It also, potentially, gives them access to information about what prescription drugs you might be taking, how much alcohol you consume, what sorts of magazines you buy, or videos you rent. Thus, while GIS and the massive data sets that support it open up new opportunities for marketing, the possibility also exists for the erosion of personal rights.

Source: This feature was adapted from *Business Geographics,* January 1996 and November/December 1995.

109

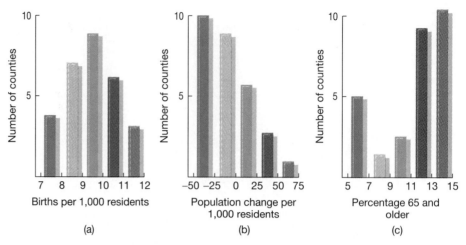

Figure 3.10 Histograms of univariate distributions Histograms allow analysis of selected characteristics of specific subgroups within a larger population. These histograms illustrate characteristics of three population subgroups for one geographical location. Figure 3.10a depicts a symmetrical distribution (i.e., evenly distributed around a midpoint) of the number of births (per 1,000 residents) among the selected counties. Figure 3.10b shows that, for those same counties, the population change per 1,000 residents is heavily skewed toward net population decline. Figure 3.10c indicates that, in terms of percentage of the population in those counties aged 65 and older, a bimodal distribution exists, with highest percentages occurring at the lowest and highest ends of the scale.

Cohort: a group of individuals who share a common temporal demographic experience.

representation of the population based on its composition according to age and sex. An age–sex pyramid is actually a histogram displayed horizontally; males are usually portrayed on the left side of the vertical axis and females on the right. Age categories are ordered sequentially from the youngest, at the bottom of the pyramid, to the oldest, at the top. By moving up or down the pyramid, one can compare the opposing horizontal bars to assess differences in frequencies for each age group. Age–sex pyramids also allow demographers to identify changes in the age and sex composition of populations. For example, an age–sex pyramid depicting Germany's population in 1989 reveals the impact of the two World Wars, especially the loss of large numbers of males of military age, and the deficit of births during those periods (Figure 3.11). Demographers call population groups like these cohorts. A **cohort** is a group of individuals who share a common *temporal* demographic experience. A cohort is not necessarily based only on age, however, and may be defined according to criteria such as time of marriage or time of graduation.

In addition to revealing the demographic implications of war or other significant events, age–sex pyramids can provide information necessary to assess the potential impacts that growing or declining populations might have. As illustrated in Figure 3.12 (see page 112), the shape of an age–sex pyramid varies depending on the proportion of people in each age cohort. The pyramid for Kenya, for example, shown in Figure 3.12a, reveals that many dependent children, ages 0 to 14, exist relative to the rest of the population. The considerable narrowing of the pyramid toward the top indicates that the population has been growing very rapidly in recent years. The shape of this pyramid is typical of peripheral countries with high birthrates and low death rates.

Serious implications are associated with this type of pyramid, however. First, in the absence of high productivity and wealth, resources are increasingly stretched to their limit to accommodate even elemental schooling, nutrition, and health care for the growing number of dependent children. Furthermore, when these children reach working age, a large number of jobs will need to be created to enable them to support themselves and their families. Also, as they form their

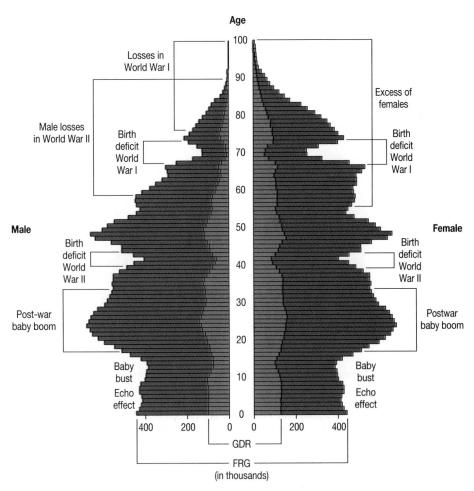

FRG = West Germany
GDR = East Germany

Figure 3.11 Population of Germany, by age and sex, 1989 Germany's population profile is that of a wealthy core country that has passed through the postwar baby boom and currently possesses a low birthrate. It is also the profile of a country whose population has experienced the ravages of two World Wars. (*Source:* J. McFalls, Jr., "Population: A Lively Introduction," *Population Bulletin,* 46(2), 1991.)

own families, the sheer number of women of childbearing age will almost guarantee that the population explosion will continue. This will be true unless strong measures are taken, such as intensive birth-control campaigns; improved education and outside opportunities for women; and modifications of cultural norms that place a high value on large family size.

In contrast, the pyramid for the United States (Figure 3.12b) illustrates the typical shape for a country experiencing a slow rate of growth. In fact, the population would not be growing much at all if it were not for substantial migration from other countries. Note that a perceptible bulge exists in this pyramid, representing the baby boom cohort. The period, immediately after the Second World War, was one of tremendous economic growth as well as political and technological change, as the United States took its place as the leading core country. With a great deal of prosperity and a rosy future, returning Americans reproduced at a rate unlike any other time in our demographic history.

While Denmark has a somewhat similar pyramid to that of the United States (Figure 3.12c), one key difference exists: Where the United States has a slow rate

Figure 3.12 Population pyramids of rapid, slow, and zero population growth (ZPG)
Population pyramids vary, depending on the age and gender structure of the population being depicted. We can derive important information about the population growth rates of different countries over time by analyzing changes in the numbers of people of each sex and age category. (*Source:* J. McFalls, Jr., "Population: A Lively Introduction," *Population Bulletin*, 46(2), 1991.)

of growth, Denmark has a zero population growth rate. Thus in Denmark, where the birthrate is 13 per thousand and the death rate is 12 per thousand, the pyramid is very columnar, hardly a pyramid at all. People are equally distributed among the cohorts, though the base is perceptibly narrower. Denmark can expect many of the same demographic challenges, including the extremely large elderly population faced by the United States, in not too many years. In both countries, however, high levels of production and wealth, combined with low birthrates, translate into a generally greater capacity to provide not only high levels of health, education, and nutrition, but also jobs as those children grow up and join the work force. Whether these opportunities are equitably distributed among individual members of the population remains to be seen.

Tables 3.2 and 3.3 and Figure 3.13 (see page 117) provide a picture of the present state and potential future impact of the baby boom cohort, the largest population cohort in U.S. demographic history. Figure 3.13 provides a series of pyramids that illustrate how the configuration changes as the boomers age. The narrower column of younger people rising below the boomer cohort in these pyramids reveals the biggest problem facing this population: a significantly smaller cohort moving into its main productive years having to support a growing cohort of aging and decreasingly productive boomers (Figure 3.14, page 118). The contemporary fight over Social Security and Medicare funding is only the tip of the iceberg with regard to this problem, however (see Feature 3.2, "Geography Matters—The Baby Boomers").

Construction and analysis of age-sex pyramids need not be restricted to the national level, however. Figure 3.15 (see page 118), for example, shows pyramids for three different census tracts within the same city. Here, at the local scale, vari-

TABLE 3.2 Baby Boomer Population Structure

While the baby boom has demographically dominated the last half of the twentieth century, its influence will begin to wane in the early decades of the new one.

Year	% of Population Who Are Baby Boomers	Age
1990	30	25–44
2000	20	35–59
2020	15	55–79
2040	7	75–85+

ations exist in the geography of population composition as informative as those constructed at the national scale. The proportional representation of each cohort, by age and sex, raises provocative questions about why such spatialized differences exist, and what the implications of the distributions might be for social services, marketing strategies, and a host of other applications. For example, a census tract with a large number of children may lead decision makers to push for provision of greater opportunities for organized sports, and perhaps a sports medicine facility. Contrastingly, a growing concentration of older people might prompt the opening of a senior citizen center and a geriatric care facility.

A critical aspect of the population pyramid is the **dependency ratio**, which is a measure of the economic impact of the young and old on the more economically productive members of the population. Traditionally, in order to assess this relation of dependency in a particular population, demographers will divide the total population into three age cohorts, sometimes further dividing those cohorts by sex. The **youth cohort** consists of those members of the population who are less than 15 years of age and generally considered to be too young to be fully active in the labor force. The **middle cohort** consists of those members of the

Dependency ratio: measure of the economic impact of the young and old on the more economically productive members of the population.

Youth cohort: members of the population who are less than 15 years of age and generally considered to be too young to be fully active in the labor force.

Middle cohort: members of the population 15 to 64 years of age who are considered economically active and productive.

TABLE 3.3 Top Ten Baby Boomer American Cities, 1995

Except for Washington, DC, and Portsmouth, NH, most of the cities baby boomers are attracted to are in the Sunbelt. As members of a cohort who are at and approaching their peak earning power, it is not surprising that baby boomers are where the fastest-growing employment bases are. It will be interesting to see whether these same cities are attractive for baby boom retirement.

City	% of Population Who Are Baby Boomers	% Change in Baby Boomer Population Since 1985
Denver, CO	32	14.4
Washington, DC	31	14.7
Houston, TX	31	15.4
Austin, TX	30	12.1
Atlanta, GA	30	19.9
San Francisco, CA	29	9.9
Dallas, TX	29	19.7
Colorado Springs, CO	29	9.6
Portsmouth, NH	28	13.1
Las Vegas, NV	27	14.4

3.2 Geography Matters

The Baby Boom and Its Impact

The baby boom generation—those individuals born between 1946 and 1964 in the United States—has been described as "one of the most powerful and enduring demographic influences on this nation." It has certainly been one of the most studied, analyzed, vilified, and sanctified. The case of the baby boom generation provides a useful way of understanding the complex of factors that shape demographic change, from political and economic factors, to social and cultural ones.

Why so many births between 1946 and 1964? Most demographers cannot give a definitive answer to this question. In fact, in the 1940s, population experts were predicting that the American population would stop growing. And although increases in births following a war are expected, the extended surge in births that followed the end of the Second World War came as quite a surprise to population experts, policymakers, government officials, and most others. Demographic transition theory would have led demographers to anticipate declining births as urbanization and "modernization" accelerated after the war. Sociological theories and predictions based on past trends were not very useful, however, in anticipating the unprecedented phenomenon of the baby boom. To understand it, then, we need to examine a whole host of factors.

Demographic Factors

Demographers insist that the baby boom not be seen as a direct or indirect result of the end of the war. In fact, although marriages (and divorces) did increase dramatically after the war, the accompanying rise in birthrates accounts only for the time from 1946 to 1947. By 1950, birthrates had actually fallen (see Figure 3.2.1). In 1951, the birthrates once again began to climb and continued to be high until 1964.

Additionally, the baby boom was not about women having more children or people having larger families. It was, instead, based on more people marrying overall and having at least two children early in the marriage. During the 1950s and early 1960s, young women also married earlier than the previous generation and had children earlier (see Figure 3.2.2).

Political and Economic Factors

In addition to examining demographic factors in order to understand the baby boom, it is also necessary to look at political and economic factors. It is significant that the end of the Second World War witnessed a phenomenal expansion in the U.S. economy. Initially stimulated by the war, it was extended by important transformations in transportation and technology. Additionally, government had expanded and created programs for education and housing that helped returning veterans start off married life with property and the opportunity for improving their economic status by attending college. Transportation—especially through the construction of the interstate highway system—helped to fuel suburbanization, which meant growth in the building industry, in automobile manufacturing, and in the manufacturing of durable goods for the home. Not surprisingly, demand for labor was high in a growing economy, and young people, most of whom had attained a higher level of education than the previous generation, were able to obtain good jobs with relatively high wages, decent benefits, and reasonable prospects for promotion.

Social and Cultural Factors

"Leave it to Beaver"—the television program of the late fifties and early sixties—is a useful popular culture mani-

Figure 3.2.1 U.S. baby boom Crude Birthrate and Total Fertility Rate (*Source:* L. Bouvier and C. DeVita, "The Baby Boom: Entering MidLife," *Population Bulletin*, 46(3), 1991.)

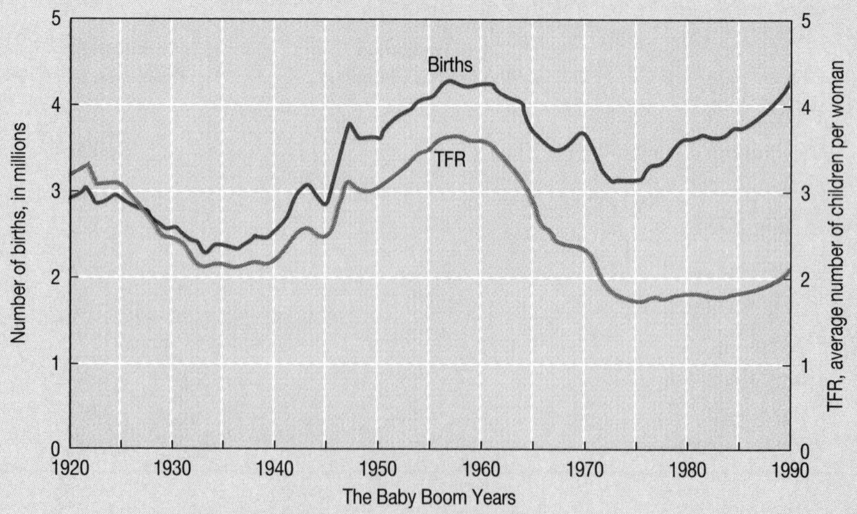

Figure 3.2.2 Comparison of number of children born to different female cohorts (*Source:* L. Bouvier and C. DeVita, "The Baby Boom: Entering MidLife," *Population Bulletin,* 46(3), 1991.)

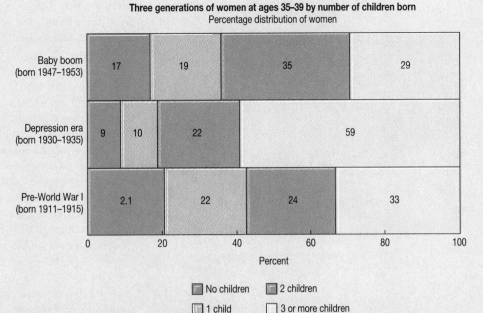

Three generations of women at ages 35–39 by number of children born
Percentage distribution of women

Baby boom (born 1947–1953): No children 17, 1 child 19, 2 children 35, 3 or more children 29

Depression era (born 1930–1935): No children 9, 1 child 10, 2 children 22, 3 or more children 59

Pre-World War I (born 1911–1915): No children 2.1, 1 child 22, 2 children 24, 3 or more children 33

Percent

No children · 1 child · 2 children · 3 or more children

festation of the "traditional American family" that re-emerged during the baby boom period. The Cleaver family, portrayed in the show, contained a gone-from-morning-till-night, at-work dad and a stay-at-home mom and two children living in a clean and safe suburban home. Ward Cleaver—the everyman, sole breadwinner—and June Cleaver—the everywoman homemaker—were very close to the "norm" of white, American, suburban living during the boom years. Though Mr. and Mrs. Cleaver are playful cultural stereotypes, it is important to note that the baby boom was confined not only to white America. Demographers point out that nearly all racial, economic, and religious segments of American society were part of the baby boom, although some segments were "more boomish" than others, most notably Catholics and African-Americans, whose fertility rates were higher than other groups.

The Impact of the Baby Boom Generation

Just as population experts have spent a great deal of energy trying to understand the many forces that brought about the baby boom, they have also expended at least an equal amount on attempting to predict the continuing impact of this very famous cohort. Indeed, a great deal of it can be directly attributed to its numerically large size compared to the generation that preceded and has followed it. Sheer numbers, in other words, have been very important in shaping the opportunities and obstacles that this cohort has confronted and will continue to face. It has affected educational institutions, government programs, job opportunities, the employment rate, voting patterns and spending patterns, not to mention its profound impact on popular culture—from fashion to

music to emotional, physical, and spiritual well-being (Figure 3.2.3, page 116). Baby boomers are now the most educated generation in American history. They have also married later and are more likely to divorce than any generation before them. Female and minority baby boomers have dramatically higher labor-force participation rates than previous generations. Indeed, women arguably have been the most important influence on the labor force in the twentieth century.

The Future

As the leading edge of this cohort enters middle age, a significant difference of opinion exists among population experts on its future impacts. Some see all these highly educated workers as having a stimulating effect on the economy in terms of improved productivity and increasing innovations. Others worry that this cohort will place unfulfillable demands on the nation's social service system. Certainly the debates about health-care reform are related to the increasing demands that this group is beginning to place on the health-care system. In addition, the pressures generated by its sheer size will certainly continue well into the first half of the twenty-first century. For example, by 2030, the oldest surviving boomers will be age 84 and the youngest will be 65. At that point, a projected 65 million people in the United States will be over the age of 65; at present there are roughly 30 million. When looking at the overall demographic portrait of the American population in 2030, we also see that more individuals over the age of 65 will be living than under the age of 18. Our understanding of the dependency load tells us that fewer young workers will exist to support the growing numbers of retirees.

115

Figure 3.2.3 Teenagers for 7up

The Impacts on Generation X: The Baby Busters

Demographers, in addition to identifying the existence of the baby boom generation, have also demonstrated the presence of a "baby bust" generation. Two ways exist to determine the extent of this cohort. One way is through the annual birthrates; if we use this measure, then this generation includes those individuals born between 1965 and 1980, when births declined below the baby boom average of 4 million births per year. A second way to determine the extent of the baby bust cohort is in terms of total fertility rates (TFR). Between 1947 and 1964, the TFR ranged between 3.1 and 3.7, well above replacement level. In 1965, however, the TFR dropped below replacement level for the first time ever in American history, and in 1976 it hit an all-time low of 1.7. Between 1977 and 1987, the TFR remained pretty stable at 1.8, and in 1990

it climbed to 2.09—close to replacement level. If we use TFR as the determining measure for the baby bust generation, then it looks as if the early 1990s mark the end of the cohort.

The baby bust generation will, of course, be the generation that has to "mop up" after the flood of boomers moves up their career ladders toward retirement and beyond. The tremendous size of the baby boom cohort is likely to affect the career and job mobility of the busters as well as labor costs, which would affect taxes, health care, and other benefit costs. These older workers will be more costly for their employers as their salaries and benefits increase in line with their seniority.

Other analysts have pointed with concern to the viability of the Social Security system as these workers age. Many worry that the system will not be able to meet its obligations without heavily taxing baby busters. In addition to the possible future implications of the low personal savings rate among baby boomers, population analysts worry that the boomers will reach retirement age with few personal resources and limited public funding to see them through a comfortable retirement.

The actual impact that the baby boom generation will have into the first half of the twenty-first century remains to be seen. What is clear, however, is that it carries the future of the nation's economic and social security along with it. Because of its sheer size, the cohort will continue to have an influence on public policy through their political participation at the ballot box and as national, regional, and local leaders and officials. Most important for the years to come, however, boomers will force to the center of political and ethical debate the issues of generational equity and the allocation of scarce resources across all cohorts.

Source: This feature was adapted from "The Baby Boom: Entering Midlife" by Leon Bouvier and Carol De Vita in *Population Bulletin*, 1991.

Old-age cohort: members of the population 65 years of age and older who are considered beyond their economically active and productive years.

population aged 15 to 64 who are considered economically active and productive. Finally, the **old-age cohort** consists of those members of the population aged 65 and older who are considered beyond their economically active and productive years. By dividing the population into these three groups it is possible to obtain a measure of the dependence of the young and old upon the economically active and the impact of the dependent population upon the independent.

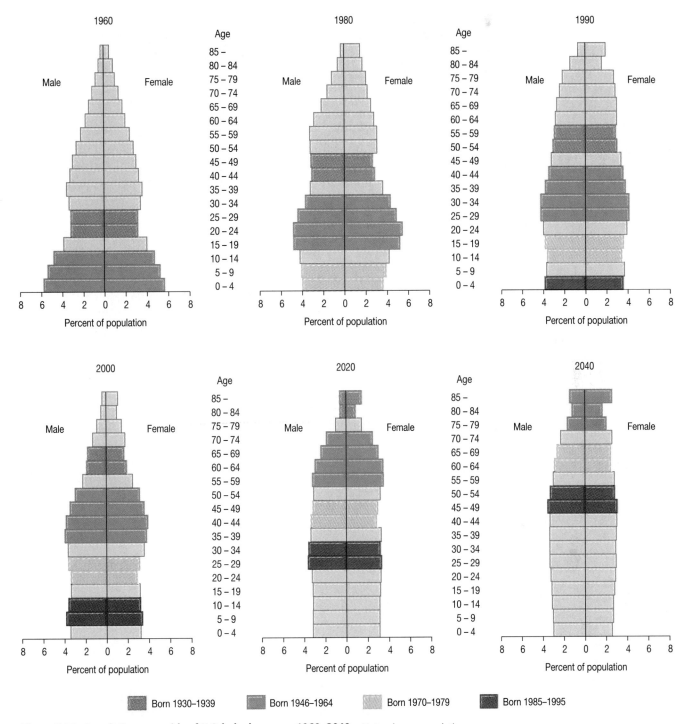

Figure 3.13 Population pyramids of U.S. baby boomers, 1960–2040 Not only are population pyramids based on past or present statistics, they also may be constructed on future estimates as well. In this series of pyramids, we can clearly see the progression of the baby boomers up the pyramid as their cohort ages. Note how the "pyramid" becomes a column by the year 2040 as birth rates remain below death rates for each cohort. Note also the significantly higher number of women compared with men in the oldest age group for this same year, reflecting the statistical tendency for women to live to older ages than men. (*Source:* L. Bouvier and C. DeVita, "The Baby Boom: Entering MidLife," *Population Bulletin*, 46(3), 1991.)

Figure 3.14 Generation X Members of Generation X, people who are currently in their late teens and twenties, are faced with the awesome burden of having to support a huge, aging baby boom population. In addition to the economic burden, many commentators have observed that they also face a cultural burden—the difficulty of establishing a distinct identity. Because so many baby boomers exist, with relatively fewer members of Generation X, preferences and styles of the baby boomers tend to dominate consumption patterns. Now in possession of a disproportionate share of the national wealth, baby boomers have a great deal of buying power, which translates into a significant impact on the music industry, the film industry, and other forms of contemporary cultural expression.

Figure 3.15 Population pyramids of Tucson census tracts, 1990
Population pyramids are as useful for analyzing populations within a single region as for an entire country. Here we see how age–sex profiles can also vary within different census tracts of the same city. The tracts show that even within a city, variation in populations can be substantial. Information like this can be very valuable in decision making and policymaking at state, regional, and local levels. (*Source:* U. S. Department of Commerce, Bureau of Census, 1990.)

Population Dynamics and Processes

In order to evaluate a different understanding of population growth and change, experts look at two significant factors: fertility and mortality. Birth and death rates, as they are also known, are important indicators of a region's level of development and its place within the world economy.

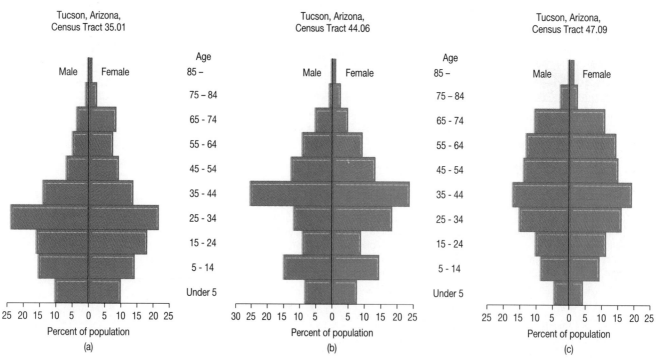

Tucson, Arizona, Census Tract 35.01

Tucson, Arizona, Census Tract 44.06

Tucson, Arizona, Census Tract 47.09

Birth, or Fertility, Rates

The **crude birthrate (CBR)** is the total number of live births in a year for every thousand people in the population. The crude birthrate is, indeed, crude, because it measures the birthrate in terms of the total population and not with respect to a particular age-specific group or cohort. For example, as of the early 1990s, the CBR of the American population was 15.48, while the birthrate for Asian Americans aged 20 to 29 was 10.3. Clearly, differences exist when we look at specific groups and age cohorts at their reproductive peak.

Although the level of economic development is a very important factor shaping the CBR, other, often equally important influences also affect CBR. In particular, it may be heavily affected by the demographic structure of the population, as graphically suggested by age–sex pyramids. In addition, an area's CBR is influenced by women's educational achievement, religion, social customs, diet and health, as well as politics and civil unrest. Most demographers also believe that the availability of birth-control methods is also critically important to a country's or region's birthrate (Figure 3.16). Figure 3.17 (see page 120) is a world map of the CBR showing high levels of fertility in most of the periphery of the world economy and low levels of fertility in the core. The highest birthrates occur in Africa, the poorest region in the world.

The crude birthrate is only one of the indicators of fertility and, in fact, is somewhat limited in its usefulness, telling very little about the potential for future fertility levels. Two other indicators formulated by population experts—the total fertility rate and the doubling time—provide more insight into the potential of a population. The **total fertility rate (TFR)** is a measure of the average number of children a woman will have throughout the years that demographers have identified as her childbearing years, approximately ages 15 through 49 (see Table 3.4). Whereas the CBR indicates the number of births in a given year, the TFR is a more *predictive* measure that attempts to portray what birthrates will be among a particular cohort of women over time. A population with a TFR of slightly higher than two has achieved replacement level fertility. This means that birthrates and death rates are approximately balanced and there is stability in the population.

Crude birthrate (CBR): ratio of the number of live births in a single year for every thousand people in the population.

Total fertility rate (TFR): average number of children a woman will have throughout the years that demographers have identified as her childbearing years, approximately ages 15 through 49.

Figure 3.16 Birth-control promotion in India
Controlling fertility has been an objective of population policy at the international level for several decades. For many years, however, birth-control programs were rejected by peripheral countries because having children is regarded as an economically sound decision. Birth-control programs coupled with improved educational and economic opportunities for women have proven to be far more effective than birth-control policies alone.

TABLE 3.4	Total Fertility Rates for Selected Countries, 1995
Country	Total Fertility Rate (TFR)
Afghanistan	6.34
Argentina	2.72
China	1.85
India	3.57
Namibia	6.46
Netherlands	1.59
Russia	1.83
United States	2.05

Doubling time: measure of how long it will take the population of an area to grow to twice its current size.

Crude death rate (CDR): ratio between the number of deaths in a single year for every thousand people in the population.

Closely related to the TFR is the doubling time of the population. The **doubling time,** as the name suggests, is a measure of how long it will take the population of an area to grow to twice its current size. For example, a country whose population increases at 1.8 percent per year will have doubled in about 40 years. In fact, world population is increasing at this rate at present. By contrast, a country whose population is increasing 3.18 percent annually will double in only 22 years—the doubling time for Kenya. Birthrates, and the population dynamics we can project from them, however, tell us only part of the story of the potential of the population for growth. We must also know the death, or mortality, rates.

Death, or Mortality, Rates

Countering birthrates and also shaping overall population numbers and composition is the **crude death rate (CDR),** the ratio between the total number of deaths in one year for every thousand people in the population. As with crude birthrates, crude death rates often roughly reflect levels of economic development—countries with low birthrates generally have low death rates (Figure 3.18).

Although often associated with economic development, CDR is also significantly influenced by other factors. A demographic structure with more men and elderly people, for example, usually means higher death rates. Other important influences on mortality include the health-care availability, social class, occupation, and even place of residence. Poorer groups in the population have higher

Figure 3.17 World crude birthrate, 1995 Crude birthrates and crude death rates are often indicators of the levels of economic development in individual countries. Compare, for example, Australia, a core country, which offers a stark contrast to statistics for Ethiopia, a very poor and underdeveloped peripheral country. (*Sources:* Reprinted with permission from Prentice Hall, from E. F. Bergman, *Human Geography: Cultures, Connections, and Landscapes,* © 1995, pp. 4–10. Data from World Resources Institute, *World Resources 1994–1995.* New York: Oxford University Press, 1994, pp. 270–71. Map projection, Buckminster Fuller Institute and Dymaxion Map Design, Santa Barbara, CA. The word Dymaxion and the Fuller Projection Dymaxion™ Map design are trademarks of the Buckminster Fuller Institute, Santa Barbara, California, ©1938, 1967 & 1992. All rights reserved.)

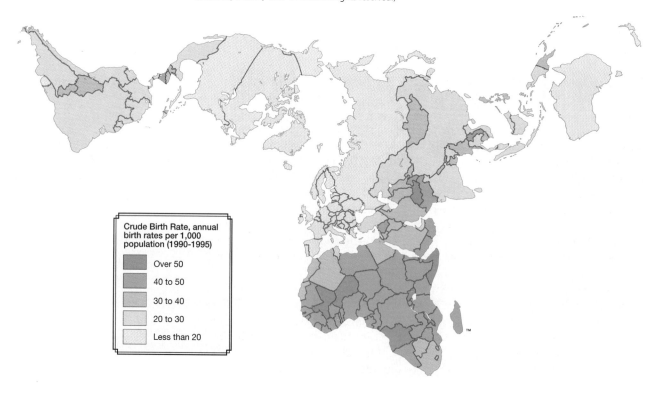

Crude Birth Rate, annual birth rates per 1,000 population (1990-1995)

	Over 50
	40 to 50
	30 to 40
	20 to 30
	Less than 20

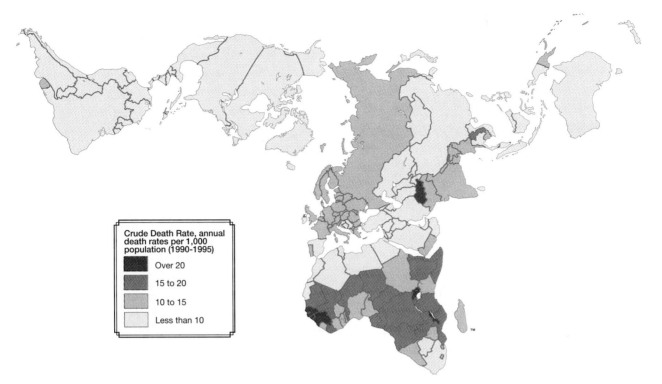

Figure 3.18 World crude death rate, 1995   The global pattern of crude death rates varies from crude birthrates. Most apparent is that the difference between highest and lowest crude death rates is relatively smaller than is the case for crude birthrates, reflecting the impact of factors related to the middle phases of the demographic transition described on page 124. (*Sources:* Reprinted with permission from Prentice Hall, from E. F. Bergman, *Human Geography: Cultures, Connections, and Landscapes,* © 1995, pp. 4–8. Data from World Resources Institute, *World Resources 1994–1995.* New York: Oxford University Press, 1994, pp. 270–71. Map projection, Buckminster Fuller Institute and Dymaxion Map Design, Santa Barbara, CA. The word Dymaxion and the Fuller Projection Dymaxion™ Map design are trademarks of the Buckminster Fuller Institute, Santa Barbara, California, ©1938, 1967 & 1992. All rights reserved.)

death rates than the middle class. In the United States, coal miners have higher death rates than school teachers. Also in the United States, it is often the case that urban areas have higher death rates than rural areas. The difference between the CBR and CDR is the rate of **natural increase**—the surplus of births over deaths— or the rate of **natural decrease**—the deficit of births relative to deaths (Figure 3.19, page 122.).

Death rates can be measured for both sex and age cohorts and one of the most common measures is the **infant mortality rate.** This figure is the annual number of deaths of infants less than one year of age compared to the total number of live births for that same year. The figure is usually expressed as number of deaths during the first year of life per 1,000 live births.

The infant mortality rate has been used by researchers as an important indicator both of a country's health-care system and of the general population's access to health care. In terms of global patterns, while infant mortality rates are high in the peripheral countries of Africa and Asia, they are low in the more developed countries of Europe and North America. Generally, this pattern reflects adequate maternal nutrition and the wider availability of health-care resources and personnel in the core regions (Figure 3.20, page 123). When patterns are examined below the global scale, at the level of countries, regions, and cities and regions within countries, infant mortality rates are not uniform. In the United States, for example, African-Americans, as well as other ethnic minorities in urban and rural areas, suffer infant mortality rates that are twice as high as the

Natural increase: difference between the CBR and CDR, which is the surplus of births over deaths.

Natural decrease: difference between CDR and CBR, which is the deficit of births relative to deaths.

Infant mortality rate: annual number of deaths of infants under one year of age compared to the total number of live births for that same year.

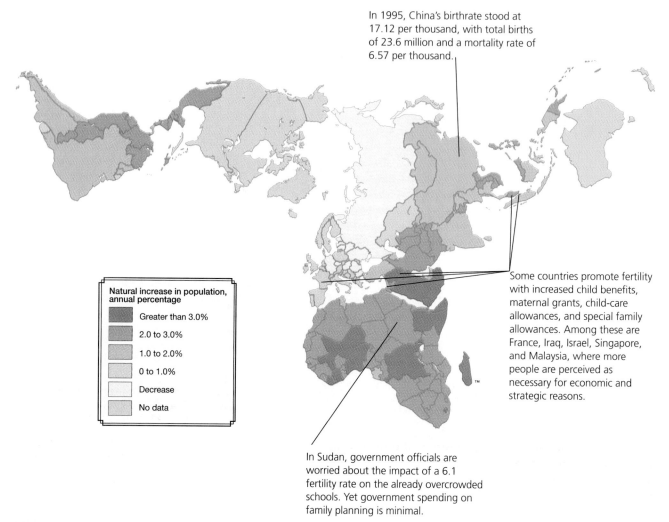

In 1995, China's birthrate stood at 17.12 per thousand, with total births of 23.6 million and a mortality rate of 6.57 per thousand.

Natural increase in population, annual percentage

Greater than 3.0%

2.0 to 3.0%

1.0 to 2.0%

0 to 1.0%

Decrease

No data

Some countries promote fertility with increased child benefits, maternal grants, child-care allowances, and special family allowances. Among these are France, Iraq, Israel, Singapore, and Malaysia, where more people are perceived as necessary for economic and strategic reasons.

In Sudan, government officials are worried about the impact of a 6.1 fertility rate on the already overcrowded schools. Yet government spending on family planning is minimal.

Figure 3.19 World rates of natural increase, 1995 As the map shows, rates of natural increase are highest in subequatorial Africa, the Near East, and parts of Asia as well as parts of South and Central America. Europe and the United States and Canada, as well as Australia and parts of central Asia and Russia have slow to stable rates of natural increase. (*Source:* Map projection, Buckminster Fuller Institute and Dymaxion Map Design, Santa Barbara, CA. The word Dymaxion and the Fuller Projection Dymaxion™ Map design are trademarks of the Buckminster Fuller Institute, Santa Barbara, California, ©1938, 1967 & 1992. All rights reserved.)

Life expectancy: average number of years an infant newborn can expect to live.

national average. In east central Europe, Bulgaria has a 12.6 per thousand infant mortality rate, while the Czech Republic has a rate of 9.7 per thousand. The point is that global patterns often mask regional and local variation in mortality rates, both for infants and for other population cohorts.

Related to infant mortality and the crude death rate is **life expectancy,** the average number of years an infant newborn can expect to live. Not surprisingly, life expectancy varies considerably from country to country, region to region, and even from place to place within cities and among different classes and racial and ethnic groups. In the United States, an infant born in 1995 can expect to live more than 76.25 years. If we begin to specify the characteristics of that infant by sex and socioeconomic characteristics, however, a great deal of variation emerges. An African-American male born in 1995 has a life expectancy of 68.8 years, while an Anglo-American male can expect to live, on average, to be 73.4 years old.

Another key factor influencing life expectancy is epidemics, which can quickly and radically alter population numbers and composition. In our times, epidemics can spread rapidly over great distances, largely because people and other disease carriers can now travel from one place to another very rapidly. Epidemics can have profound effects at various scales from the international to the local, and reflect the increasing interdependence of a shrinking globe. They may affect different population groups in different ways and, depending on the

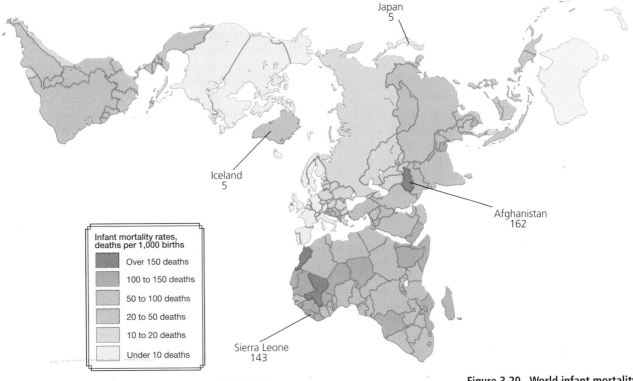

Japan
5

Iceland
5

Afghanistan
162

Infant mortality rates, deaths per 1,000 births

- Over 150 deaths
- 100 to 150 deaths
- 50 to 100 deaths
- 20 to 50 deaths
- 10 to 20 deaths
- Under 10 deaths

Sierra Leone
143

Figure 3.20 World infant mortality rates, 1995 The geography of poverty underlies the patterns shown in this map and allows us to analyze the linkages between population variables and social conditions. Infant mortality rates seem generally to parallel crude death rates, with sub-Saharan Africa reporting the highest rates. These rates reflect a number of factors, including inadequate or completely absent maternal health care as well as poor nutrition for infants. (*Sources: Hammond Atlas of the World.* 1993. New York: Oxford University Press. Map projection, Buckminster Fuller Institute and Dymaxion Map Design, Santa Barbara, CA. The word Dymaxion and the Fuller Projection Dymaxion™ Map design are trademarks of the Buckminster Fuller Institute, Santa Barbara, California, ©1938, 1967 & 1992. All rights reserved.)

quantity and quality of health and nutritional care available, may have a greater or lesser impact on different localities.

One of the most widespread epidemics of modern times is AIDS (Acquired Immunodeficiency Syndrome). The disease has come to be a serious problem in regions ranging from Southeast Asia to sub-Saharan Africa, and has affected certain populations in many core countries of Europe, as well as the United States. To date, the Islamic countries of the Middle East and North Africa, as well as South America, seem to be least affected, though lack of data makes any assessment provisional at best.

Geographers have made important contributions to the study of the diffusion of AIDS. In the United States, for example, AIDS first arose largely among male homosexuals and intravenous drug users who shared needles. Geographically, early concentrations of AIDS occurred in particular places with high concentrations of these two subpopulations, such as San Francisco, New York, Los Angeles, and southern Florida. Since first appearing, the disease has hierarchically diffused to other groups, often manifesting itself differently in each group, including hemophiliacs and other recipients of blood transfusions. It has had perhaps the most severe impact in inner-city areas, but has cropped up in every region of the United States, increasingly appearing in the male and female heterosexual population. The seriousness of the epidemic generally is revealed in the numbers: As of 1996, AIDS was one of the 10 leading causes of death in the United States.

The pattern of the disease has been different, however, in central Africa, where it has been associated with heterosexual, non–drug-using persons, and has affected both sexes equally. The geographical diffusion of AIDS in Africa has occurred along roads, rivers, and coastlines, all major transportation routes associated with regional marketing systems. The central African nations, including the Democratic Republic of Congo (formerly Zaire), Zambia, Uganda, Rwanda, and the Central African Republic, have been especially hard hit, with infection estimates as high as five million people. The impact is worst in urban areas, though no area has been immune to the disease's spread. High death rates in these countries reflect in no small part the ravages of AIDS. That these countries continue to experience high growth rates, however, is a reflection of the huge proportion of the population that is in the peak years of biological reproduction.

Although no cure has yet been found, some countries have been successful in slowing the spread of AIDS. Finland, for example, has had significant success in slowing the diffusion of HIV (Human Immunovirus, the virus that causes AIDS) through an intensive public relations campaign and the provision of top-notch health services to all its citizens. None of the sub-Saharan African countries, mired in poverty, inadequate health-care systems, and political inefficiency and often unrest, can even approach these levels of activity, however. Many Southeast Asian countries may face similar constraints and equally dismal prospects of slowing the diffusion of the disease (Figure 3.21).

Demographic Transition Theory

Many demographers believe that fertility and mortality rates are directly tied to the level of economic development of a country, region, or place. Pointing to the history of demographic change in core countries, they state that many of the economic, political, social, and technological transformations associated with industrialization and urbanization lead to a demographic transition. The **demographic transition** is a model of population change when high birth and death rates are replaced by low birth and death rates. Once a society moves from a preindustrial economic base to an industrial one, population growth slows. According to the demographic transition model, the slowing of population growth is attributable

Demographic transition: replacement of high birth and death rates by low birth and death rates.

Figure 3.21 World distribution of AIDS, 1990 AIDS is a disease that strikes the relatively young, though no age group is immune. A differentiation can be made in the incidence of the disease between core and peripheral countries. In the United States and Europe, AIDS mostly affects a young, male, homosexual population. In Africa and Asia, the disease is far less discriminating, affecting men and women and heterosexual as well as homosexual populations. (*Source:* Reprinted with permission of Simon & Schuster from *The New State of World Atlas,* 4th Ed., by M. Kidron and R. Segal. Copyright © 1991 by Michael Kidron and Ronald Segal. Maps and Grapics copyright © 1991 by Swanston Publishing Limited.)

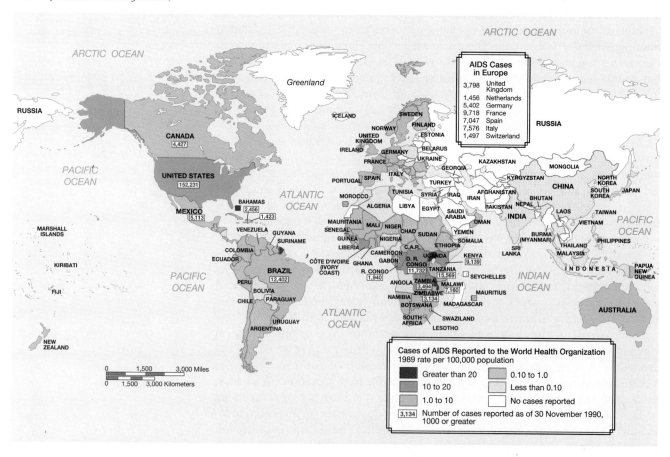

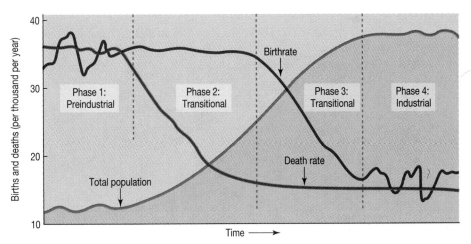

Figure 3.22 Demographic transition model The transition from a stable population based on high birth and death rates to one based on low birth and death rates progresses in clearly defined stages, as illustrated by this graph. With basic information about a country's birth and death rates and total population, it is possible to identify that country's position within the demographic transition process. Population experts disagree about the usefulness of the model, however. Many insist it is only applicable to the demographic history of core countries.

to improved economic production and higher standards of living brought about by changes in medicine, education, and sanitation.

As Figure 3.22 illustrates, the high birth and death rates of the preindustrial phase (Phase 1) are replaced by the low birth and death rates of the industrial phase (Phase 4) only after passing through the critical *transitional* phase (Phase 2) and then more moderate rates (Phase 3) of natural increase and growth. This transitional phase of rapid growth is the direct result of early and steep declines in mortality at the same time that fertility remains at levels characteristic of a place that has not yet been industrialized.

Some demographers have observed that many peripheral and semiperipheral countries appear to be stalled in the transitional phase, which has been called a "demographic trap." Figure 3.23 (see page 126) illustrates the disparity between birth and death rates for core and noncore countries. Despite a sharp decline in mortality rates, most noncore countries remain hostage to relatively high fertility rates.

The reason for this lag in declining fertility rates relative to mortality rates is that while new and more effective methods for fighting infectious diseases have been advanced, the social attitudes about the desirability of large families have yet to be affected even in the face of rapid industrialization and urbanization. Historical trends in birth and death rates and natural increase are shown in Table 3.5 for Scotland and England during their periods of demographic transition.

TABLE 3.5 Birth and Death Rates for England and Scotland, 1870–1920

The CBRs and CDRs for England and Scotland for the period between 1870 and 1920 illustrate countries moving between Phase 2 and Phase 3 of the demographic transition in which death rates are lower than birthrates. England and Scotland are clear examples of the way in which the demographic transition has been theorized to operate in the core countries. During the 50-year period covered in the table, both countries were completing their transformation as key industrial regions.

	1870	1880	1890	1900	1910	1920
Crude Birthrate						
England	35.2	35.4	32.5	29.9	27.3	22.7
Scotland	35	34.8	32.3	30.2	28.4	24
Crude Death Rate						
England	22.5	21.4	19.2	18.2	15.4	14.6
Scotland	22.1	21.6	19.2	18.5	16.6	15.3

Source: M. Anderson and D. Morse, *Scottish Demography*, Vol. 42, 1993.

Figure 3.23   World trends in birth and death rates, 1775–2050   Whereas death rates have been reduced on a global scale, for much of the periphery, birthrates continue to climb. The graph is an illustration of the impact of affluence on reproductive choices. It also portrays an optimistic view of the future where, for peripheral regions, birthrates will continue to fall as we move closer to the midpoint of the next century. For the core, the projection is that birth and death rates will stay low. (*Source:* T. Allen and A. Thomas, *Poverty and Development in the 1990s*, 1992. London: Oxford University Press.)

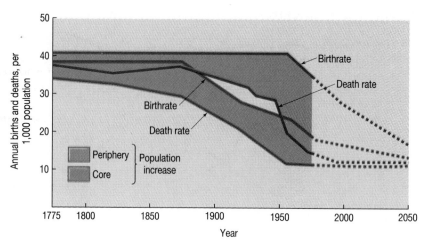

Over a roughly 50-year period, both countries were able to reduce their rates of natural increase by nearly one-third. In the fourth phase of postindustrial development (not shown in the table), birth and death rates have both stabilized at a low level, which means that population growth rates are very slow and birthrates are more likely to oscillate than death rates.

Whereas England and Scotland, early industrializers, passed through the demographic transition over a period of about 150 years beginning in the mid-nineteenth century, most peripheral and semiperipheral countries have yet to complete the transition.

Although the demographic transition model is based on actual birth and death statistics, many population geographers and other population experts increasingly question its generalizability to all places and all times. In fact, while the model adequately describes the history of population change in the core countries, it appears less useful for explaining the demographic history of countries and regions in the periphery. Such criticism has led to its declining significance for understanding population geography. Among other criticisms are that industrialization—which, according to the theory, is central to moving from Phase 2 to Phases 3 and 4—is seldom domestically generated in the peripheral countries. Instead, foreign investment seems to drive peripheral industrialization. As a result, the features of demographic change witnessed in the core countries, such as higher living standards, where industrialization was largely a result of internal capital investment, have not occurred in many peripheral countries. Other critics of the demographic transition model point to several factors undermining a demographic transition fueled by economic growth: the shortages of skilled laborers; the absence of advanced educational opportunities for all members of the population (especially women); and limits on technological advances. In other words, while demographic transition may be a characteristic experience of the core regions of the globe, it appears to have limited applicability to the periphery.

Population Movement and Migration

In addition to the population dynamics of death and reproduction, the third critical influence is the movement of people from place to place. Individuals may make far-reaching, international or intraregional moves, or they may simply move from one part of a city to another. For the most part, mobility and migration reflect the interdependence of the world-system. For example, global shifts in industrial investment result in local adjustments to those shifts as populations move or remain in place in response to the creation or disappearance of employment opportunities.

Mobility and Migration

One way to describe such movement is by the term **mobility, the ability to move, from one place to another, either permanently or temporarily.** Mobility may be used to describe a wide array of human movement ranging from a journey to work (for example, a daily commute from suburb to city or suburb to suburb) to an ocean-spanning, permanent move.

The second way to describe population movement is in terms of **migration, a** long-distance move to a new location. Migration involves a permanent or temporary change of residence from one neighborhood or settlement (administrative unit) to another. Moving from a particular location is defined as **emigration;** this process is also known as *out-migration.* Moving to a particular location is defined as **immigration** or, alternatively, as *in-migration.* Thus, a Russian who moves to Israel emigrates *from* Russia and immigrates *to* Israel. This type of move, from one country to another, is termed **international migration.** Moves may also occur within a particular country or region, in which case they are called **internal migration.** Both permanent and temporary changes of residence may occur for many reasons, but most often involve a desire for economic betterment or an escape from adverse political conditions such as war or oppression.

Governments are concerned about keeping track of both migration numbers and rates and the characteristics of the migrant populations, because these factors can have profound consequences for political, economic, and cultural conditions at national, regional, and local scales. For example, a peripheral country, such as Cuba, that has experienced substantial out-migration of highly trained professionals may find it difficult to provide needed services such as health care. Benefiting from lower labor costs are countries such as the United States, Germany, and France that have received large numbers of low-skilled in-migrants willing to work for extremely low wages. These countries may also face considerable social stress in times of economic recession when unemployed citizens begin to blame the immigrants for "stealing" their jobs or receiving welfare benefits.

Demographers have developed several calculations of migration rates. Calculation of the in-migration and out-migration rates provides the foundation for discovering gross and net migration rates for an area of study. **Gross migration** refers to the total number of migrants moving into and out of a place, region, or country. **Net migration** refers to the gain or loss in the total population of that area as a result of the migration. Migration is a particularly important concept because the total population of a country, region, or locality is dependent on migration activity as well as on birth and death rates.

Migration rates, however, provide only a small portion of the information needed to understand the dynamics of migration and its effects at all scales of resolution. In general terms, migrants make their decisions to move based on push factors and pull factors. **Push factors** are events and conditions that impel an individual to move *from* a location. They include a wide variety of possible motives, from the idiosyncratic, such as an individual migrant's dissatisfaction with the amenities offered at home, to the dramatic, such as war, economic dislocation, or ecological deterioration. **Pull factors** are forces of attraction that influence migrants to move *to* a particular location. Factors drawing individual migrants to chosen destinations, again, may range from the highly personal (such as a strong desire to live near the sea) to the very structural (such as strong economic growth, and thus relatively lucrative job opportunities).

Usually the decision to migrate is a combination of both push and pull factors, and most migrations are voluntary. In **voluntary migration,** an individual chooses to move. Where migration occurs against the individual's will, push factors can produce **forced migration.** Forced migration (both internal and international) remains a critical problem in the contemporary world. In fact, it is on the rise, growing from more than 32 million people in 1987 to more than 42 million

Mobility: the ability to move, either permanently or temporarily.

Migration: a move beyond the same political jurisdiction, involving a change of residence, either as *emigration,* a move *from* a particular location, or as *immigration,* a move *to* another location.

Emigration: a move *from* a particular location.

Immigration: a move *to* another location.

International migration: a move from one country to another.

Internal migration: a move within a particular country or region.

Gross migration: the total number of migrants moving into and out of a place, region, or country.

Net migration: the gain or loss in the total population of a particular area as a result of migration.

Push factors: events and conditions that impel an individual to move *from* a location.

Pull factors: forces of attraction that influence migrants to move *to* a particular location.

Voluntary migration: movement by an individual based on choice.

Forced migration: movement by an individual against his or her will.

in 1993. These migrants may be fleeing a region or country for many reasons, but some of the most common are war, famine (often war-induced), life-threatening environmental degradation or disaster, or governmental coercion or oppression. Among the most tragic situations are those experienced by refugees.

International Voluntary Migration

Not all migration is forced, however. Indeed, most people who migrate choose to do so (Figure 3.24). For example, the successive waves of European migration to the United States have included a preponderance of people who more or less voluntarily chose to leave their native lands. These migrants have included religious dissidents, such as the Puritans, and seekers of better economic opportunities, such as the Swedes and Norwegians. In the nineteenth century, for example, many northern European migrants actually immigrated to the United States in response to recruiting campaigns orchestrated by the railroads. Railroad owners wanted to maximize company profits by selling to immigrants the vast acreage they received from the federal government in partial payment for building railroad lines across the country.

In other situations, migration does not involve a permanent change of residence. Temporary labor migration has long been an indispensable part of the world economic order, and has at times been actively pursued by governments and companies alike. Individuals who migrate temporarily to take up jobs in other countries are known as **guest workers.** The temporary migration of North African auto workers to France is an excellent example of this process and its linkage of local, national, and international scales (see Feature 3.3, "Geography Matters—Guest Workers in the French Auto Industry"). Sending workers abroad is an important economic strategy for many peripheral and semiperipheral coun-

Guest workers: individuals who migrate temporarily to take up jobs in other countries.

Figure 3.24 Global voluntary migration, 1990 This map illustrates very complex flows of people across borders who have migrated by choice. While each of the flows represents a cluster of individual decisions, it is noteworthy that, generally speaking, the flows emanate from the periphery and are moving toward the core. Intracontinental migration, such as that in South America and Africa, represents the apparent pull of economic opportunity for residents of relatively poorer countries. (*Sources:* A. Segal, *An Atlas of International Migration.* New Providence, NJ: Hans Zell Publishers/Bowker-Saur, 1993. p. 23. Map projection, Buckminster Fuller Institute and Dymaxion Map Design, Santa Barbara, CA. The word Dymaxion and the Fuller Projection Dymaxion™ Map design are trademarks of the Buckminster Fuller Institute, Santa Barbara, California, ©1938, 1967 & 1992. All rights reserved.)

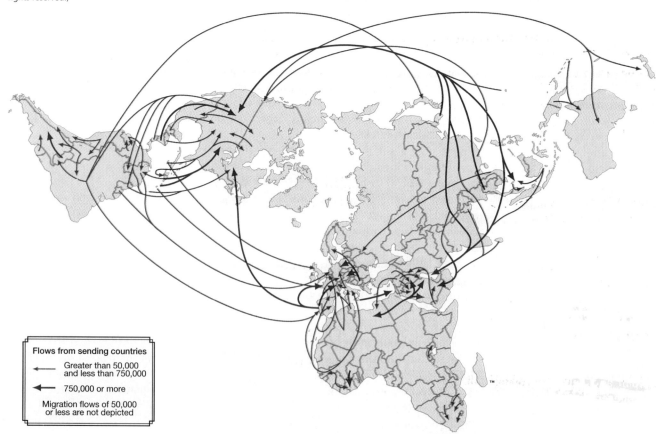

Flows from sending countries

→ Greater than 50,000 and less than 750,000

⟹ 750,000 or more

Migration flows of 50,000 or less are not depicted

tries, for it not only alleviates local unemployment, but also enables workers to send substantial amounts of money to their families at home. This arrangement helps keep the workers' families afloat and supports the dominance of the core in global economic activities. It is not always an ideal situation, however, for economic downturns in the guest worker's host country may result in a large decrease in remittances received by the home country, thus further aggravating that country's economic situation.

It is also important to know the gender of temporary workers, and the gender-based differences in the types of work performed. The Philippines, for example, has an Overseas Contract Worker (OCW) program that links foreign demand for workers to Philippine labor supply. The proportion of men and women OCWs is approximately equal, but the jobs they hold are gender-biased. Men receive most of the higher-level positions, while women are largely confined to the service and entertainment sectors. Some interesting geographical variations exist as well. Although constituting a small percentage of the total number of women working abroad, a large proportion of the female OCWs who do have professional positions (such as doctors and nurses) work in the United States and Canada. More typical of Filipina OCW experience, however, are the patterns in Hong Kong, Singapore, and Japan. In Hong Kong and Singapore, Filipinas are almost exclusively employed as domestic servants; in Japan, most work in the "entertainment industry," often synonymous with prostitution. These women in many ways have been transformed from individuals to commodities—many are even chosen by catalog. The OCW program has been criticized by feminist groups as well as human rights organizations. Many regard the OCW treatment of women as a contemporary form of slavery.

International Forced Migration

The African slave trade is a classic example of international forced migration. This migration stream was integral to European economic expansion from the seventeenth through the nineteenth centuries. The huge fortunes made in the sugar trade, for example, were largely earned on the backs of African slaves working the sugar plantations of Brazil, Guyana, and the Caribbean. Figure 3.25 (see page 130) shows the regions of Africa that acted as sources of the global slave trade from the seventeenth to the nineteenth century. Other prominent examples of international forced migration include the migration of Jews from Germany and eastern Europe preceding the Second World War and the deportation of Armenians out the Ottoman Empire after the First World War (Figure 3.26, page 131). Figure 3.27 (see page 131) shows those countries where residents have been forced from their homes by war, abuse, and fear.

Recently, many European countries, faced with rising numbers of refugees seeking political asylum, have tightened their previously liberal asylum policies. In 1993, the French National Assembly and the German Bundestag (Parliament) both voted to restrict the conditions under which asylum would be granted. Russia anticipates having to absorb several million ethnic Russians fleeing the severe economic decline, political unrest, and volatile ethnic tensions that continue to plague former republics such as Tadjikistan and Georgia.

Sudan, however, is in the contrasting position of having displaced several million of its own citizens as a result of civil war, while at the same time hosting over 750,000 refugees from neighboring countries. The situation provides a clear example of the openness of the borders that define many African States. In some cases, refugee populations remain in a host country for long periods of time. In Jordan and Lebanon, for example, some of the Palestinian refugee camps are multigenerational, dating back to the 1948 Arab–Israeli war. High fertility rates in some of these refugee camps put considerable strain on the humanitarian aid resources of both international organizations and the host countries.

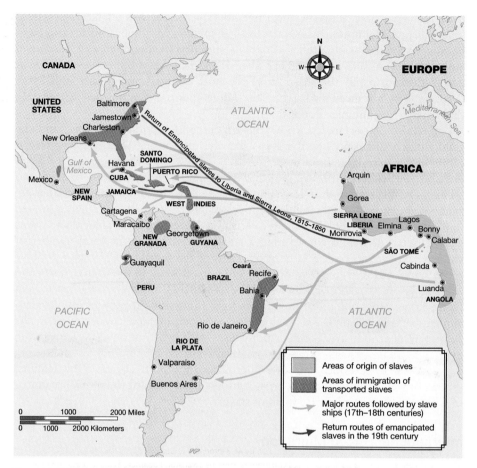

Figure 3.25 African slave trade, seventeenth through nineteenth centuries As the map illustrates, about 10 million slaves were sent from Africa to other locations in the roughly 350 years that the slave trade operated. Most of the slaves were shipped to the New World to work in French, Spanish, Dutch, Danish, and British colonies although a small number were shipped to Europe. The largest source of slaves was West Africa where the more powerful tribes sold the weaker and more passive tribe members to European buyers. (*Source:* From *The Penguin Atlas of Diasporas* by Gerard Chaliand and Jean-Pierre Rageau, translated by A. M. Berrett. Translation copyright © 1995 by Gerard Chaliand and Jean-Pierre Rageau. Used by permission of Viking Penguin, a division of Penguin Books USA Inc.)

Internal Voluntary Migration

One way to understand the geographical patterns of migration is to think in terms of waves of migration. In the United States, for example, three important and overlapping waves of internal migration over the past two centuries have altered the population geography of the country. As with a great deal of migration activity, these three major migrations were tied to broad-based political, economic, and social changes.

The first wave began in the mid-nineteenth century and increased steadily through the twentieth century. This wave has two parts. One part consisted of massive rural-to-urban migration associated with industrialization. The second part was characterized by the large movement of people from the settled eastern seaboard into the interior of the country. This westward expansion, which began spontaneously during the British colonial period, in blatant disregard for British restrictions to such expansion, became official settlement policy after the American Revolution. The federal government thereby encouraged migration

Figure 3.26 Armenian refugees Geographically located at one of the world's great crossroads, the Armenian people have experienced a long and complex history filled with conflict and subordination as well as prosperity and expansion. The twentieth century has been a particularly traumatic period in Armenian history. Circumstances emerging out of the First World War led to the massive deportation of Armenians from the Ottoman Empire and the genocide of nearly 50 percent of the Armenian population. Armenians are currently spread out throughout the globe with sizable communities on every continent and in most major cities.

Figure 3.27 Refugee-sending countries, 1990 War is certainly the most compelling factor in forcing refugee migration. This map is a veritable list of the most belligerent nations of the second half of the twentieth century. Shown are the sending countries, those whose internal situations propel people to leave. What is perhaps most distressing about this graphic is the fact that refugee populations have increased over the last decade almost exclusively in the periphery. (*Sources:* Adapted from M. Kidron and R. Segal, *The New State of World Atlas,* 4th Ed. Reprinted with the permission of Simon & Schuster from *The New State of World Atlas,* 4th Ed., by M. Kidron and R. Segal. Copyright © 1991 by Michael Kidron and Ronald Segal. Maps and Graphics copyright © 1991 by Swanston Publishing Limited. Map projection, Buckminster Fuller Institute and Dymaxion Map Design, Santa Barbara, CA. The word Dymaxion and the Fuller Projection Dymaxion™ Map design are trademarks of the Buckminster Fuller Institute, Santa Barbara, California, ©1938, 1967 & 1992. All rights reserved.)

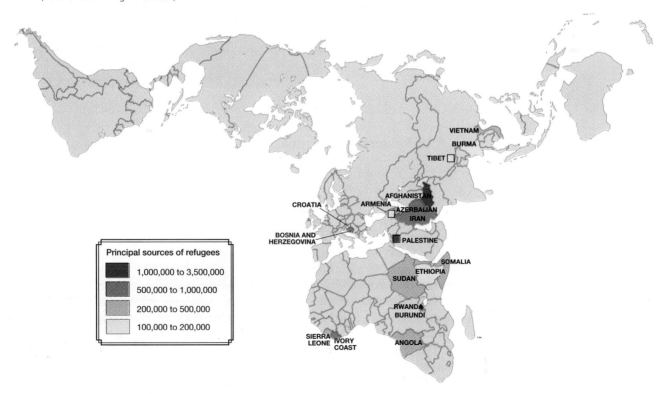

Principal sources of refugees

- 1,000,000 to 3,500,000
- 500,000 to 1,000,000
- 200,000 to 500,000
- 100,000 to 200,000

3.3 Geography Matters

Guest Workers in the French Auto Industry

The story of North African guest workers in the French automobile industry provides an excellent illustration of how international, national, and local conditions and processes interact to produce particular geographies of migration. The automobile industry in France actively recruited large numbers of North African guest workers in the 1950s and 1960s (Figure 3.3.1). Labor was needed because of low birthrates that had occurred throughout France in the 1930s and 1940s. This was not the first time France had gone abroad for workers: Previous labor shortages had brought in Belgian, Spanish, Italian, and Polish workers. However, this was the first time that non-Europeans had been intensively recruited. Some agents traveled to remote villages in Algeria, Morocco, and Tunisia and sought out illiterate males who were willing to leave their villages and families behind for work in French factories (Figure 3.3.2).

The workers constituted a cheap labor source: They worked for low wage, and did not have to be given unemployment insurance, disability insurance, or health-care benefits. They could be segregated from the French work force, and through concentration in ghetto-like residential areas, could be segregated from French society. Because they could not read and often did not speak French, there was little risk of their getting involved in union organizing. The North Africans were especially suited to auto industry needs because they were willing to withstand long hours of repetitive work on the assembly lines; could work night shifts; were easy to hire and fire; and rarely engaged in labor struggles.

Things soon became more complicated, though, for once the workers became established in France, they began bringing their families over, resulting in a substantial increase in the country's foreign-born population

Figure 3.3.1 Maghreb migrant workers As members of former French colonies, the people of Tunisia, Morocco, and Algeria were drawn to France to participate in the industrial economy. Although explicitly recruited as laborers, the Maghreb people have presented a challenge for the French government. As nationalist sentiment has grown in the Maghreb, especially in Algeria, France has been the target of terrorist bombings as retaliations against the Western cultural impacts of the colonial period.

over the course of more than a century as part of the country's expansionist strategies. The idea was to alleviate urban crowding in the East and to offset economic pressures in burgeoning eastern areas by promoting the idea of the self-sufficient farmer. Table 3.6 (see page 134) illustrates how, between 1860 and 1920, the United States was transformed from a rural to an urban society as industrialization created new jobs and as increasingly redundant numbers of agricultural workers (along with foreign immigrants) moved to urban areas to work in the manufacturing sector. Figure 3.28 (see page 134) illustrates how the demographic center of the United States moved westward in the two hundred years between 1790 and 1990 as more and more areas were colonized, largely through transportation and communications innovations, capital investment, and large internal population movements.

The second great migration wave, which began early in the 1940s and continued through the 1970s, was the massive and very rapid movement of mostly African-Americans out of the rural South to cities in the South, North, and West (Figure 3.29, page 135). Although African-Americans already formed consider-

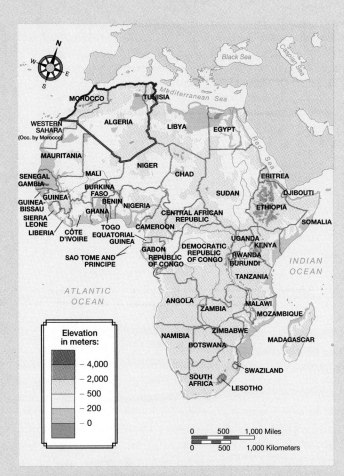

between 1967 and 1974. In addition, by the mid-1970s, a global and national economic downturn, together with mechanization of many assembly functions, resulted in decreased employment opportunities. In response, the French government, which had previously been supportive of the recruitment effort, began restricting immigration. Renault initiated a training program to teach its foreign workers French and technical skills. Nevertheless, massive unemployment occurred among the migrants, who were then replaced by French workers.

In 1977, the government began trying to repatriate the workers and their families to their home countries by offering them financial incentives to return. Some 45,000 people were repatriated, but many still remained, so the government applied further pressure in the form of housing restrictions. Massive layoffs and labor unrest persisted throughout the 1980s, however, and problems are still evident today, particularly as reflected in the xenophobic utterances from the far right of France's political spectrum.

Why do the workers remain, even in such hostile circumstances? Because going home may not be any better. For guest workers and their families, the prospect of return can be daunting: They may no longer be accepted in their local villages, and if their home countries have been experiencing severe economic and political problems, the prospects of finding any kind of job there become even dimmer. By now many second-generation North Africans (who call themselves "Beurs") live in France. Having grown up in France, their attachment to their "homeland" in North Africa may be fairly weak, and they may decide to take advantage of French laws that allow them to become French citizens upon reaching adulthood.

Figure 3.3.2 The Maghreb, North Africa The Maghreb is geographically close to Europe with push and pull factors operating simultaneously. With few jobs available in their own countries, many migrants were drawn by the possibility of economic prosperity offered through guest worker status. For some, conditions in their home countries are currently so politically volatile and economically uncertain, that return migration is not a desirable option.

able populations in cities such as Chicago and New York before the onset of this wave, the mechanization of cotton picking pushed additional large numbers of these people out of the rural areas. Tenant cotton picking was a major source of livelihood for deep-South African-Americans until mechanization reduced the number of jobs available. At the same time, pull factors attracted them to the large cities. In the early 1940s, large numbers of jobs in the defense-oriented manufacturing sector became available as other urban workers joined the war effort. This second wave of migration can be seen as part of a wider pattern of rural-to-urban migration among agricultural workers as industrialization spread globally.

After the war, a more important catalyst drove this migration: the increasing emphasis on high levels of mass consumption, which reoriented industry toward production of consumer goods and which in turn stimulated large increases in the demand for unskilled and semiskilled labor. The impact on the geography of racial distribution in the United States was profound.

The third internal migration wave began shortly after the Second World War and continues into the present. Following the end of the war and directly related

TABLE 3.6 United States Rural to Urban Population Change, 1860–1920

This table illustrates the fact that in 60 years, between 1860 and 1920, the United States went from a primarily agricultural country to a primarily industrial one. In the United States industrialization was consistently associated with urbanization: As the country industrialized, so did its population urbanize as machines replaced human labor on the farm and former farm hands and others operated machines in factories.

	1860	1870	1880	1890	1900	1910	1920
Total US Population (in millions)	31.4	38.5	50.1	62.9	75.9	91.9	105.7
Total Urban Population (in millions)	6.2	9.9	14.1	22.1	30.1	41.9	54.1
% Urban	19.8	25.7	28.2	35.1	39.7	45.7	51.2
% Increase in Total Population	-	22.7	30.1	25.5	20.7	21.0	14.9
% Increase in Urban Population	-	59.3	42.7	56.4	36.4	39.2	29.0

to the impact of governmental defense policies and activities on the country's politics and economy, the region of the United States lying below the thirty-seventh parallel—now commonly called the Sunbelt—experienced a 97.9 percent increase in population. At the same time, the Midwest and Northeast, known variously as the Snowbelt, Frostbelt, or Rustbelt, together grew by only 33.3 percent. Figure 3.30 shows a U.S. Census Bureau regional map illustrating the changes, in terms of percentages, for the years 1980 to 1990. This map shows that the western, southwestern, and southeastern states have experienced substantial population increases.

The consequences of the transformation of the U.S. economy from a goods-producing to a service-producing orientation are twofold: One, the distribution of the population has moved away from the urban manufacturing centers of the North and Midwest and toward the South and West; and, two, the movement of population has occurred out of cities into suburbs and previously rural areas.

Figure 3.28 Changing demographic center of the United States, 1790–1990 Knowing where the demographic center of a country's population is located at given times allows us to track that population's movement, growth, distribution, and concentration. Here, we can clearly perceive the expansion of the American frontier and the fact that although the East continued to be the most populated part of the country early on, the demographic center moved as more and more of the population dispersed to the West and the population of the West grew through natural increase. (*Source:* Reprinted with permission of Prentice Hall, adapted from J. M. Rubenstein, *The Cultural Landscape: An Introduction to Human Geography,* 5th ed., © 1996, p. 122.)

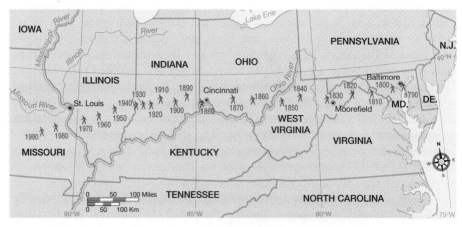

Figure 3.29 Twentieth-century African-American internal migration Many former sharecroppers migrated to the industrial cities of the northern and eastern United States to take advantage of the expansion of industrial employment. Economically motivated migration occurs at all scales from the global down to the neighborhood.

Suburbanization is the growth of population along the fringes of large metropolitan areas. The first evidence of U.S. suburbanization can be traced back to the late eighteenth century, when wealthy city-dwellers began seeking more scenic residential locations. Later, residents fled to the suburbs to get away from the new immigrants and their increasing hold over urban machine politics.

The process really took on its own life, however, with the introduction of new transportation technologies—first horse-drawn streetcars, then intraurban rail services and, finally, automobiles. Each innovation in transportation allowed

Suburbanization: growth of population along the fringes of large metropolitan areas.

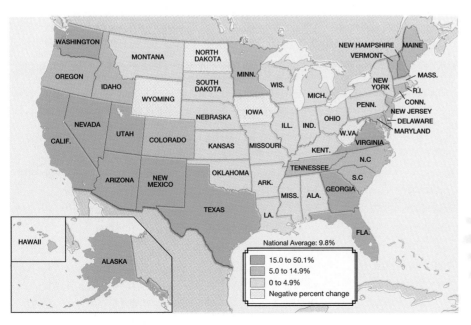

National Average: 9.8%

- 15.0 to 50.1%
- 5.0 to 14.9%
- 0 to 4.9%
- Negative percent change

Figure 3.30 U.S. state population change, 1980–1990 Mapping the degree of population change in each of the 50 states from one decennial census to the next allows us to identify which areas are experiencing the greatest gains and losses in population. Such changes have important implications, for example, on the allocation of Electoral College votes among the states and on areas undergoing marked stress (such as the Rustbelt from loss of manufacturing jobs), and so on. The recent shift in U.S. population to the Sunbelt shows up quite clearly on this map.

people to travel longer distances to and from work within the same or shorter time period. They chose to move to the suburbs in massive numbers, not in the least because the suburbs were—and are—arguably considered by many to be more healthful and better places to raise a family.

The most compelling explanation for the large-scale population shift characteristic of the third migration wave is the pull of economic opportunity. Rather than being invested in upgrading the aged and obsolescent urban industrial areas of the Northeast and Midwest, venture capital was invested in Sunbelt locations, where cheaper land and labor costs made the introduction of manufacturing and service-sector activity more profitable (Table 3.7). The 1990 census shows a decrease in the amount of in-migration to the Sunbelt, but the changes in the geography of U.S. population at the end of the twentieth century will nevertheless be almost diametrically opposed to the patterns of 150 years ago. This new population distribution illustrates the way in which political and economic transformations play an especially significant role in shaping individual choice and decision making.

In other parts of the world rural-to-urban migration trends have changed population geographies in equally dramatic ways, although local manifestations have often been quite different from those experienced in the U.S.. Shantytowns ring all major cities of the periphery, posing almost insurmountable problems to city managers for providing the most basic health, sanitation, educational, and occupational services. In some places, such as Brazil, an effort is ongoing to redirect these migrants to other locations. The Brazilian government constructed highways into the Amazon, which has drawn some migrants into the frontier area. Nevertheless, pressures on major cities like Rio de Janeiro remain overwhelming.

Another approach aimed at redressing a severe population imbalance has been taken by Indonesia. Indonesia had a 1990 population of approximately 180 million people, more than 60 percent of whom live on the islands of Java and Bali (these two islands comprise just 7 percent of the nation's land base). Because of tremendous population growth, two million new job-seekers enter the market annually, and the government has created programs aimed at inducing the people to move to other less densely populated locations.

Efforts to redistribute the population to the less densely settled islands of the nation date back to the turn of the century when the area was still under Dutch colonial rule. For many years these efforts involved offering incentives to selected landless and jobless people to move to new agriculturally based settlements in the

TABLE 3.7 United States Snowbelt/Sunbelt Population Change, 1950–1988				
This table illustrates another important economic phase in U.S. population history as the older industrial areas of the country lost their attractiveness as investment sites. Part of this transformation can be explained by the emergence of the importance of the service sector to the U.S. economy. It can also be explained by the more attractive investment opportunities in the Sunbelt—cheaper land, taxes, and labor.				
	1950–1960	**1960–1970**	**1970–1980**	**1980–1988**
Total U.S.	18.6	13.4	11.4	8.5
Snowbelt	15.4	10.6	2.3	2.6
(Northeast)	(14.1)	(10.8)	(0.8)	(3.6)
(Midwest)	(16.9)	(10.4)	(3.9)	(1.6)
South and West	22.2	16.5	21.3	13.9
(Sunbelt)	(25.9)	(18.3)	(22.5)	(15.4)
(Other areas)	(13.9)	(12.0)	(18.1)	(8.4)

Outer Islands. Since 1984, however, facing a limited development budget, the government turned to sponsoring investment in labor-intensive enterprises in established settlement areas. To date, over 1.5 million people have migrated; more than 60 percent of them have moved to one large island: Sumatra.

Internal Forced Migration

One of the best-known forced migrations in the United States was the "Trail of Tears," a tragic episode in which nearly the entire Cherokee Nation was forced to leave their once treaty-protected Georgia homelands for Oklahoma (Figure 3.31). Despite sustained legal resistance, approximately 16,000 Cherokees were forced to march across the continent in the early 1830s, suffering from drought, food scarcity, bitterly cold weather, and sickness along the way. By some estimates at least a quarter of the Cherokees died as a result of the removal.

Placed within the national and international context, the movement of Native-American populations during the nineteenth century can be seen as a response to larger political and economic forces. European populations were participating in a massive migration to the United States, and the national economy was on the threshold of an urban–industrial revolution. The eastern Native-American populations posed an obstacle to economic expansion, which itself was dependent upon geographic expansion. Growing European-American prosperity had to be secured by taking Indian land.

In more recent years, civil war, ethnic conflict, famine, deteriorating economic conditions, and political repression have produced an extraordinary series of internal forced migrations in sub-Saharan countries ranging from South Africa to the Sudan. Combined with international forced migration, the process is an important one, for 80 percent of the world population increase in the next decade will take place within the poorest countries of the world. Many of these countries have some of the highest rates of forced migration. The combination bodes ill for improving these countries' prospects for economic and political improvement.

Other recent examples of internal forced migration are provided by China and South Africa. In the late 1960s and early 1970s the government of China forcibly relocated 10 to 15 million of its citizens to rural communes, in order to enforce Chinese Communist dogma and to ease pressures arising from high

When I was a boy, I saw the white man afar off and was told he was my enemy. I could not shoot him as I would a wolf or a bear, yet he came upon me.

My horse and fields he took from me. He said he was my friend—he gave me his hand in friendship. I took it, he had a snake in the other, his tongue was forked. I am about to leave Florida forever and have done nothing to disgrace it. It was my home; I loved it, and to leave is like burying my wife and child.

Coacoochee
Seminole, 1858

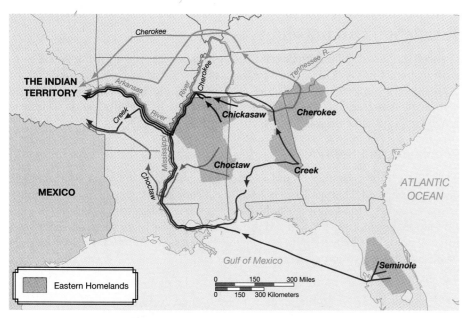

Figure 3.31 Trail of Tears, 1830s The Indian Removal Act of 1830 enabled the "voluntary" exchange of lands held by eastern Native-American tribes for territory that the federal government acquired west of the Mississippi. What the Act effectively enabled was the elimination of Indian title to land previously protected by treaties. Without government protection of their rights to the land, members of the Choctaw, Chickasaw, Cherokee, Seminole, Creek, and other eastern tribes "chose" to remove themselves further west to lands that had been acquired for them. The map shows the routes taken by these tribes as they vacated their ancestral lands for federally designated Indian Territory. (Source: *Atlas of the North American Indian* by C. Waldman. Copyright © 1995 Carl Waldman. Reprinted with permission by Facts On File, Inc., New York.)

Figure 3.32 Ethiopian famine migrants, 1984–1985 Poor climatic conditions leading to failed harvests as well as civil strife forced the migration of hundreds of thousands of Ethiopians from that country. In this case, an environmental stimulus enlarged the problems of civil war, thereby creating unbearable conditions. Most agriculturalists, as well as others, had no choice but to leave or die either from starvation or political persecution.

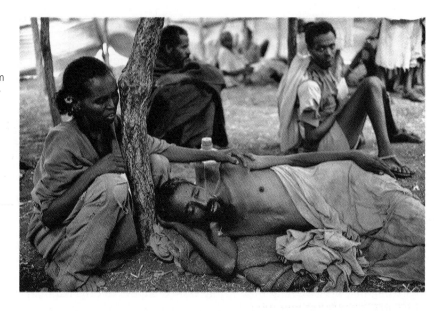

Eco-migration: population movement caused by the degradation of land and essential natural resources.

urban unemployment. The policy has since been disavowed, but the effects on an entire generation of young people were profound. Another example took place in South Africa between 1960 and 1980, when apartheid policies forced some 3.6 million blacks to relocate to government-created homelands, causing much suffering and dislocation (see "Geography Matters 9.2: Imperialism and Colonization of South Africa" for more on apartheid).

Today, forced **eco-migration**—population movement caused by the degradation of land and essential natural resources—has created a new category of refugees. In Bangladesh, for example, floodplain settlement that began occurring in the 1960s has led to severe losses of life and property, and has forced temporary relocation of huge numbers of people whenever severe flooding occurs. In Ethiopia, the 1984–1985 famine was attributed not only to drought and pest infestation, but also to governmental policies and actions that favored urban populations and rendered rural populations especially vulnerable to environmental stress (Figure 3.32). Finally, dams and irrigation projects have annually forced between one and two million people worldwide to move.

Population Debates and Policies

One big question occupies the agenda of population experts studying world population trends today: How many people can Earth sustain without depleting or critically straining its resource base? The relationship between population and resources, which lies at the heart of this question, has been a point of debate among experts since the early nineteenth century.

Population and Resources

The debate about population and resources originated in the work of an English clergyman named Thomas Robert Malthus (1766–1834), whose theory of population relative to food supply established resources as the critical limiting condition upon population growth. Malthus's theory was published in 1798 in a famous pamphlet called *An Essay on the Principle of Population*. In this tract Malthus sets up two important postulates:

- Food is necessary to the existence of human beings.
- The passion between the sexes is necessary and constant.

It is important to put the work of Malthus into the historical context within which it was written. Revolutionary changes—prompted in large part by technological innovations—had occurred in English agriculture and industry and were eliminating traditional forms of employment faster than new ones could be created. This condition led to a fairly widespread belief among wealthy members of English society that a "surplus" of unnecessary workers existed in the population. The displaced agriculturists began to be a heavy burden on charity, and so-called Poor Laws were introduced to regulate begging and public behavior.

In his treatise, Malthus insisted that "the power of the population is indefinitely greater than the power of the earth to produce subsistence." He also believed that if one accepted this premise, then a natural law would follow; that is, the population would inevitably exhaust food supplies.

Malthus was not without his critics, and influential thinkers such as William Godwin, Karl Marx, and Frederick Engels disputed Malthus's premises and propositions. Godwin argued that "there is no evil under which the human species cannot labor, that man is not competent to cure." Marx and Engels were in general agreement and insisted that technological development and an equitable distribution of resources would solve what they saw as a fictitious imbalance between people and food.

The debate about the relationship between population and resources continues to this day with provocative and compelling arguments for both positions being continually advanced. The geographer David Harvey, for example, has explored the population–resources issue in great detail. He has shown the limitations of Malthus's approach by demonstrating that by adopting a certain scientific method only one outcome is possible—a doomsday conclusion about the limiting effect of resources on population growth. By following Marx's scientific approach, however, he argues that quite different perspectives on, and solutions to, the population–resource issue can be generated. These solutions are based on human creativity and socially generated innovation, which, through technological and social change, provide opportunities for people to overcome the limitations of their environment.

Neomalthusians—people today who share Malthus's perspective—predict a population doomsday, however. They believe that growing human populations the world over, with their potential to exhaust Earth's resources, pose the most dangerous threat to the environment. Although they point out that it is the people of the core countries who consume the vast majority of resources, they and others argue that only strict demographic control everywhere will solve the problem, even if it requires severely coercive tactics.

A more moderate approach argues that people's behaviors and governmental policies are much more important factors affecting the condition of the environment and the status of natural resources than population size in and of itself. Proponents of this approach reject casting the population issue as a biological one in which an ever-growing population will create ecological catastrophe. They also reject framing it as an economic issue in which technological innovation and the sensitivities of the market will regulate population increases before a catastrophe can occur. Rather, they see the issue as a political one—one that governments have tended to avoid dealing with because they lack the will to redistribute wealth or the resources to reduce poverty, a condition strongly correlated with high fertility.

The question of whether too many people exist for Earth to sustain has bedeviled population policy makers and political leaders for most of the second half of the twentieth century. This concern has led to the formation of international agencies that monitor and often attempt to influence population change. It has also led to the organizing of a series of international conferences that have attempted to establish globally applicable population policies. The underlying assumption of much of this policy making is that countries and regions have a better chance of achieving improvement in their level of development if they can keep their population from outstripping the supply of resources and jobs.

Population Policies and Programs

Contemporary concerns about population—especially whether too many people exist for Earth to sustain—have led to the development of international and national policies and programs. A population *policy* is an official government policy designed to affect any or all of several objectives including the size, composition, and distribution of population. The implementation of a population policy takes the form of a population program. Whereas a policy identifies goals and objectives, a *program* is an instrument for meeting those goals and objectives.

Most of the international population policies of the last two decades have been directed at reducing the number of births worldwide. The instruments that have been developed to address rising fertility have largely been in the form of family-planning programs. The desire to limit fertility rates by the international population-planning community is a response to concerns about rapidly increasing global population—an increase that is being experienced significantly more in the periphery and semiperiphery than in the core countries. Accompanying this situation of imbalanced population growth between the core and the periphery is gross social and economic inequality as well as overall environmental degradation and destruction.

Figure 3.33 provides a picture of the recent history and a reasonable projection of the future of world population growth by region. The difference between core and periphery is dramatically illustrated. In mid-1992 the world contained nearly 5.5 billion people. The United Nations Population Fund projects that the world's population will increase by 90 million per year to the turn of the century. This means that by the year 2000, the world is projected to contain about 6.25 billion. Compare this figure to the fact that over the course of the entire nineteenth century less than a billion people were added to the population (Figure 3.34).

The geography of projected population growth is noteworthy. Over the next century, population growth is predicted to occur almost exclusively in Africa, Asia, and Latin America, while Europe and North America will experience very low and in some cases zero population growth. The differing rates of natural increase listed in Table 3.8 (see page 142) illustrate this point.

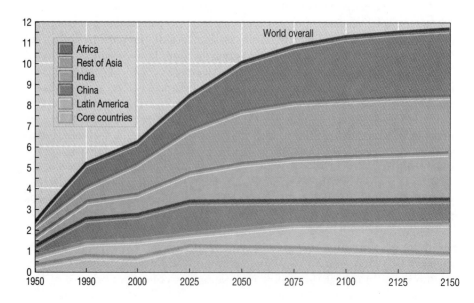

Figure 3.33 World population projection by region This graph represents a medium-variant projection, one that is in the middle of three possible scenarios, with the other two being higher or lower. In this projection, population continues to expand in the periphery, though in some regions more than in others. Africa is projected to experience the greatest growth followed by Asia, not including China, where growth is expected to level off by 2150. Less dramatic growth is expected to occur in Latin America, while in the core, population numbers remain constant or drop slightly. Though the total number of people in the world will be dramatically greater by 2150, the medium-variant forecast indicates a leveling off of world population. (*Source:* I. Hauchler and P. Kennedy (Eds.), *Global Trends: The World Almanac of Development and Peace.* New York: Continuum, 1994, p. 109.)

Figure 3.34 World population, 2020 This map provides a sense of how much the population of a country is expected to change by the year 2020. Although nearly all countries are expected to increase their populations, it is clear that some populations will grow far more dramatically than others. Notice the substantial growth expected in Saudi Arabia and Afghanistan, in contrast to the United States and Europe where little, if any, population growth will occur. Italy, for example, is expected to lose population, while the Netherlands will grow by only 5 percent.

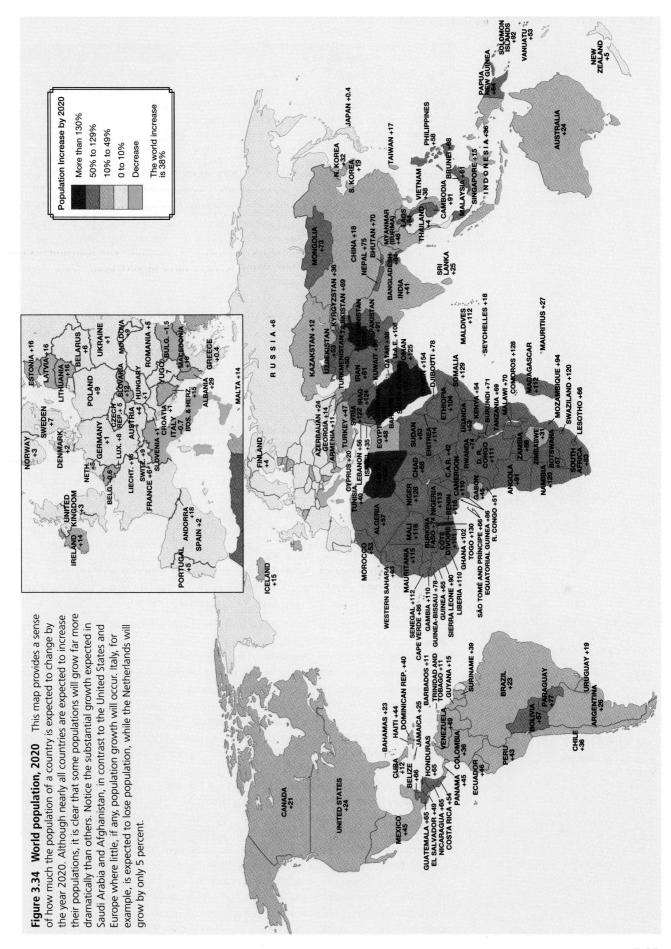

Population Increase by 2020

More than 130%
50% to 129%
10% to 49%
0 to 10%
Decrease

The world increase is 38%

141

TABLE 3.8 Global Demographic Indicators, 1994

This table illustrates the substantial population pressures emanating from the global periphery which, as of 1994, contained roughly four times the population of the core with a rate of natural increase also nearly four times as great. While Asian countries possess the largest proportion of the world's population, their rate of natural increase is considered moderate and their level of economic development—as measured in terms of GNP—higher than the average overall for the periphery. Most troubling to demographers, however, are the population dynamics of Africa, with a rate of natural increase of nearly three percent per year and a GNP of only $650 per capita.

Region	Population Mid-1994 (millions)	Natural Increase (annual %)	Birthrate (per 1,000 pop.)	Death Rate (per 1,000 pop.)	Life Expectancy (years at birth)	GNP Per Capita ($US 1992)
World	5,607	1.6	25	9	65	4,340
More Developed	1,164	0.6	12	10	75	16,610
Less Developed	4,443	1.9	28	9	63	950
Africa	700	2.9	42	13	55	650
Asia	3,392	1.7	25	8	64	1,820
Latin America and Caribbean	470	2.0	27	7	68	2,710
Europe	728	0.1	12	11	73	11,990
North America	290	0.7	16	9	76	22,840
Oceania	28	1.2	20	8	73	13,040

Source: A. Findley "Population Crises: the Malthusian Specter?" In R. J. Johnston et al. (Eds.), "Geographies of Global Change." Oxford: Blackwell, 1995, p. 156.

In 1993 the core contained 32 countries with zero population growth. By contrast, the periphery contained 39 countries with rates of natural increase of 3.0 or more. A rate of natural increase of 3.0 per year means their populations will double in approximately 24 years. Such remarkable rates of population increase and their relationship to quality of life and environment have led to a concerted and international response.

International conferences, sponsored by the United Nations to develop population policy at the global level, have been held every 10 years. Out of each conference emerged explicit population policies aimed at lowering fertility rates in the periphery and semiperiphery. The 1974 world conference was held in Bucharest, Romania, where two opposing approaches to lowering fertility were expressed. In fact, general disagreement emerged as to whether a population problem even existed. While the core countries were warning that a "population bomb" was ticking with the peripheral countries producing too many babies, the peripheral countries accused the core of advocating genocide. They argued that it wasn't the periphery but the core that was degrading the environment with its industry and heavy consumption of natural resources such as petroleum. The peripheral countries promoted the view that the most effective path to lowering fertility was through economic development, not the family-planning policies that the core was advocating. Ten years later, in Mexico City, the positions were reversed and peripheral countries were pressing for increased family-planning programs and the distribution of contraceptives, while core countries had adopted the perspective that "development is the best contraceptive."

In 1994, the most recent conference took place in Cairo. Called the International Conference on Population and Development, the name itself reflects the changes in ideas and practices that 20 years of population policy and programs had wrought. The Cairo conference, rather than focusing on the increase in global population growth rates, pointed to the fact that birthrates in almost every country on Earth are dropping—in many cases by significant amounts. Recognizing that a leveling-off of the population was possible in the foreseeable future, conference participants from both the core and the periphery were in agreement that efforts to bring down the growth rates must continue so that human numbers will peak sooner rather than later. The policy that emerged from the Cairo conference called for governments not simply to make family-planning programs available to all, but also to take deliberate steps to reduce poverty and disease; improve educational opportunities, especially for girls and women; and work toward environmentally sustainable development (Figure 3.35). The Cairo meeting was the first of the three international population conferences where nearly all the core and peripheral countries agreed on a plan for achieving population stabilization, a plan that would encourage freedom of choice in the matter of family size, population policy and program, and in development policy and practice.

The 1994 population conference recognized that the history, social and cultural practices, development level and goals, and political structures for countries and even regions within countries are highly variable and that one rigid and over-arching policy to limit fertility will not work for all. Whereas some programs and approaches will be effective for some countries seeking to cut population numbers, they will be fruitless for others.

For instance, China's family-planning policy of one child per household appears to be effective in driving down the birthrate and damping the country's overall population growth. Because so little data is available on China, however, it is not clear whether the policy is operating the same throughout the country. For instance, some population experts believe an urban bias exists. That is, while the policy is adhered to in the cities, it is disregarded in the countryside.

Family-planning regulations also include authorizing the age of marriage; offering incentives to couples who have only one child; and mandating disincentives to couples who have larger families. The disincentives are increasingly severe the larger the family. In India, family-planning policies offering free contraceptives and family-planning counseling have also had the impact of lowering the birthrate. Other countries such as Sri Lanka, Thailand, Cuba, and the Indian state of Kerala have lowered their birthrates not through regulation of family size but by establishing greater access to social resources such as health care and education, particularly for women.

Indeed, it is now a widely accepted belief among demographers and policymakers that a close relationship exists between women's status and fertility. Women who have access to education and employment tend to have fewer

Figure 3.35 International conference on population and development, 1994 The conference in Cairo had the important effect of insisting on the link between level of development and population growth. At the conference, much discussion ensued about improving the status of women as key to controlling population. It was also the case, however, that core and peripheral countries were often in disagreement about the most appropriate means to arrest the present trend of global population expansion.

children since they have less of a need for the economic security and social recognition that children are thought to provide (Figure 3.36). In Botswana, for instance, women with no formal education have, on average, 5.9 children, while those with four to six years of school have just 3.1 children. In Senegal, women with no education give birth to an average of seven children. In contrast, the average number of children born to a woman with 10 years of education drops to 3.6. The numbers are comparable for Asia and South America.

More equality between men and women inside and outside the household is also believed to have a significant impact on reducing fertility. Enabling voluntary constraints where both men and women are provided with a choice and educated about the implications of such choices, appears to be an especially successful program for small island populations such as Bali, Barbados, and Mauritius with historically high population growth. In Mauritius, in just 24 years (between 1962 and 1986), the introduction of voluntary constraints lowered the total fertility rate from 5.8 to 1.9. It is hardly any wonder, then, that the 1994 population conference in Cairo placed such a clear and well-received emphasis on (1) improving the rights, opportunities, and economic status of girls and women as the most effective way of easing global population growth and (2) rejecting coercive measures, including government sterilization quotas, that force people to violate their personal moral codes.

In order to implement the goals of improving women's (and girls') status relative to men's, it is necessary to take a broad view. Respected population experts have provided convincing evidence that an excess of female mortality characterizes much of the periphery. It is estimated that between 60 and 100 million more females would be alive today were it not for a preference for male children, which results in prenatal choices to abort females; biased health conditions; unequal nutritional provision; and female infanticide. Thus in countries such as China, for example, the apparent success of the one-child-per-household policy has been linked to horrifying abuses. The U.S. press as well as other international news media have for years published reports of how the policy has contributed to female infanticide in a culture where male children are highly valued and female children are not.

Success at damping population growth in the periphery appears to be very much tied to enhancing the possibility for a good quality of life and empowering people, especially women, to make informed choices. Both of these factors may require altering the consumption practices of populations in the core. Whether such an alteration is desirable remains to be seen.

Figure 3.36 Population policy and women and children Women are regarded as central to the success of controlling population growth. Access to education and employment security are seen as critical factors shaping a woman's decisions about how many children to have and when to have them.

Conclusion

The geography of population is directly connected to the complex forces that drive globalization. And, since the fifteenth century, the distribution of the world's population has changed dramatically as the capitalist economy has expanded, bringing new and different peoples into contact with one another and setting into motion additional patterns of national and regional migrations. When capitalism emerged in Europe in the fifteenth century, the world's population was experiencing high birthrates, high death rates, and relatively low levels of migration or mobility. Four hundred years later, birth, death, and migration rates vary—sometimes quite dramatically—from region to region, with core countries experiencing low death and birthrates and peripheral and semiperipheral countries generally experiencing high birthrates and fairly low death rates. Migration rates vary within and outside the core. These variations may be seen to reflect the level and intensity of political, economic, and cultural connectedness between core and periphery.

The example of formerly colonized peoples migrating to their ruling countries in search of work provides insights into the dynamic nature of the world economy. The same can be said of American migrants who in the 1970s and 1980s steadily left their homes in the Northeast and Midwest to participate in the employment opportunities that were emerging in the Sunbelt. Both examples show the important role that people play in acting out the dynamics of geographic variety.

In the final analysis, death rates, birthrates, and migration rates are the central variables of population growth and change. These indicators tell us much about transforming regions and places as elements in a larger world-system. Globalization has created many new maps as it has unfolded; the changing geography of population is just one of them.

Main Points

- Population geographers depend on a wide array of data sources to assess the geography of populations. Chief among these is the census, although other sources include vital records and public health statistics.

- Population geographers are largely concerned with the same sorts of questions that other population experts study, but also investigate "the why of the where": *Why* do particular aspects of population growth and change (and problems) occur *where* they do, and what are the implications of these factors for the future of these places?

- Two of the most important factors that make up population dynamics are birth and death. These variables may be examined in simple or complex ways but, in either case, the reasons for the behavior of these variables are as important as the numbers themselves.

- A third critical force in population change is the movement of populations. The forces that push populations from particular locations as well as those that pull them to move to new areas are key to understanding the resulting new settlement patterns. Population migration may not always be a matter of choice.

- Perhaps the most pressing issue facing scholars, policymakers, and other interested individuals is the one articulated at the World Population Conference held in Cairo in 1994: How many people can the world adequately accommodate with food, water, clean air, and other basic necessities necessary to the enjoyment of happy, healthy, and satisfying lives?

Key Terms

age–sex pyramid (p. 107)
agricultural density (p. 105)
baby boom (p. 106)
census (p. 98)
cohort (p. 110)

crude birthrate (CBR) (p. 119)
crude death rate (CDR) (p. 120)
crude (arithmetic) density (p. 103)

demographic transition (p. 124)
demography (p. 98)
dependency ratio (p. 113)
doubling time (p. 120)

eco-migration (p. 138)
emigration (p. 127)
forced migration (p. 127)
geodemographic analysis (p. 107)
gross migration (p. 127)

guest workers (p. 128)
histogram (p. 107)
immigration (p. 127)
infant mortality rate
    (p. 121)
internal migration (p. 127)
international migration
    (p. 127)

life expectancy (p. 122)
middle cohort (p. 113)
migration (p. 127)
mobility (p. 127)
natural decrease
    (p. 121)
natural increase
    (p. 121)

net migration (p. 127)
nutritional density (p. 105)
old-age cohort (p. 116)
pull factors (p. 127)
push factors (p. 127)
suburbanization (p. 135)
total fertility rate (TFR)
    (p. 119)

vital records (p. 99)
voluntary migration
    (p. 127)
youth cohort (p. 113)

Exercises

 ### On the Internet

In this chapter the Internet exercises will explore the relationship between globalization and population change. The Internet provides an especially valuable tool for studying this relationship because it contains extensive databases on demographic variables (such as fertility, mortality, and disease) for countries, regions, and cities worldwide. After completing these exercises you should have a better grasp of the geography of population and how globalization is shaping demographic variables and patterns.

 ### Unplugged

1. The distribution of population is a result of many factors, such as employment opportunities, culture, water supply, climate, and other physical environment characteristics. Look at the distribution of population in your state or province. Is it evenly distributed, or are the majority of people found in only a few cities? What role do you think these various factors have played in influencing where people live in your state or province? Can you think of other reasons for this distribution?

2. Immigration is an important factor contributing to the increase in the population of the United States. Chances are your great-grandparents, grandparents, parents, or even you immigrated to, or migrated within, your country of residence. Construct your family's immigration or migration history. What were some of the push and pull factors influencing your family's decision to immigrate to or migrate within your country?

3. How many stages are there in the demographic transition graph? Why does the death rate typically fall before the birthrate? List some factors that might cause the birthrate to fall. Speculate what stage your country is in.

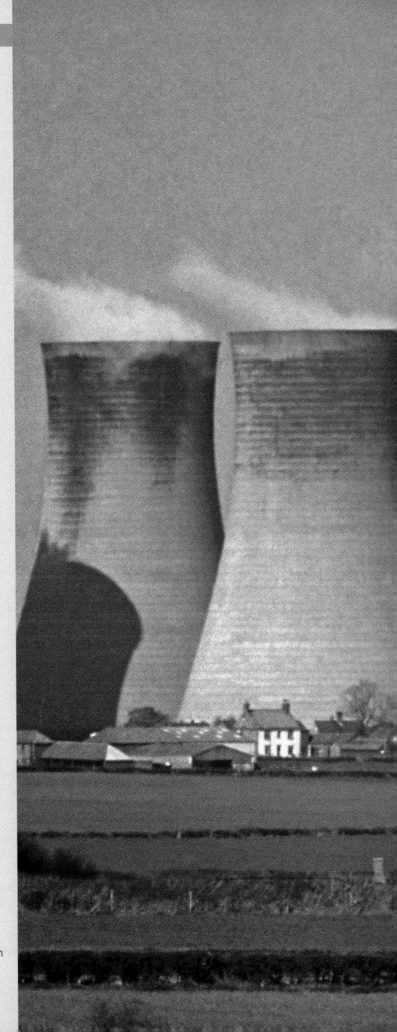

Chapter 4

Nature, Society, and Technology

Nuclear cooling towers,
East Anglia, Great Britain

In the fall of 1990, the national news media carried the story that Africanized honeybees, popularly known as killer bees, had finally crossed the U.S.–Mexico border into Texas. Taking less than 35 years to migrate from São Paulo in Brazil, the Africanized bees entered Arizona in 1993 and will continue their movement northward until climatic constraints limit their expansion by the year 2000. Given that the bees have a current range of well over 20 million square kilometers (about 7.7 million square miles), encompassing much of South America, nearly all of Central America, and a substantial portion of the southern United States, the story of the Africanized bees can be read as a cautionary tale of human interaction with nature.

Eager to strengthen its bee-keeping industry, and impatient with the European honeybees that were poorly suited to the environment, the Brazilian government introduced the bees in Brazil in the 1950s. A variant more suited to the tropics was sought, and a researcher at the University of São Paulo undertook the task of importing queens from Africa. By 1957, new colonies were well established. That same year a visitor to the experimental site apparently removed the screens at the entry to the hives intended to prevent the queens from leaving. The result was the liberation of 26 colonies of bees, which eventually flourished independently in the Brazilian environment. The migrating Africanized bees threaten to destroy the more docile European bees and the American (North, South, and Central) bee-keeping industry with them.

The history of the Africanized bees illustrates both the complex ways in which the global environmental system is connected and in which human interactions with nature can cause unforeseen consequences and require untold investments of time and capital to correct—if, in fact, they can ever be corrected. What started as a naive experiment to encourage economic growth has turned into a natural hazard for many peoples, governments, and ecosystems.

This relationship between people and the environment is perhaps the most central of all within the discipline of geography. Indeed, the discipline of geography consists of those who study natural systems and those who study human systems. In this chapter we explore the ways that society has used technology to transform and adapt to nature, together with the human and environmental impact of these technological adaptations.

Nature As a Concept

As discussed in Chapter 2, a simple model of the nature–society relation is that nature, through its awesome power and subtle expressions, limits or shapes society. This model is known as *environmental determinism*. A second model posits that society also shapes and controls nature, largely through technology and social institutions. This second model, which we will explore in this chapter, emphasizes the complexity of nature–society interactions.

Interest in the relationships among nature, society, and technology has experienced a resurgence over the last two or three decades. The single most dramatic manifestation of this interest occurred in the summer of 1992, when over 100 world leaders and 30,000 other participants attended the second Earth Summit in Rio de Janeiro (the first Earth Summit was held in Stockholm in 1972). The central focus of the agenda was to ensure a sustainable future for Earth by establishing treaties on global environmental issues such as climate change and biodiversity. An abundance of far less spectacular evidence also suggests, however, that concern for the environment has spread from the scientific community to the public and from the core to the periphery. In 1992, the George H. Gallup International Institute conducted a survey called "The Health of the Planet," polling more than 1,000 individuals in 22 nations around the world. The survey found that people in both poor and rich nations rated environmental protection as more important than economic growth and indicated that environmental problems were the most pressing of all their problems. Figure 4.1 illustrates these points.

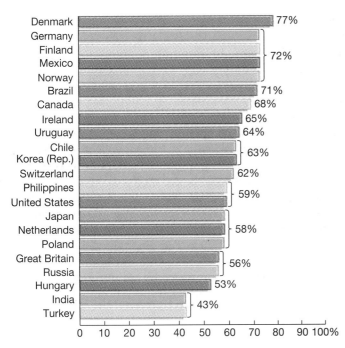

Figure 4.1 Protecting the environment or economic development? In 1992, the George H. Gallup International Institute in Princeton, New Jersey, conducted a survey of 22 major nations around the world in order to determine popular attitudes toward the environment. "The Health of the Planet Survey" was undertaken to allow citizens to register their concerns about how globalization—what they called increasing levels of economic development—is affecting their lives and their physical environments. Despite what conventional wisdom suggests—that most people are too busy trying to deal with their daily economic problems to worry about much else—people around the world are deeply concerned about the state of their physical environments. (*Source: The Health of the Planet Survey,* The George H. Gallup International Institute, 1992.)

Why is there such strong and widespread interest in the environment today? In order to answer that question it is necessary to recognize that despite repeated attempts to address environmental problems, they persist. At the beginning of the twentieth century, Gifford Pinchot, as an influential citizen and later as director of the National Conservation Commission, advocated environmental conservation. In 1962, Rachel L. Carson, in a pathbreaking book, *Silent Spring,* warned of the dangers of agricultural pesticides to ecosystems (Figure 4.2, and Figure 4.3, page 150). Yet, at the turn of the twenty-first century, the pesticide problem persists and has become a problem for peripheral as well as core countries.

Indeed, concern about pollution has been growing in the international community at least since 1972's first Earth Summit. Notably, concern was mostly focused then upon local pollution issues such as air and water pollution. Twenty years later, at the second Earth Summit in Rio, concern had advanced beyond local-level issues to embrace pollution issues that are global in scope as well as in impact, including global climate change and the destruction of Earth's rain forests.

Figure 4.2 Death from DDT DDT (or dichlorodiphenyltrichloroethane) is a chemical pesticide applied through aerial spraying. It was used widely in agriculture in the United States and other core countries in the middle decades of the twentieth century. Banned from use in the United States in 1972, it had lethal effects on populations of fish-eating birds such as osprey, cormorants, brown pelicans, and bald eagles. Despite the prohibition against the use of DDT in most core countries, countries in the periphery continue to apply it as a pesticide to improve yields. Illegal use of the pesticide in the United States as well as its airborne drift from Latin America appears to be playing a role in the continued high levels of DDT found in fish in the United States. In the early 1990s, the U.S. Environmental Protection Agency found DDT residues in 99 percent of the freshwater fish it tested. In short, pesticides do not stay in one place, and they persist in the fatty tissue of fish and animals from 2 to 15 years.

Figure 4.3 Rachel L. Carson (1907–1964) Rachel L. Carson was a professional biologist who worked for the U.S. Fish and Wildlife Service. In 1951, she wrote *The Sea Around Us,* which describes the natural history of the oceans. The book sold over two million copies and was translated into 32 languages. Her most important contribution, however, was her book *Silent Spring,* which was the first piece of critical research published on the environmental effects of pesticides. Many credit Rachel Carson as a key figure in the emergence of what is now known as the environmental movement in the United States.

Nature: a social creation as well as the physical universe that includes human beings.

Society: sum of the inventions, institutions, and relationships created and reproduced by human beings across particular places and times.

Technology: physical objects or artifacts, activities or processes, and knowledge or know-how.

In short, renewed interest in the nature–society relationship is the result of the persistence and number of environmental crises. This interest has led to attempts to rethink these relationships. In the past, technology had always emerged as the apparent solution to our environmental problems, but now the continued application of technology seems not to solve them. As a result, researchers and activists have begun to ask different questions and abandon the assumption that technology is the *only* solution.

Environmental thinkers, including a number of geographers, have begun to question the predominant Western conception of nature as well as the implicit assumptions about it that constrain our discussions, debates, and approaches. They advocate challenging the very concept of nature itself and recognizing how our social ideas about nature shape the kinds of questions we ask about it, as well as the uses to which we put it. They point out that, by understanding nature as a *social idea,* we can come to recognize that environmental crises are social phenomena as much as they are physical ones. Before asking new questions about nature, we define key terms and look at different approaches to the concept of nature. We then examine how changing conceptions of nature have translated into very different uses of, and adaptations to, it.

Nature, Society, and Technology Defined

The central concepts of this chapter—nature, society, and technology—have very specific meanings. Although we discuss the changing conceptions and understandings of nature in some detail, we hold to one encompassing position. **Nature** is a social creation as much as it is the physical universe that includes human beings—understandings of nature are the product of different times and different needs. Nature is not only an object, it is a reflection of society in that philosophies, belief systems, and ideologies shape the way people think about nature and the way they "use" it. The relationship between nature and society is two-way: Society shapes people's understandings and uses of nature at the same time that nature shapes society. The amount of shaping by society is dependent to a large extent on the state of technology and the constraints on its use at any given time.

Society is the sum of the inventions, institutions, and relationships created and reproduced by human beings across particular places and times. Nature is just one of the relationships created and reproduced by society; and the social relationship to nature varies from place to place among different social groups. In our scheme, society and nature are interrelated.

The mutual relationship between society and nature is usually mediated through technology. Knowledge, implements, arts, skills, and the socio-cultural context all are components of technology. If we accept that all of these components are relevant to technology, then we can provide a definition that has three distinguishable though equally important aspects. **Technology** is defined as:

- physical objects or artifacts (for example, the plow)
- activities or processes (for example, steelmaking)
- knowledge or know-how (for example, biological engineering).

This definition encompasses tools, applications, and understandings and enables recognition of all three as critical components of the processes and outcomes of the human production of technology. The manifestations and impacts of technology can be measured in terms of concepts such as level of industrialization and per capita energy consumption.

The definitions provided in this section reflect current thinking on the relationship between society and nature. For centuries, humankind, in its responses to the constraints of the physical environment, has been as much influenced by prevailing *ideas* about nature as by the realities of nature. In fact, prevailing ideas about nature have changed over time, as evidence from literature, art, religion, legal systems, and technological innovations make abundantly clear.

A recent attempt to conceptualize the relationship between social changes and environmental changes has emerged from concern with global environmental change. Based on the premise that individual societal changes can be both subtly and dramatically related to environmental change, a formula for distinguishing the sources of social impacts on the environment has been advanced and is now widely used. The formula is known as *I = PAT*, and it relates human population pressures on environmental resources to the level of affluence and access to technology in a society. More specifically, the formula states that *I = PAT*, where *I* (*impact* on Earth's resources) is equal to *P* (*population*) times *A* (*affluence*, as measured by per capita income) times *T* (a *technology* factor). For example, the differential impact on the environment of two households' energy use in two different countries would equal the number of people per household times the per capita income of the household times the type of technology used to provide energy for that household (Figure 4.4).

Each one of the variables in the formula—population, affluence, and technology—should be seen as complex, otherwise some erroneous conclusions can be drawn. For example, with regard to population numbers, it is generally believed that fewer people on the planet will result in fewer direct pressures on resources. Some argue, however, that the increases in world population numbers as we approach the twenty-first century are quite desirable. The argument is made that more people means more labor coupled with more potential for the emergence of innovation to solve present and future resource problems. Clearly there is no simple answer to the question of how many people is too many people.

It is also the case that the variable affluence cannot simply be assessed in terms of "less is better." Certainly, increasing affluence—a measure of per capita

Figure 4.4 Affluence differences in two households   As the *I = PAT* formula suggests, the level of affluence of households plays an important role in their impact on the global environment. Pictured here are two families, one from the core and the other from the periphery. Both families are pictured outside their homes with all of their possessions arrayed around them. The extensive range of possessions show the Icelandic family, composed of two parents and four children, to be far more affluent than the Guatemalan family, composed of two parents and three children. Iceland ranks eighth in affluence among the 183 members of the United Nations; Guatemala's rank is 114. The Icelandic family possesses two radios, one stereo, two televisions, one VCR, one home computer, two automobiles, and a private airplane. The Guatemalan family possesses one battery-operated radio and no telephone but would like to acquire a television set.

consumption multiplied by the number of consumers and the environmental impacts of their technologies—is a drain on Earth's resources and a burden on Earth's ability to absorb waste. Yet, "how much affluence is too much?" is difficult to determine. Furthermore, evidence shows that the core countries, with high levels of affluence, are more effective at protecting their environments than poor countries of the periphery. Unfortunately, it is also often the case that core countries protect their own environments by exporting their noxious industrial processes and waste products to peripheral countries. By exporting polluting industries and the jobs that go with them, however, core countries may also be contributing to a rise in the level of affluence. Given what we know about core countries, such a rise may be sufficient to sustain a set of social values that would foster attitudes and behaviors leading ultimately to protecting the environment in a new place. It is difficult to identify just when an environmental consciousness goes from being a luxury to a necessity. The role of affluence in terms of environmental impacts, is, in short, like population, difficult to assess.

Not surprisingly, the variable technology is no less complicated than either population or affluence. It is important to recognize that technologies affect the environment in a threefold way:

- through the harvesting of resources
- through the emission of wastes in the manufacture of goods and services
- through the emission of waste in the consumption of goods and services

Technology is complex simply because it can work both ways. A technological innovation can shift demand from an existing resource to a newly discovered, more plentiful one. Technology can sometimes be a solution and sometimes a problem. Such is the case with nuclear energy, widely regarded as cleaner and more efficient than coal or oil as energy sources. Producing this energy creates hazards, however, which scientists are still unable to prevent.

It is, therefore, clear that increases in human numbers, in levels of wealth, and in technological capacity are key components of social and economic progress whose impact on the environment has been extremely complex. In the last 100 years, this complexity has increasingly come to be seen as a triple-barreled threat to the quality of the natural world and the availability and quality of environmental resources. Before we look more carefully at the specific impacts of populations, affluence, and technology on nature, we need to look first at how differing social attitudes toward nature shape the human behaviors that are a basis for $I = PAT$.

Religious Perspectives on Nature

Traces on the Rhodian Shore, written by geographer Clarence Glacken in 1967, is an influential work on the history of the idea of nature. Encyclopedic in its scope, this book provides an important foundation for all who are interested in the way that nature has been understood by humankind up to the eighteenth century. Glacken demonstrated that it was the ancient Greeks who were the original source of the idea of nature, regarding it as separate from humans. Indeed, it is from the Greeks that contemporary society inherited the belief that a fundamental distinction exists between humans and nature, with the latter being defined as anything not actually fabricated by humans.

The Judeo-Christian tradition has exerted perhaps the most pervasive influence on our contemporary ideas about nature. Generally, the **Judeo-Christian perspective on nature** holds that nature was created by God and is subject to God in the same way that a child is subject to parents. "Man" (by which pre–twentieth century environmental philosophies meant males *exclusively*) was also created by God, but made in God's own image, so Man is separate from nature in this regard.

Judeo-Christian perspective on nature: the view that nature was created by God and is subject to God in the same way that a child is subject to parents.

While the Judeo-Christian idea that humans must dominate nature is perhaps the most prevalent one, it is important to remember it is not the only one. Even within the Judeo-Christian tradition, this view was opposed by thinkers such as St. Francis of Assisi, for instance. Indeed, it is important to recognize that despite the fact that human dominance over nature is a centerpiece of the Judeo-Christian worldview, both the Old and New Testaments contain ample evidence of support for a more cooperative approach to nature.

All religious traditions, in fact, involve conflictual ideas about nature. When compared in broad terms, however, they reflect some very different perspectives on humans' relation to nature. Within a **Taoist perspective on nature**, for example, a clear emphasis exists on valuing nature for its own sake, not for the utilitarian purposes to which it might be put. An ancient Chinese religion whose emergence preceded the birth of Christ by a millennium, Taoism emphasizes harmony with nature and views the natural world not as an exploitable resource but as a complex life process to be respected and appreciated.

Taoist perspective on nature: the view that nature be valued for its own sake, not for how it might be exploited.

A **Buddhist perspective on nature** teaches that nothing exists in and of itself and everything is part of a natural complex and dynamic totality of mutuality and interdependence. Humans have a special role in the totality in that they alone are capable of reflection and conscious action. It is up to human beings, therefore, not to act as if they are above nature, but to care for all of life, human and nonhuman, and to safeguard the integrity of the universe.

Buddhist perspective on nature: the view that nothing exists in and of itself and everything is part of a natural complex and dynamic totality of mutuality and interdependence.

Islam is the Muslim religion based on the teachings of the prophet Muhammad. An **Islamic perspective on nature** is closely aligned with the Judeo-Christian view, though it must be remembered that a multiplicity of views exist within all religious traditions. The sacred book of Islam, the Qur'an, supports the view that the heavens and Earth were created to serve human purposes—that humans are sovereign over the rest of creation. Where Islam departs from the Judeo-Christian view of human dominance over nature is through the teaching that authority over nature is given by Allah (God) not as an absolute right but as a test of obedience, loyalty, and gratitude to Allah. Abuse of Earth is opposed to the will of Allah; stewardship of it shows respect and fulfills the will of Allah.

Islamic perspective on nature: the view that the heavens and Earth were made for human purposes.

Indigenous religious traditions in North and South America and in Africa also conceptualize nature differently. The belief that humans are one part of a complex natural world is a somewhat simplified but certainly accurate portrayal of their ideas about the human-nature relationship. Animism, for example, was a widespread system of belief in these three continents before European contact. An **Animistic perspective on nature** is the belief that natural phenomena—both animate and inanimate—possess an indwelling spirit or consciousness. In most forms of animism, humans view themselves as extensions of animate and inanimate nature. For many native peoples, for example, humans cannot be separated from nature, and the natural cannot be separated from the supernatural. Animism embraces a sense of kinship between the natural world and the human world, as illustrated in Figure 4.5 (see page 154), a painting done by an early twentieth-century North American Iroquois Indian.

Animistic perspective on nature: the view that natural phenomena—both animate and inanimate–possess an indwelling spirit or consciousness.

Environmental Philosophies and Political Views of Nature

Henry David Thoreau (1817–1862), the American naturalist and activist, perhaps best illustrates the Western incorporation of North American Indian conceptions of nature combined with other emerging ecological approaches. Thoreau lived and studied the natural world around the town of his birth, Concord, Massachusetts, during the middle decades of the nineteenth century. He is most famous for his book *Walden*, which chronicles the two years he spent living and observing nature in solitude in a house he built on Walden Pond, a mile and a half from the village green of Concord. Thoreau represents a significant alternative to the "Man-over-nature" approach that characterized his times.

Figure 4.5 Jesse Cornplanter's "Raising of the Slain Hero" This etching by an Iroquois artist in 1908 shows the animals of the forest assisting an injured brave. The bear, an animal of great strength which was highly prized and revered by the Iroquois, holds the brave's hand while other birds and animals watch with concern.

Many people credit him as the originator of an American ecological philosophy (Figure 4.6).

Thoreau was impressed with the power of nature. He often described its unrestrained and sometimes explosive capacity, which he thought had the potential to overthrow Man's dominion if left unchecked. In contrast, he also emphasized the holism or interrelatedness of the natural world, where birds depended upon worms and fish depended upon flies, and so on, along the food chain (Figure 4.7). Most notable, however, is the fact that Thoreau regarded the natural world as an antidote to the negative effects of technology on the American landscape and the American character. Concord, where Thoreau lived, was just 20 miles west of Boston and an equal distance south of the booming milltowns of Lowell and Lawrence. Although he spent his life in a more or less rural setting, the Industrial Revolution was in full force all around him, and he was keenly aware of its impacts.

In fact, Thoreau's approach to the natural world is very much a response to the impacts on nature of the early forces of globalization. His research into the

Figure 4.6 Walden Pond, 1996 The tranquil setting of Walden Pond that Henry David Thoreau made famous is now a thing of the past. Over the years since Thoreau, it has become a popular bathing and recreational site drawing in thousands of visitors on its busiest summer weekends.

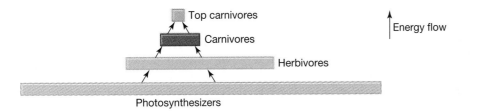

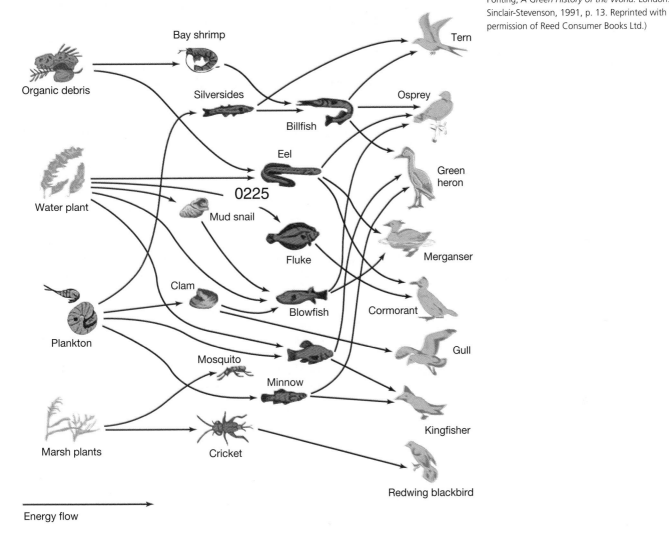

Figure 4.7 Generalized food chain
Although a contemporary schematic, this illustration of the food chain in a Long Island estuary demonstrates the nineteenth-century naturalist's view. Plants, animals, and insects are all seen to be part of a complex whole such that elimination or injury to one element affects the entire system. (*Source:* C. Ponting, *A Green History of the World.* London: Sinclair-Stevenson, 1991, p. 13. Reprinted with permission of Reed Consumer Books Ltd.)

animals and plants that surrounded Concord was an attempt to reconstruct the landscape as it had existed before colonization and massive European immigration.

Thoreau was also a primary force behind romanticism, a movement that originated in Europe. **Romanticism** is a philosophy that emphasizes interdependence and relatedness between humans and nature. In direct revolt against those who espoused a Judeo-Christian understanding of nature, the romantics believed that *all* creatures—human and otherwise—were infused with a divine presence that commanded respect and that humans were not exceptional in this scheme. Rather, their divinity issued from humble participation in the natural community.

A branch of American romanticism, known as transcendentalism, has also had an influence over contemporary understandings of nature. Transcendentalism was espoused most eloquently by Unitarian minister turned poet and

Romanticism: philosophy that emphasizes interdependence and relatedness between humankind and nature.

Transcendentalism: a philosophy in which a person attempts to rise above nature and the limitations of the body to the point where the spirit dominates the flesh.

Conservation: the view that natural resources should be used wisely, and that society's effects on the natural world should represent stewardship and not exploitation.

Preservation: an approach to nature advocating that certain habitats, species, and resources should remain off-limits to human use, regardless of whether the use maintains or depletes the resource in question.

Environmental ethics: a philosophical perspective on nature that prescribes moral principles as guidance for our treatment of it.

philosopher Ralph Waldo Emerson, a neighbor and contemporary of Thoreau. For Emerson and many members of the romantic school, **transcendentalism** is a philosophy in which a person attempts to rise above nature and the limitations of the body, to the point where the spirit dominates the flesh, where a mystical and spiritual life replaces a primitive and savage one. Thoreau and Emerson are two of the most important influences on contemporary ideas about the human–nature relationship.

Early in the twentieth century, writers like Gifford Pinchot and politicians like Theodore Roosevelt drew on the ideas of Thoreau and Emerson to advocate the "wise use" of natural resources and the conservation of natural environments. The view that nature should be conserved is one that has persisted to the present. **Conservation** is the view that natural resources should be used wisely, and that society's actions with respect to the natural world should be actions of stewardship, not exploitation. Conservation implies responsibility to future generations as well as to the natural world itself in our utilization of resources. Groups such as the Sierra Club, while also concerned with the quality of the natural world, are similar in their approach to conservationists. The Sierra Club is a well-established private institution with chapters throughout the United States. It possesses an extensive legal division that litigates cases of corporate and individual violation of federal and state environmental regulations.

For those who espouse a more radical approach to nature, conservation is seen as a passive, mainstream approach, either because it leaves intact the political and economic system that drives the exploitation of nature or because of a central belief that nature is sacred and should be preserved, not used. A more extreme position, **preservation** is an approach to nature advocating that certain habitats, species, and resources should remain off-limits to human use, regardless of whether the use maintains or depletes the resource in question. The philosophy of groups such as Earth First! is closely aligned with the preservationist perspective. Earth First!, unlike the Sierra Club, operates outside the bounds of mainstream institutional frameworks. Whereas the Sierra Club takes its opponents into the courtroom, Earth First! employs extralegal tactics—often called *ecoterrorist* tactics—such as driving spikes into trees to discourage logging. These "quick strike" actions are intended to halt what are regarded as government or corporate environmental abuses (which may, in fact, be perfectly legal though counter to the Earth First! philosophy).

Greenpeace is yet another environmental organization different from either the Sierra Club or Earth First!. Most important, Greenpeace is global in its reach, which means that both its membership and its areas of emphasis are international. Focusing on environmental polluters and combining the strategies of both the Sierra Club and Earth First!, Greenpeace utilizes oppositional tactics as well as formal international legal actions. In both its membership—with the world headquarters in Amsterdam and regional offices in most major industrial countries—as well as its objectives—halting environmental pollution worldwide—Greenpeace articulates the belief that places are interdependent and what happens in one part of the globe affects us all.

Organizations such as Earth First! and Greenpeace are practical illustrations of the new approaches to understanding human interactions with nature that have occurred in environmental philosophy since the publication of *Silent Spring* over 35 years ago. These new approaches—environmental ethics, ecofeminism, environmental justice, and deep ecology—take the view that nature is as much a physical universe as it is a product of social thought. All are new and different ways of understanding how society shapes our ideas about nature.

Environmental ethics is a philosophical perspective on nature that prescribes moral principles as guidance for our treatment of it. Environmental ethics insists that society has a moral obligation to treat nature according to the rules of moral behavior that exist for our treatment of each other. Many different approaches exist within the field, but all, more or less, advocate that what is good and right in our treatment of nature involves the same sort of basic values that guide what is good and right in our treatment of other human beings. An aspect of environmental ethics that has caused a great deal of controversy is the perspective that animals, insects, trees, rocks, and other elements of nature have rights in the same way that humans do. If the moral system of our society insists that humans are to have the right to a safe and happy life, then it is argued that the same rights should be extended to nonhuman nature.

Ecofeminism shares much of this philosophical perspective. Ecofeminists hold that patriarchy—a system of social ideas that values men more highly than women—is at the center of our present environmental malaise. Because patriarchy has equated women with nature, it has enabled the subordination and exploitation of both. The many varieties of ecofeminism range from nature-based spirituality oriented toward a Goddess (not unlike the animism of native peoples) to more political approaches that emphasize resistance and opposition to the dominant masculine models that devalue what is not male. Some ecofeminists are also environmental ethicists. Not only a movement of the core, ecofeminism has also been widely embraced in the periphery, where women are primarily responsible for the health and welfare of their families in environments that are being rapidly degraded. The unifying objective in all of ecofeminism is to dismantle the patriarchal biases in Western culture and replace them with a perspective that values both cultural and biological diversity.

> **Ecofeminism:** the view that patriarchal ideology is at the center of our present environmental malaise.

Deep ecology, which shares many points with ecofeminism, is an approach to nature revolving around two key components: self-realization and biospherical egalitarianism. Self-realization embraces the view that humans must learn to recognize that they are part of the nonhuman world, whereas biospherical egalitarianism insists that Earth, or the biosphere, is the central focus of all life and that all members of nature, human and nonhuman, deserve the same sort of respect and treatment. Deep ecologists believe that there is no absolute divide between humanity and everything else, and that a complex and diverse set of relations constitutes the universe. The belief that all things are internally related could enable society to treat the nonhuman world with respect and not simply as a source of raw materials for human use.

> **Deep ecology:** approach to nature revolving around two key components: self-realization and biospherical egalitarianism.

All of these above approaches attest to a growing concern with the implications of globalization. Acid rain, deforestation, the disappearance of species, nuclear accidents, and toxic waste have all been important stimuli for newly emerging philosophies about the preferred relationships between technology and nature within globalizing societies. While none of these philosophies provides a panacea, each has an important critique to offer. More than anything, however, each serves to remind us that environmental crises are not simple, and simple solutions will no longer suffice.

The Concept of Nature and the Rise of Science and Technology

Let us return to the point made earlier, that the most widespread ideas of nature current in Western thought—and ones that have persisted under different labels for thousands of years—are that humans are the center of all creation, and that nature in all its wildness was meant to be dominated by humans. Judeo-Christian belief insists that Man, made in the image of God, was set apart from nature and must be encouraged to control it.

While Christianity held that nature was to be dominated, that idea existed more in the religious and spiritual realm than in the political and social realm. In terms of the conduct of everyday life, it was not until the sixteenth century that Christian theology was coaxed from its isolation and conscripted to aid the goals of science. Before 1500 in Europe there existed a widely held image of Earth as a living entity such that human beings conducted their daily lives in an intimate relationship with the natural order of things. The prevailing metaphor was that of the *organism*, which emphasized interdependence among human beings and between them and Earth (Figure 4.8). Yet, even within this organic idea of nature we can find two opposing conceptions. One was of a nurturing Earth that provided for human needs in a beneficent way; the other was of a violent and uncontrollable nature that could cause general chaos in human lives. In both of these views Earth and nature were regarded as female.

Francis Bacon (1561–1626) and Thomas Hobbes (1588–1679) were English philosophers who, as prominent promoters of science and technology, were influential in changing the prevailing organic view of nature. Borrowing from the Christian ideology, they advanced a view of nature as something subordinate to Man. Bacon and Hobbes sought to rationalize benevolent nature as well as to dominate disorderly and chaotic nature.

Figure 4.8 Classes of people and the natural world These sixteenth-century woodcuts illustrate the era's view of the place of human beings in the natural world. (*Sources:* Left, New York Public Library Print Division, Hans Weitz Woodcut, and H. C. Hoover and L. H. Hoover, *De Re Metallica.* Mineola, NY: Dover Publishing, 1950, p. 337. Right, Georg Agricola Woodcut.)

This woodcut by Hans Weiditz illustrates organic society represented as a tree with peasants at the root. The second level holds artisans, journeymen, and merchants. The third level supports bishops, cardinals, nobles, and princes. On the fourth level is the king, pope, and emperor; at the top, there are two peasants, signifying that organic society depends on them for sustenance.

This woodcut by Georg Agricola shows various uses of wood and water with respect to the mining industry of the late Middle Ages and Renaissance. This scene illustrates the exploitation of nature in contrast to the view of it as a bountiful mother.

As feminist environmental historian Carolyn Merchant writes:

The change in controlling imagery was directly related to changes in human attitudes and behavior toward the earth. Whereas the nurturing earth image can be viewed as a cultural constraint restricting the types of socially and morally sanctioned human actions allowable with respect to the earth, the new images of mastery and domination functioned as cultural sanctions for the denudation of nature. Society needed these new images as it continued the process of commercialization and industrialization, which depended on activities directly altering the earth—mining, drainage, deforestation, and assarting [grubbing up stumps to clear fields]. The new activities used new technologies—lift and force pumps, cranes, windmills, geared wheels, flap valves, chains, pistons, treadmills, under- and overshot watermills, fulling mills, flywheels, bellows, excavators, bucket chains, rollers, geared and wheeled bridges, cranks, elaborate block and tackle systems, worm spur, crown, and lantern gears, cams and eccentrics, ratchets, wrenches, presses, and screws in magnificent variation and combination.[1]

Figure 4.8, the Georg Agricola woodcut, illustrates this point.

Merchant shows that, by the sixteenth and seventeenth centuries, the power of science was too great for the organic idea of nature. Subsequently, a view that nature was the instrument of Man became dominant in Western culture.

The Transformation of Earth by Ancient Humans

Although the previous discussion might suggest that Earth remained relatively unaffected by human action until well into the early modern period, the Paleolithic and Neolithic peoples altered the environment even without machines or elaborate tools. Considerable evidence exists that early humans were very active agents of change. People's perceptions of nature were usually quite influential in shaping their environmental behaviors, although many examples exist of contradictions between attitudes and actions. In this section we see that contemporary humans have inherited an environment that was significantly affected even by the practices of our very earliest ancestors.

Paleolithic Impacts

Although humans are thought to have first inhabited Earth approximately 6 million years ago, almost no evidence exists of the way in which the very earliest hominids, as they are called, used the natural world around them to survive. What is known is that their numbers were not large, and they left behind little in the way of technology or art to help us understand their relationship to nature. The earliest evidence about early people–environment relationships comes from the **Paleolithic period** (about 1.5 million years ago), a cultural period also known as the early Stone Age because this was the period when chipped stone tools first began to be used.

Hunters and gatherers living on the land in small groups, the early Stone Age people mainly foraged for wild food and killed animals and fish for their survival. Hunting under early Stone Age conditions could not support a growing population, however. It is estimated that on the African grassland, only two people

Paleolithic period: the period when chipped stone tools first began to be used.

[1]C. Merchant, *The Death of Nature.* San Francisco: Harper and Row, 1979, pp. 2–3.

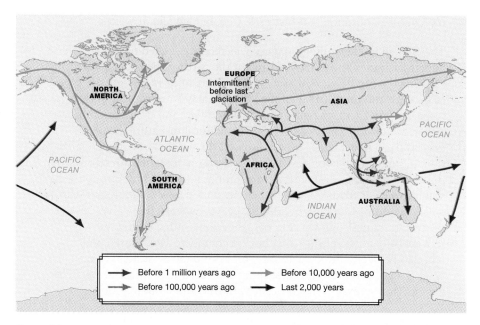

Figure 4.9   The Settlement of the World   This map shows the direction and timing of movement of humans from their point of origin in East Africa to all corners of the globe. Constant movement in search of food promoted such movement. The map represents over one million years of migration. (*Source:* C. Ponting, *A Green History of the World.* London: Sinclair-Stevenson, 1991, p. 25. Reprinted with permission of Reed Consumer Books Ltd.)

Figure 4.10   Cave Paintings   Cave paintings are an important record of the imagination of early hunter-gatherers. Some regard these paintings as clear evidence of the development of forward thinking or anticipation among humans.

could survive on the vegetation and wildlife available within about a 2.5 square kilometer (one square mile) area. To help ensure survival, early Stone Age people constantly moved over great distances, which ultimately made them a dispersed species, creating the foundation for the world's population distribution of today (see Figure 4.9).

Evidence also exists of early Stone Age tools, as well as of the importance of hunting to existence. The cave paintings of Altamira, Spain, for example, illustrate that hunting was the primary preoccupation of the Stone Age mind (Figure 4.10). Because these early peoples lived in small bands and moved frequently and in wider and wider ranges, it is tempting to conclude that they had very little impact on their environment. It does appear, however, that Stone Age people frequently used the powerful tool of fire. They used it to attract game, to herd and hunt game, to deflect predators, to provide warmth, and to encourage the growth of vegetation that would attract grazing animals like antelope and deer.

The impact of frequent and widespread fire on the environment is dramatic. Fire alters or destroys vegetation—from entire forests to vast grasslands (Figure 4.11). As can be observed in contemporary fires, the heat of the fire may reduce some vegetation to cinders, while other sorts can resist it. In those cases, fire can encourage the growth and survival of some species while eliminating others. Fire often destroys the vegetation that anchors the soil, however. Fires followed by heavy rains frequently lead to soil erosion and, in areas with a steep slope, even to the total denudation of the landscape. The use of fire by early Stone Age peoples certainly had many such impacts on the environment.

Archeologists believe that many large North American animal species disappeared around 11,000 years ago. At the end of the Pleistocene—the geologic and climatic age immediately preceding the one in which we now live—large, slow animals such as the mastodon, mammoth, cave bear, woolly rhino, and giant deer became extinct. What is so striking about this fact to archeologists is that these species constituted over two-thirds of the megafauna, or large animals, of the

Figure 4.11 Fire and its impacts on the landscape
Fires that are either allowed to burn or are unable to be brought under control can have a devastating impact on the landscape. Especially hot fires can burn not only the surface vegetation but also the organic material in the soil. The loss of these organic materials and other nutrients hinders the regeneration of vegetation. Without vegetation to anchor the soil, heavy rains can cause massive soil erosion. Shown in the photograph is Yellowstone National Park, the site of an extensive fire in 1988. Seven major, simultaneously occurring blazes burned nearly 800 thousand acres of the 2.2 million-acre park. The seven fires had different effects depending on their extent and the fuel conditions of the landscape itself. Portions of the park were severely burned, and complete vegetation recovery will take perhaps hundreds of years in some sections.

region. A great deal of controversy exists about why only these megafauna became extinct. Explanations such as climate change or large-scale natural disasters are not satisfying because neither is particularly selective.

One alternative explanation for the extinction of these megafauna is that, through their hunting and gathering practices, early Stone Age peoples were directly responsible. While it might seem implausible that such "primitive" hunting techniques could have such a devastating impact, consider the point made earlier that 2.5 square kilometers (one square mile) of vegetation and wildlife were adequate for the survival of only two people. As population sizes increased, more pressure was placed on animal populations to satisfy human food requirements. Those animals easiest to bring down were certainly the slowest and largest—the ones that could not escape quickly.

It is also the case that early Stone Age peoples had, over time, refined their killing technologies, particularly stone blades and spearheads. The double-edged Clovis point, for example, increased the likelihood of a kill rather than an injury to an animal. The **Clovis point** is a flaked, bifaced projectile whose length is more than twice its width. Used to kill large animals such as bison, the point is so named because archeologists found the projectiles (shown in Figure 4.12) in conjunction with kill sites in Clovis, New Mexico. Some Paleolithic hunters used the natural landscape to trap or kill large numbers of animals. Driving animals into canyons where they could be contained, or over cliffs where they would be killed en masse, ensured that huge numbers could be eliminated at one time (Figure 4.13, page 162). Other Paleolithic groups hunted small game, such as rabbits, using traps and small projectiles.

Clovis point: a flaked, bifaced projectile whose length is more than twice its width.

Figure 4.12 The Clovis point A sketch of the Clovis point is shown here. It is drawn from points actually found in sites in and around Clovis, New Mexico. Points were attached to wooden spears and used to hunt game. (*Source:* From *People of the Earth: An Introduction to World Prehistory,* 2nd Ed., by Brian M. Fagan. Copyright © 1977 by Little, Brown and Company, Inc. Reprinted by permission of Addison-Wesley Educational Publishers Inc.)

Neolithic Peoples and Domestication

The credit for the development of agriculture—a technological triumph with respect to nature—goes to the Neolithic peoples, also known as the late Stone Age peoples. While the divide between them and the Paleolithic peoples occurred about 10,000 years ago, the exact period is not known when Neolithic peoples shifted from hunting and gathering to cultivating certain plants and taming and herding wild animals. We have termed that period the First Agricultural Revolution (described in greater detail in Chapter 8). Climatically, we know that, for many regions of the globe, this period coincides with the end of the last Ice Age, which means that spring began slowly to occur in places that had not experienced it for thousands upon thousands of years.

It was just at this time that environmental conditions made possible the domestication of plants and animals, which also requires a sedentary lifestyle based in permanent settlements. The first domestication successes of the late Stone Age peoples were with the most docile animals (herbivores) and the hardiest plants (those with large seeds and a tolerance to drought). Once early domestication was established, it became possible for small groups of Neolithic peoples to cease to be nomads. As Chapter 2 discusses, permanent settlement then enabled further refinements in domestication. Eventually, Neolithic people achieved a truly dramatic innovation—the breeding of plants and animals to produce desired genetic characteristics such as disease resistance in plants.

The emergence of agriculture was an event that changed the course of human history, and had important environmental impacts—both negative and positive. One negative impact was the simplification of ecosystems as the multiplicity of wild species began to be replaced by fewer cultivable crops. An **ecosystem** is a community of different species interacting with each other and with the larger physical environment that surrounds it. Along with the vast number of wild species lost has gone the opportunity to understand their benefit to humans and the wider ecosystem. More positively, however, increased yields through greater control over available foodstuffs helped to improve human health and eventually increased population growth.

It is from this early period of plant domestication that we also find widespread evidence of a growing appreciation of nature through ritual, religion, and art. Human beings depended upon rain, soil fertility, and an abundance of sunlight to produce a successful harvest, and reverential attitudes toward nature appear to have been pervasive. In many places in both the Old and the New World, people worshiped Earth, Sun, Rain, and sacrificed deer and bear to them in an attempt to ensure survival.

Ecosystem: a community of different species interacting with each other and with the larger physical environment that surrounds it.

Early Settlements and Their Environmental Impacts

What is perhaps the most significant aspect of plant and animal domestication is that it eventually enabled a surplus to be produced. It also permitted the formation of human settlements where small groups—probably craftspeople and political and religious elites—were able to live off the surplus without being responsible for its production. Eventually, growing numbers of people, bolstered by increasing surpluses, were able to settle in places where water was available and the land cultivable.

Figure 4.13 Massive animal kills
Paleolithic hunters appear to have used features of the landscape to aid them in hunting large game. Archeologists believe that the mounds of skeletal remains of large animals found at various sites are evidence of this. It is not clear whether hunters and their kin were even able to consume all the animal flesh made available through such killing methods. It has been speculated that such gross killing methods may have led to the extinction of some species.

The invention of agricultural tools helped to further the domestication of plants and animals as well as multiply the early agriculturalists' impact on the landscape. Among the early tools that enabled humans a greater measure of control over nature were the sickle for harvesting wheat (Figure 4.14); the plow for preparing the soil; the yoke for harnessing draught animals (such as oxen) to pull the plow; and the wheel for grinding wheat, creating pottery, and later enabling transportation. Among other things, the wheel was also used as a pulley to draw water. In Sumer and Assyria, for example, the wheel enabled the development of large-scale irrigation systems.

Irrigation is one of the most significant ways that humans have been able to alter the limits of their environment. Throughout much of the world, in fact, agriculture could not occur without irrigation (Table 4.1). Following the success of the Fertile Crescent, agriculture diffused and new settlements emerged as a result. The food-producing minisystems of China, the Mediterranean, Mesoamerica, the Middle East, and Africa were sustained largely through irrigated agriculture. Yet, despite the existence of a vast irrigation network and a whole social structure bound up with agricultural production and attendant activities, the cities of the Mesopotamian region collapsed around 4,000 years ago. While there is no undisputed explanation for why this occurred, many researchers believe—based on archeological evidence—that it was due to environmental mismanagement. The irrigation works became clogged with salt, resulting in increasingly saline soils. To counteract the effect of salt on production, agriculturalists switched to barley, which is more salt-resistant than wheat but which ultimately resulted in a significant drop in yields. **Siltation** (the buildup of sand and clay in a natural or artificial waterway) associated with **deforestation** (the removal of trees from a forested area without adequate replanting) also occurred, filling up the deltas for nearly 200 miles. Eventually the canals filled with salt and the soils became too saline to support cultivation.

While it may seem to be a simple case of poorly informed management that led to the demise of Mesopotamian cities, increasingly saline soils currently plague agriculture in California and southwestern Arizona (Figure 4.15, page 164). And it was not only the Mesopotamians who made environmental mistakes. Other early urban civilizations, such as the Mayan in Central America and the Anasazi of Canyon de Chelly in Arizona, are also thought to have collapsed due to environmental mismanagement of water.

In the following section we examine the period of European expansion and globalization. Although many other important cultures and civilizations affected

TABLE 4.1 World Irrigated Area Since 1700	
Date (A.D.)	Area (in thousands of square kilometers/ square miles)
1700	50/19
1800	80/31
1900	480/185
1949	920/355
1959	1490/575
1980	2000/772
1981	2130/822
1984	2200/849

Source: W. Meyer, *Human Impact on the Earth,* 1990 (Cambridge: Cambridge University Press), p. 59; original source: B. G. Rozanov, V. Targulian, and D. S. Orlov, "Soils" in *The Earth Transformed by Human Action,* B. L. Turner II, W. L. Clark, R. W. Kates, J. F. Richards, J. T. Matthews, and W. B. Meyer (Eds.), 1990 (Cambridge: Cambridge University Press).

Siltation: the buildup of sand and clay in a natural or artificial waterway.

Deforestation: the removal of trees from a forested area without adequate replanting.

Figure 4.14 Wheat and flint sickle blade Perhaps the most significant factor in the spread of agriculture from Mesopotamia was the occurrence of hybrid forms of wheat, one of many wild grasses found in the area. Even before a fertile hybrid emerged, however, people were harvesting the wild forms. Sickle blades made from flint and set into a horn handle were the most common harvesting tools.

Figure 4.15 Irrigation system in the U.S. Southwest Much of the remaining cultivable land in the United States lies in dry areas such as the Southwest. Large-scale irrigation is required in these dry areas to sustain agricultural productivity. Under these circumstances, the application of water to crops is an expensive undertaking—it has to be pumped some distance and is subject to rapid evaporation during the dry, hot season of summer. Irrigation in the Southwest also contributes to the depletion of groundwater supplies. The systems that supply such irrigation are expensive to build and maintain. Much of the water delivered to the agricultural sector in the U.S. Southwest is heavily subsidized by the federal government in order to protect the producers from the negative impact that high water prices would have on the number of farms as well as overall productivity. The agricultural sector is often the largest user among all sectors (including residential, commercial, and government) in the Southwest.

the environment in the intervening period, the impacts of their technological developments were much the same as those we have already described. The period of European colonialism, however, had a profoundly different impact than preceding periods in extent, magnitude, and kind. Furthermore, it set the stage for the kinds of environmental problems contemporary society has inherited, perpetuated, and magnified.

European Expansion and Globalization

Europe provides a powerful example of how a society with different environmental attitudes was able to transform nature in ways vastly different from any in previous human history. These new attitudes drew from a newly emerging science and its contribution to technological innovation; the consolidation of the population around Judeo-Christian religious beliefs; and, most importantly, the development of a capitalist political and economic system.

Initially, European expansion was internal—largely contained within its continental boundaries. The most obvious reason for this expansion was population increase: from 36 million in 1000 to over 45 million in 1100, and over 60 million in 1200 to about 80 million by 1300 (Figure 4.16). As population continued to increase, more land was brought into cultivation. In addition, more forest land was cleared for agriculture, animals killed for food, and minerals and other resources exploited for a variety of needs. Forests originally covered upward of 90 percent of western and central Europe. At the end of the period of internal expansion, however, forested area was reduced to 20 percent.

The continental expansion of Europe ended around 1300. The bubonic plague, also known as the Black Death, had temporarily slowed population growth, while agricultural settlement had by then been extended to take up all readily available land, and then some. In England, Italy, France, and the Netherlands, for example, marshes and fens were drained, and the sea was pushed back or the water table lowered to reclaim and create new land for agriculture and settlement.

In the fifteenth century, Europe initiated its second phase of expansion—external—which changed not only the global political map but launched a 500-year period of environmental change that continues to this day. European external expansion—colonialism—was the response to several impulses, ranging from self-interest to altruism. Europeans were fast running out of land, and, as we saw in Chapter 2, "explorers" were being dispatched by monarchs to conquer new territories and enlarge their empires while collecting tax revenues from the monarch's new subjects. Many of these adventurous individuals were searching for fame and fortune or avoiding religious persecution. Behind European external expansion was also the Christian impulse to missionize and bring new souls into the kingdom of God. Other forces behind European colonialism included the need to expand the emerging system of trade, which ultimately meant increased wealth and power for a new class of people—the merchants—as well as the aristocracy.

Over the centuries, Europe came to control increasing areas of the globe. Two cases illustrate how the introduction of European people, ideologies, technologies, plant species, pathogens, and animals changed both the environments into which they were introduced as well as the societies they encountered.

Disease and Depopulation in the Spanish Colonies

Little disagreement exists among historians that the European colonization of the New World was eventually responsible for the greatest loss of human life in history. Very little disagreement exists that the primary factor responsible for that loss was disease. New World populations, isolated for millennia from the Old World, possessed immune systems that had never encountered some of the most common European diseases. **Virgin soil epidemics**—where the population at risk has no natural immunity or previous exposure to the disease within the lifetime of the oldest member of the group—were common in the so-called Columbian

Virgin soil epidemics: conditions in which the population at risk has no natural immunity or previous exposure to the disease within the lifetime of the oldest member of the group.

Figure 4.16 Growth in human numbers over the past 500,000 years The plot of the number of humans over the past half million years shows that the dramatic increase of population is a product of very recent times. Advances in health care have played an important role in helping more infants survive their first year of life as well as prolonging the lives of adults. (*Source:* A. Goudie, *The Human Impact on the Natural Environment,* 4th Ed. Cambridge: Blackwell, 1993, Fig. 1.2, p. 10.)

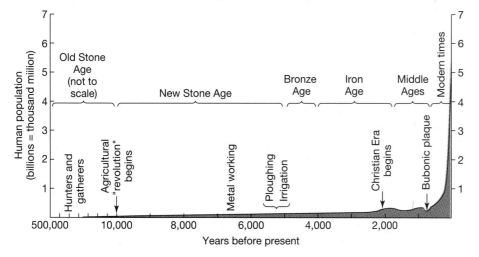

Columbian Exchange: interaction between the Old World, originating with the voyages of Columbus, and the New World.

Exchange, though the exchange in this case was mostly one-way. The **Columbian Exchange** is the interaction between the Old World (Europe) and the New World (the Americas) initiated by the voyages of Columbus. For example, diseases such as smallpox, measles, chicken pox, whooping cough, typhus, typhoid fever, bubonic plague, cholera, scarlet fever, malaria, yellow fever, diphtheria, influenza, and others were unknown in the pre-Columbian New World.

Geographer W. George Lovell has examined the role disease played in the depopulation of some of Spain's New World colonies from the point of initial contact until the early seventeenth century, using several cases to illustrate his point.[2] The first case is Hispaniola (present-day Haiti and the Dominican Republic), where Columbus's 1493 voyage probably brought influenza to the island through the introduction of European pigs carrying swine fever. Subsequent voyages brought smallpox and other diseases, which eventually led to the extinction of the island's Arawak population.

In a second example, in Central Mexico, Lovell writes of Hernán Cortés's contact with the Aztec capital of Tenochtitlán in the first decades of the sixteenth century, which led to a devastating outbreak of smallpox among a virgin soil population. A native Aztec text provides a graphic description of the disease:

> While the Spaniards were in Tlaxcala, a great plague broke out here in Tenochtitlán. It began to spread during the thirteenth month [September 30–October 19, 1520] and lasted for seventy days, striking everywhere in the city and killing a vast number of our people. Sores erupted on our faces, our breasts, our bellies; we were covered with agonizing sores from head to foot.
>
> The illness was so dreadful that no one could walk or move. The sick were so utterly helpless that they could only lie on their beds like corpses, unable to move their limbs or even their heads. They could not lie face down or roll from one side to the other. If they did move their bodies, they screamed with pain.
>
> A great many died from this plague, and many others died of hunger. They could not get up and search for food, and everyone else was too sick to care for them, so they starved to death in their beds.[3]

In a third example, Lovell also described the Jesuits' missionizing efforts in northern Mexico during a slightly later period. Because these efforts gathered dispersed groups of the population into single locations, conditions for the outbreak of disease were created. Contact with Spanish conquistadors in advance of the missionaries had already reduced native populations by perhaps 30 to 50 percent. When groups were confined to smaller areas oriented around a mission, mortality rates climbed to 90 percent of the total. Ironically, missionizing seems to have killed the Indians whose souls it was attempting to save. Eventually the disease was diffused beyond the initial area of contact as traders carried it across long distance trade routes to the periphery of the Mayan Empire in advance of the Spanish armies and missionaries. The Mayans were not defeated by European technological superiority but by the ravages of a new disease against which they possessed no natural defenses.

Demographic collapse: phenomenon of near genocide of native populations.

Lovell provides similar descriptions of disease impacts in Mayan Guatemala and the Central Andes of South America that led to devastating depopulation. The phenomenon of near genocide of native populations is called the **demographic col-**

---

[2]W. G. Lovell, "Heavy Shadows and Black Night": Disease and Depopulation in Colonial Spanish America, *Annals, Association of American Geographers*, 82, 1992.
[3]W. G. Lovell, 1992, p. 429, quoting from M. Leòn-Portilla, *The Broken Spears: The Aztec account of the conquest of Mexico*. Boston: Beacon Press, 1962, pp. 92–93.

lapse by scholars. The ecological effect of the population decline caused by the high rates of mortality was the transformation of many regions from productive agriculture to abandoned land. Many of the Andean terraces, for example, were abandoned, and dramatic soil erosion ensued. In contrast, large expanses of cleared land eventually returned to forests in areas such as the Yucatán, in present-day Mexico.

Old World Plants and Animals in the New World

A second case study of the environmental effects of European colonization involves the introduction of Old World plants and animals in the New World, and vice versa. The introduction of exotic plants and animals into new ecosystems is called **ecological imperialism,** a term now widely used by geographers, ecologists, and other scholars of the environment. The interaction between the Old and the New World resulted in both the intentional and unintentional introduction of new crops and animals on both sides of the Atlantic. Europeans brought from their homelands many plants and animals that were *exotics,* that is, unknown to American ecosystems. For example, the Spanish introduced wheat and sugarcane as well as horses, cattle, and pigs.

Ecological imperialism: introduction of exotic plants and animals into new ecosystems.

These introductions altered the environment, particularly as the emphasis on select species led to a reduction in the variety of plants and animals that constituted local ecosystems. Inadvertent introductions of hardy exotic species included rats, weeds such as the dandelion and thistle, and birds such as starlings, which crowded out the less hardy indigenous species. As with the human population, these indigenous populations of plants, birds, and animals had few defenses against European plant and animal diseases and were sometimes seriously reduced or made extinct through contact.

Contact between the Old and the New World was, however, an *exchange*—a two-way process—and New World crops and animals as well as pathogens were likewise introduced into the Old World, sometimes with devastating implications. Corn, potatoes, tobacco, cocoa, tomatoes, and cotton were all brought back to Europe; so was syphilis, which spread rapidly through the European population.

Contact between Europe and the rest of the world, though frequently violent and exploitative, was not uniformly disastrous. There are certainly examples where contact proved beneficial for both sides. The largely beneficial impacts of the Columbian Exchange were mostly knowledge-based or nutritional. Columbus's voyages (Figure 4.17, page 168) added dramatically to global knowledge, expanding understanding of geography, botany, zoology, and other rapidly growing sciences. It has been argued that the availability of American gold bullion and silver enlarged Europe's capacity for trade and may even have been indirectly responsible for creating the conditions that launched the Industrial Revolution.

The encounter also had significant nutritional impacts for both sides by bringing new plants to each. European colonization, although responsible for the extermination of hundreds of plant and animal species, was also responsible for increasing the types and amounts of foods available worldwide. It is estimated that the Columbian Exchange may have tripled the number of *cultivable* food plants in the New World. It certainly enabled new types of food to grow in abundance where they never had grown before, and it introduced animals as an important source of dietary protein. The advantages of having a large variety of food plants are several. For instance, if one crop fails, it is more than likely that another will succeed, as all plants are not subject to failure from the same set of environmental conditions.

The introduction of animals not only provided the New World with additional sources of protein but also with additional animal *power.* Before the

Figure 4.17 The voyages of Columbus, 1492–1502 These two maps show the Atlantic Ocean and the voyages of Columbus at the turn of the fifteenth century. Departing from Portugal and Spain, Columbus encountered several of the islands of the Caribbean as well as the coastal area of present-day Honduras and Venezuela. While the top map shows the tracks of the voyages across the Atlantic, the bottom one provides a closer view of the places where Columbus and his crews landed and made contact with indigenous peoples. (*Source:* From *The Penguin Atlas of the Diasporas* by Gerard Chaliand and Jean-Pierre Rageau, translated by A. M. Berrett. Translation copyright © 1995 by Gerard Chaliand and Jean-Pierre Rageau. Used by permission of Viking Penguin, a division of Penguin Books USA Inc.)

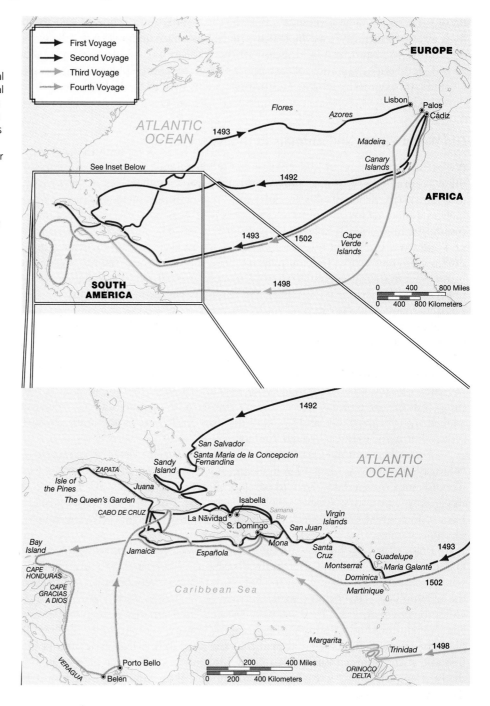

Columbian Exchange, the only important sources of animal energy were the llama and the dog. The introduction of the horse, the ox, and the ass created a virtual power revolution in the New World. These animals also, in their eventual death or slaughter, provided fibers, hides, and bones to make various tools, utensils, and a variety of coverings. Most significant in terms of its environmental impact, however, was the ox.

Land that had escaped cultivation because the indigenous digging sticks and tools were unable to penetrate the heavy soil and matted root surface became available to an ox-drawn plow. The result was that the indigenous form of intensive agricultural production (small area, many laborers) was replaced by extensive production (large area, fewer laborers). This transformation, however, was not entirely without negative impacts, such as soil destabilization and erosion.

It is also important to explore the impact of native New World peoples on their environment. The popular image of indigenous peoples living in harmony with nature, having only a minimal impact on their environment, has been shown to be flawed. In reality, different groups had very different impacts, and it is erroneous to conflate the thousands of groups into one romanticized caricature.

In New England, for example, prior to European contact, groups existed who hunted for wild game and gathered wild foods, as well as more sedentary types who lived in permanent and semipermanent villages, clearing and planting small areas of land. Hunter-gatherers were mobile, moving with the seasons to obtain fish, migrating birds, deer, wild berries, and plants. Agriculturalists planted corn, squash, beans, and tobacco, and used a wide range of other natural resources. The economy was a fairly simple one based on personal use or on barter (trading corn for fish, for example). The idea of a surplus was foreign here: People cultivated or exploited only as much land and resources as was needed to survive. Land and resources were shared in common, without concepts such as private property or land ownership. Fire was used to clear land for planting as well as for hunting. Although vegetation change did occur, it was minimal and not irreversible.

The relationship between some of the Indians of South and Central America and their environment was less harmonious, however. The Aztecs of Mexico and the Incas of Peru had developed a complex urban civilization dependent on dense populations employing intensive agricultural techniques. These groups were responsible for dramatic environmental modifications through cultivation techniques that included the irrigation of dry regions and the terracing of steep slopes. As we have seen, irrigation over several centuries resulted in the salinization of soils. In the lowland tropics, intensive agricultural practices resulted in widespread deforestation as people cut and set fire to patches of forest, planted crops, and then moved on when soil fertility declined. A surplus was key to the operations of both societies, as tribute by ordinary people to the political and religious elite was required in the form of food, animals, labor, or precious metals. The construction of the sizable Inca and Aztec Empires required the production of large amounts of building materials in the form of wood and mortar. Concentrated populations and the demands of urbanization meant that widespread environmental degradation existed prior to European contact (Figure 4.18).

Figure 4.18 Machu Picchu   Pictured is one of the most important sites of ancient Incan civilization located in highland Peru on the so-called *ceja de selva,* or eyebrow of the jungle. Machu Picchu was probably an important ceremonial center, but was also apparently a large-scale residential center as evidenced by the extensive agricultural terraces. Archeologists, geographers, and other scholars believe Machu Picchu was probably one of the last holdouts of fierce Incan resistance to the Spanish conquistadors. As the photograph suggests, natural terrain helped in the fortification of the city. Machu Picchu is one of the best preserved ruins of the Incan empire.

Human Action and Recent Environment Change

No other transition in human history has had the impact on the natural world that industrialization has. When we couple *industrialization* with its frequent companion *urbanization,* we have the two processes that, more than any others, have revolutionized human life and effected far-reaching ecological changes. For the first time in history these changes have moved beyond a local or a regional scale to affect the entire globe. In this section we will explore some of the dramatic contemporary environmental impacts that industrial technology and urbanization have produced. In doing so, we highlight the two issues most central to environmental geography today: energy use and land use change.

The Impact of Energy Needs on the Environment

Certainly the most central and significant technological breakthrough of the Industrial Revolution was the discovery and utilization of fossil fuels: coal, oil, and natural gas. Although the very first factories in Europe and the United States relied on waterpower to drive the machinery, it was hydrocarbon fuels that

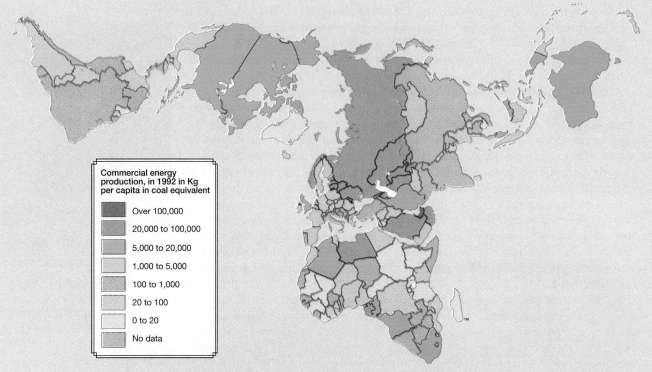

Figure 4.19 World production and consumption of energy, 1995 These paired maps provide a picture of the uneven distribution of the production and consumption of energy resources around the world. The United States is the largest producer and consumer of a range of energy resources. Notice that although the Middle East and North African countries as well as Nigeria are important producers of energy resources, their consumption (as well as that of the rest of the continent, excluding South Africa) is very low. Japan produces a negligible amount of the total of world energy

Legend (map inset):

Commercial energy production, in 1992 in Kg per capita in coal equivalent

- Over 100,000
- 20,000 to 100,000
- 5,000 to 20,000
- 1,000 to 5,000
- 100 to 1,000
- 20 to 100
- 0 to 20
- No data

provided a more constant, dependable, and effective source of power. A steady increase in power production and demand since the beginning of the Industrial Revolution has been paralleled, not surprisingly, by an increase in resource extraction and conversion.

At present, the world's population relies most heavily for its energy needs on nonrenewable energy resources, which include fossil fuels and nuclear ones, as well as renewable resources such as solar, hydroelectric, wind, and geothermal power. Fossil fuels are derived from organic materials and are burned directly to produce heat. Nuclear energy originates with isotopes, which emit radiation. Most commercial nuclear energy is produced in reactors fueled by uranium. Renewable sources of energy such as the sun, wind, water, and steam are captured in various ways and used to drive pumps, machines, and electricity generators.

The largest proportion of the world's current consumption of energy resources, 35 percent, is from oil; 24 percent is from coal; 18 percent from gas; 6 percent from hydropower (largely from dams); 5 percent from nuclear power; and 12 percent from biomass (which includes wood, charcoal, crop waste, and dung). The production and consumption of these available resources, however, is geographically uneven, as Figure 4.19 shows. Fifty percent of the world's oil supplies are from the Middle East, and most of the coal is from the Northern Hemisphere, mainly from the United States, China, and Russia. Nuclear reactors

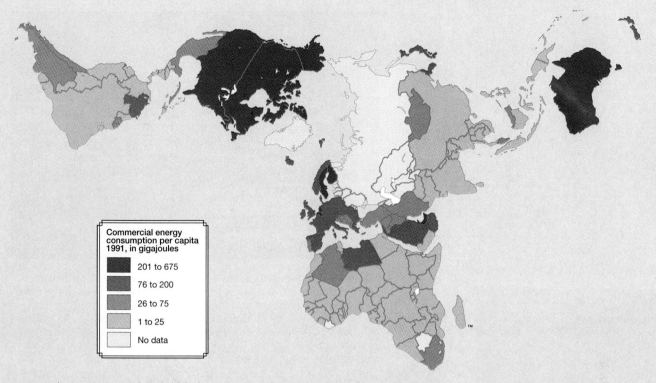

resources but consumes a relatively high share. (*Sources:* Right map, reprinted with permission of Prentice Hall, from E. F. Bergman, *Human Geography: Cultures, Connections, and Landscapes,* © 1995, p. 395. Data from the World Resources Institute, *World Resources 1994–95.* New York: Oxford University Press, 1994, pp. 334–35. Map projections, Buckminster Fuller Institute and Dymaxion Map Design, Santa Barbara, CA. The word Dymaxion and the Fuller Projection Dymaxion™ Map design are trademarks of the Buckminster Fuller Institute, Santa Barbara, California, ©1938, 1967 & 1992. All rights reserved.)

are a phenomenon of the core regions of the world. For example, France generates 90 percent of its electricity from nuclear sources.

The consumption side of energy also varies geographically. It has been estimated that at present annual world energy consumption is equal to what it took about one million years to produce naturally. In one year, global energy consumption is equal to about 1.3 billion tons of coal. What is most remarkable is that this is four times what the global population consumed in 1950, and 20 times what it consumed in 1850. And, as the $I = PAT$ formula suggests, the affluent core regions of the world far outstrip the peripheral regions in terms of energy consumption. With nearly four times the population of the core regions, however, the peripheral regions consume less than one-third of global energy expenditures. Yet consumption of energy in the peripheral regions is rising quite rapidly as globalization spreads industries, energy-intensive consumer products such as automobiles, and energy-intensive agricultural practices into regions of the world where they were previously unknown. It is projected that by the early decades of the twenty-first century, the peripheral regions will become the dominant consumers of energy (Figure 4.20, page 172).

Most important for our discussion, however, is the fact that every stage of the energy conversion process—from discovery, to extraction, processing, and utilization—has an impact on the physical landscape. In the coalfields of the world, from the U.S. Appalachian Mountains to western Siberia, mining results in a loss

Figure 4.20 Traffic in Manila, the Philippines—In urban areas of the Philippines, as in many other parts of Asia such as Bangkok and Djakarta, people spend many hours each day battling traffic. In Manila, it can take up to three hours to travel from one section of the city to another because of traffic congestion. An additional outcome of the increasing use of the automobile in the periphery is rising rates of urban air pollution. In Mexico City, air pollution from motor vehicles is so severe that respiratory ailments are increasing.

Figure 4.21 Coal mining This coal-mining operation in western Germany is typical of the surface mining technologies used to exploit shallow deposits in most parts of the world. After deposits are located, the vegetation and overburden (rocks and dirt overlaying the coal seam) are removed by bulldozers and discarded as spoil (or waste material). With the coal seam exposed, heavy equipment (such as the type pictured) is used to mine the deposit. Some countries, such as the United States, require restoration of newly mined landscapes. Sites exploited before these laws were introduced remain unrestored, however. Successfully restored surface mining sites make it difficult to tell that the area was once a mining site. Unfortunately, many such sites are in arid or semiarid areas where soil and climate prevent full restoration. In addition to substantial land disturbance, the mining and processing of coal resources often cause soil erosion as well as water and air pollution.

of vegetation and topsoil, erosion, and water pollution, acid and toxic drainage. It also contributes to cancer and lung disease in coal miners (Figure 4.21). The burning of coal is associated with relatively high emissions of environmentally harmful gases such as carbon dioxide and sulfur dioxide.

The burning of home heating oil, along with the use of petroleum products for fuel in internal combustion engines, launches harmful chemicals into Earth's atmosphere causing air pollution and related health problems. The production and transport of oil have resulted in oil spills and substantial pollution to water and ecosystems. Media images of damage to sea birds and mammals after tankers have run aground and spilled oil have shown how immediate the environmental damage can be. Indeed, the oceans are acutely affected by the widespread use of oil for energy purposes. Figure 4.22 shows how many millions of gallons of oil are spilled into the world's oceans each year. Oil drilling can also have profound environmental consequences in the form of well explosions and fires. Although by no means a common occurrence, the oil fires that occurred in the aftermath of the Persian Gulf War (see Figure 4.23) are a dramatic illustration of the impact of oil production and consumption on air quality.

Natural gas is one of the least noxious of the hydrocarbon-based energy resources because it is converted relatively cleanly. Now supplying nearly one-quarter of global commercial energy, natural gas is predicted to be the fastest growing energy source in the new century. Reserves are still being discovered, with Russia holding the largest amount—about one-third of the world's total. While regarded as a preferred alternative to oil and coal, natural gas is not produced or consumed without environmental impacts. The risk of explosions at natural gas conversion facilities is significant; leakages and losses of gas from distribution systems contribute to the deterioration of Earth's atmosphere.

At the midpoint of the twentieth century nuclear energy for civilian use was widely promoted as a clearly preferable alternative to fossil fuels. It was seen as the answer to the expanding energy needs of core countries, especially as the supply of uranium worldwide was thought to be more than adequate for cen-

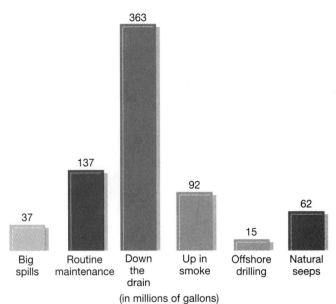

Figure 4.22 Oil poured into the oceans worldwide, 1996 Although the international press is quick to report the disastrous oil spills into the world's oceans, most of the hundreds of millions of gallons of oil that enter the seas every year come from nonaccidental sources. As this graph shows, the largest contributor to oil pollution of the oceans comes from the careless pouring of oil down municipal storm drains and the runoff from municipal and industrial wastes. Only about five percent of oil pollution in oceans is due to major tanker accidents. Tanker spills are dramatic because huge amounts of oil end up affecting fairly small areas.

turies of use. Nuclear energy was also regarded as cleaner and more efficient than fossil fuels. Although nuclear war was a pervasive threat, and there were certainly critics of nuclear energy even in the early years of its development, the civilian "atomic age" was widely seen as a triumphant technological solution to the energy needs of an expanding global economic system. It was not until serious accidents at nuclear power plants began to occur—such as at Windscale in

Figure 4.23 Smoke from the Kuwait oil fires, 1991 Following the end of the Persian Gulf War in the spring of 1991, 732 oils wells were set ablaze in Kuwait by the retreating Iraqi Army. By early summer some 550 wells were still burning. The smoke from the fires, many of which are still burning, is largely contained within the Gulf region. In addition to setting wells ablaze, Iraqi troops also released oil into the Persian Gulf. The Persian Gulf War is seen by many observers as one of the worst ecological disasters in history.

Britain and Three Mile Island in the United States—that the voices of concerned scientists and citizens began to be heard. These voices described, with incontestable evidence, the questionable aspects of nuclear energy production, such as nuclear reactor safety and the disposal of nuclear waste. The case of the meltdown of the Chernobyl nuclear reactor in 1986 illustrates both of these issues (see Feature 4.1, "Geography Matters—Chernobyl and Its Enduring Human and Environmental Impacts"). Since these accidents, many core countries have drastically reduced or eliminated their reliance on nuclear energy. Sweden, for example, has committed to eliminate its reliance on nuclear power—from which 50 percent of its current electricity use is derived—by the year 2010.

Interestingly, while the majority of core countries have begun to move away from nuclear energy because of the possibility of environmental disaster in the absence of fail-safe nuclear reactors, a few peripheral—and especially populous—countries are moving in the opposite direction (Figure 4.24). India, South Korea and China each have fledgling nuclear energy programs. So far, no accidents have been associated with nuclear energy production in the periphery.

While nuclear power problems are largely confined to the core, the periphery is not without its energy-related environmental problems. Since a large proportion of populations in the periphery rely on wood for their energy needs, as the populations have grown, so too has the demand for fuelwood. One of the most

Figure 4.24 World distribution of nuclear power, 1994   As this world map shows, most of the dependence on nuclear power is concentrated in core countries. The continents of South America and Africa contain only five nuclear reactors out of 420 reactors worldwide in 1994. Whereas some peripheral countries such as India are enthusiastic about further developing their nuclear energy production, core countries such as Sweden are phasing out dependence on nuclear power. Australia, where there is a very strong antinuclear movement, is one of the few core countries to have rejected nuclear power altogether. (*Sources:* Reprinted with permission of Prentice Hall, from J. M. Rubenstein, *The Cultural Landscape: An Introduction to Human Geography,* 5th Ed., © 1996, p. 576. Map projection, Buckminster Fuller Institute and Dymaxion Map Design, Santa Barbara, CA. The word Dymaxion and the Fuller Projection Dymaxion™ Map design are trademarks of the Buckminster Fuller Institute, Santa Barbara, California, ©1938, 1967 & 1992. All rights reserved.)

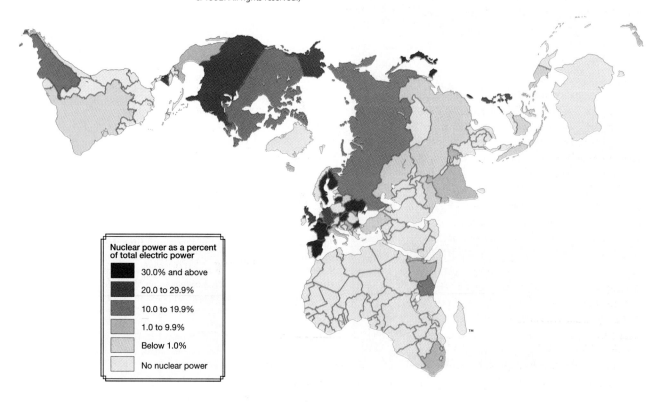

Nuclear power as a percent of total electric power

- 30.0% and above
- 20.0 to 29.9%
- 10.0 to 19.9%
- 1.0 to 9.9%
- Below 1.0%
- No nuclear power

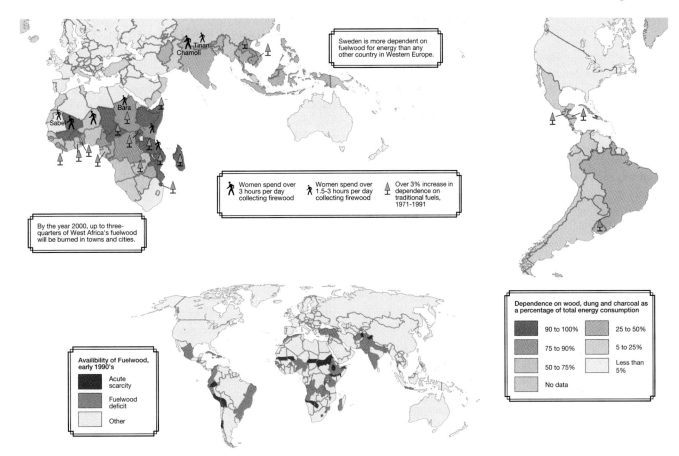

Figure 4.25 World dependence on wood, charcoal, or dung, 1991 Firewood, charcoal, and dung are considered traditional fuels, and although their availability is decreasing, dependence upon them is increasing. Dependence on traditional sources of fuel is especially high in the periphery where, in Africa, for example, they are the most important energy source for cooking and heating. Wood and charcoal, although renewable sources, are replenished at a very slow rate. Acute scarcity will be a certainty for most African households in the twenty-first century. (*Source:* Reprinted with the permission of Simon & Schuster from *The New State of World Atlas,* 4th Ed., by M. Kidron and R. Segal. Copyright © 1991 by Michael Kidron and Ronald Segal. Maps and Graphics copyright © 1991 by Swanston Publishing Limited.)

immediate environmental impacts of wood burning is air pollution. But the most alarming environmental problem related to wood burning is the rapid depletion of forest resources. With the other conventional sources of energy (coal, oil, and gas) being too costly or unavailable to most peripheral households, wood is the only alternative. The demand for fuelwood has been so great in many peripheral regions that forest reserves are being rapidly used up (Figure 4.25).

Fuelwood depletion is extreme in the highland areas of Nepal as well as in Andean Bolivia and Peru. The clearing of forests for fuelwood in these regions has lead to serious steep-slope soil erosion. In sub-Saharan Africa, where 90 percent of the region's energy needs is derived from wood, overcutting of the forests has resulted especially in denuded areas around rapidly growing cities. And although wood gathering is usually associated with rural life, it is not uncommon for city dwellers to use wood to satisfy their household energy needs as well. For example, in Niamey, the capital of Niger, the zone of overcutting is gradually expanding as the city itself expands. It is now estimated that city dwellers in Niamey must travel from 50 to 100 kilometers (31 to 62 miles) to gather wood. The same goes for inhabitants of Ouagadougou in Burkina Faso, where the average haul for wood is also over 50 kilometers.

Hydroelectric power was also once seen as a preferred alternative to the more obviously environmentally polluting fossil fuel sources. It is no exaggeration to

4.1 Geography Matters

Chernobyl and Its Enduring Human and Environmental Impacts

Geographer Peter Gould, in a short but important book called *Fire in the Rain,* provides a dramatic account of the greatest nuclear disaster the world has ever known. As he puts it:

> At 1 hour, 23 minutes, and 43 seconds after midnight on April 26, 1986, Reactor 4 at Chernobyl went into a soaring and uncontrollable chain reaction. Two seconds later the resulting steam explosion to the concrete housing blew the thousand ton "safety" cover off the top of the reactor, and spewed radioactive materials high into the night sky equal to all the atomic tests ever conducted above ground. (p. 2)

Although designed with what most nuclear experts agree were adequate safety standards and computer and human monitoring systems, the reactor failed. Scientists running a routine electrical test on one of the safety backup systems got into trouble and, as a result, overrode a series of safety regulations, leading to the explosion. While Gould tells, in gripping detail, the sequence of events that led to the meltdown of Reactor 4, what is more central to our discussion is the impact that this technological accident had on people and the environment in the region immediately surrounding the nuclear facility, in the former Soviet Union more generally, in Europe, and globally (Figure 4.1.1).

The stories of radiation sickness and eventual ghastly deaths of the facility workers, firefighters, medical personnel, and other volunteers filled the international newspapers for weeks following the meltdown (see Table 4.1.1). Less graphic and less well remembered are the invisible and enduring impacts on the population of the area surrounding the power plant as well as the natural environment.

Radiation particles entered the soil, the vegetation, the human population, and the rivers, effectively contaminating the entire food chain of the region. Secondary radiation was also a problem. In the immediate area surrounding Chernobyl, all the trees were contaminated. As the trees slowly die, and rot, and decay, radioactive material enters the physical system as "hot" nutrients. Over 3000 square kilometers (1161 square miles) of trees were brown from radiation immediately following the acci-

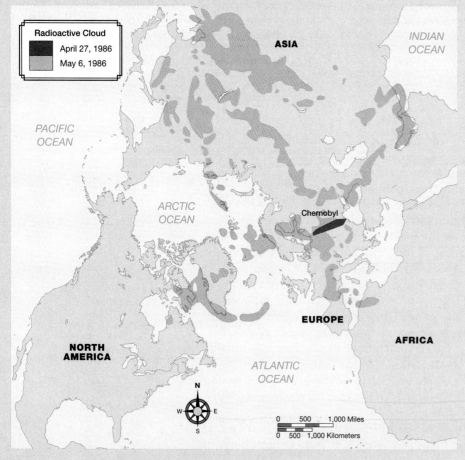

Figure 4.1.1 The geography of the initial radioactive cloud from Chernobyl While the map shows where the radioactive cloud drifted in the nine days following the accident, what is less well known is the impacts on animals months and years later. For example, 20 months after the accident, radioactive rain fell on the penguins of Antarctica. Thousands of reindeer in Lapland had to be destroyed because they were so highly contaminated. In the Lake District in England, sheep were being destroyed as late as 1994 because they were too radioactive to be sold.

TABLE 4.1.1 Grim consequences

Thirty-one people died shortly after the accident; 259 people were hospitalized with radiation sickness.

Over 650,000 military and civilian rescue and recovery workers were exposed to dangerous levels of radiation.

Over a half a million people were exposed to the radioactivity from the accident.

Four million people from in and around the area as well as northern Europe may suffer health effects from the accident.

Most children in Belarus are suffering from some form of immune deficiency disease.

The total cost of the accident will reach at least $358 billion.

dent, and it is still not clear how to decontaminate such a large area. The town of Chernobyl itself as well as the surrounding area will not be inhabitable for decades, if then (Figure 4.1.2).

An account by an American who visited Chernobyl three years after the accident says it all:

> Viktor is a haggard young engineer who was one of two people with primary responsibility for the containment of Reactor 4 after the explosion in 1986. In 60 seconds on the radioactive roof he was exposed to so [much radiation] that he cannot safely be exposed to [any] more until 1991.

Figure 4.1.2 Living in the postaccident landscape This photograph provides a sense of the immediate impact of the accident on the human landscape in and around Chernobyl immediately following as well as several years afterward. Houses were abandoned as over 100,000 residents were immediately evacuated and another 100,000 plus were evacuated several years later.

His skin turned black and came off his body. His face looks like a death mask. On a sunny March day he sent a bus for some of us staying in Kiev, two hours away. As we neared Chernobyl we sped over a road that is as clean as glass, for it is washed every other day. It goes mile after mile through deserted villages and farms and bleak gray land that has been cleared of every sign of plant and animal life. The only movement comes from an occasional curtain blowing in an empty window. The thirty square kilometer area (11.6 square miles) that is Chernobyl itself is fenced and heavily guarded. From inside the gate the landscape looks like the outside—bare, flat and dotted with deserted houses and buildings. This had been a prosperous farming area until the worst nuclear disaster in history. . . .

In this tragic wasteland, we are honored at a banquet (after washing in cold water and being scanned by Geiger counters). We filed into a formal dining room [and] after four courses and many non-alcoholic toasts to US-Soviet cooperation and peace, I asked the chief scientist how long until another accident occurs like this somewhere in the world? "Ten years at most," was the answer. We didn't make any jokes on our way back to Kiev . . . (p. 29)

Ten years after the accident, it is clear that the hardest hit is the newly independent republic of Belarus. Here 25 percent of the land is considered uninhabitable. Thousands of villages have been abandoned. The government indicates that it has spent 15 percent of its gross national product—more than $235 billion dollars over the last decade—on paying the medical bills and resettling tens of thousands of people. Despite the enormous economic and environmental costs, however, most people in southern Belarus are still reeling, 10 years later, from the psychological effects. As a recent *New York Times* journalist reported:

> People have become paralyzed with fear. They are afraid to move, afraid to stay, afraid to marry and afraid to have families. All normal life stopped here simply because there was a strong northerly wind on April 26, 1986.

Sources: Adapted from P. Gould, *Fire in the Rain: The Democratic Consequences of Chernobyl*. Cambridge: Polity Press, 1990, and M. Specter, "Ten Years Later, Through Fear, Chernobyl Still Kills in Belarus," *New York Times*, Sunday, March 31, 1996, pp. 1 and 4.

state that the wave of dam building that occurred throughout the world over the course of the twentieth century has improved the overall availability, quality, cost, and dependability of energy (Figure 4.26). Unfortunately, however, dams built to provide hydroelectric power (as well as irrigation, navigation, and drinking water supplies) for the burgeoning cities of the core and to encourage economic development in the periphery and semiperiphery have also had profound negative environmental impacts. Among the most significant of these impacts are changes in downstream flow, evaporation, sediment transport and deposition, mineral quality and soil moisture, channeling and bank scouring, aquatic biota and flora, as well as human health. Furthermore, the construction of dams also dramatically alters the surrounding terrain, often with serious consequences. For example, clearance of the forest for dam construction often leads to large-scale flooding. The felled trees are usually left to decay in the impounded waters, which become increasingly acidic. The impounded waters can also incubate mosquitoes, which carry diseases such as malaria. The remedies for such problems are difficult to determine, and many argue that new dam projects should not be undertaken without a clear sense of the complex of indirect social and environmental costs (Figure 4.27).

Figure 4.26 World distribution of hydropower, 1992   Although the great dam-building era for core countries is now largely completed, many peripheral countries, in a bid to participate more actively in the world economy, are building dams. Only a few countries are almost exclusively dependent on the hydropower produced from dams. These include Norway, Nepal, Zambia, Ghana, Paraguay, and Costa Rica. While the power produced by dams is environmentally benign, the construction of large dams can be extremely destructive of the environment and can dislocate large numbers of people. (*Source:* J. Seager, *The New State of the Earth Atlas*, 2nd Ed. New York: Simon & Schuster, 1995, p. 44–45.)

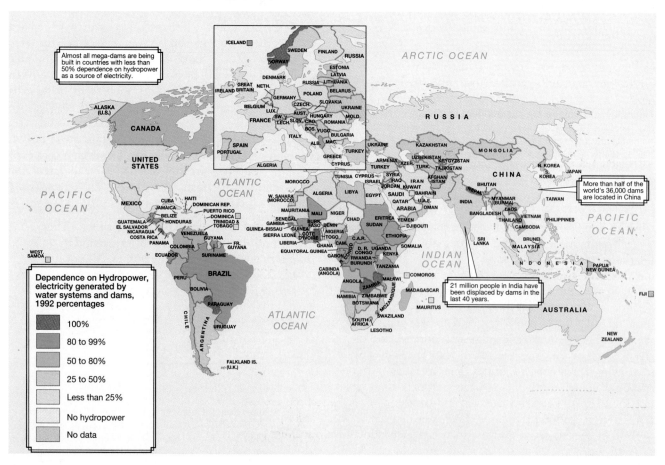

Figure 4.27 Aswan High Dam, Egypt
Completed in 1970 at a cost of $1 million, the Aswan High Dam was a significant engineering feat as well as an important symbol of Egypt's important bid for economic independence. The dam is of rock-filled construction and is 111 meters (364 feet) high. The impoundment of water caused by the dam flooded out numerous settlements along the river, creating the need for the resettlement of tens of thousands of people in both Egypt and Sudan. In addition to its human impacts, the dam also affected the natural fertilizing processes of the Nile River and flooded out the site of one important temple while restoring another to the open air.

One of the reasons hydroelectric power continues to be appealing, however, is that it produces few atmospheric pollutants as compared to fossil fuels. Indeed, coal and gas power stations as well as factories, automobiles, and other forms of transportation are largely responsible for the increasingly acidic quality of Earth's atmosphere. While people as well as other organisms naturally produce many gases, including oxygen and carbon dioxide, increasing levels of industrialization and motor vehicle use have destabilized the natural balance of such gases, leading to serious atmospheric pollution. Increasing the level of acids in the atmosphere are sulfur dioxide, nitrogen oxides, and hydrocarbons, among other gases, released into the atmosphere from motor vehicle exhaust, industrial processes, and power generation (based on fossil fuels). If these gases reach sufficient concentrations and are not effectively dispersed in the atmosphere, acid rain can result.

Acid rain is the wet deposition of acids upon Earth created by the natural cleansing properties of the atmosphere. Acid rain occurs as the water droplets in clouds absorb certain gases, which later fall back to Earth as acid precipitation. Also included under the term acid rain are acid mists, acid fogs, and smog. The effects of acid rain are widespread. Throughout much of the Northern Hemisphere, for example, forests are being poisoned and killed, and soils are becoming too acidic to support plant life. Lakes are becoming acidic in North America and Scandinavia. In urban areas, acid rain is corroding marble and limestone buildings such as the Parthenon in Athens and St. Paul's Cathedral in London as well as other historic structures in Europe. Figure 4.28 (see page 180) illustrates the global problem of acid emissions and rain in the early 1990s.

Before giving up all hope that the use of energy can ever be anything but detrimental to the environment, it is important to realize that alternatives exist to fossil fuels, hydroelectric power, and nuclear energy. Energy derived from the sun, the wind, Earth's interior (geothermal sources), and the tides has been found to be clean, profitable, and dependable. Japan, the United States, and Germany all have solar energy production facilities that have proven to be cheap and nonpolluting. Although contributing only small amounts to the overall energy supply, the production of energy from geothermal and wind sources has also been successful in a few locations around the globe. Italy, Germany, the United States, Mexico, and the Philippines all derive some of their energy production from either geothermal or wind sources.

Acid rain: the wet deposition of acids upon Earth created by the natural cleansing properties of the atmosphere.

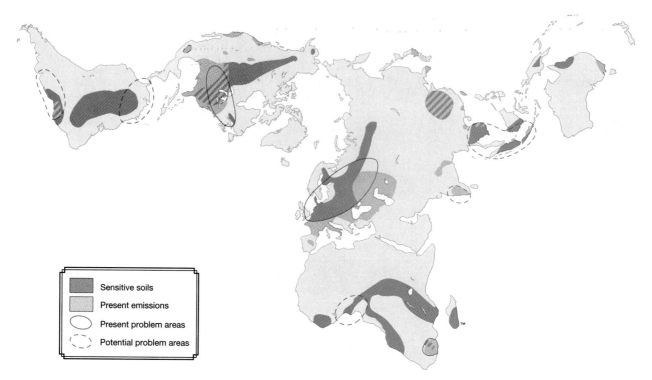

Figure 4.28 Global acid emissions, 1990 Acid emissions affect a variety of elements of the natural and the built environment. In some parts of the world, the damage to sensitive soils is especially severe. In others, acid emissions cause serious air pollution. Lakes and rivers are also affected by acid emissions resulting in fish and other wildlife kills. A large portion of acid-producing chemicals may be generated in one place but exported to another by prevailing winds. More than three-quarters of the acid deposition in Norway, Switzerland, Austria, Sweden, the Netherlands, and Finland is blown in from western and eastern Europe. (*Source:* K. Pickering and L. Owen, *An Introduction to Global Issues.* New York: Routledge, 1994, p. 124. Map projection, Buckminster Fuller Institute and Dymaxion Map Design, Santa Barbara, CA. The word Dymaxion and the Fuller Projection Dymaxion™ Map design are trademarks of the Buckminster Fuller Institute, Santa Barbara, California, ©1938, 1967 & 1992. All rights reserved.)

Monies to support the development of geothermal, wind, and tidal energy have been scarce, however, due to the lobby pressures of oil and gas companies as well as other political factors. While viable alternatives exist to traditional energy sources, the further development of these alternatives is likely to hinge on future political and economic factors.

Impacts on the Environment of Land Use Change

In addition to industrial pollution and steadily increasing demands for energy, the environment is also being dramatically affected by pressures on the land. The clearing of land for fuel, farming, grazing, resource extraction, highway building, energy generation, and war all have significant impacts on land. Land may be classified into five categories: forest, cultivated land, grassland, wetland, and areas of settlement. Geographers understand land use change as occurring in either one of two ways, conversion or modification. *Conversion* is the wholesale transformation of land from one use to another (for example, the conversion of forest land to settlement). *Modification* is an alteration of existing cover (for example, when a grassland is overlaid with railroad line or when a forest is thinned and not clear-cut). As human populations have increased and the need for land for settlement and cultivation has also increased, changes to the land have followed.

One of the most dramatic impacts of humans upon the environment is loss or alteration of forest cover on the planet as it has been cleared for millennia to make way for cultivation and settlement. Not only must forests be cleared to provide land to accommodate increases in human numbers, they are also exploited for the vast resources they contain. The approximate chronology and estimated extent of the clearing of the world's forests since preagricultural times is portrayed in Table 4.2. What the table shows is that the forested area of the world has been reduced by about 8 million square kilometers (about 3 million square miles) since preagricultural times. Rapid clearance of the world's forests has occurred either through logging, settlement, and agricultural clearing or through fuelwood cutting around urban areas.

The permanent clearing and destruction of forests is known as _deforestation_, and it is currently occurring most alarmingly in the world's rain forests. Figure 4.29 (see page 182) shows the global extent of deforestation. The United Nations Food and Agricultural Organization has estimated that rain forests globally are being destroyed at the rate of one acre (0.40 hectare) per second.

Today, rain forests cover less than 7 percent of the land surface, half of what they covered only a few thousand years ago. Destruction of the rain forests, however, is not just about the loss of trees, a renewable resource that is being eliminated more quickly than it can be regenerated. It is also about the loss of the biological diversity of an ecosystem, which translates into the potential loss of biological compounds that may have great medical value. The destruction of the rain forests is also about the destabilizing of the oxygen and carbon dioxide cycles of the forests, which may have long-term effects on global climate. Much of the destruction of the South American rain forests is the result of peripheral

TABLE 4.2 Estimated area cleared (x000km²)

Region or Country		Pre-1650	1650–1749	1750–1849	1850–1978	Total High Estimate	Total Low Estimate
North America		6	80	380	641	1,107	1,107
	H	18				288	
Central America	L	12	30	40	200		282
	H	18				925	
Latin America	L	12	100	170	637		919
	H	6	6	6		380	
Oceania	L	2	4		362		374
	H	70	180	270	575	1,095	
USSR	L	42	130	250			997
	H	204	66	146		497	
Europe	L	176	54	186	81		497
	H	974	216	596		3,006	
Asia	L	640	176	606	1,220		2,642
	H	226	80	-16	469	759	
	L	96	24	42			631
Total highest		1,522	758	1,592	4,185	8,057	
Total lowest		986	598	1,680	4,185		7,449

Source: B. L. Turner II, W. C. Clark, R. W. Kates, J. F. Richards, J. T. Mathews, and William B. Meyer, _The Earth As Transformed by Human Action: Global and Regional Changes in the Biosphere over the Past 300 Years,_ Cambridge: Cambridge University Press, 1990, p. 180.

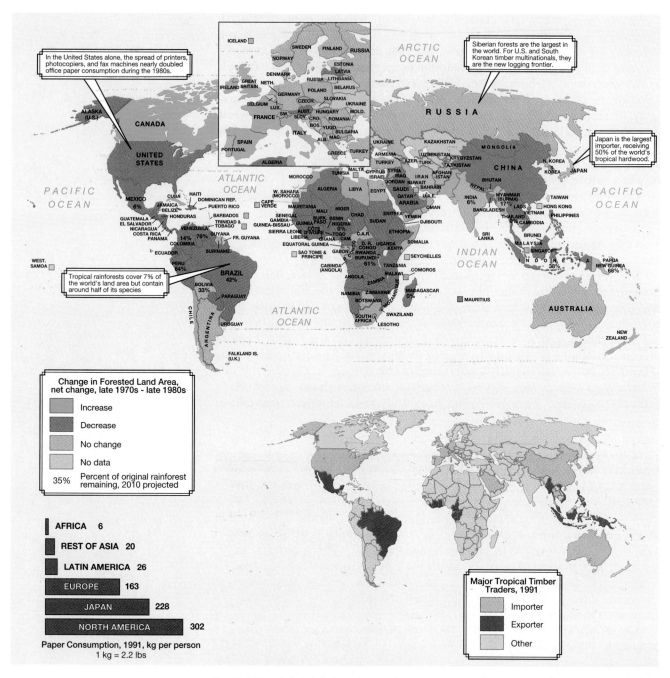

Figure 4.29 Global deforestation, 1991 The world's forests are disappearing or being reduced or degraded everywhere but especially in tropical countries. Since agriculture emerged about 10,000 years ago, human activities have diminished the world's forest resources by about one-quarter. Whereas forests once occupied about one-third of Earth's surface, they now take up about one-quarter. Playing an important role in the global ecosystem, they filter air and noise pollution; provide a habitat for wildlife; and slow down water runoff, helping to recharge streams and groundwater. They also influence climate at local, regional, and global levels. (*Source:* J. Seager, *The New State of the Earth Atlas*, 2nd Ed. New York: Simon & Schuster, 1995, p. 72–73.)

countries' attempts at economic development. Figure 4.30 illustrates this point with reference to the Bolivian Amazon rain forest. The introduction of coca production has become an important source of revenue for farmers in the region and has led to the removal of small tracts of the forest. Still, removal of the rain forest for agricultural production in Bolivia is minimal when compared to other South American countries such as Brazil and Colombia.

Figure 4.30 Coca production in the Chapare region, Bolivia The Bolivian portion of the Amazon remains relatively intact as compared with nearby Brazil. Two factors, however, may diminish its relatively pristine state. The first is the increased logging of hardwoods. The second is the production of coca for export as cocaine. Shown in this photograph are coca leaves drying on a white cloth in a cleared-out area of the rain forest. Coca is Bolivia's chief export and accounts for over half a million jobs in a largely subsistence economy. In the Chapare region there are over 100,000 hectares (247,100 acres) of coca under production.

Great geographical variability exists with respect to human impacts on the world's forests. For most of the core regions, net clearance of the forests has been replaced by regeneration. Yet for most of the periphery, clearance has accelerated to such an extent that one estimate shows a 50 percent reduction in the amount of forest cover since the early 1900s.

Cultivation is another important component of global land use, which we deal with extensively in Chapter 8. However, one or two additional points about the environmental impacts of cultivation are pertinent here. During the past 300 years, the land devoted to cultivation has expanded globally by 450 percent. In 1700, the global stock of land in cultivation took up an area about the size of Argentina. Today, it occupies an area roughly the size of the entire continent of South America. While the most rapid expansion of cropland since the mid-twentieth century has occurred in the peripheral regions, the amount of cropland has either held steady or been reduced in core regions. The expansion of cropland in peripheral regions is partly a response to growing populations and rising levels of consumption worldwide. It is also partly due to the globalization of agriculture (see Chapter 8), with some core-region production having been moved to peripheral regions. The reduction of cropland in some core regions is partly a result of this globalization, and partly the result of a more intensive use of cropland, utilizing more fertilizers, pesticides, and farm machinery, and new crop strains.

Grasslands are also used productively the world over, either as rangeland or pasture. In both cases, the land is used for livestock grazing. As Figure 4.31 shows, most grassland occurs in arid and semiarid regions which are unsuitable for farming, either because of lack of water or poor soils. Some grasslands, however, occur in more rainy regions where tropical rain forests have been removed and replaced by grasslands. Approximately 68 million square kilometers (26 million square miles) of the land surface is currently taken up by grasslands.

Human impacts on grasslands are largely of two sorts. The first has been the clearing of grasslands for other uses, most frequently settlement. As the global demands on beef production have increased, so has the intensity of use of the world's grasslands. Widespread overgrazing of grasslands has led to acute degradation of the resource. In its most severe form, overgrazing has led to desertification. **Desertification** is the degradation of land cover and damage to the soil and water in grasslands and arid and semiarid lands. One of the most severe cases of desertification has been occurring in the Sahel region of Africa since the 1970s. The degradation of the grasslands bordering the Sahara Desert, however, has not been a simple case of careless overgrazing by thoughtless herders. Severe drought, land decline, recurrent famine, and the breakdown of traditional systems for

desertification: the degradation of land cover and damage to the soil and water in grasslands and arid and semiarid lands.

[Handwritten margin notes:]
core — clearance of forests replaced by regeneration
periphery — 50% reduction since 1990's.

2 types of impacts on grasslands
① desertification overgrazing-
② settlement

Figure 4.31 African Grasslands
Shown here is a tropical grassland in Africa. Also known as savannas, these grasslands include scattered shrubs and isolated small trees and are found in areas with high-average temperatures and low-to-moderate precipitation. They occur in an extensive belt on both sides of the equator. African tropical savannas, such as the one pictured here in Kenya, contain extensive herds of hoofed animals, including gazelles, giraffes, zebras, wildebeests, and antelope.

coping with disaster all combined to create increased pressure on fragile resources, resulting in a loss of grass cover and extreme soil degradation. While the factors behind the human impacts on the Sahelian grasslands are complex, the simple fact remains that the grasslands have been severely degraded, and the potential for their recovery is still unknown (Figure 4.32).

Land included in the category "wetland" covers a wide array of types: swamps, marshes, bogs, peatlands, and the shore areas of lakes, rivers, oceans, and other water bodies. Wetlands can be associated either with salt water or fresh water (Figure 4.33). Most of Earth's wetlands are associated with the latter.

Human impacts on wetland environments have been numerous. The most widespread, however, has been the draining or filling of wetlands and their conversion to other land uses such as settlement or cultivation. One reliable estimate places the total area of the world's wetlands at about 8.5 million square kilome-

Figure 4.32 Desertification in Sub-Saharan Africa Desertification is a mounting problem in many parts of the world but especially in sub-Saharan Africa (the portion of Africa between North Africa's Sahara Desert and the five countries that make up southern Africa). This satellite photo shows the belt of desertification stretching from west to east across the widest portion of the continent. Overgrazing on fragile arid and semiarid rangelands and deforestation without reforestation are thought to be the chief causes of desertification in this part of Africa.

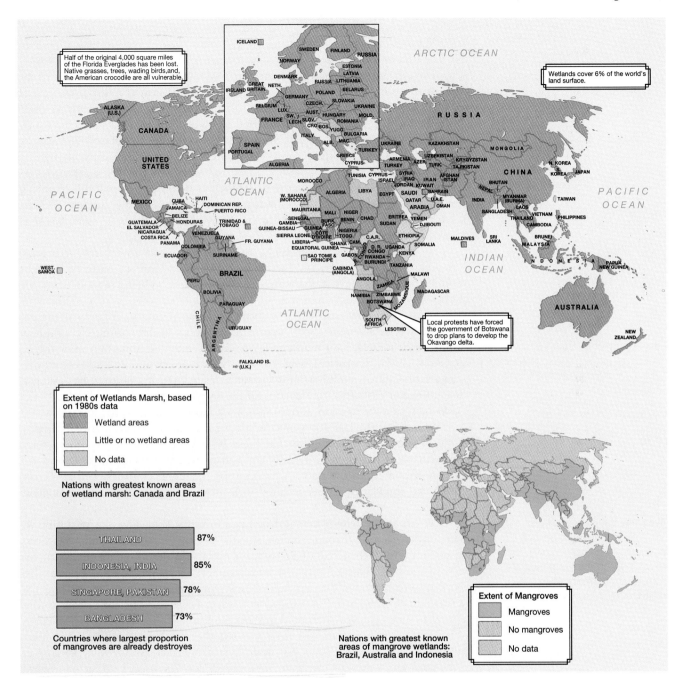

Figure 4.33 Global distribution of wetlands, 1980s   Wetlands are among the most productive ecosystems in the world. The biggest threats to wetlands are urbanization, infrastructural development, and the increasing demand worldwide for fresh water. Mangroves are a type of wetland that have been under severe assault from development and deforestation. These are swamps dominated by mangrove trees—any of 55 species of woody plants that can live partly submerged in the salty coastal wetlands. The worst destruction of the mangroves is occurring in Asia, especially in the Philippines, Thailand, Bangladesh, Indonesia, and Java. (*Source:* J. Seager, *The New State of the Earth Atlas*, 2nd Ed. New York: Simon & Schuster, 1995, p. 70–71.)

ters (3.3 million square miles) with about 1.5 million square kilometers (0.6 million square miles) having been lost to drainage or filling. In Australia, for example, all of the original 20,000 square kilometers (7,740 square miles) of wetlands have been lost to conversion. For the last 400 years or so, wetlands have been regarded as nuisances, if not sources of disease. In core countries, technological innovation made modification and conversion of wetlands possible and profitable. In San Francisco, California, for example, the conversion of wetlands in the mid-nineteenth century allowed speculators and real estate developers to extend significantly the central downtown area into the once marshy edges of the San Francisco Bay. The Gold Rush in the Sierra Nevada sent millions of tons of sediment down the rivers into the bay filling in marshes and reducing its near-shore depth. It is estimated that in 1850 the San Francisco Bay system, which includes San Pablo Bay as well as Suisun Bay, covered approximately 315 square kilometers (about 120 square miles). One hundred years later only about 125 square kilometers (about 50 square miles) remained. By the 1960s, the conversion and modification of the wetlands (as well as the effects of pollution pouring directly into the bay) had so dramatically transformed water quality and the habitats of fish, fowl, and marine life that the viability of the ecosystem was seriously threatened.

The combustion of fossil fuels, the destruction of forest resources, the damming of watercourses, and the massive change in land use patterns brought about by the pressures of globalization—industrialization being the most extreme phase—have meant that environmental problems have come to reach enormous proportions. It is now customary to speak of the accumulation of environmental problems we, as a human race, experience as *global* in dimension. Geographers and others use the term **global change** to describe the combination of political, economic, social, historical, and environmental problems with which human beings across Earth must currently contend. Very little, if anything, has escaped the embrace of globalization, least of all the environment.

Global change: combination of political, economic, social, historical, and environmental problems at the world scale.

In fact, no other period in human history has transformed the natural world as profoundly as the last 400 years. While we reap the benefits of a modern way of life, it is critical to recognize that these benefits have not been without cost. Furthermore, it has been argued that in too many cases the costs have accrued disproportionately to the poor. A growing political consciousness of these costs among the world's poor has resulted in a movement known as environmental justice.

Activists in the **environmental justice** movement consider the pollution of their neighborhoods through such elements as nearby factories and hazardous waste dumps to be the result of a structured and institutionalized inequality that is pervasive in both the capitalist core and periphery. They see their struggles as distinct from the more middle class and mainstream struggles of groups such as the Sierra Club or even Earth First! and Greenpeace. These activists have come to conceptualize their struggles as rooted in their economic status. Thus, these struggles are not about quality-of-life issues such as whether any forests will be left to hike, but about issues of sheer economic and physical survival. As a result, the answer to the questions raised by environmental justice activists must be directed toward the redistribution of economic and political resources. Such questions are not easily resolved in a court of law, but speak to more complex issues such as the nature of racism, sexism, and of capitalism as a class-based economic system.

Environmental justice: movement that reflects a growing political consciousness largely among the world's poor that their immediate environs are far more toxic than wealthier neighborhoods.

Conclusion

The relationship between society and nature is very much mediated by institutions and practices, from technology to religious beliefs. In this chapter we have seen how the society–nature relationship has changed over time, and also how the globalization of the capitalist world economy has had more of a widespread impact on attitudes and practices than any cultural or economic system that preceded it.

Ancient humans apparently displayed a reverential attitude toward the natural world, still evident among native populations in the many parts of the New

World as well as Africa and Asia. With the emergence of Judaism and, later, Christianity, humans adopted a more dominant attitude with respect to nature. The expansion of European trade, followed by colonization and eventually industrialization, broadcast worldwide the belief that humans should take their place at the apex of the natural world. It is the Judeo-Christian attitude toward nature as it was taken up by the emergence of the capitalist economic system that is the most pervasive shaper of society–nature interactions today.

Besides exploring the history of ideas about nature, this chapter has also shown that society and nature are interdependent and that events in one part of the global environmental system affect conditions in the system elsewhere. The nuclear accident in Chernobyl taught us this lesson when nuclear fallout from Ukraine was spread throughout the globe in the days following the reactor melt-down.

Finally, this chapter has also shown that events that have occurred in the past shape the contemporary state of society and nature. For example, the practices of the U.S. aerospace/defense industries in the postwar years left a legacy of danger-ously polluted drinking water in communities across the Southwest. And the fall-out from Chernobyl has been absorbed into the bodies of the survivors and even the children born after the disaster.

In short, not only has the economy been globalized but also the environment. We can now speak of a global environment where not only the people but the physical environments in which they live and work are linked in complex and essential ways.

Main Points

- Nature, society, and technology constitute a complex relationship. Our definition of nature is that while it is a physical realm, it is also a social product.

- Because we regard nature as a social product, it is important to understand the many social ideas of nature present in our society today, and especially the history of those ideas. The most prominent idea of nature in Western culture is derived from the Judeo-Christian tradition, the mainstream belief of which is that nature is an entity to be dominated by humans.

- Social relationships to nature have developed over the course of human history, beginning with the early Stone Age. The early history of humankind included people who revered nature as well as those who abused it. Urbanization and industrialization have had extremely degrading impacts on the environ-ment.

- The globalization of the political economy has meant that environmental problems are also global in their scope. Deforestation, acid rain, and nuclear fallout affect us all. Many new ways of understanding nature have emerged in the last several decades in response to these serious global crises.

Key Terms

acid rain (p. 179)

animistic perspective on nature (p. 153)

Buddhist perspective on nature (p. 153)

Clovis point (p. 161)

Columbian Exchange (p. 166)

conservation (p. 156)

deep ecology (p. 157)

deforestation (p. 163)

demographic collapse (p. 166)

desertification (p. 138)

ecofeminism (p. 157)

ecological imperialism (p. 167)

ecosystem (p. 162)

environmental ethics (p. 156)

environmental justice (p. 186)

global change (p. 186)

Islamic perspective on nature (p. 153)

Judeo-Christian perspective on nature (p. 152)

nature (p. 150)

Paleolithic period (p. 159)

preservation (p. 156)

romanticism (p. 155)

siltation (p. 163)

society (p. 150)

Taoist perspective on nature (p. 153)

technology (p. 150)

transcendentalism (p. 156)

virgin soil epidemics (p. 165)

Exercises

On the Internet

In this chapter the series of exercises will explore the relationship between globalization and environmental change. The Internet provides an especially valuable tool for studying this relationship because it contains many sites dealing with environmental issues, policies, and activism. After completing these exercises you should understand better the way in which society and technology interact to produce nature.

Unplugged

1. Many communities have begun to produce an "index of stress," which is a map of the toxic sites of a city or region. One way to plot a rudimentary map is to use the local phone book as a data source. Use the Yellow Pages to identify the addresses of environmentally harmful and potentially harmful activities such as dry cleaning businesses, gas stations, automotive repair and car care businesses, aerospace and electronic manufacturing companies, agricultural supply stores, and other such commercial enterprises where noxious chemicals may be produced or applied. Compile a map of these activities to begin to get a picture of your locale's geography of environmental stress.

2. Locate and read a natural history of the place where your college or university is located. What sorts of plants and animals dominated the landscape there during the Paleolithic period? Do any plants or animals continue to survive—in altered form or unaltered—from that period?

3. Interview a friend or classmate whose religion is significantly different from your own. Try to determine how this person's religion has shaped his or her view of nature. How different are your respective views of nature? Are the differences based on religion or personal philosophy?

Chapter 5

Mapping Cultural Identities

Young Elvis

Tempers flared in the usually genteel college town of Oxford, Mississippi, in the summer of 1995 when the Center for the Study of Southern Culture decided to sponsor an international conference on Elvis Presley. For over two decades the University of Mississippi had been host to an annual conference in honor of William Faulkner, one of the United States' best-known novelists and a native son and graduate of Ole Miss. When the Elvis Presley conference was announced, the classicists were agitated, protesting that a serious academic conference on Presley (Oxford's more famous son) was an offense to Faulkner, lovers of literature, and the townspeople as well.

The controversy caused by the Presley conference (well attended by national and international scholars alike) points to the new questions that globalization forces us to confront. What counts as culture? How do we study it? How do we come to grips with the reality that U.S. cultural practices are being exported by the media to the most remote corners of the globe? As students of culture, geographers can no longer be satisfied that attention to exotic tribes in far-flung places is sufficient to understand culture. The fact is that our own cultural practices have begun to be questioned by "outsiders" as the U.S. media extends its global reach into the villages and huts of the world. Those once seemingly exotic people are looking at and trying to understand our apparently curious rituals.

Culture has been and continues to be a central concept in geography although our understanding of its meaning and its impact have changed considerably over the last two decades. In this chapter we examine the many ways that geographers have explored the concept of culture and the insights they have gained from these explorations. We distinguish between a geographical approach to culture and the approach to culture of anthropologists, and provide background on some of the earliest approaches to cultural geography as well as the many new ways that geographers have begun to address culture.

Culture As a Geographical Process

Geographers have long been involved in trying to understand the manifestations and impacts of culture on geography *and* of geography on culture. While anthropologists are concerned with the way in which culture is created and maintained by human groups, geographers are interested in how place and space shape culture *and* the reverse—how culture shapes place and space.

Anthropologists, geographers, and other scholars who study culture, such as historians, sociologists, and political scientists, agree that culture is a complex concept. Over time, our understanding of it has been changed and enriched. A simple understanding of it is that it is a particular way of life, such as a set of skilled activities, values, and meanings surrounding a particular type of economic practice. Scholars also describe culture in terms of classical standards and esthetic excellence in opera, ballet, or literature, for example.

By contrast, the term "culture" has also been used to describe the range of activities that characterize a particular group, such as working-class culture, corporate culture, or teenage culture. Although all of these understandings of culture are accurate, for our purposes they are only partial. Broadly speaking, **culture** is a shared set of meanings that are lived through the material and symbolic practices of everyday life (Figure 5.1). The "shared set of meanings" can include values, beliefs, practices, and ideas about religion, language, family, gender, sexuality, and other important identities. Culture is often subject to reevaluation and redefinition and ultimately altered from both within and outside a particular group.

Culture is a dynamic concept that revolves around complex social, political, economic, and even historical factors. This definition of culture is part of a longer evolving tradition within geography and other disciplines such as anthropology and sociology. We will look more closely at the development of the cultural tradition in geography in the following section in which we discuss the debates surrounding culture within the discipline.

Culture: a shared set of meanings that are lived through the material and symbolic practices of everyday life.

Figure 5.1 Youth culture The term culture has been used to describe a range of practices characterizing a group. Pictured is a youth-culture group that originated in London and then spread throughout many of the core countries—punks. Hairstyle, dress, and body adornment, as well as a distinctive philosophy and music, characterize punk culture. Yet culture is more than just the physical, distinguishing aspects of a group. It is also a way in which groups derive meaning and attempt to shape the world around them.

Figure 5.2   Yard shrine in the U.S. Southwest   Yard shrines are a ubiquitous feature of Roman-Catholic Mexican-American communities in the U.S. Southwest. Their purpose is to show respect and devotion to the Virgin Mary as well as to appeal for favors. Yard shrines are a public display of religiosity outside of the formal bounds of church-based worship and a particular characteristic of a regional culture group.

For much of the twentieth century geographers, like anthropologists, have focused most of their attention on *material culture* as opposed to its less tangible symbolic or spiritual manifestations. Thus, while geographers have been interested in religion as an object of study, for a long time they have largely confined their work to examining its material basis. For example, they have explored the spatial extent of particular religious practices (the global distribution of Buddhism) and the expression of religiosity (the appearance of yard shrines in the U.S. Southwest, Figure 5.2). In the last 20 or so years, the near-exclusive focus on material cultural practices has changed—driven by the larger changes that are occurring in the world around us.

As with agriculture, politics, and urbanization, globalization has also had complex effects on culture. Terms such as "world music" and "international television" are a reflection of the sense that the world has become a very small place, indeed, and people everywhere are sharing aspects of the same culture through the widespread influence of television and other media such as radio. Yet, as pointed out in Chapter 2, while powerful homogenizing global forces are certainly at work, the world has not become uniform so that place no longer matters. With respect to culture, just the opposite is true. Place matters more than ever in the negotiation of global forces, as local forces confront globalization and translate it into unique place-specific forms. An example is given below (see Feature 5.1, "Geography Matters—STAR TV in Asia") that illustrates how the global spread of Western popular music has been received in Asia.

The place-based interactions that occur between culture and global political and economic forces are at the heart of cultural geography today. **Cultural geography** focuses on the way in which space, place, and landscape shape culture at the same time that culture shapes space, place, and landscape. As such, cultural geography demarcates two important and interrelated parts. *Culture* is the ongoing process of producing a shared set of meanings, while *geography* is the dynamic setting within which groups operate to shape those meanings and in the process form an identity and act. Geography in this definition can be as small as the micro space of the body and as large as the macro space of the globe.

Cultural geography: how space, place, and landscape shape culture at the same time that culture shapes space, place, and landscape.

Figure 5.3   Carl Sauer (1889–1975)   Born in the U.S. Midwest of German-immigrant parents, Carl Sauer spent his career as a geographer at the University of California, Berkeley. He rejected environmental determinism as a way of understanding human geography and emphasized the uniqueness of landscape through the impact of both cultural and physical processes.

Building Cultural Complexes

Geographers have long been interested in the interactions among people and culture, and among space, place, and landscape. One of the most influential individuals in this regard was Carl Sauer, a geographer who taught at the University of California. Sauer was largely responsible for creating the "Berkeley school" of cultural geography (Figure 5.3). He was particularly interested in trying to understand the material expressions of culture by focusing on

5.1 Geography Matters

STAR TV in Asia

Asia's first television service, STAR TV, beams five channels of entertainment from Saudi Arabia to Shanghai. It emanates from Hong Kong, hardly a coincidence in a world where globalization is the norm, for Hong Kong is an international city that has learned well how to operate and prosper within the global economy. STAR TV has produced an astonishing revolution in entertainment, broadcasting Arrested Development's latest tunes to Tibet, and *Santa Barbara* into Burma, where even Karen rebels lay down their AK-47s to watch the latest episode (Figure 5.1.1). Asian sports teams receive coverage on STAR TV's sports channels, and news of disturbances in Bangkok, in 1992, played on the system's news channel. Chinese and Taiwanese citizens may be locked in a political struggle of ill feelings, but they share laughter thanks to STAR's Chinese programming.

Figure 5.1.1   American television in Asia The communications industry is perhaps the most important medium for deploying core cultural practices to the periphery. American television shows are routinely exported throughout the world, providing a common point of reference for diverse groups of people. Shown here are a group of men watching an American TV show through the window of an electronics store in Mumbai, India.

STAR TV reflects the growing affluence of Asian customers who have prospered in the global economy (Figure 5.1.2). It broadcasts to 38 countries, which together comprise a huge population—and potential market—of 2.8 billion people (Figure 5.1.3). In just 10 of

Figure 5.1.2   Hollywood stars advertising Asian products The Hollywood film industry has been particularly effective in penetrating foreign markets. This is very much the case in Japan, where American movie stars are routinely used to promote Japanese products.

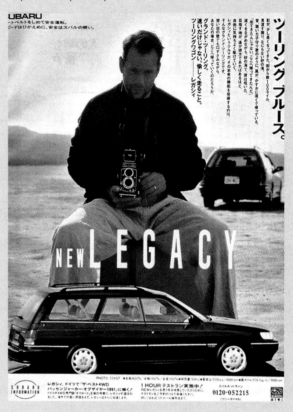

Cultural landscape: a characteristic and tangible outcome of the complex interactions between a human group and a natural environment.

their manifestations in the landscape. This interest came to be embodied in the concept of the **cultural landscape,** a characteristic and tangible outcome of the complex interactions between a human group—with its own practices, preferences, values, and aspirations—and a natural environment. Sauer differentiated the cultural landscape from the natural landscape. He emphasized that the former was a "humanized" version of the latter such that the activities of humans resulted in an identifiable and understandable alteration of the natural environment. Figure 5.4 (see page 194) illustrates the idea through a listing of the differences between a natural and a cultural landscape.

For roughly five decades, interest in culture within geography largely followed Sauer's important work. His approach to the cultural landscape was ecological, and his many published works reflect his interest in trying to understand the myriad ways that humans transformed the surface of Earth. In his own words:

those countries, STAR TV already reaches 11 million households, which account for up to 45 million viewers.

But what kind of culture is STAR TV transmitting? The music arm of the organization, STAR MTV, is heavily dominated by Western music and entertainers, leading some Asians to worry about Western cultural imperialism. The company is aware of the concern, and also promotes indigenous Asian music, although the music must be pan-Asian in nature and easily accessible to its multilingual, multicultural audience. The company could promote much more Asian music, but the sheer number of types is a problem. As the owner of the company states, "Imagine you're a person in Saudi Arabia and we broadcast to you a parochial Taiwanese song. You would think it was noise, and vice versa for Saudi Arabian music." What, then, do all of Star TV's customers have in common? Core-country entertainers like Madonna!

Source: Description of STAR TV adapted from L. Fitzpatrick, "Does Asia Want My MTV," *Hemispheres,* July 1993, pp. 21–22.

Figure 5.1.3 Potential Asian broadcast audience Language may present something of an obstacle to broad-based marketing by STAR TV. The potential Asian market includes 55 countries where over 10 times that number of languages are spoken. While sports and music broadcasts may be able to rise above the limits of language, news and talk shows depend more closely on audience intelligibility. Thus, lighter forms of broadcasting may reach a wider audience, while more language-dependent programs may be more limited in their range.

The cultural landscape is fashioned from a natural landscape by a cultural group. Culture is the agent, the natural area is the medium, the cultural landscape is the result. Under the influence of a given culture, itself changing through time, the landscape undergoes development, passing through phases, and probably reaching ultimately the end of its cycle of development. With the introduction of a different—that is an alien—culture, a rejuvenation of the cultural landscape sets in, or a new landscape is superimposed on remnants of an older one.[1]

[1]C. Sauer, "The Morphology of Landscape" in J. Leighly (Ed.), *Land and Life: Selections from the Writings of Carl Ortwin Sauer.* Berkeley, CA: University of California Press. 1964, pp. 315–350.

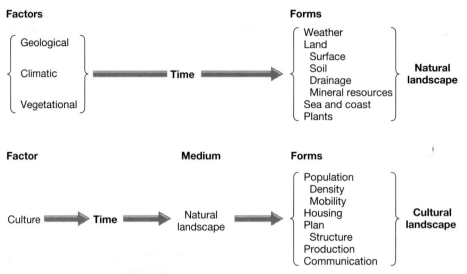

Figure 5.4 Sauer's cultural landscape This figure provides a graphic representation of the way in which the natural landscape and cultural landscapes are transformed. Physical features as well as climate factors shape the natural landscape. Cultural practices also have an important impact upon it. The results of cultural factors are cultural forms such as population distributions and patterns and housing. Over time, people, through culture, reshape the natural landscape to meet their needs. (*Source:* Adapted from C. Sauer, "The Morphology of Landscape" in J. Leighly (Ed.), *Land and Life: Selections from the Writings of Carl Ortwin Sauer.* Berkeley, CA: University of California Press. 1964, pp. 315–350.)

Historical geography: the geography of the past.

Genre de vie: a functionally organized way of life that is seen to be characteristic of culture groups.

Figure 5.5 H. C. Darby (1909–1992) H. C. Darby argued that historical geography is an essential foundation for the study of all human geography. His own studies of past geographies were undertaken through cross-sectional analyses of archival materials published in a series of Domesday Geographies of England.

In Europe, geographers interested in human interactions with the landscape produced slightly different approaches. For example, in Great Britain, the approach to understanding the human imprint on the landscape was given the term "historical geography," while in France it was conceptualized as *genre de vie.* **Historical geography,** very simply defined, is the geography of the past. Its most famous practitioner was H. C. Darby, who attempted to understand "cross-sections" or sequences of evolution, especially of rural landscapes (Figure 5.5). *Genre de vie,* a key concept in Vidal de la Blache's approach to cultural geography in France, referred to a functionally organized way of life that was seen to be characteristic of a particular culture group (Figure 5.6). *Genre de vie* centered on the livelihood practices of a group, which were seen to shape physical, social, and psychological bonds. Although emphasizing some landscape components over others or giving a larger or smaller role to the physical environment, all of these approaches placed the cultural landscape at the heart of their study of human—environment interactions.

H. C. Darby most successfully implemented his historical approach to cultural geography and landscape by developing a geography of *The Domesday Book.* William the Conqueror ordered *The Domesday* compiled in 1085 so that he could have a list of his spoils of war. The book provides a rich catalog of the ownership of every tract of land in England and of the conditions and contents of the lands at that time. For geographers like Darby, such data are invaluable for reconstructing past landscapes.

Vidal de la Blache, on the other hand, emphasized the need to study small, homogeneous areas in order to uncover the close relationships that exist between people and their immediate surroundings. He constructed complex descriptions of preindustrial France that demonstrated how the various *genres de vie* emerged from the possibilities and constraints posed by local physical environments. Subsequently, he wrote about the changes in French regions brought on by industrialization, observing that regional homogeneity was no longer the unifying element. Instead, the increased mobility of people and goods had produced new, more complex geographies wherein previously isolated *genres de vie* were being integrated into a competitive industrial economic framework. Anticipating the widespread impacts of globalization, de la Blache also recognized how people in places struggled to mediate the big changes that were transforming their lives.

Geographers have also been interested in understanding specific aspects of culture ranging from single attributes to complex systems. One simple aspect of culture of interest to geographers is the idea of special traits, which include such things as distinctive styles of dress, dietary habits, or styles of architecture. A **cultural trait** is a single aspect of the complex of routine practices that constitute a particular cultural group (Figure 5.7). For example, dietary law for Muslims prohibits the consumption of pork. This avoidance may be said to be a cultural trait of Muslim people. Additionally, in their religious iconography, Muslims have prohibitions against displaying human faces. This, too, is considered a cultural trait. While understanding the geographical dynamics of cultural traits such as pork avoidance has been of importance to geographers, an interest has also existed to learn how traits come together to form larger frameworks for living in the world. Ultimately, cultural traits are not necessarily unique to one group, and understanding them is only one aspect of the complexity of culture. For instance, there are certainly other cultural groups (such as Hindus) that avoid pork in their diet.

Many cultures also recognize the passage from childhood into adulthood with a celebration or ceremony. Called **rites of passage,** these are ceremonial acts, customs, practices, or procedures that recognize key transitions in human life—birth, menstruation, and other markers of adulthood such as sexual awakening and marriage (Figure 5.8, page 196).

In Roman Catholicism, the passage of 12-year-old boys and girls into adulthood is celebrated with a religious ceremony known as confirmation. In this ceremony, the confirmed chooses a new name to mark this important spiritual transition. Similarly, Jews also mark the passage of adolescent boys and girls into adulthood with separate religious ceremonies: a *bar mitzvah* for boys and a *bat mitzvah* for girls. Although the marking of the passage into adulthood is a trait that is celebrated by both groups, neither exhibits the trait in exactly the same way. In fact, this and other traits always occur in combination with others. The combination of traits characteristic of a particular group is known as a **cultural complex.** The avoidance of pork, the celebration of *bar* and *bat mitzvahs,* and other dietary, religious, and social practices constitute the cultural complex of Judaism, although it is important to note that even within the cultural complex of Judaism variation exists among regions and sects.

Figure 5.6 Paul Vidal de la Blache (1845–1919) Vidal de la Blache was a founder of the *Annales de Geographie,* an influential academic journal that fostered the idea of human geography as the study of people–environment relationships. His most long-lasting conceptual contribution was *genre de vie,* which is the lifestyle of a particular region reflecting the economic, social, ideological, and psychological identities imprinted on the landscape.

Cultural trait: a single aspect of the complex of routine practices that constitute a particular cultural group.

Rites of passage: the ceremonial acts, customs, practices, or procedures that recognize key transitions in human life such as birth, menstruation, and other markers of adulthood such as marriage.

Cultural complex: combination of traits characteristic of a particular group.

Figure 5.7 Maori architecture The indigenous peoples of the islands of New Zealand, the Maoris have a distinctive and symbolic architecture. Different aspects of design are meant to convey different meanings about reverence for a higher power and people's place in the universe. The meaning systems of culture groups can be conveyed not only through ritual practices but through architecture.

Figure 5.8 A coming-of-age ceremony, Apache reservation The Apache Indians recognize the passage of female children into adulthood during adolescence. A corn ceremony initiates a young girl into womanhood, signaling to other members of the community that an important transition has occurred. Such rites of passage are not uncommon among many of the world's cultures. Some non-Western cultures, for example, send adolescent boys away from the village to experience an ordeal—ritual scarring or circumcision, for example—or to meditate in extended isolation on the new roles they must assume as adults. After an extended absence from the social group, these youth return transformed and ready to lay down their previous childish occupations.

Cultural region: the area(s) within which a particular cultural system prevails.

Another concept key to traditional approaches in cultural geography is the cultural region. Although a **cultural region** may be quite extensive or very narrowly described and even discontinuous in its extension, it is the area within which a particular cultural system prevails. A cultural region is an area where certain cultural practices, beliefs, or values are more or less practiced by the majority of the inhabitants.

TABLE 5.1 Salt Lake City, Utah: What's Hot/What's Not

Certain dietary preferences and prohibitions as well as an orientation to family emerge in the results of a product analysis of Salt Lake City, where large numbers of Mormons reside. In addition to theme parks and national television shows, politics are also distinctly family oriented. Salt Lake City and Utah constitute a distinct cultural region where certain religious beliefs and family values are practiced by a majority of the inhabitants.

What's Hot	What's Not
Theme parks	Fashion
The Bible	Gourmet cooking
RVs	Walking
Coach	*American Detective*
Roseanne	*In Living Color*
Family Handyman	*New Yorker*
GMC Safaris	BMW 6/7s
Mitsubishi minivans	Jaguar XJ6s
Chevy Geos	Mercedes-Benz 560s
Conservatives	Liberals
Pro-lifers	Gay rights

Source: Michael J. Weiss, *Latitudes and Attitudes.* Boston: Little, Brown and Co., 1994, p. 180.

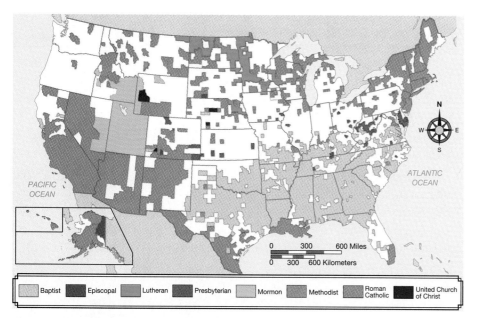

| Baptist | Episcopal | Lutheran | Presbyterian | Mormon | Methodist | Roman Catholic | United Church of Christ |

Figure 5.9 Mormon population distribution This map shows a concentration of Mormons in states of the western United States. Driven from communities in the East, Mormons made their way westward in the nineteenth century to escape persecution and establish free-standing communities. Their settlement of areas of the West was largely successful. In a state like Utah, which is dominated by Mormon inhabitants, a great deal of government support exists for the Mormon way of life. (*Source:* Reprinted with permission from Prentice Hall, from J. M. Rubenstein, *The Cultural Landscape: An Introduction to Human Geography*, 5th Ed., © 1996, p. 203. Adapted from Douglas W. Johnson, Paul R. Picard, and Bernard Quinn, *Churches and Church Membership in the United States.* Bethesda, MD: Glenmary Research Center.)

For example, the state of Utah is considered to be a Mormon cultural region because, as Table 5.1 as well as Figure 5.9 show, the population of the state is dominated by people who practice the Mormon religion and presumably adhere to its beliefs and values.

Cultural Systems

Broader than the cultural complex concept is the concept of a **cultural system,** a collection of interacting components that taken together shape a group's collective identity. A cultural system includes traits, territorial affiliation, and shared history as well as other more complex elements such as language. In a cultural system, it is possible for internal variation to exist in particular elements at the same time that broader similarities lend coherence. For example, Christianity unites all Protestant religions, yet the practices of particular denominations— Lutherans, Episcopalians, and Quakers—vary. And, while Mexicans, Bolivians, Cubans, and Chileans exhibit variations in pronunciation, pitch, stress, and other aspects of vocal expression, they all speak Spanish. This means they share a key element of a cultural system (which, for these nationalities, also includes Roman Catholicism and a Spanish colonial heritage).

Cultural system: a collection of interacting elements that taken together shape a group's collective identity.

Geography and Religion

Two key components of a cultural system for most of the world's people are religion and language. **Religion** is a belief system and a set of practices that recognizes the existence of a power higher than humans. Although religious affiliation is on the decline in some parts of the world's core regions, it still acts as a powerful shaper of daily life from eating habits and dress codes to coming-of-age rituals and death ceremonies in both the core *and* the periphery. And, like language,

Religion· belief system and a set of practices that recognize the existence of a power higher than humans.

TABLE 5.2 Changes in Strength of Major Religions, 1970–1985, Caused by Natural Population Increase and by Religious Conversion				
Religion	Natural	Conversion	Total	Rate
Christianity	21.4	0.2	21.6	1.64
Islam	17.1	0.1	17.2	2.74
Nonreligious	9.3	8.0	17.3	2.76
Hinduism	12.1	-0.2	11.9	2.03
Buddhism	5.1	-0.9	4.2	1.67
Chinese folk religion	3.4	-5.1	-1.6	-0.08
Atheism	2.4	0.5	3.0	1.66
New-religions	1.8	0.1	2.0	2.28
Tribal religion	2.4	-2.2	0.2	0.21
Judaism	0.2	-0.01	0.2	1.09
Sikh	0.3	0.03	0.4	2.94
Shamanism	0.3	-0.6	-0.3	-0.41
Confucianism	0.09	-0.05	0.04	0.98
Bahai	0.08	0.04	0.1	3.63
Shinto	0.05	-0.1	-0.06	-1.66
Jain	0.07	-0.01	0.06	2.00
Total world population	**76.4**	**76.4**	**0**	**1.93**

Source: C. Park, 1994. *Sacred Worlds.* London: Routledge, 1994, p. 131.

religious beliefs and practices change as new interpretations are advanced or new spiritual influences are adopted (Table 5.2).

The most important influence on religious change has been conversion from one set of beliefs to another. From the onset of globalization in the fifteenth century, religious missionizing—propagandizing and persuasion—and the conversion of non-Christian souls were key elements. In the 500 years since the onset of the Columbian Exchange, conversion of all sorts has escalated throughout the globe. In fact, since 1492, traditional religions have become dramatically dislocated from their sites of origin through missionizing and conversion as well as diaspora and emigration. Whereas missionizing and conversion are deliberate efforts to change the religious views of a person or peoples, diaspora and emigration involve the involuntary and voluntary movement of peoples who bring their religious beliefs and practices to their new locales.

Diaspora: a spatial dispersion of a previously homogeneous group.

Diaspora is a spatial dispersion of a previously homogeneous group. The processes of global political and economic change that led to the massive movement of the world's populations over the last five centuries has also meant the dislodging and spread of the world's many religions from their traditional sites of practice. Religious practices have become so spatially mixed that it is a challenge to present a map of the contemporary global distribution of religion that reveals more than it obscures. This is because the global scale is too gross a level of resolution to portray the wide variation that exists among and within religious practices. Figure 5.10 identifies the contemporary distribution of what are considered by religious scholars to be the world's "major religions" because they contain the largest number of practitioners globally. As with other global-scale representations, the map is useful in that it helps to present a generalized picture.

Figure 5.11 (see page 200) identifies the source areas of four of the world's major religions and their diffusion from those sites over time. This global-scale map also distorts the place-based variations within those religions, promoting a

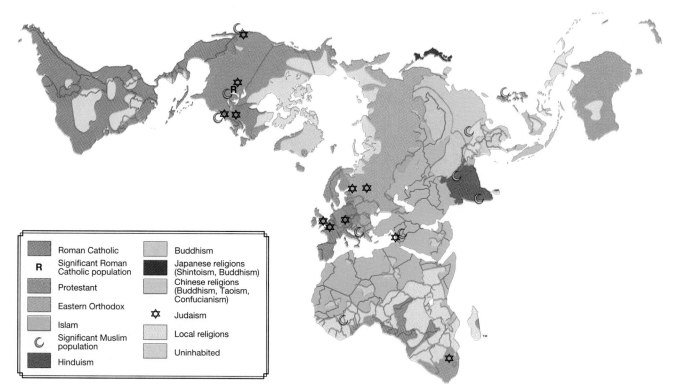

Figure 5.10 World distribution of major religions The map shows a non-nuanced picture of the world's major religions. Most of the world's peoples are members of one of these religions. What is not evident are the local variations in practices as well as the many other different religions that are practiced worldwide. (*Source:* Map projection, Buckminster Fuller Institute and Dymaxion Map Design, Santa Barbara, CA. The word Dymaxion and the Fuller Projection Dymaxion™ Map design are trademarks of the Buckminster Fuller Institute, Santa Barbara, California, ©1938, 1967 & 1992. All rights reserved.)

perspective that a Christian is a Christian or a Buddhist is a Buddhist no matter where they are located. The reality is that dramatic variations exist among and between the place-based practices of Christianity, Islam, Buddhism, and Hinduism, not to mention the hundreds of other religions that also exist and are spreading. Yet, as with the previous map, this one also helps to provide a broad sense of the distribution of religious practice globally. Contrast the world-scale maps in Figures 5.10 and 5.11 with Figure 5.12 (see page 200), which shows just how much variation exists in an apparently homogeneous area when you train your lens to a finer scale of resolution.

Perhaps what is most interesting about the present state of the geography of religion is how during the colonial period religious missionizing and conversion flowed from the core to the periphery. In the current postcolonial period, however, the opposite is becoming true. For example, the fastest-growing religion in the United States today is Islam, and the core countries are where Buddhism is making the greatest number of converts. While Pope John Paul has been the most widely traveled pontiff in Roman Catholic history, the same can be said for the Tibetan Bhuddist religious leader, the Dalai Lama, who is also a tireless world traveler for Buddhism. The Pope's efforts are mostly directed at maintaining Roman Catholic followers and attempting to dissuade their conversion to other religions, such as evangelicalism in the United States and Latin America. The Dalai Lama is promoting conversion to Buddhism by carrying its message to new places, especially in the core (Figure 5.13, page 200).

To provide a dynamic picture of the way in which religion is ever changing, we illustrate the point that religious beliefs and practices are altered through larger

Figure 5.11 Origin areas and diffusion of four major religions The world's major religions originated in a fairly small region of the world. Judaism and Christianity began in present-day Israel and Jordan. Islam emerged from western Arabia. Buddhism originated in India, and Hinduism in the Indus region of Pakistan. The source areas of the world's major religions are also the cultural hearth areas of agriculture, urbanization, and other key aspects of human development.

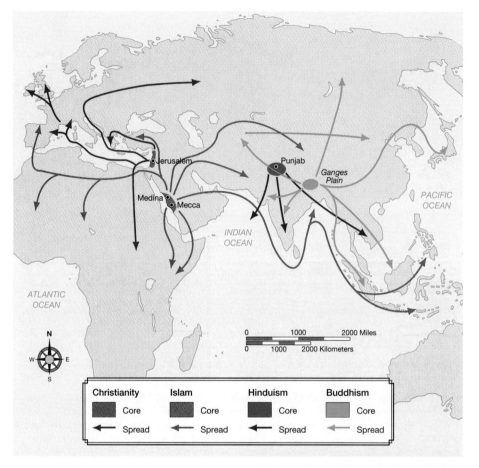

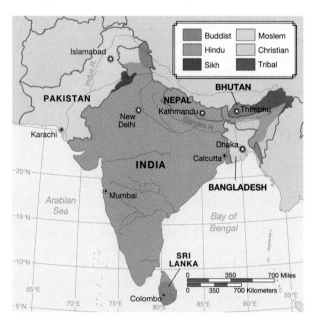

Figure 5.12 Regional religious diversity in South Asia This map of South Asia provides a sense of how varied religious practices are when seen at scales below that of the globe. Where the global scale provides a sense of homogeneity within certain regions, these maps show just how much variability exists within countries and at more local levels. (Source: Reprinted with permission of Prentice Hall, from E. F. Bergman, *Human Geography: Cultures, Connections, and Landscapes,* © 1995, p.280.)

Figure 5.13 The Dalai Lama China's refusal to acknowledge an independent Tibet has sent the Dalai Lama on numerous international tours to broadcast the plight of Tibetans who are experiencing extreme persecution. The practices of Buddhism are increasingly being accepted by Westerners over the teachings of Christianity and Judaism. High-profile practitioners such as Richard Gere have helped to give Buddhism cachet.

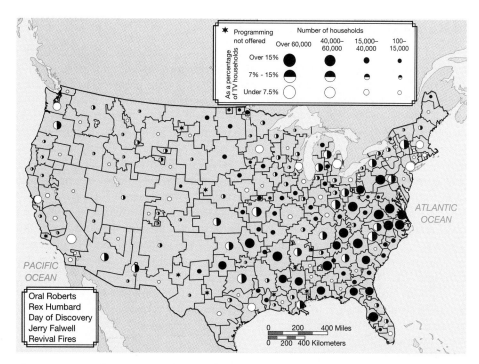

Figure 5.14 Sunday viewing of the top five televangelism programs, 1973 The map of televangelism watchers shows that audiences are drawn from the traditional U.S. Bible Belt. This region is so named because the inhabitants are overwhelmingly Protestant fundamentalists who proclaim a strict adherence to the teachings of the Bible. (*Source:* C. Park, *Sacred Worlds.* New York: Routledge, 1994, p. 91.)

processes of globalization (see Feature 5.2, "Geography Matters—Globalization and Religion in the Americas"). The celebration of the Roman Catholic mass, for example, varies noticeably from country to country in Europe, where it has been shaped by forces within the national and local culture such as dialect, regional history, musical traditions, esthetics of art and architecture, livelihood practices, political history, and many other factors.

One other impact of globalization upon religious change occurs by conversion through the electronic media. The rise of television evangelism, or *televangelism*, especially in the United States, has meant the conversion of large numbers of people to Christian fundamentalism. Figure 5.14 shows the Sunday audiences for the five leading independent religious television programs in the United States, illustrating a loosely formed televangelism cultural region that stretches from Virginia and northern Florida in the east and from the Dakotas to central Texas in the west.

Geography and Language

Geographers have also been interested in understanding other aspects of cultural systems such as languages. Language is an important focus for study because it is a central aspect of cultural identity. Without language, cultural accomplishments could not be transmitted from one generation to the next. The distribution and diffusion of languages tells much about the changing history of human geography and the impact of globalization on culture. Before looking more closely at the geography of language and the impacts of globalization upon the changing distribution of languages, however, it is necessary to become familiar with some basic vocabulary.

Language is a means of communicating ideas or feelings by means of a conventionalized system of signs, gestures, marks, or articulate vocal sounds. In

Language: a means of communicating ideas or feelings by means of a conventionalized system of signs, gestures, marks, or articulate vocal sounds.

Religion is a dynamic cultural practice that has been dramatically affected by the processes of globalization. A contemporary map of the world distribution of religions tells us much about the "frozen" where of religion, but it tells us nothing about how that distribution came about or about the important differences that exist within the same religion as it is practiced in different parts of the world. Maps at lower scales of resolution begin to reveal how religious variations exist within regions, but, again, they are silent about the processes behind the pattern. It is possible, however, to identify the diffusion of religion in ways that are similar to the diffusion of languages.

An excellent illustration of the global forces behind the changing geography of religion is through Columbian contact with the New World. Before Columbus and later Europeans reached the continents of North and South America, the people living there practiced, for the most part, various forms of animism and related rituals. They viewed themselves holistically, as one part of the wider world of animate and inanimate nature. They used religious rituals and charms to guide and enhance the activities of everyday life-as well as the more extreme situations of warfare. Shamanism, in which spiritually gifted individuals are believed to possess the power to control preternatural forces, is one important aspect of the belief system that existed among Native American populations at the time of European contact (Figure 5.2.1).

European contact with the New World was, from the beginning, accompanied by Christian missionizing efforts directed at changing the belief systems of the aboriginal peoples and converting them to what the missionizers believed to be "the one, true religion" (Figure 5.2.2 and 5.2.3). Religion, especially for the Spanish colonizing agents, was especially important in integrating Native Americans into the feudal system.

Figure 5.2.1 Pre-Columbian religions in North America   Before European contact, Native Americans had developed a range of religious practices, as shown on the map. Religious traditions based on agrarian practices diffused from South to North, while those religious traditions based on hunting diffused from North to South. (*Source: Atlas of the North American Indian* by Carl Waldman, p. 58. Copyright © 1995 Carl Waldman. Reprinted with permission by Facts on File, Inc., New York.)

Figure 5.2.2 Post-Columbian religious resistance in North America   European efforts to Christianize Native Americans were met with strenuous resistance by some tribes. This map illustrates the most well-known examples of rebellion against missionizing. While many native religions were wiped out, others were hybridized, merging elements of native practices with Christianity. (*Source: Atlas of the North American Indian* by Carl Waldman, p. 59. Copyright © 1995 Carl Waldman. Reprinted with permission by Facts on File, Inc., New York.)

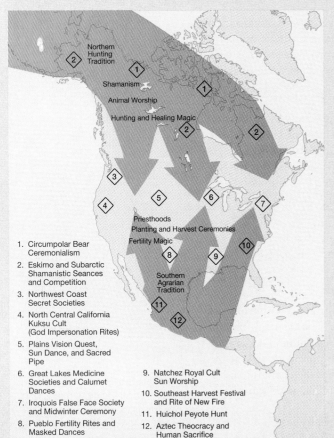

1. Circumpolar Bear Ceremonialism
2. Eskimo and Subarctic Shamanistic Seances and Competition
3. Northwest Coast Secret Societies
4. North Central California Kuksu Cult (God Impersonation Rites)
5. Plains Vision Quest, Sun Dance, and Sacred Pipe
6. Great Lakes Medicine Societies and Calumet Dances
7. Iroquois False Face Society and Midwinter Ceremony
8. Pueblo Fertility Rites and Masked Dances
9. Natchez Royal Cult Sun Worship
10. Southeast Harvest Festival and Rite of New Fire
11. Huichol Peyote Hunt
12. Aztec Theocracy and Human Sacrifice

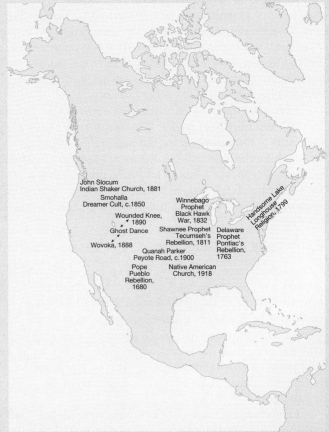

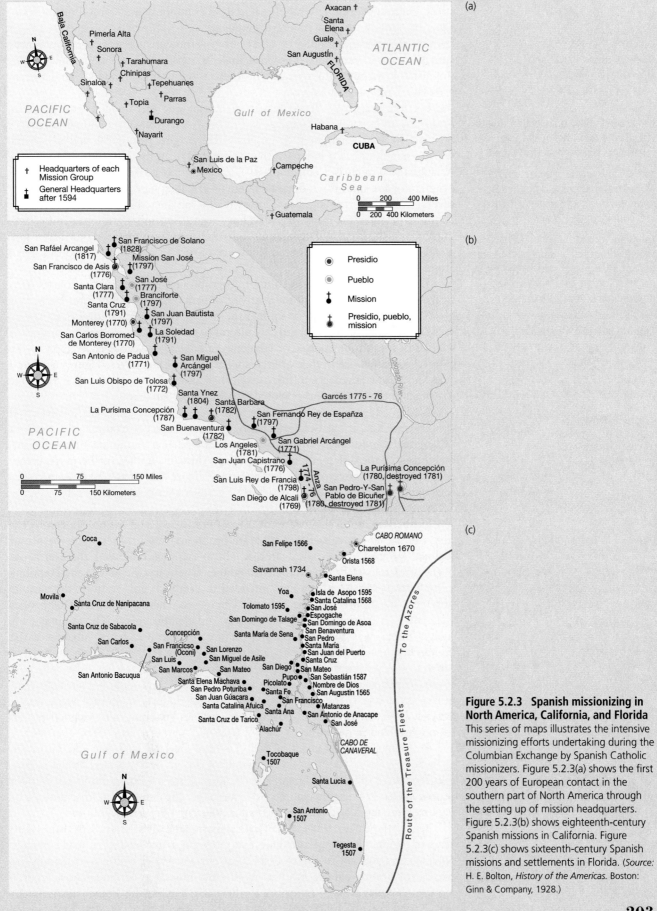

(a)

(b)

(c)

Figure 5.2.3 Spanish missionizing in North America, California, and Florida

This series of maps illustrates the intensive missionizing efforts undertaking during the Columbian Exchange by Spanish Catholic missionizers. Figure 5.2.3(a) shows the first 200 years of European contact in the southern part of North America through the setting up of mission headquarters. Figure 5.2.3(b) shows eighteenth-century Spanish missions in California. Figure 5.2.3(c) shows sixteenth-century Spanish missions and settlements in Florida. (*Source:* H. E. Bolton, *History of the Americas.* Boston: Ginn & Company, 1928.)

203

Dialects: regional variations in standard languages.

Language family: a collection of individual languages believed to be related in their prehistorical origin.

Language branch: a collection of languages that possess a definite common origin but have split into individual languages.

Language group: a collection of several individual languages that are part of a language branch, share a common origin, and have similar grammar and vocabulary.

Cultural hearths: the geographic origins or sources of innovations, ideas, or ideologies.

short, communication is symbolic, based on commonly understood meanings of signs or sounds. Within standard languages (also known as official languages because they are maintained by offices of government such as education and the courts), regional variations, known as **dialects,** exist. Dialects emerge and are distinguishable through differences in pronunciation, grammar, and vocabulary that are place-based in nature.

For the purposes of classification, languages are divided into families, branches, and groups. A **language family** is a collection of individual languages believed to be related in their prehistorical origin. About 50 percent of the world's people speak a language that originated from the Indo-European family. A **language branch** is a collection of languages that possess a definite common origin but have split into individual languages. A **language group** is a collection of several individual languages that are part of a language branch and that share a common origin in the recent past and have relatively similar grammar and vocabulary. Spanish, French, Portuguese, Italian, Romanian, and Catalan are a language *group,* classified under the Romance *branch* as part of the Indo-European language *family.*

Traditional approaches in cultural geography have identified the source areas of the world's languages and the paths of diffusion of those languages from their places of origin. Carl Sauer identified the origins of certain cultural practices by way of the term "cultural hearth." **Cultural hearths** are the geographic origins or sources of innovations, ideas, or ideologies. *Language* hearths are a subset of the cultural hearth concept; they are the source areas of languages. Figures 5.15 and 5.16 show how languages diffused from their hearth areas to form a remarkable new map of the contemporary global distribution of language.

We can add even more dynamism to the geography of language, however, and demonstrate how it has been affected by globalization. For example, the plethora of languages and dialects in any one region may make communication and commerce among the different language speakers difficult. It may, furthermore, create problems for the governing of a population (Figure 5.17, page 206). As a result, regional languages may become extinct or go into a serious state of decline (Figure 5.18, page 206).

Extinction and decline were the case in France. As early as the sixteenth century, the highly centralized French State began actively to discourage the use of regional languages and dialects in official transactions. Following the overthrow of the monarchy in France in the late eighteenth century, the new French republic more actively advanced a policy intended to establish unity among the various provinces by suppressing the regional languages. The multiplicity of languages was seen as a barrier to stable democracy and egalitarianism. The argument for such a policy was that free people—that is, people no longer subjects of the monarch—must speak the same language (north-central, or Parisian, French) in order to unify France and promote democracy as a way of life. After all, how could the people create and operate a government if they could not speak to each other? As a result, the regional languages and dialects of France went into a decline, hastened by official government policies spanning an extended period from Napoleon to deGaulle.

An example is Provençal, a dialect of the Occitan language spoken in the Provence region of southern France, which has been nearly erased by the dictum that French be the official language of France. Currently, only about one million people speak Provençal. Figure 5.19 (see page 206) shows the distribution of regional languages and dialects in France before the French Revolution.

Yet, in the early 1980s, the French government, in a reversal of nearly 400 years of regional language and dialect suppression, reassessed its official language policies. Responding to the emergence of the European Union and fearing the erosion of French culture more generally, the French government now regards regional languages as a treasure. In the early 1990s, the French government

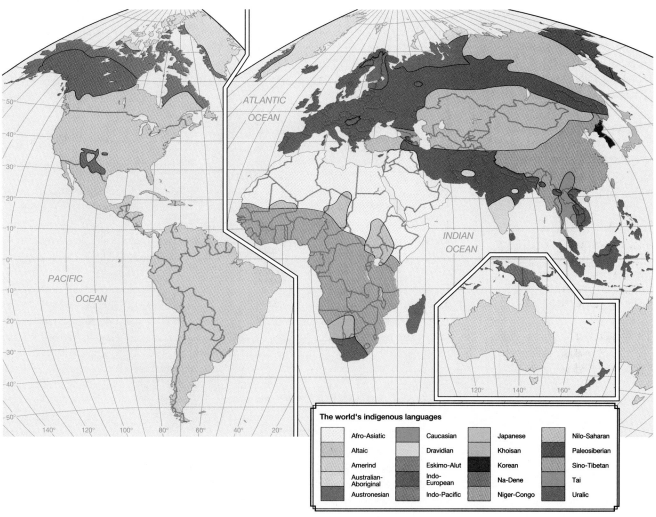

The world's indigenous languages

Afro-Asiatic	Caucasian	Japanese	Nilo-Saharan
Altaic	Dravidian	Khoisan	Paleosiberian
Amerind	Eskimo-Alut	Korean	Sino-Tibetan
Australian-Aboriginal	Indo-European	Na-Dene	Tai
Austronesian	Indo-Pacific	Niger-Congo	Uralic

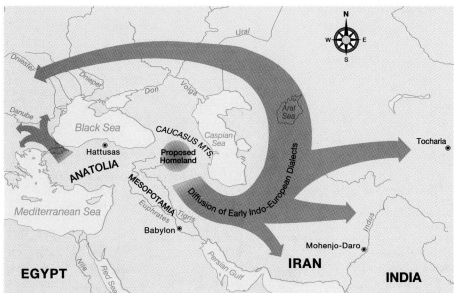

Figure 5.16 The diffusion of the Indo-European language family Although scholars are not certain where the Indo-European language arose, the evidence suggests that it blossomed in the portion of Asia that lies between the Tigris-Euphrates Valley, the Black Sea, and the Caspian Sea. As the language diffused west, east, and south, it mutated into the multiplicity of languages that today make up the Indo-European language family. (*Source:* H. de Blij, *Human Geography: Culture, Society and Space,* 4th Ed. New York: John Wiley & Sons, 1993, p. 277.)

Figure 5.15 World distribution of major languages and major language families Classifying languages by family and mapping their occurrence across the globe provide insights about human geography. For example, we may discover interesting cultural linkages between seemingly disparate cultures widely separated in space and time. We may also begin to understand something about the nature of population movements across broad expanses of time and space. (*Source:* Reprinted with permission from Prentice Hall, E. F. Bergman, *Human Geography: Cultures, Connections, and Landscapes,* © 1995, p. 240. Western Hemisphere adapted from *Language in the Americas* by Joseph H. Greenberg with the permission of the publishers, Stanford University Press. 1987 by the Board of Trustees of the Leland Stanford Junior University. Eastern Hemisphere adapted with permission from David Crystal, *Encyclopedia of Language,* New York: Cambridge University Press, 1987.)

In 1492, the inaugural year of the European contact with the New World, the linguist Nebrijia presented to Queen Isabella of Spain the following comment regarding language and power:

Your Majesty, language is a perfect instrument of rule, a means toward eternal conquest and internal suppression of undesired languages. Language was always a companion of power and it will remain its companion in the future.

Figure 5.17 Language and rule
(*Source:* I. Hauchler and P. Kennedy,. *Global Trends.* New York: Continuum, 1994, p. 93.)

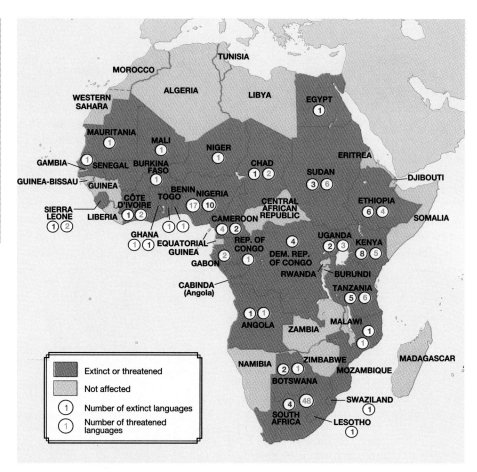

Figure 5.18 Extinct and threatened languages in Africa   It is not absolutely certain how many languages are currently being spoken worldwide, but the estimates range between 4,200 and 5,600. While some languages are being created through the fusion of an indigenous language with a colonial language such as English or Portuguese, indigenous languages are mostly dying out. This is true throughout the Americas and Asia as well as in Africa, as shown in this map.

Figure 5.19 The languages of France in 1789 On the eve of the French Revolution, language diversity in France was not so dissimilar from other European regions that were consolidating into States. Whereas a multiplicity of local languages and dialects prevailed before the emergence of a strong central State, many governments created policies to eliminate them. Local languages made it difficult for States to collect taxes, enforce laws, and teach new citizens. (*Source:* D. Bell, "Lingua Populi, Lingua Dei," *American Historical Review,* 1995, p. 1406.)

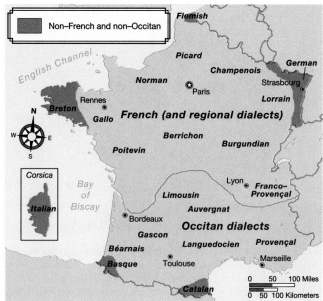

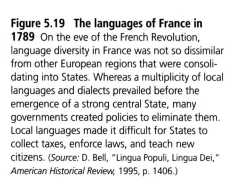

began financing bilingual education in state schools in regions with indigenous languages. Now, the revival of regional languages is viewed by the government as a way of moderating the homogenizing forces of globalization. This is especially true in places such as Provence, where high-technology industries and foreign capital are transforming the landscape as well as local practices. While still predominantly agricultural, Provence is increasingly attracting computer and other light manufacturing firms. As a result, farmlands are being converted to industrial parks, and villages are expanding into towns as workers migrate to take up jobs in the region.

In short, where official languages are put into place, indigenous languages may eventually be lost. Yet the actual unfolding of globalizing forces—like official languages—works differently in different places and in different times. The overall trend appears to be toward the loss of indigenous language (and other forms of culture), as Figure 5.20 shows. It is also important to recognize, however, that religion, language, and other forms of cultural identity can also be used as a means of challenging the political, economic, cultural, and social forces of globalization as they occur in France, Spain (the Basque Separatist Movement), Canada (the Quebecois Movement), and other countries.

Figure 5.20 Tongue-tied   Another way to examine the geography of language is to consider the proportion of people around the world whose mother tongue is not their country's official language. In places like China, where the dominant language is often not the mother tongue, problems of simple communication may pose considerable challenges. In many cases, the disparity between the official language and the mother tongue is the result of colonization. (*Sources:* Adapted from M. Kidron and R. Segal, *The New State of World Atlas,* 4th Ed.

Reprinted with the permission of Simon & Schuster from *The New State of World Atlas,* 4th Ed., by M. Kidron and R. Segal. Copyright © 1991 by Michael Kidron and Ronald Segal. Maps and Graphics copyright © 1991 by Swanston Publishing Limited. Map projection, Buckminster Fuller Institute and Dymaxion Map Design, Santa Barbara, CA. The word Dymaxion and the Fuller Projection Dymaxion™ Map design are trademarks of the Buckminster Fuller Institute, Santa Barbara, California, ©1938, 1967 & 1992. All rights reserved.)

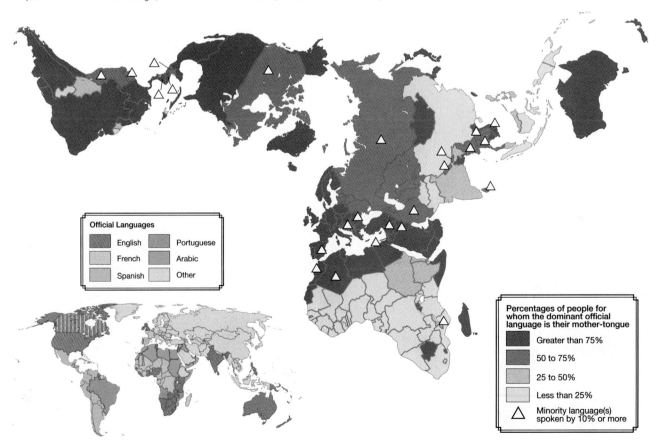

Official Languages
- English
- French
- Spanish
- Portuguese
- Arabic
- Other

Percentages of people for whom the dominant official language is their mother-tongue
- Greater than 75%
- 50 to 75%
- 25 to 50%
- Less than 25%
- △ Minority language(s) spoken by 10% or more

Cultural Nationalism

Cultural nationalism: an effort to protect regional and national cultures from the homogenizing impacts of globalization, especially the penetrating influence of U.S. culture.

The protection of regional languages is part of a larger movement in which geographers and other scholars have become interested. The movement known as **cultural nationalism** is an effort to protect regional and national cultures from the homogenizing impacts of globalization, especially from the penetrating influence of U.S. culture. Figures 5.21 and 5.22 provide a picture of one widespread aspect of U.S. culture—television. In addition to television, U.S. products also travel widely outside of U.S. borders (Figure 5.23, page 210). While many products of U.S. culture are welcomed abroad, many are not. France, for example, has been fighting for years against the "Americanization" of its language.

Nations can respond to the totalizing forces of globalization and the spread of U.S. culture in any number of ways. Some groups may attempt isolationism as a way of sealing themselves off from undesirable influences. Other groups may attempt to legislate the flow of foreign ideas and values, the case with some Muslim countries.

Figure 5.21 Programmed views, 1983 This map illustrates the proportion of television programs imported as a proportion of total screentime. The United States, the former Soviet Union, China, India, Bhutan, and Japan are the only countries who import no foreign television programs. For the former Soviet Union, China, India, and Bhutan the explanation is that the countries at the time the statistics were gathered were closed. For Japan, language has acted as an important barrier. For many other countries around the world, however, the importing of television broadcasting outstrips the production of domestically produced broadcasts. The United States is also a very large exporter of feature films, as the inset to the right of the main map shows. (*Source:* Adapted from M. Kidron and R. Segal, *The New State of World Atlas,* 4th Ed. Reprinted with the permission of Simon & Schuster from *The New State of World Atlas,* 4th Ed., by M. Kidron and R. Segal. Copyright © 1991 by Michael Kidron and Ronald Segal. Maps and Graphics copyright © 1991 by Swanston Publishing Limited. Map projection, Buckminster Fuller Institute and Dymaxion Map Design, Santa Barbara, CA. The word Dymaxion and the Fuller Projection Dymaxion™ Map design are trademarks of the Buckminster Fuller Institute, Santa Barbara, California, ©1938, 1967 & 1992. All rights reserved.)

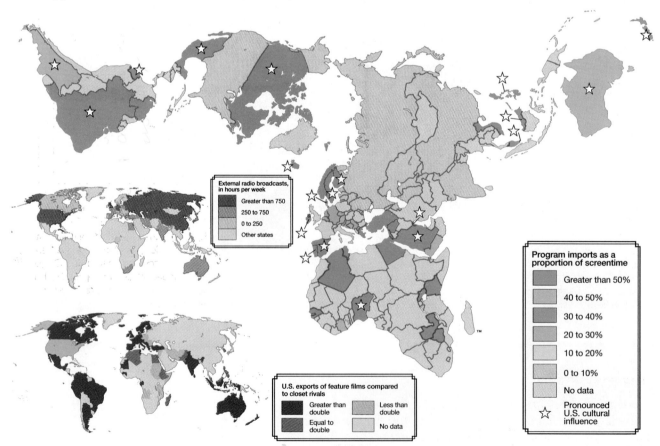

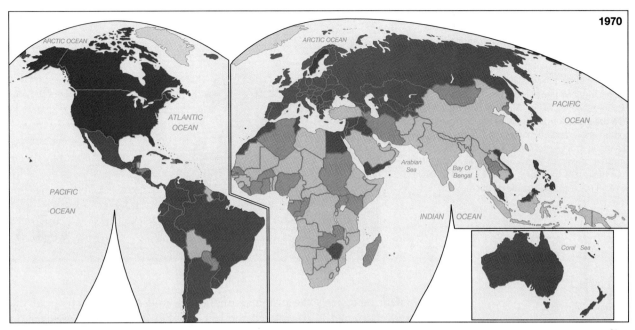

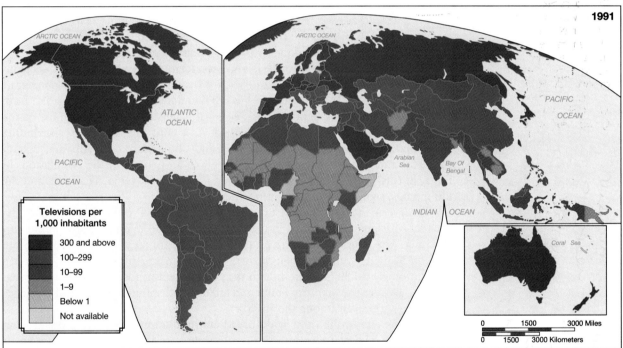

Figure 5.22 World distribution of TV sets, 1970–1991 The ability to receive television broadcasts no matter who is exporting or producing them, of course, depends on access to a television set. In the periphery, especially in some of the wealthier South American countries such as Brazil and Uruguay, approximately 200 television sets exist per 1,000 people. In most of Africa, however, fewer than 10 sets exist for every 1,000 people. In the core, in countries such as the United States and Japan, there is on average one color television set per household as well as a high level of ownership of videocassette recorders (77 percent in the United States, 111 percent in Japan). (*Source:* Reprinted with permission from Prentice Hall, J. M. Rubenstein, *The Cultural Landscape: An Introduction to Human Geography,* 5th ed., © 1996, p. 273)

Islam and Islamism

After Christianity, Islam possesses the next largest number of adherents worldwide—about 1 billion. A Muslim is a member of the community of believers whose duty is obedience and submission (*islam*) to the will of God. As a revealed

Figure 5.23 Seri tribal woman and boombox Pictured in this photograph is a Seri tribal woman from Sonora, Mexico. She is dressed in traditional garments and carries a boombox. Although their total population was probably never more than a few thousand, the Seri now number only about 500. They make their livelihood from hunting, gathering, fishing, and craft production for tourism. Living along the coastal desert of the Gulf of California, the Seri have had a great deal of contact with non-Seri peoples and have adopted and modified many of the items that have been exchanged in these contacts. While boomboxes are not an unusual consumer item among the Seri, it is interesting to note that other traditions, in this case, dress, have changed very little through outside contact.

[Handwritten marginalia:]
religiously and politically unified state
↑ model society protects the
purity of Islamic religion & it's
Islamism is not Islamic fundamentalism; it's
anticolonial, antimperial, and anti-core political
movement. - resist globalization

Islamic fundamentalism - describes a desire
for strict adherence to the fundamentals
of the religious system

religion, Islam recognizes the prophets of the Old and New Testaments of the Bible, but Muhammad is considered the last prophet and God's messenger on Earth. The Qur'an, the principal holy book of the Muslims, is considered to be the word of God as revealed to Muhammad beginning in about 610 A.D. The map in Figure 5.24 shows the relative distribution of Muslims throughout the globe; Figure 5.25 shows the heartland of Islamic religious practice.

As the two figures show, the Islamic world includes very different societies and regions from Southeast Asia to Africa. Muslims comprise over 85 percent of the populations of Afghanistan, Algeria, Bangladesh, Egypt, Indonesia, Iran, Iraq, Jordan, Pakistan, Saudi Arabia, Senegal, Tunisia, Turkey, and most of the newly independent republics of Central Asia and the Caucasus (including Azerbaijan, Turkmenistan, Uzbekistan, and Tajikistan). In Albania, Chad, Ethiopia, and Nigeria, Muslims make up 50 to 85 percent of the population. In India, Burma (Myanmar), Cambodia, China, Greece, Slovenia, Thailand, and the Philippines, significant Muslim minorities also exist.

Disagreement over the line of succession from the Prophet Muhammad, the founder of Islam, has resulted in the emergence of two main sects, the Sunni and the Shi'a. The majority of Iran's 60 million people follow Shi'a, the official State religion of the Islamic Republic of Iran, founded in 1979. The majority of Iraq's population is also Shi'a, even though the government headed by Saddam Hussein is Sunni. Besides the majority Sunni and Shi'a, small splinter groups and branches also exist, especially among the Shi'a.

Perhaps one of the most widespread cultural counterforces to globalization has been the rise of Islamism, more popularly, although *incorrectly*, known as Islamic fundamentalism. Whereas fundamentalism is a term that describes the desire to return to strict adherence to the fundamentals of a religious system, Islamism is an anticolonial, anti-imperial, and overall anticore political movement. The latter is a more accurate and general description of a movement within Muslim countries to resist the core forces of globalization—namely modernization and secularization. Not all Muslims are Islamists, although Islamism is the most militant movement within Islam today.

The basic intent of Islamism is to create a model of society that protects the purity and centrality of Islamic precepts through the return to a universal Islamic State—a State that would be religiously *and* politically unified. Islamists object to modernization because they believe the corrupting influences of the core place the rights of the individual over the common good. They view the popularity of Western ideas as a move away from religion to a more secular (nonreligious) society. Islamists desire to maintain religious precepts at the center of State

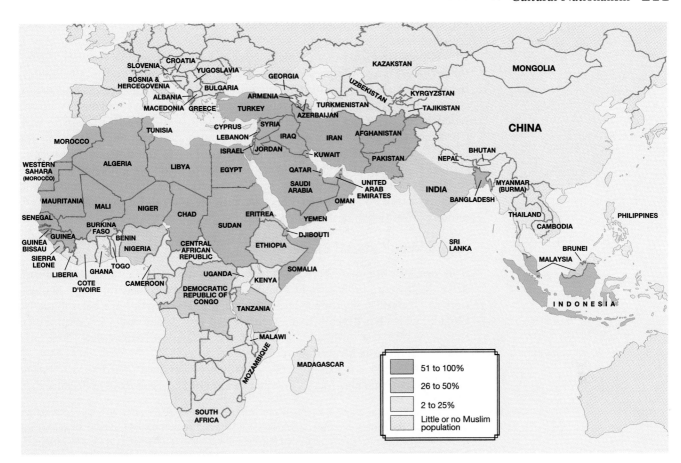

Figure 5.24 Muslim World The diffusion of Islam is much more widespread than many people realize. Like the Spanish colonial effort, the rise and growth of Muslim colonization was accompanied by the diffusion of the colonizers' religion. The distribution of Islam in Africa, Southeast Asia, and South Asia that we see today testifies to the broad reach of Muslim cultural, colonial, and trade activities. (*Source:* D. Hiro, *Holy Wars.* London: Routlege, 1989.)

Figure 5.25 Cultural hearth of Islam Islam has reached into most regions of the world, but the heart of the Muslim culture remains the Middle East, the original cultural hearth. It is also in this area that Islamism is most militant. (*Source:* D. Hiro, *Holy Wars.* London: Routlege, 1989.)

actions, such as reintroducing principles from the sacred law of Islam into State constitutions.

As popular media reports make clear, no other movement emanating from the periphery is as widespread and has had more of an impact politically, militarily, economically, and culturally than Islamism. Yet Islamism—a radical and sometimes militant movement—should not be regarded as synonymous with the practices of Islam any more generally than Christian fundamentalism is with Christianity. Islam is not a monolithic religion, and even though all accept the basic creeds, specific practices may vary according to the different histories of countries, nations, and tribes. Some expressions of Islam are moderate and allow for the existence and integration of Western styles of dress, food, music, and other aspects of culture, while others are extreme and call for the complete elimination of Western things and ideas (Figure 5.26).

Maintaining Cultural Borders

It is not just the periphery, however, that is resisting cultural imperialism. Australia, Britain, France, and Canada also have formally attempted to erect barriers to U.S. cultural products. The most aggressive moves have been made by Canada, which has developed an extensive and very public policy of cultural protection against the onslaught of U.S. music, television, magazines, films, and other art and media forms (Figure 5.27).

In early 1995, for example, the Canadian government levied an 80 percent excise tax against Time, Inc.'s Canadian version of *Sports Illustrated* because they

Figure 5.26 The adherence to Islamic principles of dress and consumption of Western products vary from country to country and even between regions within countries, oftentimes being most marked between rural and urban areas. What this means in practice is variety rather than uniformity.

This urban street scene in Cairo illustrates how some individuals in Cairo adhere to strict principles of dress.

It is also the case that in Egypt, and especially in Cairo, more liberal and Westernized attitudes toward Islamic principles also exist as is shown in this urban street scene. Increasingly, however, certain groups in Egypt are pressing—sometimes through terrorism—for a stricter adherence to Islamic principles at the formal political level as well as on the street.

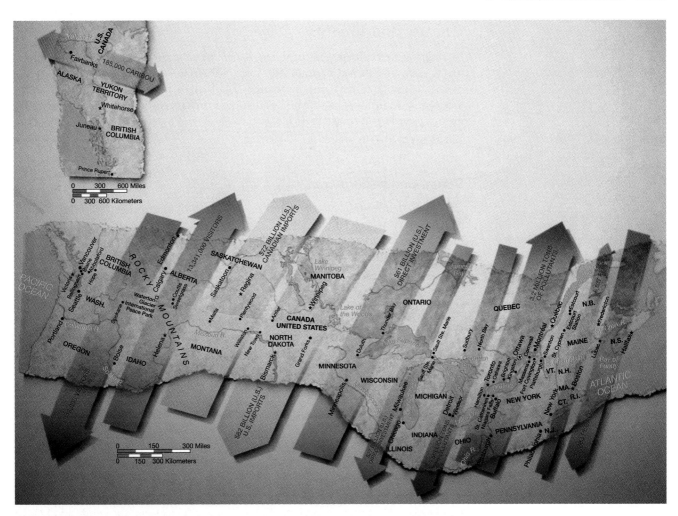

Figure 5.27 United States–Canada border This map of the border shows just how close many of the major Canadian and U.S. cities are. As cities are often the sources of cultural and technological innovation and diffusion, it is not surprising that Canada is experiencing a flood of U.S. culture over the border either through people or media. It should be pointed out, however, that the flow is not one way. The United States also receives cultural products from Canada. (*Source:* Christopher A. Klein/ National Geographic Image Collection.)

did not think it was Canadian enough. The authorities complained that too many of the articles were directed at U.S. sports issues and not enough at Canadian ones.

Other government bodies, such as the National Film Board of Canada and the Canadian Radio and Television Commission, are also active in monitoring the media for the incursion of U.S. culture. For example, 30 percent of the music on Canadian radio must be Canadian. Nashville-based Country Music TV was discontinued from Canada's cable system in the early 1990s and replaced with a Canadian-owned country music channel.

Besides regulating how much and what type of U.S. culture can travel north across the border, the Canadian government also sponsors a sort of "affirmative action" grant program for its own culture industries. Television programs about Canadian life, such as *Liberty Street,* a drama about twenty-something Canadians, are being subsidized by the government. The internationally famous music group, Crash Test Dummies, produced its first album with the help of a $40,000 grant from the Canadian government. In short, the story surrounding cultural nationalism is that the struggle to control cultural production is an intense one. This is especially true for Canada, which continues to struggle to establish an independent identity beyond the shadow of the United States.

Culture and Identity

In addition to exploring cultural forms such as religion and language and movements such as cultural nationalism, geographers have increasingly begun to ask questions about other forms of identity. This interest largely has to do with the fact that certain long-established and some more recently self-conscious cultural groups have begun to use their identities as a way of asserting political, economic, social, and cultural claims.

Sexual Geographies

Sexuality: set of practices and identities that a given culture considers related to each other and to those things it considers sexual acts and desires.

One such identity that has captured the attention of cultural geography is sexuality. **Sexuality** is a set of practices and identities that a given culture considers related to each other and to those things it considers sexual acts and desires. One of the earliest and most effective examples of the geographic study of sexuality is the examination of the spatial expression of prostitution. Research on prostitution in California found that sex work—as it is now more currently known—is spatially differentiated based upon the target clientele as well as systems of surveillance. Paralleling the different classes of men that female sex workers service, those oriented to an upper-income clientele perform their work in the homes, hotels, and private areas of upper- and middle-class society. Female sex workers serving a lower-class clientele, however, tend to operate within more public spaces or to bring clients to their dwellings or hotel rooms in the locales within which they "street walk" (Figure 5.28).

Not surprisingly, female sex workers at the lower end of the economic spectrum are more likely to be subject to police surveillance and other forms of public scrutiny as their advertising and work activities are largely carried out in public places. In contrast, those oriented toward a wealthier clientele have, in most cases, little or no obvious public identity. The distinction should be clear with respect to geography. The lower-class "red light" districts of most cities such as 42nd Street in New York and the Barri Xines in Barcelona are widely known to residents. But those sex workers that serve a higher-income clientele do not conform to segregated work spaces. This difference was made clear in the highly publicized trial of Heidi Fleiss, who allegedly procured sex workers for some of the biggest stars in Hollywood. Although Fleiss, obviously, did not escape the notice of police—she is currently in jail—the sex workers and the clients she represented have mostly remained anonymous to the public. As these examples illustrate, sexuality is also an identity that is cross-cut with other identities like class and certainly gender, race, and ethnicity.

More typical of more contemporary work on sexuality in geography, is that which explores the spatial constraints on homosexuality and the ways in which gays and lesbians respond to and reshape them. Two particular areas of research have emerged: gay and lesbian consumerism and political action.

The open expression of sexual preference in consumption has emerged in a phenomenon called "pink spending." Use of the word "pink" is a direct reference to the pink triangle Hitler required homosexuals to wear in Nazi concentration camps. Consumer support by gays and other consumers of openly gay businesses has enabled the establishment of identifiably gay spaces in the form of shopping districts and neighborhoods. These openly gay spaces have enabled gay communities in many of the core countries to gain significant political power in cities, regions, and even throughout the country. The gay-pride parades that occur in several major U.S. cities each year are also a reminder of the demographic and political power of gays and lesbians (Figure 5.29, page 216). They indicate attempts by homosexuals to resist the dominant ideology about sexuality and to occupy the spaces of U.S. cities on an equal footing with heterosexuals.

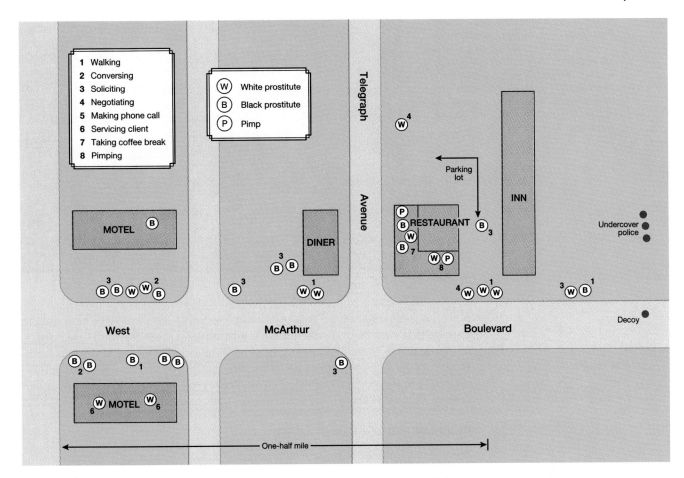

Figure 5.28 Map of prostitution This is a map of nocturnal activity among prostitutes, pimps, and customers at the "Meatrack," a center of street prostitution in Oakland, California. For working-class prostitutes, their trade is plied on the streets usually in known "red-light" districts either through their face-to-face solicitation or that of their pimps. This contrasts with higher-class prostitutes whose interactions with customers are most often arranged by someone else and who interact with customers in spatially dispersed settings. (*Source:* R. Symanski, *The Immoral Landscape: Female Prostitution in Western Societies.* Toronto: Butterworth-Heinemann Ltd., 1981, p. 19.)

Given the emphasis that the forces of globalization have placed on commodities and consumption in the contemporary world, it is not surprising that gays and lesbians constitute a market for advertisers, retailers, clubs, and restaurants. Rejecting tourist destinations that may emphasize nuclear-family fun or romantic heterosexual vacation spots, gays and lesbians can consume different and specially marketed alternative destinations. Global gay tourism guides explicitly market places like Amsterdam in Europe, San Francisco in the United States, and Rio de Janeiro in Brazil as "gay capitals" offering alternative places of tourism for gay consumers.

A second approach to homosexuality has coalesced around AIDS political activism and so-called "queer tactics" employed by such groups as ACTUP (AIDS Coalition To Unleash Power), GMFA (Gay Men Fighting AIDS), and Queer Nation. Members of these groups use direct forms of political action—more radically called *cultural terrorism*—to redefine what can be public in public space. These activists inhabit the public space in radically new ways through action meant to challenge the complacency of public agencies and conservative opponents. ACTUP, for example, has been militant in exposing the private sexual preferences of high-profile public figures. Alternatively, Queer Nation seeks a

Figure 5.29 Gay pride parade in San Francisco The gay pride parades in San Francisco and other cities across the United States provide an opportunity for gays, lesbians, and bisexuals to challenge the mainstream ideas about them and about heterosexuality by taking over spaces usually occupied by mainstream cultural, political, and economic pursuits.

playful questioning of rigid constructions of sexuality such as gay versus straight, arguing that such rigid distinctions are misleading. Queer Nation's direct forms of political action include kissing, holding hands, and dressing in drag in mainstream spaces like city streets and shopping malls. All of these political actions centered around sexuality are explicit challenges to conventional ideas about definitions of what is acceptable behavior in public and private spaces.

Ethnicity and the Use of Space

Ethnicity: a socially created system of rules about who belongs and who does not belong to a particular group based upon actual or perceived commonality.

Ethnicity is another way in which geographers are exploring cultural identity. **Ethnicity** is a socially created system of rules about who belongs to a particular group based upon actual or perceived commonality such as language or religion. A geographic focus on ethnicity is an attempt to understand how it shapes and is shaped by space and how ethnic groups use space with respect to mainstream culture. For cultural geographers, territory is also a basis for ethnic group cohesion (see Chapter 9 for more on territory). For example, cultural groups—ethnically identified or otherwise—may be spatially segregated from the wider society in ghettos or ethnic enclaves (see Chapter 11). Or these groups may use space to declare their subjective interpretations about the world they live in and their place in it. The use of the city streets by many different cultural groups demonstrates this point (Figure 5.30). Nineteenth-century immigrants to the U.S. city used the streets to broadcast their ideas about life in their adopted country (Figure 5.31).

Ethnic parades in the nineteenth century—such as St. Patrick's Day for the Irish and Columbus Day for the Italians—were often very public declarations about the stresses that existed between and among classes, cultures, and generations.

At the height of late nineteenth-century immigration, the Irish and many other immigrant groups such as the Italians, Greeks, Poles, and Slavs were largely shunned and vilified by the host society, and were relegated to low-paying jobs and poor housing. Publicly ridiculed in newspaper and other print venues, these groups took to the streets to reinterpret the city's public spaces, even if only for a day. Released from a strict work routine of 10 to 15 hours a day, participants and spectators could use the parade to promote an alternative reality of pride and festivity, among other things. Participants acquired a degree of power and autonomy that was not possible in their workaday lives. Because of their festive and extraordinary nature, nineteenth-century parades temporarily helped to change the world in which they occurred.

Figure 5.30 Indian parade, New York City
Contemporary ethnic groups in U.S. cities have also adopted the parade as a forum. Many groups must constantly confront prejudice and stereotyping on a daily basis at work and in their neighborhoods. The parades are an opportunity to promote pride and an alternative picture of what it means to belong to an ethnic group.

The same can be said of cultural parades in the twentieth century. The St. Patrick's Day parade is still a highly politically charged event whether held in Boston, New York, Chicago, or elsewhere where nontraditional Irish groups appealing for a place in the procession have recently been denied access. This has been the case with Irish-American gay and lesbian groups in Boston and New York in the 1990s. Proffering their own interpretations of "Irishness," these groups have been turned away by the more mainstream interpreters of the term. Such a confrontation between ethnicity and sexuality also highlights how difficult it is to separate cultural identity into distinct categories. In different places, for different historical reasons, the complex combinations of cultural identities of race, class, gender, and sexual preference result in unique and sometimes powerful expressions.

Race and Place

Prevailing ideas and practices with respect to race have also been used to understand the shaping of places and responses to these forces.[2] **Race** is a problematic classification of human beings based on skin color and other physical

Race: a problematic classification of human beings based on skin color and other physical characteristics.

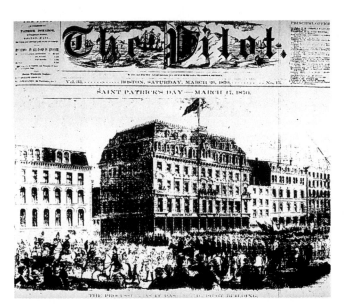

Figure 5.31 St. Patrick's Day parade, Boston, 1870 The use of the streets to broadcast messages about ethnicity has a long tradition in the U.S. city. St. Patrick's Day parades were actually invented in the United States as a way for immigrant Irish to promote themselves as respectable and hardworking people. It is important to note, however, that St. Patrick's Day parades were then and are now often exclusionary, meaning only certain members of the community have been allowed to participate and demonstrate their understandings of "Irishness." (Source: Boston Pilot, 1870.)

characteristics. *Biologically* speaking, however, no such thing as race exists within the human species. Yet, consider the categories of race and place that correspond to "Chinese" and "Chinatown." Powerful Western ideas about Chinese as a racial category enabled the emergence and perpetuation of Chinatown as a type of landscape found throughout many North American cities (Figures 5.32). In this and in other cases, the visible characteristics of hair, skin, and bone structure made race into a category of difference that was (and still is) widely accepted and often spatially expressed.

The mainstream approach to neighborhood is to see it as a spatial setting for systems of affiliation more or less chosen by people with similar skin color. Recently, cultural geographers have begun to overturn this approach and to see neighborhoods as spaces that affirm the dominant society's sense of identity. For example, from the perspective of white society, nineteenth-century Chinatowns were the physical expression of what set the Chinese apart from whites. The distinguishing characteristics revolved around the way the Chinese looked, what they ate, their non-Christian religion, their opium consumption, gambling habits, and other "strange" practices. Place—Chinatown—maintained and manifested differences between white and Chinese society. Furthermore, place con-

Figure 5.32 Chinatowns   Chinatowns are features of most major North American cities.

Chinatown, New York City—While they have been portrayed as being voluntarily occupied by Chinese immigrants, Chinatowns can also be seen in a different way. They are also landscapes of exclusion and racism created by the prejudices of the dominant society.

Chinatown, Vancouver, 1907—The marginalization of the Chinese population of Vancouver, British Columbia, had strong racial and ethnic undercurrents that surfaced in illustrations such as this one, published in 1907. Notice how the white settlements are pictured as airy, light, and single-family, whereas the Chinese domiciles are labeled as "warrens" "infested" by thousands of Chinese.

[2]Adapted from K. Anderson, "The Idea of Chinatown: The Power of Place and Institutional Practice in the Making of a Racial Category," *Annals of the Association of American Geographers* 77(4), 1987, pp. 580–598.

tinues to be a mechanism for creating and preserving local systems of racial classification and for containing geographical difference within defined geographical confines. The homelands of South Africa, discussed in Feature 9.2 in Chapter 9, are also an illustration of the interaction of race and place, though at a much larger scale.

Gender, Class, and Vulnerability

Gender is an identity that has received a great deal of attention by cultural geographers within the last two to three decades. **Gender** is a category reflecting the social differences between men and women rather than the anatomical differences that are related to sex. As with other forms of identity, gender implies a socially created difference in power between groups. In the case of gender, the power difference advantages males over females and is not biologically *determined*, but socially and culturally *created*. As with other forms of identity, class position can intensify the power differences among and between groups. Furthermore, the implications of these differences are played out differently in different parts of the world.

For example, Nigeria, part of the Sahelian region of central Africa, has been experiencing a severe drought that has lasted over many years (Figure 5.33). The

Gender: the social differences between men and women rather than the anatomical differences that are related to sex.

Figure 5.33 Hausaland in Northern Nigeria and Southern Niger This map of Hausaland shows the region of the Hausa people as straddling the political boundary between Nigeria and Niger. Standard political maps do not show the existence of Hausaland because it ceased to exist as a political entity at the beginning of the twentieth century when it was divided between the French and the British. On both sides of the boundary, however, Hausa people make their living farming millet and sorghum in the sandy soil of the Sahelian desert. (*Source:* Map, W. Miles, *Hausaland.* Ithaca: Cornell University Press, 1994. Used by permission of the publisher, Cornell University Press.)

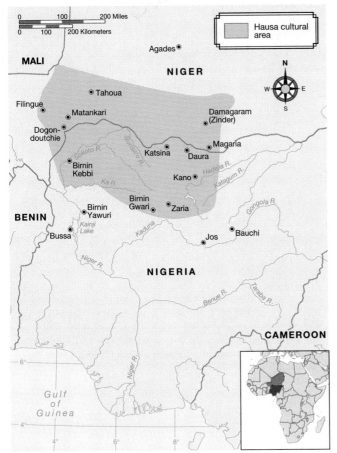

Gender systems that identify men as more valuable than women create situations of increased vulnerability to drought and famine for women and children in Sahelian Africa. The additional constraints of Islamic law may also work against women's ability to withstand the vicissitudes of natural disasters.

rural populations have been particularly hard hit and, without adequate water supplies, have been unable to grow enough food for subsistence. The result has been widespread famine in the rural areas. One of these areas is Hausaland in northern Nigeria where, because of the particularities of the gender and class systems, poor women, relative to their male counterparts, are especially vulnerable to malnourishment, undernourishment, and even starvation. Because of the gender system in Hausaland, women are more likely than men not to have access to food resources.[3] Because of the class system in Hausaland, peasant and working-class women are more likely than their male counterparts as well as higher-class women born into the urban and rural aristocracy to lack access to money resources to purchase food.

In Hausaland as elsewhere, vulnerability to drought is linked to the inability to prepare for its likelihood; to make mitigating adjustments when it occurs; and to develop new strategies, after a drought occurs, to prepare for future ones. *Gender vulnerability* refers to the fact that contemporary gender relationships among the Hausa make it more difficult for women to cope and adjust to drought situations than men. For instance, in Muslim households men eat before women and children, who are then allowed to consume whatever is left. In periods of sustained drought and famine, it is routinely the case that after men have eaten, there is no food left for women and children.

Numerous other factors affect gender vulnerability to drought among the Hausa. As Table 5.3 indicates, these include lower pay for all forms of women's work; restrictions on education for women; restrictions on direct access to public space; women's dependence on child labor; and lack of job opportunities outside of the private sphere for women. In Hausa society, Islamic law operates variably

TABLE 5.3 Factors Increasing Women's Vulnerability (Relative to Men) to Drought
High Effect
Low remuneration for all forms of women's work
Less ownership and control of the means of production
Restrictions on education
Competitive disadvantage
Medium Effect
Seasonal effects
Spatial restriction on ability to sell labor power
Dependence on child labor
Loss of parental rights through divorce
Lack of job opportunities in the public sphere
Discrimination in inheritance laws
Summary divorce
Inattention of government programs
Low or Unknown Effect
Virilocal (male-centered) marriage

Source: R. A. Schroeder, "Gender Vulnerability to Drought: A Case Study of the Hausa Social Environment," National Hazards Research Applications Information Center Working Paper No. 58. Boulder, Colorado: University of Colorado, 1987.

---

[3] Adapted from R. A. Schroeder, "Gender Vulnerability to Drought: A Case Study of the Hausa Social Environment," NHRAIC Working Paper No. 58. Boulder: University of Colorado, 1987.

TABLE 5.4	Factors Increasing Peasant Women's Vulnerability (Relative to Women in Ruling Classes) to Drought
High Effect	
Contributions to household subsistence	
Losses through sale of assets in distress	
Lack of investment capital	
Greater vulnerability to seasonal effects	
Smaller dowries	
Medium Effect	
Less reciprocal protection through extended family ties	
More completely bound by domestic responsibilities	
Less access to child labor	
Low or Unknown Effect/Low Reliability of Data	
Inability to educate children	
The early marriage of daughters	
Restriction on form of capital accumulation due to small compound size	

Source: R. A. Schroeder, "Gender Vulnerability to Drought: A Case Study of the Hausa Social Environment," National Hazards Research Applications Information Center Working Paper No. 58. Boulder, Colorado: University of Colorado, 1987.

among different geographically and socially located groups of women. While all of the factors listed in Table 5.4 are significant with respect to gender relations, they may operate more or less intensively on different *classes* of women.

In short, the power differentials of class mean that not all Muslim women in Hausaland experience the same level of vulnerability to drought. Table 5.4 outlines the factors that increase Muslim peasant women's vulnerability to drought relative to Muslim women in the ruling classes. In fact, women in the ruling classes tend to escape most aspects of drought vulnerability through their husbands' or their own access to economic resources. Peasant and other poor women experience few, if any, such mitigating factors, however.

Culture and the Physical Environment

While interest in culture and the built environment has become prominent among geographers over the last several decades, a great deal of attention continues to be paid to culture and the physical environment. As with Sauer's original concept of the cultural landscape, geographers continue to focus their attention on people's relationships to the natural world and how the changing global economy disrupts or shapes those relationships. In this section we look at two related but distinct ways of understanding the relationship between culture and the natural environment—cultural ecology and political ecology.

Cultural Ecology

Cultural ecology is the study of the relationship between a cultural group and its natural environment. Cultural ecologists study the material practices (food production, shelter provision, levels of biological reproduction) as well as the nonmaterial practices (belief systems, traditions, social institutions) of cultural groups. Their aim is to understand how cultural processes affect adaptation to the environment. Whereas the traditional approach to the cultural landscape focuses on *human impacts* on the landscape or its form or history, cultural ecologists seek to explain how cultural processes affect *adaptation* to the environment.

Cultural ecology: the study of the relationship between a cultural group and its natural environment.

Cultural adaptation: the complex strategies human groups employ to live successfully as part of a natural system.

Cultural adaptation involves the complex strategies human groups employ to live successfully as part of a natural system. Cultural ecologists recognize that people are components of complex ecosystems and that the way they manage and consume resources is shaped by cultural beliefs, practices, values, and traditions as well as larger institutions and power relationships.

The cultural ecology approach incorporates three key points:

- Cultural groups and the environment are interconnected by systemic interrelationships. Cultural ecologists must examine how people manage resources through a range of strategies to comprehend how the environment shapes culture and vice versa.

- Cultural behavior must be examined as a function of the cultural group's relationship to the environment through both material and nonmaterial cultural elements. Such examinations are conducted through intensive fieldwork.

- Most studies in cultural ecology investigate food production in rural and agricultural settings in the periphery in order to understand how change affects the relationship between cultural groups and the environment.[4]

These three points illustrate the way in which cultural geographers go about asking questions, collecting data, and deriving conclusions from their research. They also show how cultural ecology is both similar to and different from Sauer's approach to the cultural landscape, described at the beginning of the chapter. While each shares an emphasis on culture, in cultural ecology the cultural processes of particular groups have come to take center stage rather than the imprint that culture makes upon the landscape. As a result, cultural ecologists look at food production, demographic change and its impacts on ecosystems, and ecological sustainability. Additionally, the scale of analysis is not on culture areas or culture regions, but on small groups' adaptive strategies to a particular place or setting.

The impact of Spanish agricultural innovations on the culture of the indigenous people of the Central Andes region of South America (an area encompassing the mountainous portions of Peru, Bolivia, and Ecuador) presents an excellent case study in cultural ecology.[5] The transformation of Andean culture began when Pizarro arrived in Peru from Spain in 1531 and set about vanquishing the politically, technologically, and culturally sophisticated Inca empire. The Spaniards brought with them not only domestic plants and animals (mainly by way of Nicaragua and Mexico) but also knowledge about how to fabricate the tools they needed and a strong sense of what was necessary for a "civilized" life.

By 1620, however, the indigenous Andean people had lost 90 percent of their population and had been forced to make significant changes in their subsistence lifestyles (an illustration of demographic collapse as discussed in Chapter 4). The Inca empire, with its large population base, had once engaged in intensive agriculture practices, including building and maintaining irrigation systems, terracing fields, and furrowing hillsides. With the severe drop in population and consequent loss of labor power, the survivors turned to pastoralism because herding requires less labor than intensive agriculture. Ultimately, it was the introduction of Old World domesticated animals that had the greatest impact on the Central Andes (Figure 5.34).

Figure 5.34 Bolivian herders, Lake Titicaca Sheep are a nonindigenous animal to the Andes but have been widely adopted since the Columbian Exchange. Sheep are well adapted to the high altitudes and provide wool and meat. Shown are young girls of the Lake Titicaca region of Bolivia returning at day's end from herding.

[4]K Butzer, "Cultural Ecology," in G. L. Gaile and C. J. Wilmot (Eds.), *Geography in America.* Columbus, OH: Merrill Publishing Co., 1989, p. 192.
[5]Adapted from D. Gade, "Landscape, System, and Identity in the Post-Conquest Andes," *Annals of the Association of American Geographers* 82(3), 1992, pp. 461–477.

Of the range of animals the peasants could have incorporated into their agricultural practices (including cattle, oxen, horses, donkeys, pigs, sheep, goats, rabbits, and turkeys), however, only a few animals were widely adopted. Sheep were by far the most important introduction and were kept by Andean peasants as early as 1560. In many areas, sheep herding soon displaced the herding of indigenous animals such as llamas and alpaca. By the seventeenth century, at elevations below 3500 meters (11,550 feet), sheep herding had been fully integrated into peasant economies and practices.

Adoption of sheep herding was facilitated by several factors. Sheep wool was oilier than that produced by native animals, a quality that made wool clothing more water-resistant. Sheep also provided a source of meat, tallow, and manure for farm plots and could become an important source of food to a family in times of flood, frost, or drought. Sheep had a higher fertility level and a lower mortality rate than native herd animals and did not require large inputs of labor to manage. Finally, unlike crops, sheep could be marched to market on their own feet—no small advantage in the rugged terrain of the Andes.

Pigs and goats also proved popular because they fit well into available niches of the peasant economy. Rabbits, on the other hand, even though they produced more meat, never replaced the native guinea pigs. Guinea pigs retained a high cultural value as they continued to be a featured food at native ceremonies celebrating life-cycle events and curing rites. Likewise cows, though valued by the Spanish colonists, never became important to indigenous ways of life. Cows did not do well at higher altitudes or on steep terrain, and it was difficult to find appropriate fodder for them during the dry season. They also constituted a high risk, for they were an expensive investment, and loss or theft created economic hardship for the owner.

The pattern of selective adoption among Central Andean peasants also extended to plants. Of the approximately two dozen crops they could have adopted, Andean peasants adopted only about half of them. Peasants based their planting decisions on usefulness, on environmental fit, and on competition from other plants. For example, of the various grain crops that were available (rye, barley, oats, and wheat), Andean peasants adopted only wheat and barley. Andean peasants began to cultivate these grains in the highlands as early in the 1540s.

Both wheat and barley found a good ecological "fit" within the Central Andes, and could be integrated into the fallow cycles that the peasants had long practiced. These crops also had the advantage of supplementing the natives' array of foods because they complemented—rather than competed with—cultivation of indigenous crops such as maize and potatoes (Figure 5.35).

By the 1590s, a "bundle" of Spanish cultural traits had been integrated into the Central Andean rural culture complex, creating a hybridized rural culture. The hybridized culture—and cultural landscape—combined a much-simplified version of Spanish material life with important (though altered) Incan practices of crop growing, herding, agricultural technology, and settlement patterns. That this hybrid culture complex remains identifiable today, even after four centuries and in the face of contemporary globalizing forces, is due to a combination of factors: the peasants' strong adherence to custom; their geographic isolation; and the poverty of their circumstances.

Following the three points outlined earlier, cultural ecologists have been able to understand complex relationships between cultural groups and their environment, showing how choices are shaped by both culture and environmental conditions. Some critics have argued, however, that this conceptual framework of cultural ecology leaves out other intervening influences of the relationship between culture and the environment: the impact of the political and economic institutions and practices.

Figure 5.35 Andean potatoes in the marketplace The Andean region boasts hundreds of varieties of potatoes. The potato is one of the New World plants that was introduced into Europe following the Columbian Exchange. The sole reliance on the potato in nineteenth-century Irish agriculture led to disaster, however. The Great Potato Famine of the 1840s caused widespread mortality and migration. While many died, many others migrated to the United States, reducing the country's population to levels from which it still has not recovered.

Political Ecology

During the 1980s cultural ecologists began moving away from a strict focus on a particular group's interactions with the environment and placed that relationship within a wider context. The result is political ecology, the merging of political economy with cultural ecology. **Political ecology** stresses that human–environment relations can be adequately understood only by reference to the relationship of patterns of resource use to political and economic forces. Just as with the study of agriculture, industrialization, urbanization, and comparable geographical phenomena, this perspective requires an examination of the impact of the State and the market on the ways in which particular groups utilize their resource base.

Political ecology incorporates the same human–environment components analyzed by cultural ecologists. Yet, because political ecologists frame cultural ecology within the context of political and economic relationships, political ecology is seen to go beyond what cultural ecologists seek to understand.

A case study of the banana industry on St. Vincent, an island in the Caribbean, illustrates this difference (Figure 5.36).[6] Agriculturalists in St. Vincent have shifted to banana production for *export* as they have increasingly abandoned local food production, and without recognizing the impacts of politics and

Political ecology: approach to cultural geography that studies human–environment relations through the relationships of patterns of resource use to political and economic forces.

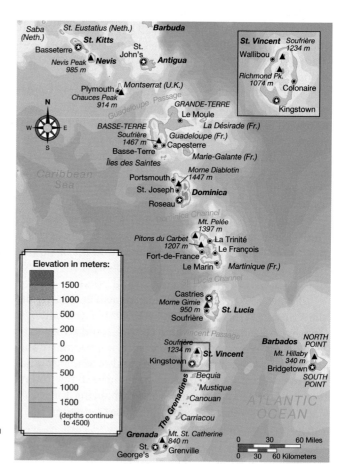

Figure 5.36 St. Vincent and the Grenadines St. Vincent and the Grenadines are part of the island chain of the Lesser Antilles in the Caribbean Sea. The total population is about 115,000, occupying an area of 150 square miles.

[6]Adapted from L. Grossman, "The Political Ecology of Banana Exports and Local Food Production in St. Vincent, Eastern Caribbean," *Annals of the Association of American Geographers* 83(2), 1993, pp. 347–367.

the wider economy, it would be impossible to understand why. Disincentives and incentives have both played a role. *Disincentives* to maintain local food production include marketing constraints, the quality of available land, and the need to use land for other purposes. *Incentives* to produce for export include state subsidies to export-oriented agriculture and access to credit for banana producers, as well as a strong British market for Caribbean bananas. As a result, local food production, although faced with the same environmental conditions as banana production, does not enjoy the same political and economic benefits. Because production for export is potentially more lucrative and an economically safer option and, to some extent because of changing dietary practices, local food production is a less attractive option for agriculturalists.

As the St. Vincent case illustrates, the political ecology approach provides a framework for understanding how the processes of the world economy affect local cultures and practices. The St. Vincent case also indicates how State policies and practices and economic demand in the global economy shape local decision making. Furthermore, local cultural practices (especially dietary) are being abandoned as people develop a taste and preference for low-cost and convenient imported agricultural commodities (including flour and rice). Unfortunately, however, production for export also opens up the local economy to the fluctuations of the wider global economy. Recent changes in European Union policy on banana imports, for example, are likely to have negative effects on banana production in St. Vincent.

Globalization and Cultural Change

The discussion of cultural geography in this chapter has raised one central question: How has globalization changed culture? We have seen that it affects different cultural groups differently and that different groups respond in different ways to these changes. With so much change occurring for so long, however, we must still ask ourselves what impact, overall, globalization is having on the multiplicity of culture groups that inhabit the globe.

Anyone who has ever traveled between major world cities—or, for that matter, anyone who has been attentive to the backdrops of movies, television news stories, and magazine photojournalism—will have noticed the many familiar aspects of contemporary life in settings that, until recently, were thought of as being quite different from one another. Airports, offices, and international hotels have become notoriously alike, and their similarities of architecture and interior design have become reinforced by near-universal dress codes of the people who frequent them. The business suit, especially for males, has become the norm for office workers throughout much of the world. Jeans, T-shirts, and sneakers, meanwhile, have become the norm for both young people and those in lower-wage jobs. The same automobiles can be seen on the streets of cities throughout the world (though sometimes they are given different names by their manufacturers); the same popular music is played on local radio stations; and many of the movies shown in local theaters are the same (*The Flintstones, Jurassic Park, Terminator 2, Forrest Gump,* and *Independence Day,* for example). Some of the TV programming is also the same—not just the music videos on MTV and STAR TV, but CNN's news, major international sports events, drama series like *Dallas,* and comedy series like *Cheers* and *Friends.* The same brand names also show up in stores and restaurants: Coca-Cola, Perrier, Carlsberg, Nestlé, Nike, Seiko, Sony, IBM, Nintendo, and Microsoft, to list just a few. Everywhere there is Chinese food, pita bread, pizza, classical music, rock music, and jazz.

It is these commonalities that provide a sense of familiarity among the inhabitants of the "fast world" that we described in Chapter 2. From the point of view of cultural nationalism, the "lowest common denominator" of this familiarity is

Figure 5.37 Benetton in Romania Even the newly democratized country of Romania has a Benetton's as shown by the green Benetton's sign in the window of the building in the foreground right. Although hugely expensive, frequenting shops like Benetton's and restaurants like McDonald's is a sign of status and personal prosperity in Romania and other recently communist countries.

often seen as the culture of fast food and popular entertainment that emanates from the United States. Popular commentators have observed that cultures around the world are being Americanized, or "McDonaldized," which represents the beginnings of a single global culture that will be based on material consumption, with the English language as its medium (Figure 5.37).

There is certainly some evidence to support this point of view, not least in the sheer numbers of people around the world who view *Sesame Street,* drink Coca-Cola, and eat in McDonald's franchises or similar fast-food chains. Meanwhile, U.S. culture is increasingly embraced by local entrepreneurs around the world. Travel writer Pico Iyer, for example, describes finding dishes called "Yes, Sir, Cheese My Baby" and "Ike and Tuna Turner" in a local buffeteria in Guanzhou, China.[7] It seems clear that U.S. products are consumed as much for their symbolism of a particular way of life as for their intrinsic value. McDonald's burgers, along with Coca-Cola, Hollywood movies, rock music, and NFL and NBA insignia have become associated with a lifestyle package that features luxury, youth, fitness, beauty, and freedom.

The economic success of the U.S. entertainment industry has helped reinforce the idea of an emerging global culture based on Americanization. In 1996, the entertainment industry was a leading source of foreign income in the United States, with a trade surplus of $23 billion. Similarly, the United States transmits much more than it receives in terms of the sheer volume of cultural products. In 1995, the originals of over half of all the books translated in the world (more than 20,000 titles) were written in English. In terms of international flows of everything from mail and phone calls to press-agency reports, television programs, radio shows, and movies, a disproportionately large share originates in the United States.

---

[7]P. Iyer, *Video Nights in Kathmandu: Reports from the Not-So-Far East.* London: Black Swan, 1989.

Neither the widespread consumption of U.S. and U.S.-style products nor the increasing familiarity of people around the world with global media and international brand names, however, adds up to the emergence of a single global culture. Rather, what is happening is that processes of globalization are exposing the inhabitants of both the fast world and the slow world to a common set of products, symbols, myths, memories, events, cult figures, landscapes, and traditions. People living in Tokyo or Tucson, Turin or Timbuktu may be perfectly familiar with these commonalities without necessarily using or responding to them in uniform ways. Audience research on *Dallas,* for example, has shown that viewers from different cultural backgrounds interpret the program in different ways, depending on their own cultural contexts. Because of the particular conceptions of kinship and interpersonal relations held by Australian aboriginals, for example, their interpretations of episodes of *Dallas* involve meanings that are quite different from those intended by the program's makers.

Equally, it is important to recognize that cultural flows take place in all directions, not just outward from the United States. Think, for example, of European fashions in U.S. stores; of Chinese, Indian, Italian, Mexican, and Thai restaurants in U.S. towns and cities; and of U.S. and European stores selling exotic craft goods from the periphery.

A Global Culture?

The answer to the question "Is there a global culture?" then, must be no. While an increasing familiarity exists with a common set of products, symbols, and events (many of which share their origins in U.S. culture of fast food and popular entertainment), these commonalties become configured in different ways in different places, rather than constitute a single global culture. The local interacts with the global, often producing hybrid cultures. Sometimes, traditional, local cultures become the subject of global consumption; sometimes it is the other way around. This is illustrated very well by the case of two suqs (linear bazaars) in the traditional medieval city of Tunis, in North Africa. Both suqs radiate from the great Zaytuna Mosque, which always constituted the geographic focal point of the old city. One suq, which was once *the* suq, leads from the mosque to the gateway that connects the medieval core to the French-built new city, where most tourists tend to stay. The second sets off at right angles to another exit from the formerly walled city.

> The first suq now specializes in Tunisian handicrafts, "traditional" goods, etc. It has kept its exotic architecture and multicolored colonnades. The plaintive sound of the ancient nose flute and the whining of Arabic music provide background for the European tourists in their shorts and T-shirts, who amble in twos and threes, stopping to look and to buy. Few natives, except for sellers, are to be seen. The second suq, formerly less important, is currently a bustling madhouse. It is packed with partially veiled women and younger Tunisian girls in blouses and skirts, with men in knee-length tunic/toga outfits or in a variety of pants and shirts, with children everywhere. Few foreigners can be seen. The background to the din is blaring rock and roll music, and piled high on the pushcarts that line the way are transistor radios, watches, blue jeans (some prewashed), rayon scarves, Lux face and Omo laundry soaps.[8]

---

[8]J. Abu-Lughod, "Going Beyond Global Babble," in A. D. King (Ed.), *Culture, Globalization, and the World-System.* Basingstoke, England: Macmillan, 1991, p. 132.

Conclusion

Culture is a complex and exceedingly important concept within the discipline of geography. A number of approaches exist to understand culture. It may be understood through a range of elements and features from single traits to complex systems. Cultural geography recognizes the complexity of culture and emphasizes the roles of space, place, and landscape and the ecological relationships between cultures and their environment. It distinguishes itself from other disciplinary approaches providing unique insights that reveal how culture shapes the worlds we live in at the same time that the worlds we inhabit shape culture.

Two of the most universal forms of cultural identity are religion and language. Despite the secularization of many people in core countries, religion is still a powerful form of identity, and it has been used to mediate the impacts of globalization. Resulting from globalization have been dramatic changes in the distribution of the world's religions as well as interaction among and between religions. Most remarkable has been the fact that religious conversion to religions of the periphery is now under way in the core.

While the number of languages that exist worldwide are threatened by globalization, some cultures have responded to the threat by providing special protection for regional languages. The 500-year history of globalization has resulted in the steady erosion of many regional languages and heavy contact and change in the languages that persist. It appears recently, however, that some governments are taking action to protect official and regional languages against the onslaught of globalization. Not only religion and language, however, are at risk from globalization, but other forms of cultural expression such as art and film are as well.

Cultural geographers are also interested in understanding how culture shapes groups' adaptations to the natural environment. The aim of cultural ecology is to understand how the availability of resources and technology, as well as value and belief systems, shapes the behaviors of culture groups as active modifiers of, and adapters to, the natural environment. Recently, geographers have begun to pay attention to the role of politics and the wider economy in understanding the relationship between adaptive strategies and the natural world. This approach is known as political ecology.

Different groups in different parts of the world have begun to use cultural identities such as gender, race, ethnicity, and sexuality as a way of buffeting the impacts of globalization on their lives. It is also the case that when the impacts of globalization are examined at the local level, some groups suffer more harm or appreciate the benefits more than others. The unevenness of the impacts of globalization and the variety of responses to it indicate that the possibility of a monolithic global culture wiping out all forms of difference is limited.

Main Points

- While culture is a central, complex concept in geography, it may be thought of as a way of life involving a particular set of skills, values, and meanings.

- Geographers are particularly concerned about how place and space shape culture and, conversely, how culture shapes place and space. They recognize that culture is dynamic, and is contested and altered within larger social, political, and economic contexts.

- Like other fields of contemporary life, culture has been profoundly affected by globalization. However, globalization has not produced a homogenized culture so much as it has produced distinctive impacts and outcomes in different societies and geographical areas as global forces come to be modified by local cultures.

- Contemporary approaches in cultural geography seek to understand the role played by politics and the economy in establishing and perpetuating cultures, cultural landscapes, and global patterns of culture traits and culture complexes.

- Cultural geography has been broadened to include analysis of gender, class, sexuality, race, ethnicity, stage in the life-cycle, and so on, in recognition of the fact that important differences can exist within as well as between cultures.

- Cultural ecology, an offshoot of cultural geography, focuses on the relationship between a cultural group and its natural environment.

- Political ecologists also focus on human–environment relations, but stress that relations at all scales from the local to the global are intertwined with larger political and economic forces.

Key Terms

cultural adaptation (p. 222)
cultural complex (p. 195)
cultural ecology (p. 221)
cultural geography (p. 191)
cultural hearths (p. 204)
cultural landscape (p. 192)

cultural nationalism (p. 208)
cultural region (p. 196)
cultural system (p. 197)
cultural trait (p. 195)
culture (p. 190)
dialects (p. 204)
diaspora (p. 198)

ethnicity (p. 216)
gender (p. 219)
genre de vie (p. 194)
historical geography (p. 194)
language (p. 201)
language branch (p. 204)
language family (p. 204)

language group (p. 204)
political ecology (p. 224)
race (p. 217)
religion (p. 197)
rites of passage (p. 195)
sexuality (p. 214)

Exercises

On the Internet

In this chapter, the Internet exercises will explore the relationship between culture and geography. The Internet provides an especially valuable tool for studying this relationship because it provides current information on the impacts of globalization on people in places. After completing these exercises you will have a clearer sense of the reach and impacts of globalization on the cultural practices of different groups of people in different parts of the world.

Unplugged

1. The world's cultures have been represented to the English-speaking public for over 100 years through the pages of *National Geographic* magazine. The thousands upon thousands of photographs have provided armchair travelers with a rear window on the world. One way to observe changes in cultural representations is to evaluate the composition and framing of those glossy photographic images that are the hallmark of the magazine. You will need to use your university or college library as a resource for this exercise. Your task is to identify a country or a region and explore the kinds of images that have been published in *National Geographic* to represent it over at least a 50 year period. In evaluating the images, consider that when we look at anything, we are always looking at the relationship between the object of our gaze and ourselves.

A standard format exists for looking at and talking about photographs. Consider the following:

Angle—How does the angle affect the photograph? (The angle influences the composition and the content of an image.)

Framing—What might have been visible beyond the edge of the picture? (Framing is a process of inclusion and exclusion that also affects composition and content.)

Light—How does the light contribute to the feeling conveyed by the photograph? (Light shows details and creates shadows and sets the mood of an image.)

All of these elements affect the way in which the content of the photograph is conveyed to the viewer. Using these variables, examine the photos for the ways in which a culture you have chosen is represented over time. How are women represented? How are gender relations? What activities are men involved in? Children? How similar or different are other cultures pictured relative to your own? If similarities exist, do they change over the time period? How has globalization affected the culture you are examining?

2. Ethnic identity is often expressed spatially through the existence of neighborhoods or business areas

dominated by members of a particular group. One way to explore the spatial expression of ethnicity in a place is to look at newspapers over time. In this exercise you are expected to look at ethnic change in a particular neighborhood over time. You can do this by using your library's holdings of local or regional newspapers. Examine change over at least a four-decade period. To do this you must identify an area of the city in which you live or some other city for which your library has an extensive newspaper collection. You should trace the history of an area you know is now occupied by a specific ethnic group. How long has the group occupied that area? What aspects of the group's occupation of that area have changed over time (school, church, or sports activities, or the age of the households)? If different groups have occupied the area, what might be the reasons for the changes?

3. College and university campuses generate their own cultural practices and ideas that shape behaviors and attitudes in ways that may not be so obvious at first glance. For this exercise you are asked to observe a particular practice that occurs routinely at your college or university. For example, fraternity and sorority initiations are important rituals of college life, as are sports events and even class discussions. Observe a particular cultural practice that is an ordinary part of your life at college. Who are the participants in this practice? What are their levels of importance? Are there gender, age, or status differences in the carrying out of this ritual or practice? What are the time and space aspects of the practice? Who controls its production? What are the intended outcomes of the practice? How does the practice or ritual contribute to the maintenance or disruption of order in the larger culture?

Chapter 6

Interpreting Places and Landscapes

Shark Dreaming,
Australian Aboriginal art

The aboriginal peoples of Australia have produced beautiful and fascinating images of their landscape known as Dreamings. The Dreamings reflect a religious world-view as well as a kind of fundamental law instructing all aboriginal tribes on how to interact with the surrounding environment. Dreamings are also held to involve ancestral beings who remain present within the belief systems and everyday practices of aboriginal life. At places where ancestral beings lived and died and left their special imprint on the land, secret rituals and ceremonies occur. The aboriginals believe that ancestral beings as well as plants, animals, and natural forces such as lightning and water were responsible for the creation of physical features and ensuring the productivity of the land and the availability of water. In the process of creating the world, these ancestral beings traveled across the land along clearly defined routes, known as the Dreamtime Tracks or Songlines. These Dreamtime Tracks connect places and the sacred events that occurred there.

Aboriginal Dreamings are just one expression of the human relationship to the environment. In the case of the Australian Aborigines, that relationship is deeply religious and is passed down from one generation to the next through song, dance, stories, and paintings. Every initiated male is responsible for the Dreaming that connects his family to their inherited land. These men exercise their Dreaming responsibilities by learning and teaching their associated stories, songs, and dances, as well as painting the image of the Dreaming. Nothing less than cultural maps of the artist's family territory, the paintings appear to be abstract. In actuality, however, the combination of dots, circles, lines, and symbols is actually representational of the artist's attachment or feelings for the landscape. It has even been said that the Australian desert looks very similar to the aboriginal Dreamings when viewed from an aircraft.

While aboriginal knowledge of the environment is culturally based, certainly other sources for environmental knowledge and understanding exist to the world at large, from direct individual-level experience to scientific research. Contrast, for example, the aboriginal Dreamings with a NASA-generated satellite image of Earth from space. In this chapter we explore the relationships between people, landscape, and place in order to assess how individuals and groups experience their environments, create places, and find meaning in the landscapes they create.

Behavior, Knowledge, and Human Environments

In addition to understanding how the environment shapes and is shaped by people, geographers also seek to identify how it is *perceived* and *understood* by people. Arguing that there is an interdependence between people and places, geographers explore how individuals and groups acquire knowledge of their environments and how this knowledge shapes their attitudes and behaviors. Some geographers have focused their research on natural hazards as a way of addressing environmental knowledge, while others have tried to understand how people ascribe meaning to landscape and places. In this chapter we take the key geographical concepts of place, space, and landscape and explore the ways in which people understand them, create them, and operate within them.

In their attempts to understand environmental perception and knowledge, geographers share a great deal with other social scientists but especially with psychologists. Human cognition and behavior have always been at the center of psychology. What makes *environmental* knowledge and behavior uniquely geographical is its relation to both the environmental context and the humans who struggle to understand and operate within it. Much of what we as humans know about the environment we live in is learned through both direct and indirect experience. Our environmental knowledge is also acquired through a filter of personal and group characteristics such as race, gender, stage of the life cycle, religious beliefs, or even where we live.

For instance, children have their own geographies and display interesting and distinct relationships to the physical and cultural environment. How do children acquire knowledge about their environments? What kind of knowledge do they acquire and how do they use it? What role do cultural influences play in the process? What happens when larger social, economic, and environmental changes take place? Geographer Cindi Katz traveled to rural Sudan to find answers to these questions. Working with a group of 10-year-old children in a small village, she sought to discover how they acquired environmental knowledge. What she also learned, through this work, was how the transformation of the agriculture of the region changed both the children's relationships to their families and community, and also their perceptions of nature.[1] In this Sudanese village, as in similar communities elsewhere in the periphery, children are important contributors to subsistence activities, especially planting, weeding, and harvesting. These villagers are strict Muslims and thus have strict rules about what female members of the community are allowed to do and where they are allowed to go. In this region, many of the subsistence activities that required leaving the family compound were customarily delegated to male children. In fact, within the traditional subsistence culture, boys predominated in all agricultural tasks except planting and harvesting, and were responsible for herding livestock as well. Many boys (and occasionally girls) were also responsible for fetching water and helping to gather firewood. By contrast, both boys and girls collected seasonal foods from lands surrounding the village. Work and play were often mixed together, and play, as well as work, provided a creative means for acquiring and using environmental knowledge, and for developing a finely textured sense of their home area (Figure 6.1, page 234).

What happens when globalization reshapes the agricultural production system, as it did in the village when irrigated cash crop cultivation was introduced? A Sudanese government development project transformed the agriculture of the region from subsistence-level livestock raising and cultivation of sorghum and sesame to cultivation of irrigated cash crops such as cotton. The new cash-crop regime, which required management of irrigation works, application of fertilizers, herbicides, and pesticides, and more frequent weeding, actually required children—as well as adults—to work longer and harder. Parents were often forced to keep their children out of school because many of the tasks had to be done during the school term.

Loss of forests required children to range farther afield to gather fuelwood, and to make gathering trips more frequently (Figure 6.2, page 234). Soon, wealthier households began buying wood rather than increasing demands on their children or other household members. For the children of more marginalized households, the selling of fuelwood, foods, and other items provided a new means for earning cash to support their families, but also placed increasing demands on their energies and resourcefulness, and changed their whole experience of their world.

For the children of this village, globalization (in the form of the transition to cash-crop agriculture) changed their relationships with their environments and with their futures. The kinds of skills the children had learned for subsistence production were no longer useful for cash-crop production. As they played less and worked longer hours in more specialized roles, their experience of their environment became narrowed. As their roles within the family changed, they attended school less and learned less about their world through formal education. As a result, their perceptions of their environment changed, along with their values and attitudes toward the landscape and the place they knew as home.

---

[1]C. Katz, "Sow What You Know: The Struggle for Social Reproduction in Rural Sudan," *Annals of the Association of American Geographers*, 81, 1991, pp. 488–514.

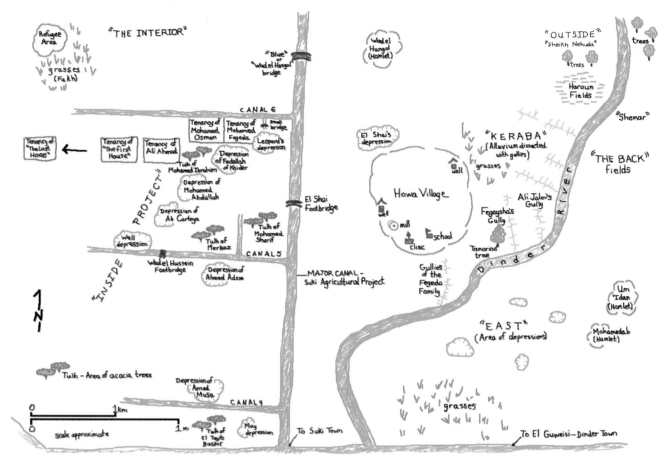

Figure 6.1 Shepherd's map This map, drawn by a 10-year-old boy, shows the area over which the sheep are herded. It illustrates the detailed environmental knowledge Sudanese children possess of the landscape that surrounds their village. The village is an Islamic one, and norms determine the kinds of tasks in which boys and girls can participate. Only boys are allowed to tend sheep, allowing a very particular environmental knowledge about grazing areas and water availability. (*Source:* C. Katz, "Sow What You Know: The Struggle for Social Reproduction in Rural Sudan," *Annals of the Association of American Geographers,* 81, 1991, pp. 488–514.)

Today, many of the children face a considerable challenge, for there simply are not enough agricultural jobs to go around. Many of the boys—especially those whose parents are tenant farmers and lacking status—will be left with few options but to become agricultural wage laborers or to seek nonagricultural work outside the village. The girls, trained largely to assume their mothers' household roles, face another challenge—how to take over the agricultural tasks previously performed

Figure 6.2 Sudanese children working and playing The introduction of an agricultural development project brought the forces of globalization very directly to this small village in Sudan. The project dramatically changed both work and play for children and created a gap between the environmental knowledge they were learning and the knowledge they would need to survive under different economic circumstance. In this photograph, two 10-year-old girls are shown gathering firewood for domestic use. The wood is of very poor quality, and the girls must range far to procure it. These branches were gathered in and around the irrigation canal, which for all members of the village is a risky environmental site. The waters in the canals contain the parasitic blood flukes that cause schistosomiasis.

by the boys, but without the advantage of knowing how to do those tasks or use the proper tools. Traditional relationships within village society have been undermined by globalization, and traditional opportunities to acquire important environmental knowledge are being lost to new generations of children.

Landscape As a Human System

The foundation for geographers' interest in landscape was Carl Sauer's concept of the cultural landscape, described in Chapter 5. Since 1925, when Sauer advocated the study of the cultural landscape as a uniquely geographical pursuit, new generations of geographers have been expanding the concept. Generally speaking, landscape is a term that means different things to different people. For some, the term brings to mind the design of formal gardens and parks as in landscape architecture. For others, landscape signifies a bucolic countryside or even the organization of plantings around residences and public buildings. For still others, landscape calls to mind the artistic rendering of scenery as in landscape painting.

What Is Landscape?

Contemporary geographers think of landscape as a comprehensive product of human action such that every landscape is a complex repository of society. It is a collection of evidence about our character and experience, our struggles and triumphs as humans. To understand better the meaning of landscape, geographers have developed different categories of landscape types based on the elements contained within them.

Ordinary landscapes (or **vernacular landscapes,** as they are sometimes called) are the everyday landscapes that people create in the course of their lives together. From parking lots and trailer parks to tree-shaded suburbs and the patchwork fields of midwestern farms, these are landscapes that are lived in and changed and that in turn influence and change the perceptions, values, and behaviors of the people who live and work in them.

Symbolic landscapes, by contrast, stand as representations of particular values or aspirations that the builders and financiers of those landscapes want to impart to a larger public. For example, the neoclassical architecture of the federal government buildings in Washington, DC, along with the streets, parks, and monuments of the capital, constitute a symbolic landscape intended to communicate a sense of power, but also of democracy in its imitation of the Greek city-state (Figure 6.3, page 236). "Main Street, U.S.A." is a symbol—an idea—derived from the small towns that sprang up in places like Illinois and Iowa in tandem with the growth of the nation's railroad network (Figure 6.4, page 237). Today this symbolic landscape is memorialized not only in literature but also in Disneyland, where the focal point of the amusement park is a carefully scaled-down version of this culturally important American landscape. Geographers also speak about "landscapes of power" such as military outposts, and "landscapes of despair" such as homeless encampments and **derelict landscapes.** The latter are ones that have experienced abandonment, misuse, disinvestment, or vandalism.

Geographers now recognize that many layers of meaning are embodied in the landscape, meanings that can be expressed and understood differently by different social groups at different times. Put another way, many cultural landscapes exist in any single place. These landscapes reflect the lives of ordinary people as well as the more powerful, and they reflect their dreams and ideas as well as their material lives.

One outcome of this shift to individual subjectivity in the study of landscape—also known as the humanistic approach—was the emergence of the study of environmental perception, which pointed out that different people

Ordinary landscapes (vernacular landscapes): the everyday landscapes that people create in the course of their lives.

Symbolic landscapes: representations of particular values and/or aspirations that the builders and financiers of those landscapes want to impart to a larger public.

Derelict landscapes: landscapes that have experienced abandonment, misuse, disinvestment, or vandalism.

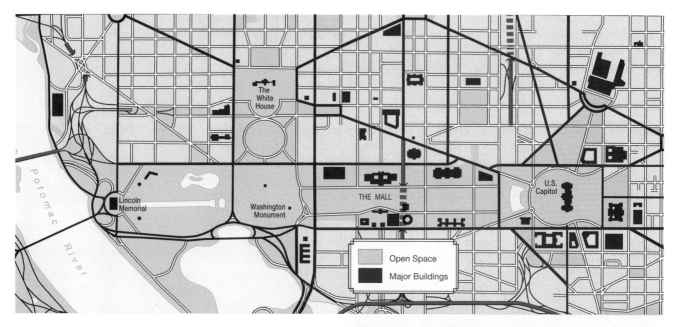

Figure 6.3 Lincoln Memorial and the Mall The Lincoln Memorial, like the other public buildings, monuments, and statuary around the Mall in Washington, DC, is intended to convey a sense of sobriety, authority, and the power of democratic principles. The Mall itself has been designed to communicate pomp and distinction. The wide entry to the Lincoln Memorial and the seated statue of Lincoln, inside, are meant to inspire awe and appreciation for democratic principles and the wisdom of Lincoln as a democratic leader.

Humanistic approach: places the individual—especially individual values, meaning systems, intentions, and conscious acts—at the center of analysis.

comprehend the landscape differently. The **humanistic approach** in geography is one that places the individual—especially individual values, meaning systems, intentions, and conscious acts—at the center of analysis. As the Sudanese example above suggests, children's perceptions of their worlds are different from those of their parents, and girls see their world differently from boys, even in the same family.

Environmental perception, and its close relative behavioral geography, are interdisciplinary, drawing together geographers, landscape architects, psychologists, architects, and others. Professionals in these disciplines conduct investigations into individuals' preferences in landscapes, into how people construct cognitive images of their worlds, and into how they find (or fail to find) their way around in various settings. Recall from Chapter 1 that cognitive images are representations of the world that can be called to mind through the imagination. The humanistic approach's focus on the perceptions of individuals is an important counterweight to the tendency to generalize at higher levels of aggregation. Nevertheless, some critics argue that such research has limited utility because individual attitudes and views do not necessarily add up to the views held by a group or a society.

Environmental perception scholars, and humanistic geographers more generally, work at larger levels of aggregation to try to address this conundrum. Since

Figure 6.4 Main Street, Disneyland The entire structure of Disneyland is organized around Main Street, U.S.A., a replica of an imaginary early twentieth-century small town. It is situated at the entrance to the park, and from it all other amusements can be reached. Main Street, U.S.A., is the spatial mediator between "America" and all other culture represented in the park. Importantly, Main Street is not supposed to represent any particular Main Street but a hype- real one in which exist none of the problems experienced in real American cities and towns.

the 1970s, however, their approach has been overshadowed by orientations that explore both the role of larger forces, such as culture, gender, and the State, and the ways in which these forces enhance or constrain people's lives. The traditional notion—that humans create culture and, through culture, create landscape—has consequently been replaced. Much recent cultural geographical work conceptualizes the relationship of people and the environment as interactive, not one-way, and emphasizes the role that landscapes play in shaping and reinforcing human practices. This most recent conceptualization of landscape is more dynamic and complex than the one Carl Sauer advanced, and it encourages geographers to look outside their own discipline—to anthropology, psychology, sociology, and even history—to understand fully its complexity.

Landscape As Text

Such a dynamic and complex approach to understanding landscape is based on the conceptualization of **landscape as text,** by which we mean that, like a book, landscape can be read and written by groups and individuals. This approach departs from traditional attempts to systematize or categorize landscapes based on the different elements they contain. The landscape-as-text view holds that landscapes do not come ready-made with labels on them. Rather, there are "writers" who produce landscapes and give them meaning, and there are "readers" who consume the messages embedded in landscapes. The messages embedded in landscape can be read as signs about values, beliefs, and practices, though not every reader will take the same message from a particular landscape (just as people may differ in their interpretation of a passage from a book). In short, landscapes both produce and communicate meaning, and one of our tasks as geographers is to interpret those meanings. Later in this chapter, in the section on coded spaces, we provide an extended example of the "writing" and "reading" of two landscapes: the shopping mall and the capital city of Brazil, Brasilia. First, however, we must establish how places and spaces are given meaning by individuals and by different social and cultural groups.

Landscape as text: the idea that landscapes can be read and written by groups and individuals.

Place Making/Place Marketing

Some social scientists believe that humans, like many other species, have an innate sense of territoriality. The concept of **territoriality** refers to the persistent attachment of individuals or peoples to a specific location or territory. The concept is

Territoriality: the persistent attachment of individuals or peoples to a specific location or territory.

Ethology: the scientific study of the formation and evolution of human customs and beliefs.

important to geographers because it can be related to fundamental place-making forces. The specific study of people's sense of territoriality is part of the field of **ethology,** the scientific study of the formation and evolution of human customs and beliefs. The term is also used to refer to the study of the behavior of animals in their natural environments. According to ethologists, humans carry genetic traits produced by our species' need for territory. Territory provides a source of physical safety and security, a source of stimulation (through border disputes), and a physical expression of identity. These needs add up to a strong territorial urge that can be seen in the microgeography of people's behavior: claims to space in reading rooms or on beaches, for example, and claims made by gangs to neighborhood turf (Figure 6.5). Ethologists argue that the territorial urge also can be observed in cases where people become frustrated because of overcrowding. They become stressed and, in some circumstances, begin to exhibit aggressive or deviant behavior. Crowding has been linked by ethologists and environmental psychologists to everything from vandalism and assault to promiscuity, listlessness, and clinical depression.

While such claims are difficult to substantiate, as is the whole notion that humans have an inborn sense of territoriality, the idea of territoriality as a product of *culturally* established meanings is supported by a large body of scientific evidence. Some of this evidence comes from the field of proxemics. **Proxemics** is the study of the social and cultural meanings that people give to personal space. These meanings make for unwritten territorial rules (rather than a biological urge) that can be seen in the microgeography of people's behavior. It has been established, for example, that people develop unwritten protocols about how to claim space. One common protocol is simply regular use (think of people's habits in classroom seating arrangements). Another is through the use of spatial markers such as a newspaper or a towel to fix a space in a reading room or on a beach. There are also bubbles, or areas, of personal space that we try not to invade (or allow to be invaded by others). Varying in size and shape according to location and circumstance, these bubbles tend to be smaller in public places and in busier and more crowded situations; they tend to be large among strangers and in situations involving members of different social classes; and they tend to vary from one social class or cultural group to another.

Proxemics: the study of the social and cultural meanings that people give to personal space.

At larger spatial scales, territoriality is mostly a product of forces that stem from social relations and cultural systems. This dimension of territoriality underpins a great deal of human geography. All social organizations and the

Figure 6.5 Graffiti as territorial markers
Graffiti are used by neighborhood-based youth groups to establish and proclaim their identity. Some graffiti, such as these, also function as simple territorial markers that help to stake out their "turf" in high-density environments where there exist few other clues about claims to territory.

individuals that belong to them are bounded at some scale or another through formal or informal territorial limits. Many of them—nations, corporations, unions, clubs—actually claim a specific area of geographic space under their influence or control. In this context, territoriality can be defined as any attempt to assert control over a specific geographic area to achieve some degree of control over other people, resources, or relationships. Territoriality also is defined as any attempt to fulfill socially produced needs for identity, defense, and stimulation. Territoriality covers many different phenomena, from the property rights of individuals and private corporations, to the neighborhood covenants of homeowners' associations; the market areas of commercial businesses; the heartlands of ethnic or cultural groups; the jurisdictions of local, state, and national governments; and the reach of transnational corporations and supranational organizations.

Territoriality thus provides a means of accomplishing three social and cultural needs:

- The regulation of social interaction
- The regulation of access to people and resources
- The provision of a focus and symbol of group membership and identity

Territoriality does so because, among other things, it facilitates classification, communication, and enforcement. We can classify people and/or resources in terms of their location in space much more easily than we can classify them in relation to personal or social criteria. All that is necessary to communicate territory is a simple marker or sign that constitutes a boundary. This, in turn, makes territory an efficient device for determining whether or not people are subject to the enforcement of a particular set of rules and conditions.

Territoriality also gives tangible form to power and control, but does so in a way that directs attention away from the personal relationships between the controlled and the controllers. In other words, rules and laws become associated with particular spaces and territories rather than with particular individuals or groups. Finally, territoriality allows people to create and maintain a framework through which to experience the world and give it meaning. Bounded territories, for example, make it easier to differentiate "us" from "them."

Sense of Place

The bonds established between people and places through territoriality allow people to derive a pool of shared meanings from their lived experience of everyday routines. People become familiar with one another's vocabulary, speech patterns, gestures, humor, and so on. Often, this carries over into people's attitudes and feelings about themselves and their locality. When this happens, the result is a self-conscious sense of place. The concept of a **sense of place** refers to the feelings evoked among people as a result of the experiences, memories, and symbolism that they associate with a given place. It can also refer to the character of a place as seen by outsiders: its unique or distinctive physical characteristics and/or its inhabitants.

For *insiders,* this sense of place develops through shared dress codes, speech patterns, public comportment, and so on. It also develops through familiarity with the history and symbolism of particular elements of the physical environment—the birthplace of someone notable, for example, the location of some particularly well-known event, or the expression of community identity through community art (Figure 6.6, page 240). Sometimes, it is deliberately fostered by the construction of symbolic structures such as monuments and statues. Often, it is a natural outcome of people's familiarity with one another and their surroundings. Because of this consequent sense of place, insiders feel at home and "in place."

Sense of place: feelings evoked among people as a result of the experiences and memories that they associate with a place, and to the symbolism that they attach to it.

Figure 6.6 Community art
Community art can provide an important element in the creation of a sense of place for members of local communities. This example is from the Mission District of San Francisco.

Consider the sense of place evoked by one writer describing a London neighborhood, Ladbrooke Grove, where he had lived for a time:

> Like many impoverished areas in big cities, [Ladbrooke Grove] is picturesque in the sun, and Americans walk the length of the street market in Portobello Road snapping it with Kodaks; but on dull days one notices the litter, the scabby paint, the stretches of torn wire netting, and the faint smell of joss-sticks competing with the sickly-sweet odour of rising damp and rotting plaster. Where the area shows signs of wealth, it is in the typically urban non-productive entrepreneurism of antique shops and stalls. Various hard-up community action groups have left their marks: a locked shack with FREE SHOP spraygunned on it, and old shoes and sofas piled in heaps around it; a makeshift playground under the arches of the motorway with huge crayon faces drawn on the concrete pillars; slogans in whitewash, from SMASH THE PIGS to KEEP BRITAIN WHITE. The streets are crowded with evident isolates: a pair of nuns in starched habits, a Sikh in a grubby turban, a gang of West Indian youths, all teeth and jawbones, . . . limp girls in flaky Moroccan fleeces, macrobiotic devotees with transparent parchment faces, mongrel dogs, bejeaned delivery men, young mothers in cardigans with second hand [strollers].[2]

Another writer with an eye for the distinctive character of places describes his regular walk from his apartment in New York's Greenwich Village to the midtown restaurants on the East Side (Figure 6.7), where he likes to eat:

> To reach the French restaurants I have to pass from my house through a drug preserve just to the east of Washington Square. Ten years ago junkie used to sell to junkie. . . . In the morning stoned men lay on park benches, or in doorways; they slept immobile under the influence of the drugs, sometimes having spread newspapers out on the pavement as mattresses. . . . The dulled heroin addicts are now gone, replaced by addict-dealers in cocaine. The cocaine dealers are never still, their arms are jerky, they pace and pace; in their electric nervousness, they radiate more danger than the old stoned men. . . .
>
> The last lap of my walk passes through Murray Hill. The townhouses here are dirty limestone or brownstone; the apartments have no imposing entrance lobbies. There is a uniform of fashion in Murray Hill: elderly women in black silk dresses and equally elderly men sporting pencil-thin mustaches and malacca canes, their canes visibly decades old. This is a quarter of the old elite in New York. . . . Near to the Morgan Library is B. Altman's, an enormous store recently closed which was regularly open in the evenings so that people could shop after work. One often saw women, of the sort who live nearby in Gramercy Park, shopping for sheets there; the sheet-shoppers had clipped the advertisement for a white sale out of the newspaper and still carried it in their unscuffed calf handbags; they were hardworking, thrifty.[3]

For *outsiders*, such details add up to a sense of place only if they are distinctive enough to evoke a significant common meaning for those with no direct experience of them. For most outsiders the sights and sounds of Ladbrooke Grove are probably not strong enough to elicit much of a sense of place.

[2]J. Raban, *Soft City*. London: Fontana, 1975, pp. 169–170.
[3]R. Sennett, *The Consciousness of the Eye*. New York: Knopf, 1990, pp. 123–124.

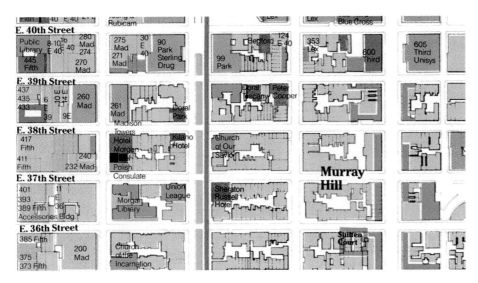

Figure 6.7 Murray Hill, New York City—This map extract shows part of Richard Sennett's walk from Greenwich Village to his favorite restaurants in the midtown section of the East Side. As this extract suggests, few distinctive landmarks exist in this section of the city. What gives Murray Hill a sense of place for Sennett is its people. To Sennett's eye, the squares, streets, apartment houses, townhouses, bars, stores, and institutions along his walk are given character and meaning by people, including the "elderly women in black silk dresses and equally elderly men sporting pencil-thin mustaches and malacca canes." (*Source:* Extract from *Midtown Manhattan,* Williams Real Estate Co. and Incentra International, Inc., 1991.)

Manhattan, New York, on the other hand, is one setting that does carry a strong sense of place to outsiders—many people feel a sense of familiarity with the skyline, busy streets, and distinctive commercial districts that together symbolize the heart of the American business world (Figure 6.8). The monumental core of Washington, DC, is another place that evokes a strong sense of place among outsiders—many people have a sense of familiarity with the buildings and landmarks that symbolize the nation's capital.

Experience and Meaning

The interactions between people and places raise some fundamental questions about the meanings that people attach to their experiences: How do people process information from external settings? What kind of information do they use? How do new experiences affect the way they understand their worlds? What meanings do particular environments have for individuals? How do these meanings influence behavior? The answers to these questions are by no means complete, but it is clear that people not only filter information from their environments through neurophysiological processes, but also draw on personality and culture to produce cognitive images of their environment, pictures or representations of the world that can be called to mind through the imagination (Figure 6.9, page 242). Cognitive images are what people see in the mind's eye when they think of a particular place or setting.

Two of the most important attributes of cognitive images are that they both simplify and distort real-world environments. Research on the ways in which people simplify the world through such means has suggested, for example, that many people tend to organize their cognitive images of particular parts of their world in terms of the several simple elements:

Paths: The channels along which they and others move; for example, streets, walkways, transit lines, canals

Edges: Barriers that separate one area from another; for example, shorelines, walls, railroad tracks

Districts: Areas with an identifiable character (physical and/or cultural) that people mentally "enter" and "leave"; for example, a business district or an ethnic neighborhood

Nodes: Strategic points and foci for travel; for example, street corners, traffic junctions, city squares

Landmarks: Physical reference points; for example, distinctive landforms, buildings, or monuments

Figure 6.8 Manhattan's Financial District The landscape of Wall Street and the surrounding financial district of New York City is composed of clean, crisp lines and an imposing skyline. The image of the financial district, which is often portrayed in films and advertising, conveys a sense of the confidence and authority of American corporate capitalism, as well as the role of the city as the world's financial center.

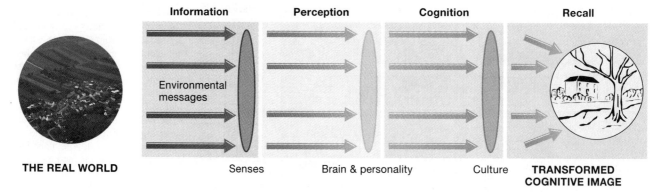

Information   Perception   Cognition   Recall

THE REAL WORLD

Senses   Brain & personality   Culture   TRANSFORMED COGNITIVE IMAGE

Environmental messages

Figure 6.9 The formation of cognitive images   People form cognitive images as a product of information about the real world, experienced directly and indirectly, and filtered through their senses, their brain, and personality, and the attitudes and values they have acquired from their cultural background. (*Source:* Diagram, R. G. Gollege and R. J. Simpson, *Analytical Behavioral Geography.* Beckenham: Croom Helm, 1987, Fig. 3.2, p. 3.)

For many people, individual landscape features may function as more than one kind of cognitive element. A freeway, for example, may be perceived as both an edge and a path in a person's cognitive image of a city. Similarly, a railroad terminal may be seen as both a landmark and a node (see Figure 6.10).

Distortions in people's cognitive images are partly the result of incomplete information. Once we get beyond our immediate living area, there are few spaces that any of us know in complete detail. Yet our worlds—especially those of us in the fast world who are directly tied to global networks of communication and knowledge—are increasingly large in geographic scope. As a result, these worlds must be conceived, or cognized, without many direct stimuli. We have to rely on fragmentary and often biased information from other people, from books, maga-

Figure 6.10 Cognitive image of Boston   This map was compiled by Kevin Lynch, one of the pioneer researchers into cognitive images, from interviews with a sample of Boston residents. Lynch found that the residents of Boston tended to structure their cognitive images of the city with the same elements as one another. He produced ingenious maps such as this one to demonstrate the collective "mental map" of the city, using symbols of different boldness to indicate the proportion of respondents who had mentioned each element. (*Source:* K. Lynch, *The Image of the City.* Cambridge, MA: M.I.T. Press, 1960, p. 146.)

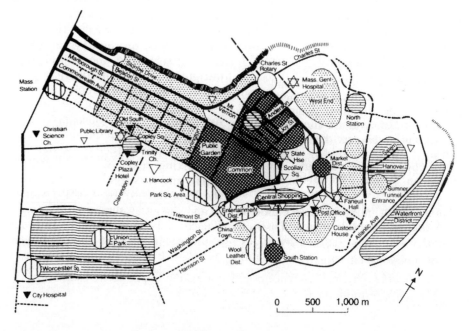

zines, television, and the Internet. Distortions in cognitive images are also partly the result of our own biases. What we remember about places; what we like or dislike; what we think is significant; and what we impute to various aspects of our environments all are a function of our own personalities, our past experiences, and the cultural influences to which we have been exposed.

Images and Behavior

Cognitive images are compiled, in part, through behavioral patterns. Environments are "learned" through experience. Meanwhile, cognitive images, once generated, influence behavior. In the process of these two-way relationships, cognitive images are constantly changing. Each of us also generates—and draws on—different kinds of cognitive images in different circumstances.

Elements such as districts, nodes, and landmarks are important in the kinds of cognitive images that people use to orient themselves and to navigate within a place or region. The more of these elements an environment contains and the more distinctive they are, the more legible it is to people and the easier it is to get oriented and navigate. In addition, the more first-hand information people have about their environment and the more they are able to draw on secondary sources of information, the more detailed and comprehensive their images will be.

This phenomenon is strikingly illustrated in Figure 6.11 (see page 244), which shows the collective image of Los Angeles as seen by three different groups: the residents of Westwood, an affluent neighborhood; Avalon, a poor, inner-city neighborhood; and Boyle Heights, a poor, immigrant neighborhood. The residents of Westwood have a well-formed, detailed, and comprehensive image of the entire Los Angeles basin. At the other end of the socioeconomic spectrum, however, residents of the black ghetto neighborhood of Avalon, near Watts, have a vaguer image of the city, structured only by the major east–west boulevards and freeways, and dominated by the gridiron layout of streets between Watts and the city center. The Spanish-speaking residents of Boyle Heights—even less affluent, less mobile, and somewhat isolated by language— have an extremely restricted image of the city. Their world consists of a small area around Brooklyn Avenue and First Street, bounded by the landmarks of City Hall, the bus depot, and Union Station.

The importance of these images goes beyond people's ability simply to navigate around their environments. The narrower and more localized people's images are, for example, the less they will tend to venture beyond their home area. Their behavior becomes circumscribed by their cognitive imagery, in a kind of self-fulfilling prophecy. People's images of places are also important in shaping particular aspects of their behavior. Research on shopping behavior in cities, for example, has shown that customers do not necessarily go to the nearest store or the one with the lowest prices; they also are influenced by the configuration of traffic, parking, and pedestrian circulation within their imagery of the retail environment. The significance of this has clearly not been lost on the developers of shopping malls, who always provide extensive space for free parking and multiple entrances and exits.

In addition, shopping behavior, like many other aspects of behavior, is influenced by people's values and feelings. A district in a city, for example, may be regarded as attractive or repellent, exciting or relaxing, fearsome or reassuring or, more likely, a combination of such feelings. As with all other cognitive imagery, such images are a product of a combination of direct experience and indirect information, all filtered through personal and cultural perspectives. Images such as these often exert a strong influence on behavior. Returning to the example of consumer behavior, one of the strongest influences on shopping patterns relates to the imagery evoked by retail environments—something else that has not escaped the developers of malls, who spend large sums of money to establish the right atmosphere and image for their projects.

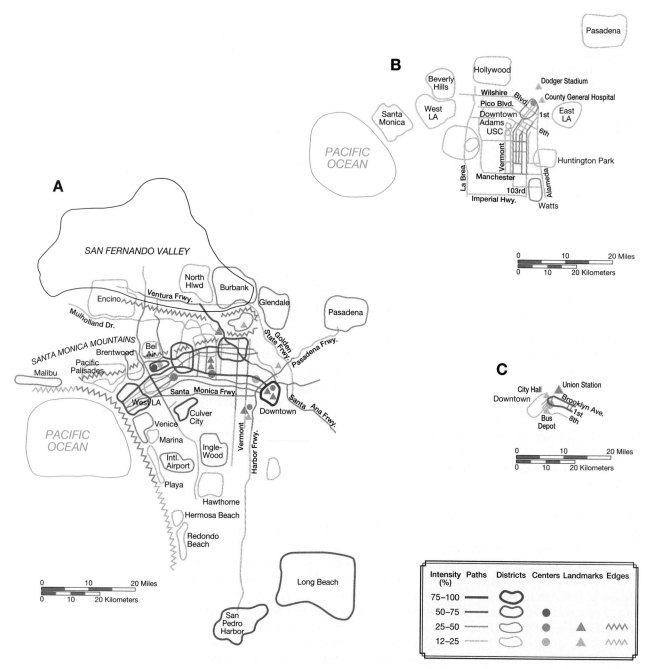

Figure 6.11 Images of Los Angeles Here are images (A) as seen by residents of Westwood (an affluent, high-status neighborhood), (B) as seen by residents of Avalon (a poor, inner-city neighborhood), and (C) as seen by residents of Boyle Heights (a Spanish-speaking neighborhood). Note that all three are drawn to the same scale. (*Source:* P. Orleans, in R. Downs and D. Stea (Eds.), *Image and Environment.* Chicago: Aldine, 1973, pp. 120–122.)

While shopping behavior is one narrow example of the influence of place imagery on behavior, other examples can be drawn from every aspect of human geography and at every spatial scale. The settlement of North America, for example, was strongly influenced by the changing images of the Plains and the West. Indeed, the images associated with different regions and localities continue to shape settlement patterns. People draw on their cognitive imagery in making decisions about migrating from one area to another. Figure 6.12 shows the com-

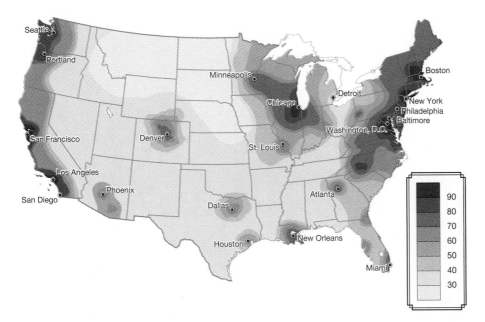

Figure 6.12 Preference map of the United States This map illustrates collective preferences for cities in the coterminous United States as places in which to live and work, as expressed by architecture students at Virginia Tech in 1996. This isoline map is a generalization based on the scores the students gave to the 150 largest cities in the country. The higher the score, the stronger the preference for living and working there.

posite image of the United States held by a group of Virginia Tech students, based on the perceived attractiveness of cities and states as places in which to live.

Another example of the influence of cognitive imagery on people's behavior is the way that people respond to environmental hazards such as floods, droughts, earthquakes, storms, and landslides, and come to terms with the associated risks and uncertainties. Some people tend to change the unpredictable into the knowable by imposing order where none really exists (resorting to folk wisdom about weather, for example), while others deny all predictability and take a fatalistic view. Some tend to overestimate both the degree and the intensity of natural hazards, while others tend to underestimate them. These differences point to another important dimension of behavior: people's attitudes to risk taking. Figure 6.13 (see page 246) illustrates diagrammatically the perceptions of individuals with different attitudes to risk taking. While the reckless person (A), for example, exaggerates gains and discounts losses, the poor person (C) exaggerates both. The cautious person (B) overestimates loss and undervalues gain. In reality, individuals adopt different risk-taking attitudes to different situations, making their decisions difficult to predict through models based on "rational" behavior.

Finally, one aspect of cognitive imagery exists that is of special importance in modifying people's behavior: the sentimental and symbolic attributes ascribed to places. Through their daily lives and through the cumulative effects of cultural influences and significant personal events, people build up affective bonds with places. They do this simultaneously at different geographic scales: from the home, through the neighborhood and locality, to the national State. The tendency for people to do this has been called topophilia. **Topophilia** literally means "love of place." Geographers use it to describe the complex of emotions and meanings associated with particular places that, for one reason or another, have become significant to individuals. The result is that most people have a home area, home town, or home region for which they have a special attachment or sense of identity and belonging.

Topophilia: the emotions and meanings associated with particular places that have become significant to individuals.

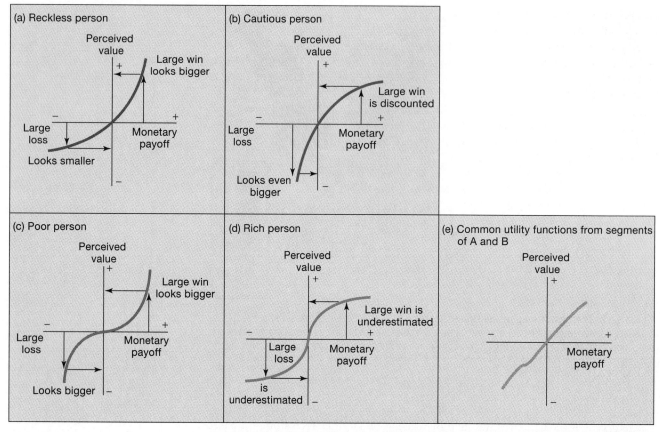

Figure 6.13 Individual utility functions and attitudes to risk taking Cautious people (B) tend to overestimate the consequences of a loss, while underestimating the consequences of a gain. For reckless people (A), the opposite is true. Many people are not especially cautious or reckless, their utility functions being somewhere between these two extremes (E). Attitudes to risk taking can also depend on people's level of affluence. Poor people (C) tend to exaggerate both large gains and large losses, while rich people (D) tend to underestimate them. (*Source:* R. G. Golledge and R. J. Stimpson, *Analytical Behavioral Geography.* Beckenham: Croom Helm, 1987, Figure 2.3, p. 48.)

Place Marketing

Economic and cultural globalization has meant that places and regions throughout the world are increasingly seeking to influence the ways in which they are perceived by tourists, businesses, media firms, and consumers. As a result, places are increasingly being reinterpreted, reimagined, designed, packaged, and marketed. Through place marketing, sense of place has become a valuable commodity, and culture has become an important economic activity. Furthermore, culture has become a significant factor in the ability of places to attract and retain other kinds of economic activity. Seeking to be competitive within the globalizing economy, many places have sponsored extensive makeovers of themselves, including the creation of pedestrian plazas, cosmopolitan cultural facilities, festivals, and sports and media events. An increasing number of places have also set up home pages on the Internet containing maps, information, photographs, guides, and virtual spaces in order to promote themselves in the global marketplace for tourism and commerce. Meanwhile, the question of who does the reimagining and cultural packaging, and on whose terms, can become an important issue for local politics.

Central to place marketing is the deliberate manipulation of material and visual culture in an effort to enhance the appeal of places to key groups. These groups include the upper-level management of large corporations; the higher-skilled and better-educated personnel sought by expanding high-technology industries; wealthy tourists; and the organizers of business and professional conferences and other income-generating events. In part, this manipulation of culture depends on promoting traditions, lifestyles, and arts that are locally rooted (see Feature 6.1, "Visualizing Geography—Place Marketing and Economic Development"); in part, it depends on being able to tap into globalizing culture through new cultural amenities and specially organized events and exhibitions. Some of the most widely adopted strategies for the manipulation and exploitation of culture include funding for facilities for the arts; investment in public spaces; the re-creation and refurbishment of distinctive settings like waterfronts and historic districts; the expansion and improvement of museums (especially with blockbuster exhibitions of spectacular cultural products that attract large crowds and can be marketed with commercial tie-ins); and the designation and conservation of historic landmarks.

The Dutch city of Amsterdam provides a good example of how investment in the arts can provide a catalyst for economic development. The construction in the late 1980s of an arts and civic complex, the Stopera, in a declining neighborhood in the eastern part of the central area of the city led to the recovery of the whole area within less than a decade. By the early 1990s, nearly 2,000 people were working in the Stopera itself, and its presence had attracted bookstores, record and magazine stores, restaurants, cafes, and specialized food stores. Altogether, the neighborhood experienced a 40 percent increase in the number of shops between 1988 and 1992. A number of small businesses have also been attracted into the area because of its new atmosphere, and increasing numbers of tourists, who previously avoided the area, now seek it out.

New York City also provides a good example of place marketing through investment in public space. Bryant Park, on 42nd Street in Manhattan (Figure 6.14), has become a small but celebrated component of the city's attempt to clean up its image and attract tourists and business investment. After a period of decline, disuse, and daily occupation by vagrants and drug dealers, the park was taken over by the Bryant Park Restoration Corporation, a nonprofit business association of local property owners and their major corporate tenants. This group redesigned and renovated the park; installed new food services; hired private security guards; and organized a series of cultural events, including the showing, twice a year, of the fashion collections of New York designers (held in marquees set up inside the park). The result has been described as "pacification by cappuccino." The previous users of the space have been displaced, the image of the whole area around Bryant Park has been radically altered, and investments are once again flowing in.

Examples of the re-creation and refurbishment of distinctive settings like waterfronts and historic districts can be found in many cities. The popularity of waterfront redevelopments can be traced to the success of Harborplace in Baltimore (Figure 6.15, page 250). Harborplace is a waterfront complex that was redeveloped from a semiderelict wharf and docks area. Built in 1980, it has become one of the best-known examples of the kind of large, mixed-use developments that industrial cities have sponsored in an effort to bolster their image and attract tourists and shoppers. By 1995, Harborplace was attracting more than 30 million visitors a year. Not surprisingly, this kind of success has encouraged other cities to re-present themselves through major waterfront developments. Other examples include South Street Seaport in New York, the Marketplace and Harborwalk in Boston, and Darling Harbour in Sydney, Australia. The largest and most ambitious of all such developments, however, is in London's Docklands (Figure 6.16, page 250). The remaking of the Docklands in the 1980s was a deliberate attempt by Prime Minister Margaret Thatcher's government not simply to market this part of London to global tourists and investors, but to sell the whole idea of the United Kingdom as a rejuvenated, postindustrial economy.

Figure 6.14 Bryant Park, New York City
Bryant Park abuts the main building of the New York City Public Library in the heart of central Manhattan. Before renovation it was usually occupied by homeless people and drug dealers. Now, through the renovation efforts of a nonprofit volunteer group, the park has been transformed into a fashionable site for meeting and being seen. In its new incarnation, it is not clear whether Bryant Park is a public space or a combination of public and private. Or perhaps the question is: public for whom? The homeless are certainly no longer welcome to linger among the other, more affluent visitors.

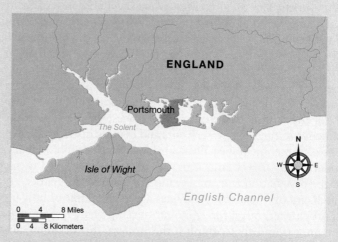

The legacy of Portsmouth's naval history has allowed the city to develop a heritage industry that has helped to make up for the loss of jobs in the naval dockyard that was the mainstay of the city for so long. HMS *Victory,* Nelson's flagship at the Battle of Trafalgar (see Figure 2.2.4), had been a long-standing attraction in the Royal dockyard. In 1979, a volunteer organization recovered the *Mary Rose,* a Tudor warship that had been sunk just off Portsmouth in 1545, during an engagement with a French invasion fleet. It has since been restored and is a major attraction in the dockyard's Heritage Area. In 1988 HMS *Warrior,* built in 1860 as Britain's first ironclad battleship, was added to the dockyard's collection of major tourist attractions.

Portsmouth, England, with its superb natural harbor, has been England's chief naval port and dockyard since the reign of Henry VIII. At the height of the British Empire, the growth of the dockyard fueled the development of the city, helping to create a specialized economy based around marine engineering. With the decline of the British Empire after the 1960s, the British navy was much reduced in size, and Portsmouth's dockyard functions shrank dramatically. After a period of economic adjustment, the city has begun to remake its image, exploiting its past to market itself as a distinctive conference center and tourist site.

The *Mary Rose*

HMS *Victory*

Throughout the second half of the nineteenth century and the first half of the twentieth century, Portsmouth's economy included a lucrative resort and recreation function. When international travel became more affordable in the 1960s, many British vacationers chose the warmer beaches of the Mediterranean, and Portsmouth's tourist industry began to decline sharply. Since the 1980s, Portsmouth has repositioned itself as a different kind of resort, drawing on the city's naval history in order to attract overnight visitors traveling to and from France on the cross-Channel car ferries that now operate from an unused part of the naval dockyard. In 1996, the city formed a partnership with neighboring municipalities to redevelop Portsmouth Harbour. The £90 million ($135 million) redevelopment project will be substantially funded by proceeds from the National Lottery and will include a Baltimore-style wharf with a conference center, shops, cafes, hotels, and viewing terraces for maritime events. The centerpiece of the project will be a spectacular entrance to the harbor with a 452-foot (165-meter) tower, a 410-foot (150-meter) water arch lit at night by lasers, and 20-foot high (8-meter) water curtains that will serve as giant projection screens. In 1996, the city attracted over 4.2 million tourists; the harbor redevelopment is expected to increase that figure to around 5.6 million by the year 2000.

18 July, 1545: Henry VIII's fleet, in Portsmouth Harbour, fends off an attack by the French fleet.

Sixteenth- and seventeenth-century coastal fortifications, formerly boarded up and fenced off as eyesores, have been landscaped in order to take advantage of their historic interest. In one formerly abandoned fort, Southsea Castle, a museum of Tudor life has been established, exploiting the city's link with Henry VIII, who first developed the British navy into a significant military force.

Southsea Castle

Entrance to the *Mary Rose* exhibit, in the Heritage Area of Portsmouth naval base

During World War II, Portsmouth was the main port of departure for Allied forces invading German-held France on D-Day, June 6, 1944. In 1982, the city opened the D-Day Museum, reinforcing the city's emerging identity as a specialized heritage site. A few years later, the closure of a Royal Marine barracks provided the opportunity for the development of another specialized military museum.

D-Day, June 6, 1944

Royal Marines Museum, Portsmouth

Restored naval cutter, in the Mast Pond, part of the Heritage Area within Portsmouth naval base

HMS *Warrior*

D-Day Museum, Portsmouth

Figure 6.15 Harborplace, Baltimore
The developer of this reclaimed waterfront, James Rouse, sought to provide a setting that is attractive to daytime office workers from the city's nearby downtown; to visitors to the adjacent convention center and hotel; and to tourists visiting the nearby aquarium. Rouse provided a new waterfront promenade with decorative paving, benches, and streetlights; a foodcourt offering a variety of ethnic, prepared, and fresh foods from restaurants, fast-food counters, stalls, and pushcarts; and a total of 142,000 square feet of shopping space.

Examples of the re-creation and refurbishment of historic districts and settings are even more widespread—so widespread, in fact, that they have become known as a mainstay of the heritage industry. This industry, based on the commercial exploitation of the histories of peoples and places, is now worldwide, as evidenced by the involvement of the Economic, Social, and Cultural Organization of the United Nations (UNESCO) in identifying places for inclusion on World Heritage Lists. In countries such as the United Kingdom, with a high density of historic districts and settings, place marketing relies heavily on the heritage industry. In the United Kingdom, more than 90 million people pay to view about 650 historic properties each year, and millions more visit free-of-charge heritage sites such as cathedrals, field monuments, remote ruins, and no-longer-useful industrial waterways. In 1996, more than 170 million tourists visited designated heritage sites in the United Kingdom, generating about $36 billion on entry fees, retail sales, travel, and hotels.

One important aspect of the influence of the heritage industry on spaces, places, and landscapes is through the tendency for historic districts and settings to be re-created, imitated, simulated, and even reinvented according to commercial considerations rather than principles of preservation or conservation. As a result, contemporary landscapes contain increasing numbers of inauthentic settings. These are as much the product of contemporary material and visual culture as they are of any cultural heritage. Particularly influenced by images and sym-

Figure 6.16 London's Docklands
Once the commercial heart of Britain's empire, employing over 30,000 dockyard laborers, London's extensive Docklands fell into a sharp decline in the late 1960s because of competition from specialized ports using new container technologies. In 1981 the London Docklands Development Corporation was created by the central government and given extensive powers to redevelop the derelict dock areas. The Docklands are now recognized as the largest urban redevelopment scheme in the world, with millions of square feet of office and retail space, and substantial amounts of new housing.

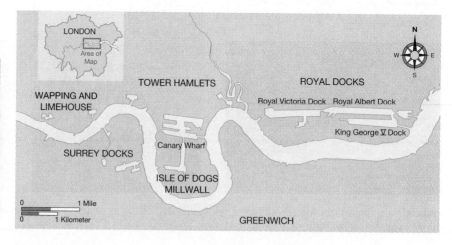

18 July, 1545: Henry VIII's fleet, in Portsmouth Harbour, fends off an attack by the French fleet.

Sixteenth- and seventeenth-century coastal fortifications, formerly boarded up and fenced off as eyesores, have been landscaped in order to take advantage of their historic interest. In one formerly abandoned fort, Southsea Castle, a museum of Tudor life has been established, exploiting the city's link with Henry VIII, who first developed the British navy into a significant military force.

Southsea Castle

Entrance to the *Mary Rose* exhibit, in the Heritage Area of Portsmouth naval base

During World War II, Portsmouth was the main port of departure for Allied forces invading German-held France on D-Day, June 6, 1944. In 1982, the city opened the D-Day Museum, reinforcing the city's emerging identity as a specialized heritage site. A few years later, the closure of a Royal Marine barracks provided the opportunity for the development of another specialized military museum.

D-Day, June 6, 1944

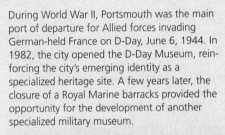

Royal Marines Museum, Portsmouth

HMS *Warrior*

D-Day Museum, Portsmouth

Restored naval cutter, in the Mast Pond, part of the Heritage Area within Portsmouth naval base

Figure 6.15 Harborplace, Baltimore
The developer of this reclaimed waterfront, James Rouse, sought to provide a setting that is attractive to daytime office workers from the city's nearby downtown; to visitors to the adjacent convention center and hotel; and to tourists visiting the nearby aquarium. Rouse provided a new waterfront promenade with decorative paving, benches, and streetlights; a foodcourt offering a variety of ethnic, prepared, and fresh foods from restaurants, fast-food counters, stalls, and pushcarts; and a total of 142,000 square feet of shopping space.

Examples of the re-creation and refurbishment of historic districts and settings are even more widespread—so widespread, in fact, that they have become known as a mainstay of the heritage industry. This industry, based on the commercial exploitation of the histories of peoples and places, is now worldwide, as evidenced by the involvement of the Economic, Social, and Cultural Organization of the United Nations (UNESCO) in identifying places for inclusion on World Heritage Lists. In countries such as the United Kingdom, with a high density of historic districts and settings, place marketing relies heavily on the heritage industry. In the United Kingdom, more than 90 million people pay to view about 650 historic properties each year, and millions more visit free-of-charge heritage sites such as cathedrals, field monuments, remote ruins, and no-longer-useful industrial waterways. In 1996, more than 170 million tourists visited designated heritage sites in the United Kingdom, generating about $36 billion on entry fees, retail sales, travel, and hotels.

One important aspect of the influence of the heritage industry on spaces, places, and landscapes is through the tendency for historic districts and settings to be re-created, imitated, simulated, and even reinvented according to commercial considerations rather than principles of preservation or conservation. As a result, contemporary landscapes contain increasing numbers of inauthentic settings. These are as much the product of contemporary material and visual culture as they are of any cultural heritage. Particularly influenced by images and sym-

Figure 6.16 London's Docklands
Once the commercial heart of Britain's empire, employing over 30,000 dockyard laborers, London's extensive Docklands fell into a sharp decline in the late 1960s because of competition from specialized ports using new container technologies. In 1981 the London Docklands Development Corporation was created by the central government and given extensive powers to redevelop the derelict dock areas. The Docklands are now recognized as the largest urban redevelopment scheme in the world, with millions of square feet of office and retail space, and substantial amounts of new housing.

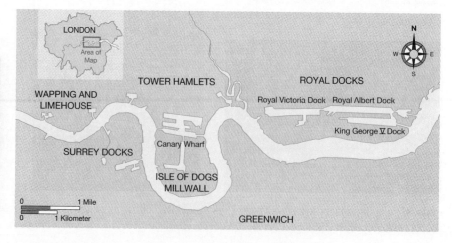

Figure 6.17 Old Tucson, Arizona The boundaries between the heritage industry and the leisure and entertainment industries have become increasingly blurred, with the result that a great deal of investment has been channeled toward the creation of inauthentic "historic" settings whose characteristics owe as much to movies and popular stereotypes as to the realities of the past.

bols derived from movies, advertising, and popular culture, examples include the re-creation of the Wild West in the fake cowboy town of Old Tucson in Arizona (Figure 6.17); the Bavarian village created in Torrance, California; and the simulation of America in the American Adventure leisure center in Derbyshire, England. The town of Helen, Georgia, boasts a full-scale reconstruction of a Swiss mountain village, complete with costumed staff and stores selling Swiss merchandise. In Japan, an artificial village called British Hills, complete with church, pub, and school, has been constructed north of Tokyo.

Coded Spaces

As we discussed in an earlier section of this chapter, landscapes are embedded with meaning—which can be interpreted differently by different people and groups. In order to interpret or read our environment, however, we need to understand the language in which it is written. We must learn how to recognize the signs and symbols that go into the making of landscape. The practice of writing and reading signs is known as **semiotics**.

Semiotics: the practice of writing and reading signs.

Semiotics in the Landscape

Semiotics proposes the view that innumerable signs are embedded or displayed in landscape, space, and place, sending messages about identity, values, beliefs, and practices. The signs that are constructed may have different meanings for those who produce them and those who read, or interpret, them. Some signs are so subtle as to be recognizable only when pointed out by a knowledgeable observer; others may be more readily available and more ubiquitous in their spatial range. For example, semiotics enables us to recognize that college students, by the way they dress, send messages to each other and the wider world, about who they are and what they value. For some of us, certain types, such as the "jock" or the "tree-hugger," are readily identifiable their clothes, hairstyle, or footwear, by the books they carry, or even the food they eat.

Semiotics, however, is not only about the signs that people convey by their mode of dress. Messages are also deployed through the landscape and are embedded in places and spaces. Consider the very familiar landscape of the shopping mall. Although there is certainly a science to the size, scale, and marketing of a mall based on demographic research as well as environmental and architectural analysis, more exists to the mall than these concrete features. The placement and mix of stores and their interior design, the arrangement of products within stores, the amenities offered to shoppers, and the ambient music all combine to send signals to the consumer about style, taste, and self-image (Figure 6.18). Called by some "palaces of consumption," malls are complex semiotic sites, directing important signals not only about what to buy but also about who should shop there and who should not.

Lest you underestimate the importance of shopping to life in the United States, the following statistics should convince you otherwise. By 1990, 36,650 shopping centers existed in the United States, accounting for more than $725 billion in sales, or 55 percent of retail sales excluding automobiles. After time spent at home and at work or school, Americans spend more time shopping than any other activity. For Americans, shopping is the second most popular leisure activity after television watching. Indeed, recent market research shows that many Americans prefer shopping to sex.[4] There persists for a great many Americans, however, an explicit disdain for shopping and the commercialism and materialism that precede and accompany it. Thus, shopping is a complicated activity that is full of ambivalence. It is not surprising, therefore, that developers have promoted shopping as a kind of tourism. The mall is a "pseudoplace" meant to encourage one sort of activity—shopping—by projecting the illusion that something else besides shopping (and spending money) is actually going on. Because of their important and complex function, malls are places with rich semiotic systems expressed through style, themes, and fantasy. While some malls are intended to convey their messages to particular subsets of the population such as the very wealthy, or teenagers (even within the same mall space), it is possible to send different messages to different consumers. The South Coast Plaza in Orange County, California, possesses a kind of sociocultural geology, where the lowest level of the mall—which contains a Sears and a J.C. Penney—is a landscape with lower-middle-class semiotics; the

Figure 6.18 Shopping mall
Developers of shopping malls know that consumer behavior is heavily conditioned by the spatial organization and physical appearance of retail settings. As a result, they find it worthwhile to spend large sums creating what they consider to be the appropriate atmosphere for their projects.

[4]J. Levine, "Lessons from Tysons Corner," *Forbes,* April, 30, 1990, pp. 186–7.

more elaborate and pricey upper levels—with stores such as Gucci and Cartier—is a landscape more consistent with affluence. Thus, even within one place different spaces send different messages to different consumers.

However complex the messages that malls send, one message is consistent across class, race, gender, age, ethnicity, and other cultural boundaries: consumption, a predominant aspect of globalization. Indeed, malls are the late twentieth-century's spaces of consumption, where just about every aspect of our lives has become a commodity. Consumption—or shopping—defines who we are more than ever before, and what we consume sends signals about who we want to be. Advertising and the mass media tell us what to consume, equating ownership of products with happiness, a good sex life, and success in general. Within the space of the mall these signals are collected and re-sent. The architecture and design of the mall are an important part of the semiotic system shaping our choices and molding our preferences. As architectural historian Margaret Crawford writes:

> All the familiar tricks of mall design—limited entrances, escalators placed only at the end of corridors, fountains and benches carefully positioned to entice shoppers into stores—control the flow of consumers through the numbingly repetitive corridors of shops. The orderly processions of goods along endless aisles continuously stimulates the desire to buy. At the same time, other architectural tricks seem to contradict commercial consideration. Dramatic atriums create huge floating spaces for contemplation, multiple levels provide infinite vistas from a variety of vantage points, and reflective surfaces bring near and far together. In the absence of sounds from the outside, these artful visual effects are complemented by the "white noise" of MUZAK and fountains echoing across enormous open courts.[5]

Malls, condominium developments, neighborhoods, university campuses, and any number of other possible geographic sites possess codes of meaning. By linking these sites with the forces behind globalization, it is possible to interpret them and understand the implicit messages they contain. And it is certainly not necessary to restrict our focus to sites in the core.

Consider Brasilia, the capital city of Brazil. As early as independence from Portugal in 1822, Brazilian politicians began suggesting that a new capital be established on the central plateau in the undeveloped interior of the country. Although authorized in 1899, the establishment of a federal district and a new capital city did not occur until the mid-twentieth century. In 1957 the Brazilian Congress approved the proposed construction of the new capital, and in 1960 the government was officially transferred to Brasilia. A symbol of the taming of the wild interior of the country and the conquest of nature through human ingenuity, Brasilia is also a many-layered system of signs conveying multiple and frequently contradictory messages. Interestingly, Brasilia, intended to symbolize a new age in Brazilian history, was also literally an attempt to construct one. That is why its plan and its architecture are so self-consciously rich with messages meant to transform Brazilian society through a new and radical form of architecture. To launch both the idea and the reality of a "city in the wilderness," the Brazilian government sponsored a design contest hoping to encourage the development of a new vision for the new capital (and by extension a new society). The winner of the contest was engineer Lucio Costa. His original plan, submitted to an international jury, was a simple sketch of a series of three crosses, each more elaborate than the previous one (see Figure 6.19).

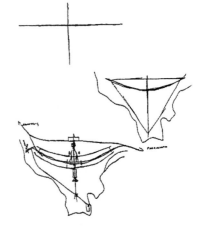

Figure 6.19 Sketches of Lucio Costa's crosses A cross symbolizes a sacred place in Christianity and other religions. By using a cross to designate the location and orientation of Brasilia, Costa was suggesting a holy origin for Brasilia, the new city in the wilderness. (*Source:* J. Holston, *The Modernist City: An Anthropological Critique of Brasilia.* Chicago: University of Chicago Press, 1989, p. 63.)

[5]M. Crawford, "The World in a Shopping Mall," in M. Sorkin (Ed.), *Variations on a Theme Park.* New York: Noonday Press, 1992, p. 14.

In a semiotic reading of Costa's plan as well as of the city, which is now more than 35 years old, the sign-of-the-cross ordering of the plan is an intentional use of a well-known mark. First, in its graphic form, the crossed axes represent the cross of Christianity. This aspect of the sign is important because it suggests that Brasilia was to be built on a sacred site, an important endorsement for the founding of a new capital. The sign of the cross, as the centerpiece of the plan, also makes an important semiotic connection to two ideal types found in both ancient and contemporary city planning and founding. Anthropologist John Holston writes:

> The first is considered one of the earliest pictorial representations of the idea of city: the Egyptian hieroglyph of the cross within the circle, itself an iconic sign standing for "city." . . . The second is the diagram of the *templum* of ancient Roman augury, a circle quartered by the crossing of two axes. It represents a space in the sky or on earth marked out . . . for the purpose of taking auspices. Hence it signifies a consecrated place, such as a sanctuary, asylum, shrine, or temple.[6]

Many other observers of the plan and the completed city have said that it resembles an airplane. This observation seems especially astute with respect to Figure 6.20. The plan shows that the residential districts were to be located along the north and south wings and administrative government offices on the part corresponding with the fuselage. The commercial district was to be constructed at the intersection of the wings and the fuselage with a cathedral and museum along the monumental axis. Like the sign of the cross, the significance of an airplane is obvious. Politicians and planners envisioned Brasilia as both the engine

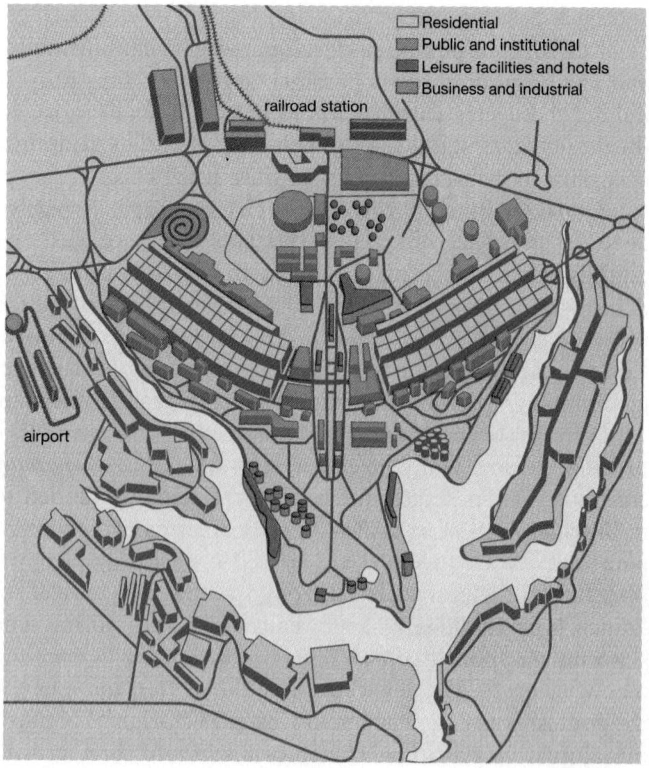

Figure 6.20 Costa's master plan for the new capital of Brasilia   In 1957 Lucio Costa submitted his entry to the Master Plan competition for the new capital of Brazil on five medium-sized cards, each containing a set of freehand sketches and some commentary. Three essential elements comprised his plan: a cross created by the intersection of two highway axes; an equilateral triangle superimposed on this cross that defined the geographical area of the city; and two terraced embankments and a platform. The more detailed rendition of the plan Costa drew for Brazil's new capital, illustrated here, reveals how he thought the various activities and material forms of city life should be geographically arranged within the city. (*Source:* B. Marshall (Ed.), *The Real World.* London: Marshall Editions/Houghton Mifflin, 1991, p. 171.)

[6]J. Holston, *The Modernist City: An Anthropological Critique of Brasilia.* Chicago: University of Chicago Press, 1989, p. 71.

and the symbol of the rapid modernization of the country. The image of an airplane in flight was an exciting, soaring, uplifting, and speedy means of achieving modernization.

All of Brasilia's major public buildings were designed by the internationally famous Brazilian architect Oscar Niemeyer (Figures 6.21). The residential axes were designed with clusters of apartment buildings, each cluster surrounding a set of recreational facilities, school buildings, and shopping areas. The University of Brasilia was also part of the early vision of the city, as was the creation of a lake and the official home of the president of the country, the *Palacio de Alvorada*, or the Palace of Dawn.

The architecture of Brasilia, like that of other federal districts—Canberra in Australia and Washington, DC, in the United States—contains both subtle and more explicit messages about the strength and purpose of government there. The exclusively Modernist architectural style of Brasilia (as opposed to the neoclassical style of Washington, DC, for example) was intended to convey the Modernist utopian ideal of technological progress and a democratic and egalitarian society. Niemeyer's architectural designs were conceived to transform colonial Brazilian society by projecting modernity and innovation through bold new urban images. A photo of the president's residence is shown in Figure 6.22. As the plan for the city indicates, the palace is far removed from the heart of the city. This spatial distance can be read as a social distance between the ruler and the ruled. It also suggests that the executive branch is elevated over the legislative—as the residences of legislators are located within the confines of the city. The name itself, the Palace of Dawn, can be interpreted as suggesting the optimism of a new day for democracy dawning through Brasilia.

Like the city, the palace was intended to reflect the new society it was to inaugurate. In the palace, one can see both the real and the imaginary counterpoised. In its present form, the city itself truly embodies this tension, because egalitarianism was never achieved. The real builders of the city—the migrant workers who came to construct the city and live out their own dreams there—

Figure 6.21 The landscape of Brasilia The landscape of Brasilia, as designed and constructed by architect Oscar Niemeyer, is that of a dynamic, modern world city. Esthetic beauty, a sense of permanence, the expression of power, and the idiosyncrasies of the individual creative process are among the primary elements of this designed landscape.

Figure 6.22 Palacio de Alvorada, Brasilia The official residence of the president of Brazil combines the homogenized international style of "glass box" architecture with culturally distinctive artistic elements such as the sweeping colonnades pictured here.

have been relegated to ever-growing *favelas* (or squatter settlements) surrounding the city. Thus while the original plan and architecture were intended to send a message about the aspirations of the new Brazil, the contemporary image of the city contradicts those dreams with the harsh realities of poverty and inequality.

The examples of the American shopping mall and the federal district of Brasilia illustrate the way that landscapes can be read, or decoded, by interpreting the signs and symbols they project. Not all the signs are consistent, even when planners and designers have complete control over their projects, because readers do not always interpret signs in ways the creators intended. In both cases, social and political realities can disrupt the plan and insert very different messages.

Sacred Spaces

Religious places can also be read and decoded. Indeed, most religions designate certain places as sacred, often because a special event occurred there. Sites are often designated as sacred in order to distinguish them from the rest of the landscape, which is considered ordinary or profane. Sacred spaces are special because they are the sites of intense or important mystical or spiritual experiences.

Sacred space: an area recognized by individuals or groups as worthy of special attention as a site of special religious experiences or events.

Sacred space includes those areas of the globe recognized by individuals or groups as worthy of special attention because they are the sites of special religious experiences and events. Sacred space does not occur naturally; rather, it is assigned sanctity through the values and belief systems of particular groups or individuals. Geographer Yi-Fu Tuan insists that what defines the sacredness of a space goes beyond the obvious shrines and temples. Sacred spaces are simply those that rise above the commonplace and interrupt ordinary routine.

In almost all cases, sacred spaces are segregated, dedicated, and hallowed sites that are maintained as such generation after generation. Believers—including mystics, spiritualists, religious followers, and pilgrims—recognize sacred spaces as being endowed with divine meaning. The range of sacred spaces includes sites as different as the temple of Kyichu LhaKang the holiest spot for Bhutanese Buddhists, and the Black Hills of South Dakota, the sacred mountains of the Lakota Sioux (Figures 6.23 and 6.24).

It is often the case that religious followers are expected to journey to especially important sacred spaces to renew their faith or to demonstrate devotion. A *pilgrimage* is a journey to a sacred space, and a *pilgrim* is a person who undertakes such a journey. In India, many of the sacred pilgrimage sites for Hindus are concentrated along the seven sacred rivers: the Ganges, the Yamuna, the Saraswati, the Naramada, the Indus, the Cauvery, and the Godavari. The Ganges is India's holiest river, and many sacred sites are located along its banks (Figure

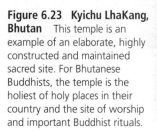

Figure 6.23 Kyichu LhaKang, Bhutan This temple is an example of an elaborate, highly constructed and maintained sacred site. For Bhutanese Buddhists, the temple is the holiest of holy places in their country and the site of worship and important Buddhist rituals.

6.25). Hindus visit sacred pilgrimage sites for a variety of reasons, including to seek a cure for sickness, wash away sins, or fulfill a promise to a deity.

Perhaps the most well-known pilgrimage is the *hajj*, the obligatory once-in-a-lifetime journey of Muslims to Mecca. For one month every year, the city of Mecca in Saudi Arabia swells from its base population of 150,000 to over 1,000,000 as pilgrims from all over the world journey to fulfill their obligation to pray in the city and receive the grace of Allah. Figure 6.26 (see page 258) shows the principal countries that send pilgrims to Mecca.

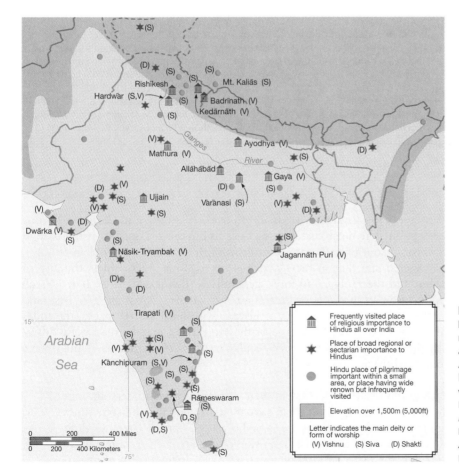

Figure 6.25 Sacred sites of Hindu India
India's many rivers are holy places within Hindu religion, so it is not surprising that sacred sites are located along the country's many riverbanks. Apparently, those shrines closer to the rivers are holier than those farther away. (*Source:* Reprinted with permission from Prentice Hall, from J. M. Rubenstein, *The Cultural Landscape: An Introduction to Human Geography,* 5th ed., © 1996, p. 216. From Ismail Ragi al Farugi and David E. Sopher, *Historical Atlas of the Religions of the World.* New York: Mcmillan, 1974.)

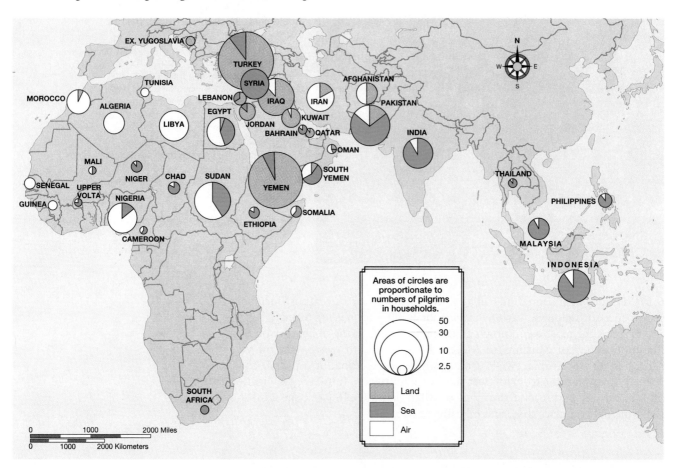

Figure 6.26 Source areas for pilgrims to Mecca Islam requires that every adult Muslim perform the pilgrimage to Mecca at least once in a lifetime. This obligation is deferred for four groups of people: those who cannot afford to make the pilgrimage; those who are constrained by physical disability, hazardous conditions, or political barriers; slaves and those of unsound mind; and women without a husband or male relative to accompany them. The pattern of actual pilgrimages to Mecca (which is located close to the Red Sea coast in Saudi Arabia) suggests a fairly strong distance-decay effect, with most traveling relatively short distances from Middle Eastern Arab countries. More distant source areas generally provide smaller numbers of pilgrims, though Indonesia and Malaysia are notable exceptions. (*Source:* C. C. Park, *Sacred Worlds.* London: Routledge, 1994, p. 268.)

Pilgrimages to sacred sites are made all over the world, and Christian Europe is no exception. The most visited sacred site in Europe is Lourdes, at the base of the Pyrenees in southwest France, not far from the Spanish border (Figure 6.27). Another sacred site that attracts pilgrims throughout the world is the city of Jerusalem, and the Holy Land more generally, which is visited by Jews, Greek Orthodox, Roman Catholics, Protestant Christians, Christian Zionists, and many other religious followers. As is the case with most sacred spaces, the codes that are embedded in the landscape of the Holy Land may be read quite differently by different religious and even secular visitors. Two students of pilgrimage have observed that:

Each group brings to Jerusalem their own entrenched understandings of the sacred; nothing unites them save their sequential—and sometimes simultaneous—presence at the same holy sites. For the Greek Orthodox pilgrims, indeed, the precise definition of the site itself is largely irrelevant; it is the icons on display which are the principal focus of attention.

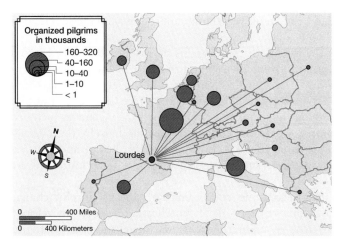

Figure 6.27 Source areas for pilgrims to Lourdes This map shows the points of origin of European, group-organized pilgrims to Lourdes in 1978. These represent only about 30 percent of all pilgrims to Lourdes, most of whom travel to the shrine on their own. Improved transportation (mainly by train) and the availability of organized package trips have contributed to a marked increase in the number of pilgrims visiting it. Many of the five million pilgrims who visit the town each year do so in the hope of a miraculous cure for medical ills at a grotto where the Virgin Mary is said to have appeared before 14-year-old Bernadette Soubirous in a series of 18 visions in the year 1858. (*Source:* C. C. Park, *Sacred Worlds.* London: Routledge, 1994, p. 268.)

For the Roman Catholics, the site is important in that it is illustrative of a particular biblical text relating to the life of Jesus, but it is important only in a historical sense, as confirming the truth of past events. Only for the Christian Zionists does the Holy Land itself carry any present and future significance, and here they find a curious kinship with indigenous Jews.[7]

That sacred spaces, like other cultural products, can convey different messages to different readers is an insight generated by a new generation of theorists known as postmodernists. We turn to consider their approach in the next section.

Postmodern Spaces

Since the 1980s, many commentators on cultural change have noted a broad shift in cultural sensibilities that seems to have permeated every sphere of creative activity, including art, architecture, advertising, philosophy, clothing design, interior design, music, cinema, novels, television, and urban design. This shift, broadly characterized as a shift from Modernity to Postmodernity, has involved both avant-garde and popular culture. It seems to have originated in parts of the world's core countries and is currently spreading throughout the rest of the world. The shift to postmodern cultural sensibilities is of particular importance to cultural geography because of the ways in which changed attitudes and values have begun to influence place making and the creation of landscapes.

Modernity and Postmodernity

Throughout the world, the philosophy of Modernity has been one of the major influences on the interdependencies among culture, society, space, place, and landscape for more than a century. **Modernity** is a forward-looking view of the world that emphasizes reason, scientific rationality, creativity, novelty, and progress. Its origins can be traced to the European Renaissance and the emergence of the world-system of competitive capitalism in the sixteenth century, when scientific discovery and commerce began to displace sociocultural views of the world that were backward-looking—views that emphasized mysticism, romanticism, and fatalism. These origins were consolidated into a philosophical movement during the eighteenth century, when the so-called Enlightenment

Modernity: a forward-looking view of the world that emphasizes reason, scientific rationality, creativity, novelty, and progress.

---

[7]J. Eade and M. Sallnow (Eds.), *Contesting the Sacred.* London: Routledge, 1991, p. 14.

established the widespread belief in universal human progress and the sovereignty of scientific reasoning.

At the beginning of the twentieth century, this philosophy developed into a more widespread intellectual movement. Around the turn of the twentieth century, there occurred a series of sweeping technological and scientific changes that not only triggered a new round of spatial reorganization (see Chapter 2) but also transformed the underpinnings of social and cultural life. These changes included the telegraph, the telephone, the X-ray, cinema, radio, the bicycle, the internal combustion engine, the airplane, the skyscraper, relativity theory, and psychoanalysis. Universal human progress suddenly seemed to be a much more realistic prospect.

Nevertheless, the pace of economic, social, cultural, and geographic change was unnerving, and the outcomes uncertain. The intellectual response, developed among a cultural avant-garde of painters, architects, novelists, and photographers, was a resolve to promote Modernity through radical changes in culture. These ideas were first set out in the "Futurist Manifesto," published in 1909 in the Paris newspaper *Le Figaro* in the form of a letter from the Italian poet Filippo Marinetti. Gradually, the combination of new technologies and radical design contributed to the proliferation of landscapes of Modernity. Among the most striking of these were Modernist urban landscapes, from Helsinki to Hong Kong and from New York to Nairobi (Figure 6.28). Indeed, in a general sense almost

Figure 6.28 Urban landscapes of Modernity These photographs reflect the pervasive influence of Modernist architecture in the skylines of the central areas of large cities throughout the world's core regions.

Sydney, Australia

Toronto, Ontario, Canada

Paris, France

all of the place making and landscapes of the early and mid-twentieth century are the products of Modernity.

Throughout this period, a confident and forward-looking Modernist philosophy remained virtually unquestioned, with the result that places and regions everywhere were heavily shaped by people acting out their notions of rational behavior and progress. Rural regions, for example, bore the imprint of agricultural modernization. The hedgerows of traditional European field patterns were torn up to make way for landscapes of large, featureless fields in which heavy machinery could operate more effectively (Figure 6.29, page 262). At a different scale, peripheral countries sought to remake traditional landscapes through economic modernization. Economic development and social progress were to be achieved through a modern infrastructure of highways, airports, dams, harbors, and industrial parks (Figure 6.30, page 262).

Postmodernity is a view of the world that emphasizes an openness to a range of perspectives in social inquiry, artistic expression, and political empowerment. Postmodernity is often described in terms of cultural impulses that are playful, superficial, populist, pluralistic, and spectacular. Many commentators see it as the result of a reconfiguration of sociocultural values that has taken place in tandem with the reconfiguration of the political economy of the core countries of the world. In this context, Postmodernity has been described as the "cultural clothing" of the postindustrial economy.

Postmodernity abandons Modernity's emphasis on economic and scientific progress and, instead, focuses on living for the moment. Above all, Postmodernity is consumption-oriented. This has made for sociocultural environments in which the emphasis is not so much on ownership and consumption as such but, rather, on the possession of particular combinations of things and on the style of consumption. Postmodern society has been interpreted as a "society of the spectacle," in which the symbolic properties of places and material possessions have assumed unprecedented importance. The stylistic emphases of Postmodernity include eclecticism, decoration, parody, and a heavy use of historical and vernacular motifs—all rendered with a self-conscious stylishness. In architecture, for example, the Modernist aphorism "Less is more" is itself parodied: "Less is a bore." Since the mid-1970s, Postmodernity has been manifest in many aspects of life, from architecture, art, literature, film, and music to urban design and planning. It has begun to be reflected in the landscapes of some places and regions, especially in more affluent settings. Some of the most striking of these Postmodern landscapes are to be found in the redeveloped waterfronts, revitalized downtown shopping districts, and neotraditional suburbs of major cities (Figure 6.31, page 263).

Postmodernity: a view of the world that emphasizes an openness to a range of perspectives in social inquiry, artistic expression, and political empowerment.

Globalization and Postmodernity

As we saw in Chapter 2, economic globalization has brought about a generalization of forms of industrial production, market behavior, trade, and consumption. This economic interdependence is also tied in to several other dimensions of globalization, all of which have reinforced and extended the commonalities among places. Three of these dimensions are especially important. First, mass communications media have created global culture markets in print, film, music, television, and the Internet. Indeed, the Internet has created an entirely new *kind* of space, cyberspace, with its own "landscape" (or technoscape), and its own embryonic cultures (see Feature 6.2, "Geography Matters—Core and Periphery in Cyberspace"). The instantaneous character of contemporary communications has also made possible the creation of a shared, global consciousness from the staging of global events such as LiveAid, the Olympic Games, and the World Cup. Second, mass communications media have diffused certain values and attitudes toward a wide spectrum of sociocultural issues, including citizenship, human

Figure 6.29 Rural landscape of Modernity
This photograph of rural East Anglia, Great Britain, shows a landscape that is the product of agricultural modernization. Urban land uses—commuter homes and the cooling towers of a power station, in this case—have encroached into the countryside. In addition, many of the traditional hedgerows in much of East Anglia have been torn own and small land holdings have been consolidated in order to create a 'prairie' landscape of large fields in which modern agricultural equipment can operate efficiently.

rights, child rearing, social welfare, and self-expression. Third, international legal conventions have increased the degree of standardization and level of harmonization not only of trade and labor practices but also of criminal justice, civil rights, and environmental regulations.

These commonalities have been accompanied by an increased importance of material consumption within many cultures. This is where globalization and Postmodernity meet, with each reinforcing the other. People's enjoyment of material goods now depends not only on their physical consumption or use. It is also linked to the role of material culture as a social marker. A person's home, automobile, clothes, reading, viewing, eating and drinking preferences, and choice of vacations are all increasingly to be interpreted as indicators of that person's social distinctiveness and sense of style. This means that there must be a continuous search for new sources of style and distinctiveness. The wider the range of foods, products, and ideas from around the world—and from past worlds—the greater the possibilities for establishing such style and distinctiveness.

Given that material consumption is so central to the repertoire of symbols, beliefs, and practices of Postmodern cultures, the "culture industries"—advertising, publishing, communications media, and popular entertainment—have become important shapers of spaces, places, and landscapes. Because the symbolic meanings of material culture must be advertised (in the broadest sense of the word) in order to be shared, advertising (in its narrower sense) has become a key component of contemporary culture and place making. In addition to stimu-

Figure 6.30 Peripheral landscape of Modernity Throughout many of the world's peripheral regions, large-scale infrastructure projects, such as the Uribante Caparo hydro-electric dam project in the Venezuelan Andes, provide striking elements of Modernity in otherwise traditional landscapes.

Figure 6.31 Landscape of Postmodernity This photograph shows part of Seaside, Florida, a small community that has become famous as an example of neotraditional planning, whereby designers attempt to reproduce the ambience, appearance, and serenity of small-town neighborhoods of the past.

lating consumer demand, advertising has always had a role in teaching people how to dress, how to furnish a home, and how to signify status through groupings of possessions.

In the 1970s and 1980s, however, the emphasis in advertising strategies shifted away from presenting products as newer, better, more efficient, and more economical (in keeping with Modernist sensibilities) to identifying them as the means to self-awareness, self-actualization, and group stylishness (in keeping with Postmodern sensibilities). Increasingly, products are advertised in terms of their association with a particular lifestyle rather than in terms of their intrinsic utility. Many of these advertisements deliberately draw on international or global themes, and some entire advertising campaigns (for Coca-Cola, Benetton, and American Express, for example) have been explicitly based on the theme of cultural globalization. Many others rely on stereotypes of particular places or kinds of places (especially exotic, spectacular, or "cool" places) in creating the appropriate context or setting for their product. Images of places therefore join with images of global food, architecture, pop culture, and consumer goods in the global media marketplace. Advertisements both instruct and influence consumers not only about products but also about spaces, places, and landscapes.

One result of these trends is that contemporary cultures rely much more than before not only on material consumption but also on *visual* and *experiential* consumption: the purchase of images and the experience of spectacular and distinctive places, physical settings, and landscapes. Visual consumption can take the form of magazines, television, movies, World Wide Web sites, tourism, window shopping, people-watching, or visits to galleries and museums. The images, signs, and experiences that are consumed may be originals, copies, or simulations.

The significance of the increased importance of visual consumption for place making and the evolution of landscapes is that settings such as theme parks, shopping malls, festival marketplaces, renovated historic districts, museums, and galleries have all become prominent as centers of cultural practices and activities. The number of such settings has proliferated, making a discernible impact on metropolitan landscapes. The design of such settings, however, has had an even greater impact on metropolitan landscapes. Places of material and visual consumption have been in the vanguard of Postmodern ideas and values, incorporating eclecticism, decoration, a heavy use of historical and vernacular motifs, and spectacular features in an attempt to create stylish settings that are appropriate to contemporary lifestyles.

One interesting aspect of the increasing trend toward the consumption of experiences is the emergence of restaurants as significant cultural sites. Restaurants often represent a synthesis of the global and the local, and they can be

6.2 Geography Matters

Core and Periphery in Cyberspace

The rapid growth of the Internet (Figure 6.2.1) has brought a massive global immigration into cyberspace. In the mid-1990s, one million new electronic citizens were being added to the Internet each month; by the year 2000, there may well be half a billion people with personal access to the Internet. Culture is fundamentally based on communication, and in cyberspace we have an entirely new form of communication: uncensored, multidirectional, written, visual, and aural.

At face value, the Internet represents the leading edge of the globalization of culture. In broad terms, the culture propagated by the Internet is very much core-oriented. With about 90 percent of all Internet communication being conducted in English, the Internet portends a global culture based on English as the universal world language. With its origins in affluent Western educational establishments and corporations, the Internet also carries a heavy emphasis on core-area cultural values such as novelty, spectacle, fashionability, material consumption, and leisure. By its very nature, the Internet empowers individuals (rather than social groups or institutions), allowing millions of people to say whatever they want to each other, free (for the first time in history) from State control.

It is unlikely, however, that the Internet will simply be a new medium through which core-area values and culture are spread. Itself an agent of cultural change, the Internet is a unique space with its own landscape of Web pages, dial-in bulletin boards, and multiuser dungeons inhabited by technoyuppies, chipheads, and info-surfers. A vocabulary of Internet slang is already finding its way into everyday usage. Electronic mail is developing its own distinctive syntax and stream-of-consciousness style. A proliferation of electronic magazines (more than 750 "E-zines" were listed in 1996) includes many that are self-consciously attempting to be in the vanguard of pop culture, while commenting critically on that culture. While E-zines on narrow or avant-garde subjects can reach large audiences on the Internet, they would not be commercially viable in other media. In addition, they offer audio and video clips, instant links to advertisers' home pages, and the opportunity for interactive discussions among authors, editors, and readers. As such, they are potentially important vehicles for the spread of participatory democracy.

Other elements of the landscapes of cyberspace include virtual shops and catalogs, virtual peep shows, teleclinics, and special-interest networks. Internet shopping has begun to change the nature of consumerism: Virtual catalogs make price and product comparisons quick and cost free. Surfing the Internet has become an important new form of recreational activity (with voyeurism and cybersex currently heading the list of most frequently visited categories). Internet connectivity is creating virtual communities by drawing together people from different places, and social and cultural backgrounds into specialized, common-interest chat groups and news groups. Last, but not least, the Internet has become a vast resource of knowledge and information that is both liberating and empowering.

These changes, however, are likely to be highly uneven in their impact because of uneven accessibility to, and use of, Internet connections. In global terms, this unevenness broadly reflects core–periphery patterns of affluence and poverty (Table 6.2.1). With the growth of the Internet, the fast world has gotten faster. The United States, for example, with less than 5 percent of the world's population, had almost 64 percent of the Internet hosts in 1996, giving it an "Internet Quotient" of 13.2. A quotient of greater than 1.0 means that a country has more Internet hosts than would be expected on the basis of its population alone. Other parts of the world that had a high Internet Quotient in 1996 include Australia, New Zealand, and much of northwestern Europe (especially the Scandinavian countries). In contrast, most of the world's periphery and semiperiphery occupies a disproportionately small part of cyberspace. In much of the slow world, poor accessibility to cyberspace is compounded by the discouragingly high cost of Internet connections. In China, for example, the cost of simply opening an Internet account in 1996 was equivalent to two months' salary on a middle-class income.

Figure 6.2.1 Worldwide Growth of Internet Hosts—An Internet host is any computer system connected to the Internet, via full- or part-time, direct or dial-up connections. In January 1987 there were just 5,089 Internet hosts; by January 1996 there were 9,472,000.

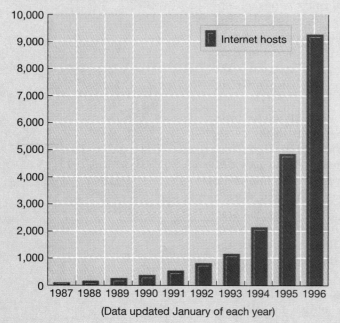

(Data updated January of each year)

The Internet also allows some new twists on old core–periphery relationships—relationships in which the periphery and its inhabitants are systematically disadvantaged or exploited. Consider, for example, the World Wide Web site that appeared in 1996 on prostitution in Havana, Cuba. This Web site, operating from a host server in Finland, made no bones about its function: as a tourist guide to the re-emerging sex market in Havana. Written in English by Italian correspondents, the site provided detailed information on locations, specializations, and prices, together with tips on how to handle the local police. While undoubtedly a boost for Havana's sex industry, the overall result is to intensify the economic and cultural domination of the periphery by the core.

Unevenness in the accessibility and use of the Internet is also compounded by language. Non–English speakers are at a disadvantage not simply because most communications on the Internet are currently in English; they must understand English to use the "search engines" necessary even to find information in their own languages. Search engines, in addition, often cannot handle accented letters. French- and German-language search engines were established in 1996, but for other language groups the Internet is linguistically inaccessible. One reaction to this is simply to not use the Internet. Thus, in Japan, where affordability is not a serious issue, only 10 percent of all corporate offices and 9 percent of all personal computers had Internet connections of some sort in 1996 (compared with 42 percent and 52 percent, respectively, in the United States). At a press conference to reveal plans for an information infrastructure in 1996, the Japanese telecommunications minister was asked his E-mail address. Not only did he not have one, he did not even know what an electronic mailbox was.

Another reaction is to resist the cultural globalization associated with cyberspace, using it only selectively. The French and French-Canadian authorities, already sensitive about the influence of English and English-language popular culture on their own, have actively sought ways of allowing Francophone *cybernautes* to use the Internet without submitting to English. In addition, the French government has subsidized an all-French alternative to the Internet: *Minitel,* an on-line videotext terminal that plugs in to French telecommunication networks. A free Minitel terminal is available to anyone who stops by a France Telecom office.

In much of Asia, the Internet's basic function as an information-exchange medium clashes with local cultures in which information is a closely guarded commodity. Whereas many American World Wide Web sites feature lengthy government reports and scientific studies as well as lively debates about government policy, comparable Asian sites typically offer little beyond public relations materials from government agencies and corporations. In puritanical Singapore, political leaders, worried that the Internet will undermine morality, have taken to reading private E-mail as part of an all-out effort to beat back the menace of on-line pornography. Chinese authorities have been afraid that the Internet will foment political rebellion, so officials have limited access to ensure that the Chinese portion of the Internet can easily be severed from the world in the event of political unrest. The reluctance of major Asian organizations to put important information on their Web sites—along with the need for Westerners to use special software to read any local language documents that do exist—has resulted in a largely one-way flow of information, from America to Asia.

TABLE 6.2.1 Share of global Internet hosts, selected countries, 1996

Country	Number of Internet Hosts	A Percentage of Total Hosts	B Percentage of Total Population	(A/B) "Internet Quotient"
Finland	208,502	2.2	0.09	24.4
United States	6,053,402	63.7	4.82	13.2
Norway	88,356	0.9	0.08	11.3
Australia	309,562	3.2	0.32	10.0
New Zealand	53,610	0.57	0.06	9.4
Canada	372,891	3.9	0.50	7.8
Switzerland	85,844	0.9	0.13	6.9
Netherlands	174,888	1.8	0.28	6.4
United Kingdom	451,750	4.8	1.00	4.7
Japan	269,327	2.8	2.30	1.2
Brazil	20,113	0.21	2.83	0.07
Russia	14,320	0.15	2.67	0.06
China	2,146	0.02	21.9	0.0009
India	788	0.008	17.0	0.0004

powerful cultural symbols in their own right. The dining experience in a particular restaurant can be an important symbolic good. By the same token, the social standing or celebrity status of customers contributes to the value of the dining experience offered by a restaurant. Restaurants themselves can be both theater and performance—particularly in big cities like New York and Los Angeles, where underpaid or out-of-work actors, dancers, and other artists form an important part of the restaurant labor force. Restaurants bring together a global and a local labor force (immigrant owners, chefs, and waiters, as well as locals) and clientele (tourists and business travelers as well as locals). Finally, restaurant design also contributes to a city's visual style as architects and interior designers, restaurant consultants, and restaurant-industry magazines diffuse global trends adapted to local styles.

While the idea of the emergence of a single global culture is too simplistic, we must acknowledge that the postmodern emphasis on material, visual, and experiential consumption means that many aspects of contemporary culture transcend local and national boundaries. Furthermore, many of the residents of the fast world are world travelers—either directly or via the TV in their living room—so that they are knowledgeable about many aspects of others' cultures. This contributes to **cosmopolitanism,** an intellectual and esthetic openness toward divergent experiences, images, and products from different cultures.

Cosmopolitanism: an intellectual and esthetic openness toward divergent experiences, images, and products from different cultures.

Cosmopolitanism is an important geographic phenomenon because it fosters a curiosity about all places, peoples, and cultures, together with at least a rudimentary ability to map, or situate, such places and cultures geographically, historically, and anthropologically. It also suggests an ability to reflect upon, and judge esthetically between, different places and societies. Furthermore, cosmopolitanism allows people to locate their own society and its culture in terms of a wide-ranging historical and geographical framework. For travelers and tourists, cosmopolitanism encourages both the willingness and the ability to take the risk of exploring off the beaten track of tourist locales. It also develops in people the skills needed to be able to interpret other cultures, and to understand what visual symbolism is meant to represent.

Conclusion

Geographers study the interdependence between people and places and are especially interested in how individuals and groups acquire knowledge of their environments, and how this knowledge shapes their attitudes and behaviors. People ascribe meanings to landscapes and places in many ways, and they also derive meanings from the places and landscapes they experience. Different groups of people experience landscape, place, and space differently. For instance, the experience that rural Sudanese children have of their landscapes and the way in which they acquire knowledge of their surroundings differ from how middle-class children in an American suburb learn about and function in their landscapes. Furthermore, both landscapes elicit a distinctive sense of place that is different for those who live there and those who visit or see the place as outsiders.

As indicated in previous chapters, the concepts of landscape and of place are central to geographic inquiry. They are the result of intentional and of unintentional human action, and every landscape is a complex reflection of the operations of the larger society. Geographers have developed categories of landscape to help distinguish the different types that exist. Ordinary landscapes such as neighborhoods and drive-in movie theaters are ones that people create in the course of their everyday lives. By contrast, symbolic landscapes represent the particular values and aspirations that the developers and financiers of those landscapes want to impart to a larger public. An example is Mount Rushmore in the Black Hills of South Dakota, designed and executed by sculptor Gutzon Borglum. Chiseling the heads of George Washington, Thomas Jefferson,

Theodore Roosevelt, and Abraham Lincoln into the granite face of the mountain, he intended to construct an enduring landscape of nationalism in the wilderness.

More recently, geographers have come to regard landscape as a text, something that can be written and read, rewritten and reinterpreted. The concept of landscape as text suggests that more than one author of a landscape can exist, and different readers may derive different meanings from what is written there. The idea that landscape can be written and read is further supported by the understanding that the language in which the landscape is written is a code. To understand the significance of the code is to understand semiotics, or the language in which the code is written. The code may be meant to convey many things, including a language of power or of playfulness; a language that elevates one group above another; or a language that encourages imagination or religious devotion and spiritual awe.

The global transition from Modernity to Postmodernity has altered cultural landscapes, places, and spaces differently as individuals and groups have struggled to negotiate the local impacts of this widespread shift in cultural sensibilities. The shared meanings that insiders derive from their place or landscape have been disrupted by the intrusion of new sights, sounds, and smells as values, ideas, and practices from one part of the globe have been exported to another. The Internet and the emergence of cyberspace have meant that new spaces of interaction have emerged that have neither distinct historical memory attached to them nor well-established sense of place. Because of this, cyberspace carries with it some unique possibilities for cultural exchange. It remains to be seen, however, whether access to this new space will be truly open or whether the Internet will become another landscape of power and exclusion.

Main Points

- In addition to understanding how the environment shapes (and is shaped by) people, geographers seek to identify how it is *perceived* and *understood* by people.

- Different cultural identities and status categories influence the ways in which people experience and understand their environments as well as how they are shaped by and able to shape them.

- Landscape serves as a kind of archive of society. It is a reflection of our culture and our experiences. Like a book, landscape is a text that is both written by individuals and groups and read by them as well.

- The language in which a landscape is written is a kind of code. The code or codes are signs that direct our attention toward certain features and away from others.

- The written code of landscape is also known as semiotics. Codes signify important information about landscapes such as whether they are sacred or profane, accessible or off-limits, or oriented toward work or play.

- The emergence of the most recent phase of globalization has occurred in parallel with a transition from Modernity to Postmodernity.

- Modernity as a historical period embraced scientific rationality and optimism about progress.

- Postmodernity, the name for the contemporary period, revolves around an orientation toward consumption and emphasizes the importance of multiple perspectives.

Key Terms

cosmopolitanism (p. 266)
derelict landscapes (p. 235)
ethology (p. 238)
humanistic approach
    (p. 236)
landscape as text (p. 237)
Modernity (p. 259)
ordinary landscapes
    (vernacular landscapes)
    (p. 235)
Postmodernity
    (p. 261)
proxemics (p. 238)
sacred space (p. 256)
semiotics (p. 251)
sense of place (p. 239)
symbolic landscapes
    (p. 235)
territoriality (p. 237)
topophilia (p. 245)

Exercises

On the Internet

The Internet offers some useful examples of the ways in which places and regions are being "constructed" through the images that are presented in the home pages of various states and cities in different parts of the world. The Internet exercises for this chapter explore the idea of the construction of place, place making and place marketing, and the existence of core–periphery differences in cyberspace. After completing the Internet exercises for this chapter, you should have a better understanding of the idea of a sense of place and the ways in which people's sense of place can be influenced.

Unplugged

1. Write a short essay (500 words, or two, double-spaced, typed pages) that describes, from your own personal perspective, the sense of place that you associate with your home town or county. Write about the places, buildings, sights, and sounds that are especially meaningful to you.

2. Now write another short essay about the sense of place evoked by the same town or county, but this time write as if you were an outside observer, emphasizing the distinctive features of the place as they would seem interesting or significant to others.

3. Draw up a list of the top 10 places in the United States in which you would like to live and work; then draw up a list of the bottom 10. How do these lists compare with the map of student preferences shown in Figure 6.12? Why might your preferences be different from theirs?

4. On a clean sheet of paper, and without reference to maps or other materials, draw a detailed sketch map of the town or city in which you live. When you have finished, turn to p. 242 and compare your sketch to Figure 6.10. Does your sketch contain nodes? Landmarks? Edges? Districts? Paths? How does "your" city compare to the "real" city?

The Geography of Economic Development

Pearl River Delta, Guandong China

In the first half of the 1990s, many of China's regions went through an unexpected transformation. In Guangdong province, near Hong Kong, for example, industrial growth approached 50 percent a year. Once known mainly for its abundant and high-quality long-grain rice, Guangdong itself now imports fragrant jasmine rice from Thailand. Along Guangdong's Pearl River Delta it is difficult to photograph one of its classic rice paddies without capturing a construction site as well. Incomes and consumer spending in Guangdong have suddenly skyrocketed. In Shandong province, 1,500 kilometers (930 miles) north, conditions are also changing. Traditionally a region of peasant agriculture and a modest prosperity based on winter wheat and summer corn, it is now beginning to join in China's industrial surge. In the tiny village of Xishan, on the plain of the Yellow River, a slaughterhouse, a pharmaceuticals plant, and a furniture factory now employ workers who until recently were peasant farmers. "Not even a single villager grows grain now," boasts Shan Chujie, the village leader. "We're not country bumkins here."[1] As individual regions develop new specializations, China itself is changing from a self-sufficient agricultural producer to a more diversified and more affluent country that is beginning to both export substantial amounts of manufactures and import substantial amounts of food.

Economic development is always an uneven geographic phenomenon, and countries and regions can follow various pathways to development. Explaining how and why these processes occur is an important aspect of human geography.

What "Economic Development" Means

Economic development is often discussed in terms of levels and rates of change in prosperity, as reflected in bottom-line statistical measures of productivity, incomes, purchasing power, and consumption. Increased prosperity is only one aspect of economic development, however. For human geographers and other social scientists, the term "economic development" is used to refer to processes of change involving the nature and composition of the economy of a particular region as well as increases in the overall prosperity of a region. These processes can involve three types of changes:

- Changes in the structure of the region's economy (for example, a shift from agriculture to manufacturing)
- Changes in forms of economic organization within the region (for example, a shift from socialism to free-market capitalism)
- Changes in the availability and use of technology within the region

Economic development is also expected to bring with it some broader changes in the economic well-being of a region. The most important of these are changes in the capacity of the region to improve the basic conditions of life (through better housing, health care, and social welfare systems) and to improve the physical framework, or infrastructure, on which the economy rests.

The Unevenness of Economic Development

Geographically, the single most important feature of economic development is that it is *uneven*. At the global scale, this unevenness takes the form of core–periphery contrasts within the evolving world-system (see Chapter 2). These global core–periphery contrasts are the result of a competitive economic system

---

[1]"Collectives Make a Comeback," *Wall Street Journal*, March 10, 1995, p. A1.

that is heavily influenced by cultural and political factors (see Feature 7.1, "Geography Matters—How Politics and Culture Modify the Economics of Development"). The core regions within the world-system—currently, the tripolar core of North America, Europe, and Japan—have the most diversified economies, the most advanced technologies, the highest levels of productivity, and the highest levels of prosperity. They are commonly referred to as *developed regions* (though processes of economic development are, of course, continuous, and no region can ever be regarded as fully developed). Other countries and regions—the periphery and semiperiphery of the world-system—are often referred to as *developing* or *less-developed*. Indeed, the nations of the periphery are often referred to as LDCs (less-developed countries). Another popular term for the global periphery, developed as a political label but now synonymous with economic development, is *Third World*. As we saw in Chapter 2, this is a term that had its origins in the early Cold War era of the 1950s and 1960s, when the newly independent countries of the periphery positioned themselves as a distinctive political bloc, aligned neither with the First World of developed, capitalist countries nor with the Second World of the Soviet Union and its satellite countries.

Global Core–Periphery Patterns At this global scale, levels of economic development are usually measured by economic indicators such as gross domestic product and gross national product. **Gross domestic product (GDP)** is an estimate of the total value of all materials, foodstuffs, goods, and services that are produced by a country in a particular year. To standardize for countries' varying sizes, the statistic is normally divided by total population, which gives an indicator, *per capita* GDP, that provides a good yardstick of relative levels of economic development. **Gross national product (GNP)** includes the value of income from abroad—flows of profits or losses from overseas investments, for example. In making international comparisons, GDP and GNP can be problematic because they are based on each nation's currency. Recently, it has become possible to compare national currencies based on *purchasing power parity (PPP)*. In effect, PPP measures how much of a common "market basket" of goods and services each currency can purchase locally, including goods and services that are not traded internationally. Using PPP-based currency values to compare levels of economic prosperity usually produces lower GDP figures in wealthy countries and higher GDP figures in poorer nations, compared with market-based exchange rates. Nevertheless, even with this compression between rich and poor, economic prosperity is very unevenly distributed across nations.

As Figure 7.1 (see page 272) shows, most of the highest levels of economic development are to be found in northern latitudes (very roughly, north of 30°N), which has given rise to another popular shorthand for the world's economic geography: the division between the "North" (the core) and the "South" (the periphery). Viewed in more detail, the global pattern of per capita GDP (measured in the "international dollars" of PPP) in 1992 is a direct reflection of the core–semiperiphery–periphery structure of the world-system. In almost all of the core countries of North America, northwestern Europe, and Japan, annual per capita GDP (in PPP) exceeds $15,000; in some, it exceeds $20,000. The only other countries that approach these levels are semiperipheral countries such as Hong Kong and Singapore, where annual per capita GDP in 1992 was around $15,000.

In the rest of the world—the periphery—annual per capita GDP (in PPP) typically ranges between $750 and $7,000. The gap between the highest per capita GDPs ($23,220 in the United States, $21,631 in Switzerland) and the lowest ($504 in Chad, $604 in Guinea) is huge. The gap between the world's rich and poor is also getting wider rather than narrower (Figure 7.2, page 273). In 1970, the average GDP per capita of the 10 poorest countries in the world was just

Gross domestic product (GDP): an estimate of the total value of all materials, foodstuffs, goods, and services that are produced by a country in a particular year.

Gross national product (GNP): similar to GDP, but in addition includes the value of income from abroad.

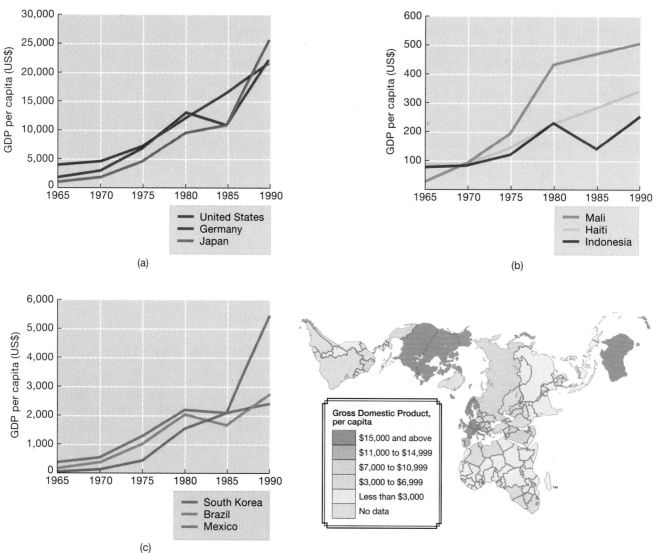

(a)

(b)

(c)

Figure 7.1 Gross domestic product (GDP) per capita GDP per capita is one of the best single measures of economic development. This map, based on 1992 data, shows the tremendous gulf in affluence between the core countries of the world economy—the United States, Japan, and Germany, for example—with an annual per capita GDP (in PPP "international" dollars) of more than $20,000 and peripheral countries like Rwanda, Mali, and Vietnam, where annual per capita GDP was less than $1,000. In semiperipheral countries like South Korea, Brazil, and Mexico, per capita GDP ranged between $5,000 and $10,000. The global average per capita GDP in 1992 was $5,336. (*Source:* Map projection, Buckminster Fuller Institute and Dymaxion Map Design, Santa Barbara, CA. The word Dymaxion and the Fuller Projection Dymaxion™ Map design are trademarks of the Buckminster Fuller Institute, Santa Barbara, California, ©1938, 1967 & 1992. All rights reserved.)

one-fiftieth of the average GDP per capita of the 10 most prosperous countries. By 1990, the relative gap had doubled: The average of the top 10 was 100 times greater than the average of the bottom 10.

Development and Gender Equality One way the core–periphery pattern is reflected in indicators that measure economic development is in terms of *gender equality.* The United Nations Development Programme has established a gender-sensitive development index based on employment levels, wage rates, adult literacy, years of schooling, and life expectancy. According to this index, in no country are women better off than men. In many core countries, the score for

women is between 85 and 95 percent of the score for men, with the affluent and liberal countries of Scandinavia and Australasia (that is, Australia and New Zealand) coming closest to gender equality. Canada, which is ranked first on an overall (gender-blind) indicator of development, slips to eighth place in a gender-sensitive ranking because Canadian women have lower employment and wage rates than men. In affluent but less progressive core countries (such as Italy, Japan, and Switzerland), the indicator score for women is barely 80 percent of the score for men. As with so many patterns in human geography, however, the most striking difference is between core countries and peripheral countries. In much of Africa and Latin America, the scores for women are only between 55 and 65 percent of those for men.

These patterns reflect deep-seated cultural positions rather than women's actual contributions to economic development. Women are, in fact, playing a central and increasing role in processes of development and change in the global economy. In many peripheral countries, women constitute the majority of workers in the manufacturing sector created by the new international division of labor (Figure 7.3). In others, it is women who keep households afloat in a world economy that has resulted in localized recession and intensified poverty. On average, women earn 30 to 40 percent less than men for the same work. They also tend to work longer hours than men: 12 to 13 hours a week more (counting both paid and unpaid work) in Africa and Asia.

Regional Patterns Inequality in economic development often has a regional dimension. Initial conditions are a crucial determinant of regional economic performance. Scarce resources, a history of neglect, lack of investment, and concentrations of low-skilled people all combine to explain the lagging performance of

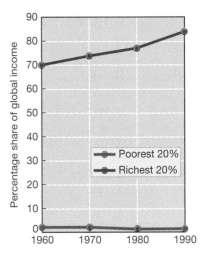

Figure 7.2 Long-term trends in per capita GDP This graph shows the steady divergence in international economic prosperity between the richest and poorest of the world's population. In 1960, the richest 20 percent of the world's population accounted for 70.2 percent of global income, while the poorest 20 percent accounted for 2.3 percent: a ratio of 30 to 1. By 1980, the ratio had increased to 45 to 1, and by 1990 it was 64 to 1.

Figure 7.3 Women in development The changing global economy has placed unprecedented demands on women. Much of the industrialization in peripheral countries depends on female labor, because wage rates for women are significantly lower than those for men. In times of rapid economic change and adjustment, women are typically called on to help sustain household incomes, in addition to their traditional household responsibilities. When employed, women are often more vulnerable than men, disproportionately concentrated in occupations with little or no job security and few opportunities for advancement.

7.1 Geography Matters

How Politics and Culture Modify the Economics of Development

Within the modern world-system, the territorial framework of political States provides a competitive economic system within which each State attempts to insulate itself as much as possible from the rigors of the world market, while attempting to turn the market to its own advantage. In terms of world-systems theory, the most important feature of this competitive system is that it has ensured that the modern world has not become a global world-empire; that is, that no single core area has been able permanently to dominate the rest of the world. In terms of human geography, the significance of the competitive system lies in the way that States influence the trajectory of change of individual countries, regions, and places within the international division of labor. States attempt to consolidate their advantages within the world market (and to overcome their disadvantages) in three main ways:

- Through their ability to organize territorial expansion and control, either through diplomacy or through military strength. This organization provides a simple and direct means of accessing and controlling resources. Where territorial expansion and control are contested, however, conflict and war may result.

- Through their ability to distort markets to the advantage of domestic producers and/or consumers. This is routinely achieved through (1) tariffs (taxes on imports) and trade quotas that restrict the access of foreign producers to domestic markets; (2) financial incentives (tax breaks, for example) to stimulate and encourage domestic firms; and (3) control of currency exchange rates, domestic interest rates, and money supply in order to manipulate production and consumption.

- Through their ability to develop an infrastructure that helps to mobilize national resources. This includes both the physical infrastructure of roads, railways, harbors, airports, dams, and so on, and the social infrastructure of legal codes, educational systems, public health care, and social welfare services.

Another important characteristic of the world-system is that it is inherently uneven, not just in its structures of economic and political power but also in the degree to which individual places and regions are incorporated into the system. The periphery, for example, is not uniform, even though its regions do share some important attributes of economic and political subordination. Individual countries, regions, and places have their own histories, their own specific attributes, and their own integrity as geographical entities.

Much of this unevenness is a result of the history and geography of the modern world-system, which we reviewed in Chapter 2. Three main factors have been important. First, differences in local and regional cultures have resulted in different reactions to the economic and political logic of the world-system. These reactions range from acceptance and cooperation through resistance and insurrection. A great deal, of course, depends on cultural compatibility with the values of materialism, individualism, competitiveness, popular democracy, and so on that are attached to the market system that drives the core. Thus, for example, Islamic fundamentalists in Iran see the capitalism of the world-system as decadent and corrupt. As a result, Iran has been only tenuously incorporated into the world-system since 1979, when a popular revolution installed a Shi'a Islamic government dedicated to economic and cultural insularity.

The second factor at work has been the geographical spread of the modern world-system, which took place under the auspices of nation-states whose strengths and objectives have varied. As a result, patterns of incorporation into the world-system have been equally varied. In some instances (for example, North America and Australia), colonization almost completely displaced indigenous societies. In others (Brazil and South Africa for example), colonization was imposed on subjugated peoples. In some cases (Egypt and India) incorporation was deep and widespread; in others (Cambodia and Libya) it was patchy and superficial. In some instances (Brazil and Malaysia) incorporation was motivated by economic opportunity; in others it was motivated simply by political territoriality—preempting a rival power, or securing a militarily strategic but economically unimportant location. Somalia's location, for example, made it strategically important for controlling access to the Indian Ocean and the Persian Gulf, with the result that the United States and the former Soviet Union vied for influence in this country that was otherwise unimportant to core countries.

The third factor modifying economic development has been that some States have responded to the unevenness and inequalities inherent to the modern world-system by pursuing alternatives to trade-based capitalism. The most important of these have been the socialist political economies of the former Soviet Union, its Eastern European satellites, and Cuba, and the communist political economy of the People's Republic of China. These alternatives have been presented to the world as being based on political beliefs—an opting-out of the world-system in favor of the pursuit of egalitarian sociocultural ideals. They can also be interpreted, however, as tactical alternatives—independent attempts to maintain sovereignty over territory; to distort and manipulate markets in order to achieve domestic economic efficiency; and to develop a modern infrastructure as a means of catching up with (and perhaps eventually joining) the core.

certain areas. In some regions, for example, initial disadvantages are so extreme as to constrain the opportunities of individuals born there. A child born in the Mexican state of Chiapas, for example, has much bleaker prospects than a child born in Mexico City. The child from Chiapas is twice as likely to die before age five; less than half as likely to complete primary school; and 10 times as likely to live in a house without access to running water. On reaching working age, he or she will earn 20 to 35 percent less than a comparable worker living in Mexico City, and 40 to 45 percent less than one living in northern Mexico.

Other examples of regional inequality can be found throughout the world. Gansu, with an income per capita 40 percent below the national average, is one of the poorest and most remote regions in China. With poor soils highly susceptible to erosion, low and erratic rainfall, and few off-farm employment opportunities, a high proportion of its inhabitants live in poverty. Chaco Province, in Argentina, another example, has a GDP per capita that is only 38 percent of the national average. Low educational attainment and lack of infrastructure, especially roads, explain much of this deviation.[2] Even when growth in other parts of a country's economy is strong, relative and absolute regional poverty can persist for long periods, however. The decline of the sugar economy in the seventeenth century, for example, pushed northeastern Brazil into a decline from which it has never fully recovered. In the United States the fortunes of coal-mining West Virginia waned with the collapse of the coal industry and the increased importance of oil and gas in energy production. It remains one of the poorest regions in the United States to this day. In Thailand, rapid development has failed to reach the northern hill people. Fewer than 30 percent of their villages have schools, and only 15 percent of the hill people can read and write. Their average annual income is less than a quarter of the country's overall GNP per capita.

At this regional scale, as at the global scale, levels of economic development often exhibit a fundamental core–periphery structure. The most useful single indicator of levels of economic development at subnational scales is per capita income (though it must be recognized that, in many countries, data for this basic economic indicator are not very accurate). As Figure 7.4 (see page 276) shows, within-country core–periphery contrasts are evident throughout the world: within core countries such as France and the United States; within semiperipheral countries such as South Korea; and in peripheral countries such as China and India.

Resources and Technology

These patterns of economic development are the result of many different factors. One of the most important is the availability of key resources such as cultivable land, energy sources, and valuable minerals. Unevenly distributed across the world, however, are both key resources and just as important, the *combinations* of energy and minerals crucial to economic development (Figure 7.5, page 277). A lack of natural resources can, of course, be remedied through international trade (Japan's success is a prime example of this). For most countries, however, the resource base remains an important determinant of development.

A high proportion of the world's key industrial resources—basic raw materials and sources of energy—are concentrated in Russia, the United States, Canada, South Africa, and Australia. The United States, for example, has 42 percent of the world's *known* resources of hydrocarbons (oil, natural gas, and oil

---

[2]These examples are taken from *World Development Report 1995*, New York: Oxford University Press for the World Bank, 1995, p. 46.

Figure 7.4 **Regional inequality in incomes** These maps show the extent to which regions within countries vary from the average national income. Most national governments try as a matter of policy to maintain a relatively uniform level of income across their national territory. Nevertheless, core–periphery patterns with significant income gradients are common.

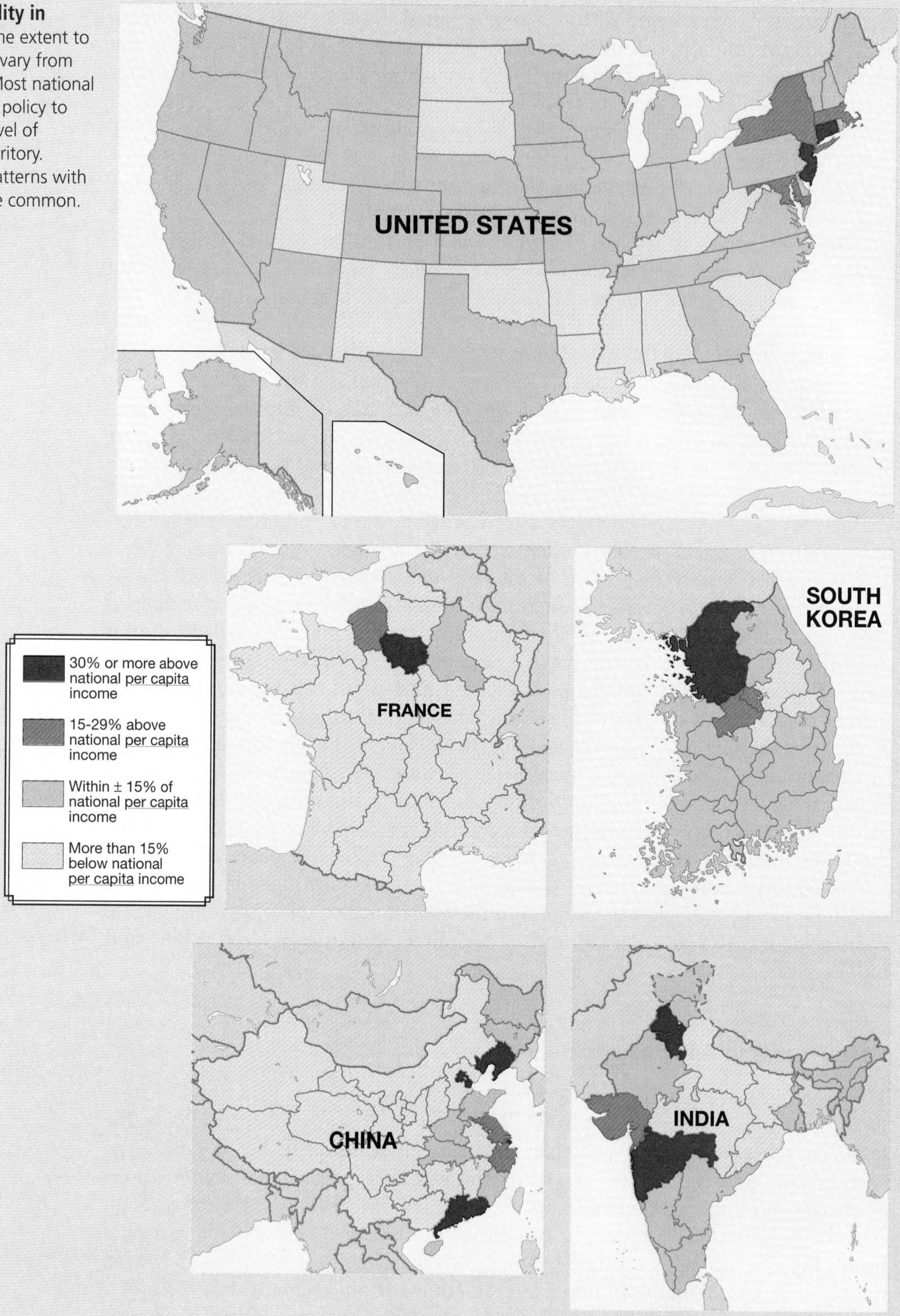
UNITED STATES
FRANCE
SOUTH KOREA
CHINA
INDIA
30% or more above national per capita income
15-29% above national per capita income
Within ± 15% of national per capita income
More than 15% below national per capita income

shales), 38 percent of the lignite ("brown coal," used mainly in power stations—Figure 7.6, page 278), 38 percent of the molybdenum (used in metal alloys), 21 percent of the lead (used chiefly for batteries, gasoline, and construction), 19 percent of the copper (used mainly for electrical wiring and components, and for coinage), 18 percent of the bituminous coal (used mainly for fuel in power sta-

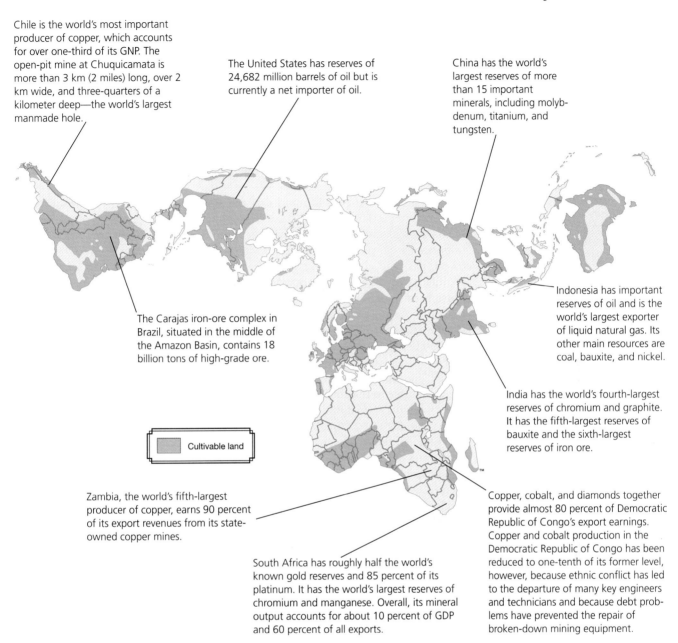

Chile is the world's most important producer of copper, which accounts for over one-third of its GNP. The open-pit mine at Chuquicamata is more than 3 km (2 miles) long, over 2 km wide, and three-quarters of a kilometer deep—the world's largest manmade hole.

The United States has reserves of 24,682 million barrels of oil but is currently a net importer of oil.

China has the world's largest reserves of more than 15 important minerals, including molybdenum, titanium, and tungsten.

The Carajas iron-ore complex in Brazil, situated in the middle of the Amazon Basin, contains 18 billion tons of high-grade ore.

Indonesia has important reserves of oil and is the world's largest exporter of liquid natural gas. Its other main resources are coal, bauxite, and nickel.

India has the world's fourth-largest reserves of chromium and graphite. It has the fifth-largest reserves of bauxite and the sixth-largest reserves of iron ore.

Cultivable land

Zambia, the world's fifth-largest producer of copper, earns 90 percent of its export revenues from its state-owned copper mines.

Copper, cobalt, and diamonds together provide almost 80 percent of Democratic Republic of Congo's export earnings. Copper and cobalt production in the Democratic Republic of Congo has been reduced to one-tenth of its former level, however, because ethnic conflict has led to the departure of many key engineers and technicians and because debt problems have prevented the repair of broken-down mining equipment.

South Africa has roughly half the world's known gold reserves and 85 percent of its platinum. It has the world's largest reserves of chromium and manganese. Overall, its mineral output accounts for about 10 percent of GDP and 60 percent of all exports.

Figure 7.5 The uneven distribution of the world's known resources Some countries, like the United States, are fortunate in having a broad resource base of energy, minerals, and cultivable land, which allows for many options in economic development. Many countries have a much narrower resource base, however, and must rely on the exploitation of one major resource as a means to economic development. Countries that have few natural resources must pursue pathways to development that are based on manufacturing or services, relying on the profits from exports to pay for imported energy resources, foodstuffs, and raw materials. (*Source:* Map projection, Buckminster Fuller Institute and Dymaxion Map Design, Santa Barbara, CA. The word Dymaxion and the Fuller Projection Dymaxion™ Map design are trademarks of the Buckminster Fuller Institute, Santa Barbara, California, ©1938, 1967 & 1992. All rights reserved.)

tions, and in the chemical industry), and 15 percent of the zinc. Russia has 68 percent of the vanadium (used in metal alloys), 50 percent of the lignite, 38 percent of the bituminous coal, 35 percent of the manganese, 25 percent of the iron, and 19 percent of the hydrocarbons. The biggest single exception to the concentration of key resources in these countries is presented by the vast oil fields of the

Figure 7.6 Lignite mining in Germany
Lignite, or brown coal, is formed from peat under moderate geological pressure. About 45 percent of the world's known coal reserves are lignitic, but many of these reserves have not been exploited because lignite is inferior to bituminous coal in heating value, storage stability, and other properties. In some areas, however, scarcity of alternative fuels has led to extensive development, especially in Germany, where many lignite beds lie close to the surface and are of great thickness, sometimes more than 30 meters (100 feet). They are thus easily worked, and the cost of production is low.

Technology systems: clusters of interrelated energy, transportation, and production technologies that dominate economic activity for several decades at a time.

Middle East. It is an exception that has enabled formerly peripheral countries like Saudi Arabia to become wealthy, and that has made the region especially important in international politics.

The concentration of known resources in just a few countries is largely a result of geology, but it is also partly a function of countries' political and economic development. Political instability in much of postcolonial Africa, Asia, and Latin America has seriously hindered the exploration and exploitation of resources. On the other hand, the relative affluence and strong political stability of the United States has led to a much more intensive exploration of resources. We should bear in mind, therefore, that Figure 7.5 reflects only the currently *known* resource base.

We should also bear in mind that the significance of particular resources is often tied to particular technologies. As technologies change, so resource requirements change, and the geography of economic development is "rewritten." One important example of this was the switch in industrial energy sources from coal to oil, gas, and electricity early in the twentieth century. When this happened, coalfield areas like central Appalachia found their prospects for economic development on indefinite hold, while oil field areas like West Texas suddenly had potential. Another example was the switch in the manufacture of mass-produced textiles from natural fibers like wool and cotton to synthetic fibers in the 1950s and 1960s. When this happened, many farmers in the U.S. South, for example, had to switch from cotton to other crops.

Regions and countries that are heavily dependent on one particular resource are vulnerable to the consequences of technological change. They are also vulnerable to fluctuations in the price set for their product on the world market. These vulnerabilities are particularly important for countries such as Bolivia, Chile, Guyana, Liberia, Mauritania, Sierra Leone, Surinam, and Zambia, whose economies are especially dependent on nonfuel minerals.

Technological innovations in power and energy, transportation, and manufacturing processes have been important catalysts for changes in the pattern of economic development. They have allowed a succession of expansions of economic activity in time and space; as a result, many existing industrial regions have grown bigger and more productive. Industrial development has also spread to new regions whose growth has become interdependent with the fortunes of others through a complex web of production and trade. Each major cluster of technological innovations tends to create new requirements in terms of natural resources as well as labor forces and markets. The result is that each major cluster of technological innovations—called "technology systems"—has tended to favor different regions, and different kinds of places. **Technology systems** are clusters of interrelated energy, transportation, and production technologies that dominate economic activity for several decades at a time—until a new cluster of improved technologies evolves. What is especially remarkable about these technology systems is that they have come along at about 50-year intervals. Since the beginning of the Industrial Revolution, we can identify four of them:

1790–1840: early mechanization based on water power and steam engines; the development of cotton textiles and ironworking; and the development of river transport systems, canals, and turnpike roads

1840–1890: the exploitation of coal-powered steam engines; steel products; railroads; world shipping; and machine tools

1890–1950: the exploitation of the internal combustion engine; oil and plastics; electrical and heavy engineering; aircraft; radio and telecommunications

1950– : the exploitation of nuclear power, aerospace, electronics, and petrochemicals; and the development of limited-access highways and global air routes

A fifth technology system, still incomplete, began to take shape in the 1980s with a series of innovations that are now being commercially exploited:

1990– : the exploitation of solar energy, robotics, microelectronics, biotechnology, advanced materials (fine chemicals, thermoplastics, for example), and information technology (digital telecommunications and geographic information systems, for example)

Each of these technology systems has rewritten the geography of economic development as it has shifted the balance of advantages between regions (Figure 7.7, page 280).

The Economic Structure of Countries and Regions

The relative share of primary, secondary, tertiary, and quaternary economic activities determines the *economic structure* of a country or region. **Primary activities** are those concerned directly with natural resources of any kind; they include agriculture, mining, fishing, and forestry. **Secondary activities** are those that process, transform, fabricate, or assemble the raw materials derived from primary activities, or that reassemble, refinish, or package manufactured goods. Secondary activities include, for example, steelmaking, food processing, furniture making, textile manufacturing, automobile assembly, and garment manufacturing. **Tertiary activities** are those involving the sale and exchange of goods and services; they include warehousing, retail stores, personal services such as hairdressers, commercial services such as accounting and advertising, and entertainment. **Quaternary activities** are those dealing with the handling and processing of knowledge and information. Examples include data processing, information retrieval, education, and research and development (R & D).

As Figure 7.8 (see page 282) shows, the economic structure of much of the world is dominated by the primary sector (that is, primary activities such as agriculture, mining, fishing, and forestry). In much of Africa and Asia, between 50 and 75 percent of the labor force is engaged in primary-sector activities. In contrast, the primary sector of the world's core regions is typically small, occupying only 5 or 10 percent of the labor force. The secondary sector is much larger in the core countries and in semiperipheral countries, where the world's specialized manufacturing regions are located (Figure 7.9, page 283). The tertiary and quaternary sectors are significant only in the most affluent countries of the core. In the United States, for example, the primary sector in 1992 accounted for less than 4 percent of the labor force; the secondary sector for about 25 percent; the tertiary sector for just over 50 percent; and the quaternary sector for 21 percent of the labor force.

Stages of Development and Geographical Divisions of Labor

Variations in economic structure—according to primary, secondary, tertiary, or quaternary activities—reflect *geographical divisions of labor.* Geographical divisions of labor are national, regional, and locally based economic specializations that have evolved with the growth of the world-system of trade and politics (see

Primary activities: economic activities that are concerned directly with natural resources of any kind.

Secondary activities: economic activities that process, transform, fabricate, or assemble the raw materials derived from primary activities, or that reassemble, refinish, or package manufactured goods.

Tertiary activities: economic activities involving the sale and exchange of goods and services.

Quaternary activities: economic activities that deal with the handling and processing of knowledge and information.

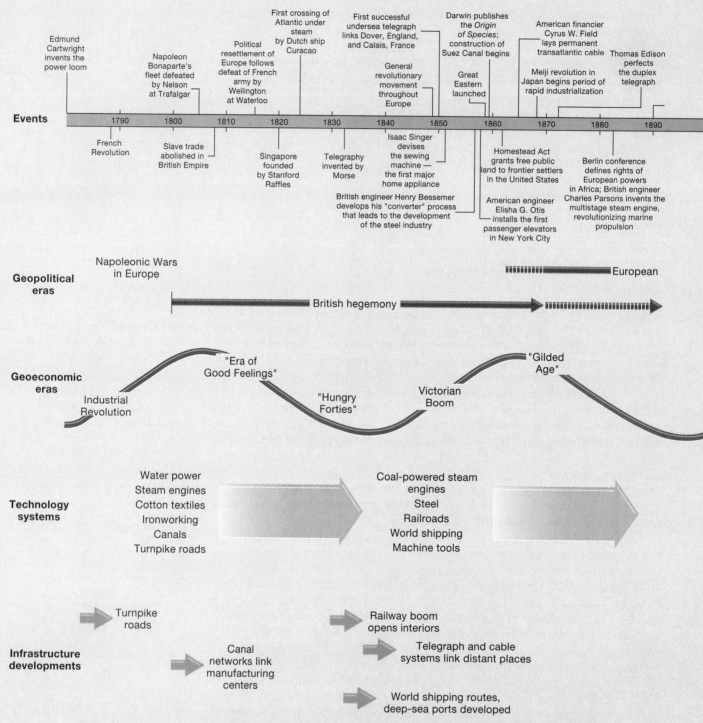

Events

Edmund Cartwright invents the power loom

French Revolution

Napoleon Bonaparte's fleet defeated by Nelson at Trafalgar

Slave trade abolished in British Empire

Political resettlement of Europe follows defeat of French army by Wellington at Waterloo

First crossing of Atlantic under steam by Dutch ship Curacao

Singapore founded by Stanford Raffles

Telegraphy invented by Morse

First successful undersea telegraph links Dover, England, and Calais, France

General revolutionary movement throughout Europe

Isaac Singer devises the sewing machine — the first major home appliance

British engineer Henry Bessemer develops his "converter" process that leads to the development of the steel industry

Darwin publishes the *Origin of Species*; construction of Suez Canal begins

Great Eastern launched

Homestead Act grants free public land to frontier settlers in the United States

American engineer Elisha G. Otis installs the first passenger elevators in New York City

American financier Cyrus W. Field lays permanent transatlantic cable

Meiji revolution in Japan begins period of rapid industrialization

Berlin conference defines rights of European powers in Africa; British engineer Charles Parsons invents the multistage steam engine, revolutionizing marine propulsion

Thomas Edison perfects the duplex telegraph

1790 1800 1810 1820 1830 1840 1850 1860 1870 1880 1890

Geopolitical eras

Napoleonic Wars in Europe

European

British hegemony

Geoeconomic eras

"Era of Good Feelings"

"Gilded Age"

Industrial Revolution

"Hungry Forties"

Victorian Boom

Technology systems

Water power
Steam engines
Cotton textiles
Ironworking
Canals
Turnpike roads

Coal-powered steam engines
Steel
Railroads
World shipping
Machine tools

Infrastructure developments

Turnpike roads

Canal networks link manufacturing centers

Railway boom opens interiors

Telegraph and cable systems link distant places

World shipping routes, deep-sea ports developed

Figure 7.7 Technological change and economic development The Industrial Revolution at the end of the eighteenth century was driven by a technology system based on water power and steam engines, cotton textiles, ironworking, river transport systems, and canals. Since then several more technology systems have occurred, each of which has opened up new geographic frontiers and rewritten the geography of economic development as it has shifted the balance of advantages between regions. Overall, the opportunities for development brought by each new technology system have been associated with distinctive economic epochs and long-wave fluctuations in the overall rate of change of prices in the economy.

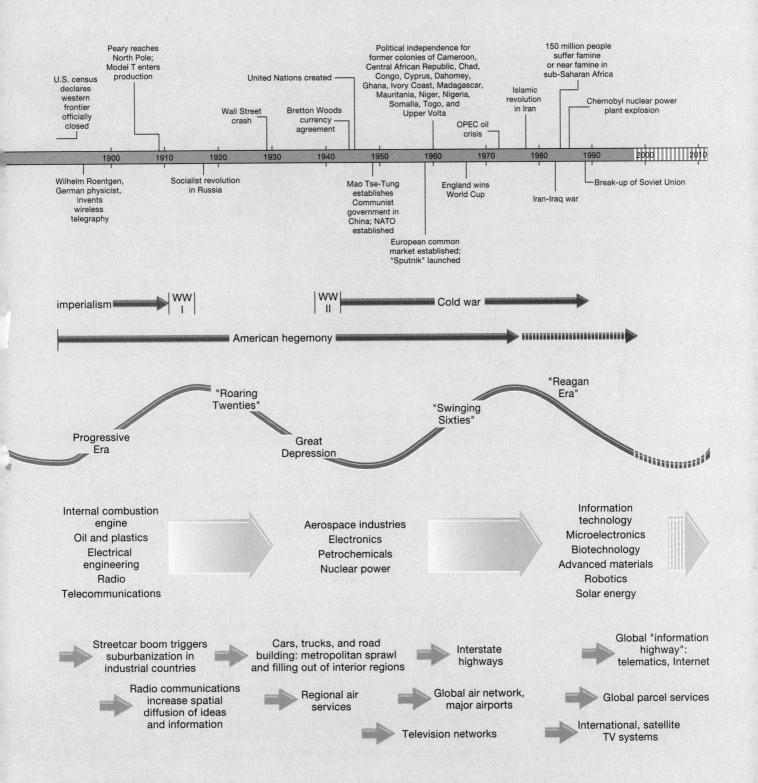

U.S. census
declares
western
frontier
officially
closed

Peary reaches
North Pole;
Model T enters
production

Wall Street
crash

Bretton Woods
currency
agreement

United Nations created

Political independence for
former colonies of Cameroon,
Central African Republic, Chad,
Congo, Cyprus, Dahomey,
Ghana, Ivory Coast, Madagascar,
Mauritania, Niger, Nigeria,
Somalia, Togo, and
Upper Volta

OPEC oil
crisis

Islamic
revolution
in Iran

150 million people
suffer famine
or near famine in
sub-Saharan Africa

Chernobyl nuclear power
plant explosion

1900   1910   1920   1930   1940   1950   1960   1970   1980   1990   2000   2010

Wilhelm Roentgen,
German physicist,
invents
wireless
telegraphy

Socialist revolution
in Russia

Mao Tse-Tung
establishes
Communist
government in
China; NATO
established

European common
market established;
"Sputnik" launched

England wins
World Cup

Iran-Iraq war

Break-up of Soviet Union

imperialism   WW I   WW II   Cold war

American hegemony

Progressive
Era

"Roaring
Twenties"

Great
Depression

"Swinging
Sixties"

"Reagan
Era"

Internal combustion
engine
Oil and plastics
Electrical
engineering
Radio
Telecommunications

Aerospace industries
Electronics
Petrochemicals
Nuclear power

Information
technology
Microelectronics
Biotechnology
Advanced materials
Robotics
Solar energy

Streetcar boom triggers
suburbanization in
industrial countries

Cars, trucks, and road
building: metropolitan sprawl
and filling out of interior regions

Interstate
highways

Global "information
highway":
telematics, Internet

Radio communications
increase spatial
diffusion of ideas
and information

Regional air
services

Global air network,
major airports

Global parcel services

Television networks

International, satellite
TV systems

281

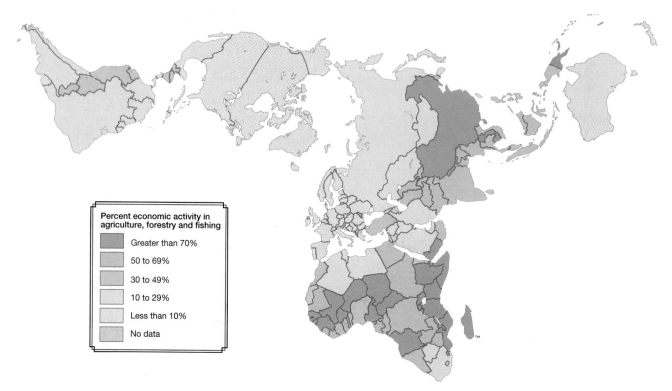

Figure 7.8 The geography of primary economic activities Primary economic activities are those concerned directly with natural resources of any kind. They include agriculture, mining, forestry, and fishing. The vast majority of the world's population, concentrated in China, India, Southeast Asia, and Africa, is engaged in primary economic activities. This map shows the percentage of the labor force in each country that was engaged in primary employment in 1995. In some countries, including China, primary activities account for more than 70 percent of the work force. In contrast, primary activities always account for less than 10 percent of the labor force in the world's core countries, and often for less than 5 percent. (*Source:* Map projection, Buckminster Fuller Institute and Dymaxion Map Design, Santa Barbara, CA. The word Dymaxion and the Fuller Projection Dymaxion™ Map design are trademarks of the Buckminster Fuller Institute, Santa Barbara, California, ©1938, 1967 & 1992. All rights reserved.)

Chapter 2), and with the locational needs of successive technology systems. They represent one of the most important dimensions of economic development. For instance, countries whose economies are dominated by primary-sector activities tend to have a relatively low per capita GDP. The exceptions to this are oil-rich countries such as Saudi Arabia, Qatar, and Venezuela. Where the geographical division of labor has produced national economies with a large secondary sector, per capita GDP is much higher (as, for example, in Argentina and Korea). The highest levels of per capita GDP, however, are associated with economies that are *postindustrial:* economies where the tertiary and quaternary sectors have grown to dominate the work force, with smaller but highly productive secondary sectors.

This overall relationship between economic structure and levels of prosperity makes it possible to interpret economic development in terms of distinctive *stages.* Each region or country, in other words, might be thought of as progressing from the early stages of development, with a heavy reliance on primary activities (and relatively low levels of prosperity), through a phase of industrialization and on to a "mature" stage of postindustrial development (with a diversified economic structure and relatively high levels of prosperity). This, in fact, is a commonly held view of economic development, and one that has been conceptualized by a prominent economist, W. W. Rostow (Figure 7.10, page 284). Rostow's

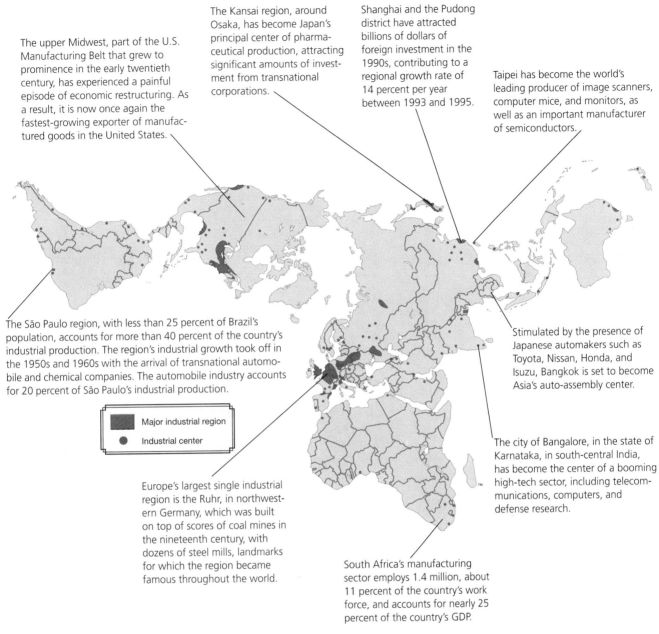

The upper Midwest, part of the U.S. Manufacturing Belt that grew to prominence in the early twentieth century, has experienced a painful episode of economic restructuring. As a result, it is now once again the fastest-growing exporter of manufactured goods in the United States.

The Kansai region, around Osaka, has become Japan's principal center of pharmaceutical production, attracting significant amounts of investment from transnational corporations.

Shanghai and the Pudong district have attracted billions of dollars of foreign investment in the 1990s, contributing to a regional growth rate of 14 percent per year between 1993 and 1995.

Taipei has become the world's leading producer of image scanners, computer mice, and monitors, as well as an important manufacturer of semiconductors.

The São Paulo region, with less than 25 percent of Brazil's population, accounts for more than 40 percent of the country's industrial production. The region's industrial growth took off in the 1950s and 1960s with the arrival of transnational automobile and chemical companies. The automobile industry accounts for 20 percent of São Paulo's industrial production.

Stimulated by the presence of Japanese automakers such as Toyota, Nissan, Honda, and Isuzu, Bangkok is set to become Asia's auto-assembly center.

The city of Bangalore, in the state of Karnataka, in south-central India, has become the center of a booming high-tech sector, including telecommunications, computers, and defense research.

Major industrial region

Industrial center

Europe's largest single industrial region is the Ruhr, in northwestern Germany, which was built on top of scores of coal mines in the nineteenth century, with dozens of steel mills, landmarks for which the region became famous throughout the world.

South Africa's manufacturing sector employs 1.4 million, about 11 percent of the country's work force, and accounts for nearly 25 percent of the country's GDP.

Figure 7.9 The geography of secondary economic activities Secondary economic activities are those that process, transform, fabricate, or assemble raw materials, or that reassemble, refinish, or package manufactured goods. As this map shows, the world's largest and most productive manufacturing regions are located in the core regions of Europe, North America, and Japan. Important concentrations of manufacturing industry are also located in semiperipheral countries such as South Korea, Mexico, and Brazil, but the increasing globalization of manufacturing means that patterns are subject to rapid change. (*Source:* Map projection, Buckminster Fuller Institute and Dymaxion Map Design, Santa Barbara, CA. The word Dymaxion and the Fuller Projection Dymaxion™ Map design are trademarks of the Buckminster Fuller Institute, Santa Barbara, California, ©1938, 1967 & 1992. All rights reserved.)

model, like those of most economists, rests on certain simplifying assumptions about the world. As we have seen, however, the real world is highly differentiated, not just in terms of natural resources, but in terms of demographics, culture, and politics. The assumptions in Rostow's model fit the experience of some parts of the world, but not all of it. In reality, a variety of pathways exist to development, as well as a variety of different processes and outcomes of development.

Figure 7.10 Stages of economic development This diagram illustrates a model of economic development that is based on the idea of successive stages of development. Each stage is seen as leading to the next, though different regions or countries may take longer than others to make the transition from one stage to the next. According to this view, first put forward by the economist W. W. Rostow, places and regions can be seen as following parallel courses within a world that is steadily modernizing. Late starters will eventually make progress, but at speeds determined by their resource endowments, their productivity, and the wisdom of their people's policies and decisions.

This view of the world has become very widespread, especially as applied to the experience and prospects of different countries, as intended by Rostow's model. It is, however, too simplistic to be of much help in understanding human geography. The reality is that places and regions are interdependent. The fortunes of any given place are tied up with those of many others, and increasingly so. Rostow's model perpetuates the myth of "developmentalism": the idea that every country and region will eventually make economic progress toward "high mass consumption" provided that they compete to the best of their ability within the world economy. But the main weakness of developmentalism is that it is simply not fair to compare the prospects of late starters to the experience of those places, regions, and countries that were among the early starters. For these early starters, the horizons were clear: free of effective competition, free of obstacles, and free of precedents. For the late starters, the situation is entirely different. Today's less-developed regions must compete in a crowded field while facing numerous barriers that are a direct consequence of the success of some of the early starters.

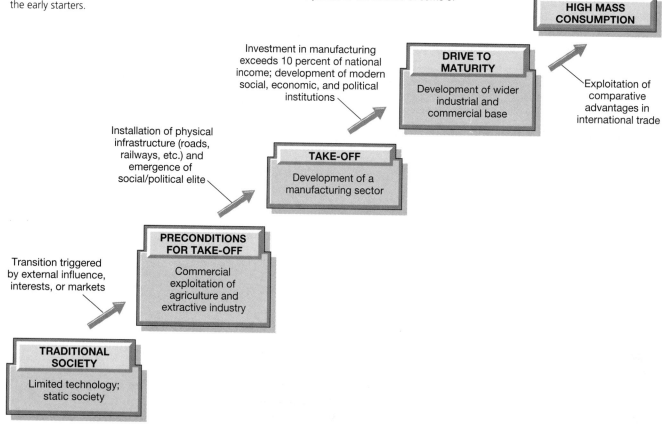

Everything in Its Place: Principles of Location

As geographers have sought to understand and explain local patterns of economic development, they have uncovered some fundamental principles that shape and influence decisions involving the location of economic activities. To the extent that regularity and predictability exist in economic geography, they stem from the logic of these principles. The geography of economic development is the cumulative outcome of decisions guided by fundamental principles of commercial and industrial location and interaction. As location decisions are played out in the real world, so distinctive geographic linkages and spatial structures emerge.

Principles of Commercial and Industrial Location

Locational decisions in commercial and industrial life are subject to a number of key factors:

- The relative importance of accessibility to whatever *material inputs* are involved (for example, raw materials, energy).
- The relative importance of the availability of *labor* with particular skills
- The relative importance of *processing* costs; these include the cost of land and buildings, machinery and hardware, software, maintenance, wages and salaries, utility bills, and local taxes.
- The relative pull of the *market* for the product or service, which depends on the importance of being near customers.
- The relative *transfer* costs that would be accrued at alternative locations. Transfer costs involve the costs not only of transporting inputs from various sources and of transporting outputs to markets, but also of insuring, storing, unloading, and repacking raw materials and finished products.
- The influence of cultural and institutional factors that channel certain activities away from some locations and toward others. The most important of these are *government policies* of one kind or another. It is quite common, for example, for local governments to offer tax breaks to companies in order to attract investments that will result in the creation of new jobs in the area.
- The influence of *behavioral considerations* that stem from the objectives and constraints affecting individual decision makers.

The importance and influence of these factors vary according to the type of activity involved. In retailing activities, for example, proximity to specific consumer markets is almost always of paramount importance—*almost* always, since some retailing occurs through mail order or telemarketing, in which case accessibility to a geographically scattered market becomes even more important. For small retailers, behavioral factors are likely to involve the personal values and priorities of business owners—the owner's attachment to a particular neighborhood, for example, or the owner's desire to locate his or her business at an upscale address. Large, corporate retailers, in contrast, will seek to maximize utility and minimize uncertainty through the extensive use of market research and geographic information systems (GIS—see Chapter 3).

Locational decisions regarding manufacturing activities, on the other hand, depend a great deal on the attributes of the inputs (that is, raw materials or components) and outputs (products) of particular industries. Where heavy, bulky raw materials are used to produce high-value products, it follows that proximity to sources of inputs is likely to be the most important locational factor. Steel production, for example, uses large quantities of iron ore, limestone, coking coal, and water. These basic raw materials lose both bulk and weight as raw steel is produced. When such weight-losing raw materials are involved in production and when these raw materials are to be found in or near one place, it is most economical to locate the manufacturing plant at that place—for why transport unnecessary weight? The industrial region around Sheffield, England (Figure 7.11, page 286), for example, was established on exactly this locational logic.

Where the manufacturing process adds significant bulk or weight to a product (as, for example, in brewing), proximity to markets is more likely to be the overriding factor. Many different inputs to different industries, exist, however, each with different attributes (weight, bulk, form, perishability, fragility) and different degrees of availability across geographic space. Some inputs (for example, bauxite, sulfur, zinc) are relatively uncommon and very localized; others (water, for example) are much more generally available. How can we sort this all out in

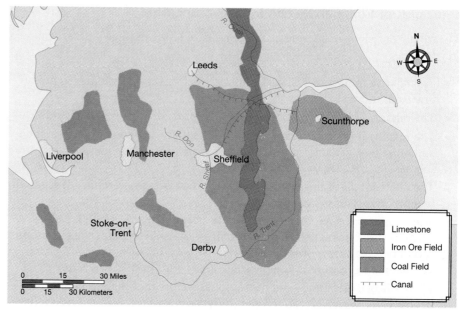

Figure 7.11 The Sheffield manufacturing region The growth of Sheffield was largely a result of the locational "pull" of the weight-losing inputs needed for iron and steel manufacturing: limestone, coal, iron ore, and water. Initially, iron ore came from deposits of clayband ironstone found amid nearby coal deposits; water came from the River Sheaf, and magnesian limestone from outcrops a few miles to the east. As the steel industry grew, the broad valley of the River Don provided sites for large steelworks, which drew on new coal mines in the extensive coalfield to the east; on high-grade iron-ore deposits found in seams 6 to 9 meters (20 to 30 feet) thick near Scunthorpe; and on limestone from the same sources to the east. In Sheffield itself, a multitude of firms sprang up to use high-quality iron and steel in the manufacture of everything from anchors, cutlery, files, and nails to needles, pins, and wire.

terms of generalizations and principles that shed light on fundamental patterns and tendencies in spatial organization?

Economic Interdependence: Agglomeration Effects

We can begin by recognizing that, in the real world, the various factors of commercial and industrial location all operate within complex webs of *functional interdependence*. These webs include the relationships between different kinds of industries, between different kinds of stores, and between different kinds of offices. They are based on linkages and relationships that tend to follow certain principles. Among the most important of these, in terms of human geography, are principles of agglomeration that influence the locational patterns of economic activities.

Agglomeration is the clustering together of functionally related activities—the way, for example, high-tech firms are clustered in Silicon Valley (the area between Santa Clara and San Jose, California). **Agglomeration effects** are the cost advantages that accrue to individual firms because of their locations within such a cluster. These advantages are sometimes known as external economies. **External economies** are cost savings that result from advantages that are derived from circumstances beyond a firm's own organization and methods of production. Thus, for example, it pays a wire-making factory to locate in close proximity to a steel mill not just in order to save on transporting the steel for the wire but also to save on the cost of re-heating the steel and to take advantage of the mill's experienced labor force.

Such situations lead to complex linkages among local economic activities. In relation to any given industry or firm, *backward linkages* are those that develop

Agglomeration effects: cost advantages that accrue to individual firms because of their location among functionally related activities.

External economies: cost savings that result from circumstances beyond a firm's own organization and methods of production.

with suppliers. In terms of our example, the wire-making factory has a backward linkage with the steel mill. The steel mill, in turn, will have backward linkages with its own suppliers: firms that supply raw materials, machinery, specialized maintenance services, and so on. For the steel mill, the wire-making factory is one of a number of *forward linkages* with firms using its output. Others might include can-making firms, tube-making firms, and cutlery firms. In addition, these firms will likely have their own forward linkages. The wire-making factory, for example, will supply local assemblers, finishers, packagers, and distributors. Together, backward and forward linkages often create a threshold of activity large enough to attract **ancillary activities**. Ancillary activities are those such as maintenance, repair, security, and haulage services that serve a variety of industries. Figure 7.12 summarizes the kinds of linkages that typically develop around a steel production plant.

Where external economies and local economic linkages are limited to firms involved in one particular industry, they are known as **localization economies**. These economies are cost savings that accrue to particular industries as a result of clustering together at a specific location. Examples include the sharing of a pool of labor with special skills or experience; supporting specialized technical schools; joining together to create a marketing organization or a research institute; and drawing on specialized subcontractors, maintenance firms, suppliers,

Ancillary activities: activities such as maintenance, repair, security, and haulage services that serve a variety of industries.

Localization economies: cost savings that accrue to particular industries as a result of clustering together at a specific location.

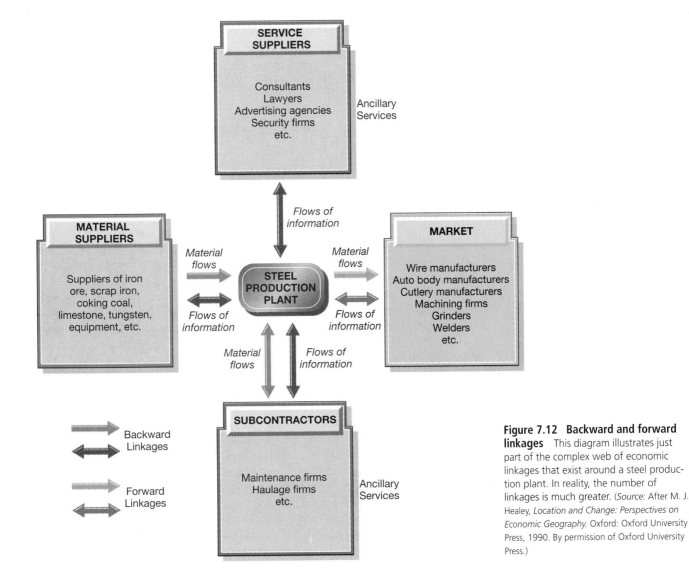

Figure 7.12 Backward and forward linkages This diagram illustrates just part of the complex web of economic linkages that exist around a steel production plant. In reality, the number of linkages is much greater. (*Source:* After M. J. Healey, *Location and Change: Perspectives on Economic Geography.* Oxford: Oxford University Press, 1990. By permission of Oxford University Press.)

Figure 7.13 Recording companies in Los Angeles Major and independent recording companies provide a good example of localization economies, whereby external economies and economic linkages have been established around a particular industry—recording studios for popular music, in this case. The map shows a central cluster of firms in Hollywood, with two subsidiary clusters in Burbank and Santa Monica. (*Source:* A. J. Scott, "The Craft, Fashion, and Cultural-Products Industries of Los Angeles. *Annals of the Association of American Geographers,* 86, 1996, Figure 2, p. 313.)

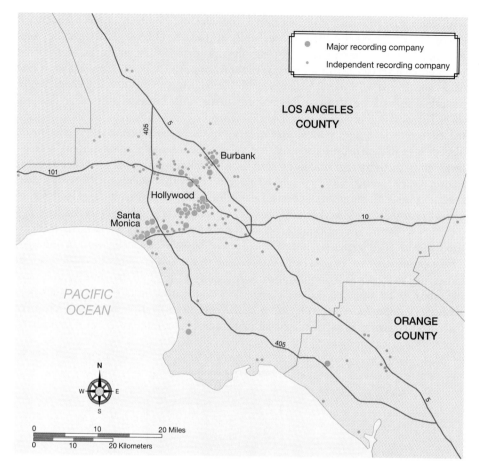

distribution agents, and lawyers. Where such advantages lead to a reputation for high-quality production, agglomeration will be intensified, since more producers will want to be able to cash in on the reputation. Among the many examples are the electronics and software industries of Silicon Valley; recording companies in Los Angeles (Figure 7.13); Sheffield steel and steel products (especially cutlery); the U.S. auto industry (Detroit); Swiss watches (concentrated in the towns of Biel, Geneva, and Neuchâtel); Belgian lace (Bruges, Brussels, Mechelen); English worsted (Bradford, Huddersfield); English china (Stoke-on-Trent); Irish linen (Athlone); and French perfume (Grasse).

Three main ways exist in which external economies can be derived. The first is through external *economies of scale.* By clustering together, firms can collectively support ancillary activities that can only operate efficiently in settings where enough different customers exist to ensure a continuous demand. Accessibility to these ancillary services, meanwhile, helps to make all firms more efficient. An example is the way that specialized engineering, maintenance, and repair firms support, and are sustained by, clusters of manufacturing plants. This is what happens in Sheffield and Detroit. Another, very different example is the way that clusters of offices and managerial activity can support, and gain from the presence of, specialized legal services, advertising services, and financial services. This is what happens in Manhattan, New York, and London.

The second source of external economies is through the atmosphere that results from the clustering of functionally related activities. Over time, a community of interest develops among producers, government agencies, research institutions, and specialized services, and this helps to generate *information* and to facilitate the flow of information. Close association also brings out rivalries that breed innovation. Frequent, easy contact among producers also helps minimize

uncertainty. Thus, for example, small firms in industries in which demand is unpredictable (because it is subject to fashion, for example, or to rapidly changing technologies) tend to agglomerate in order to be able to share information as quickly as possible. Examples include women's clothing (in Paris [Figure 7.14] and Milan), biotechnology (around Washington, DC), and computer software (in Silicon Valley).

The third source of external economies is the fixed social capital that is generated by clusters of activity. Fixed social capital constitutes the **infrastructure** of society, the underlying framework of services and amenities needed to facilitate productive activity. It comes from a mixture of public and private investment, and includes roads, highways, railroads, schools, hospitals, shopping centers, and recreational and cultural amenities. Agglomeration means that the costs of providing this infrastructure can be shared, and that more extensive and sophisticated infrastructures can be supported. The more extensive and sophisticated the infrastructure, the more advantages accrue to the producers—more efficient transportation, more specialized education, more attractive environments for key workers, and so on.

Providing firms in many industries with opportunities for external economies are urban settings, with large pools of labor containing a wide range of skills; intensively developed infrastructures; specialized education and training facilities; research institutions; and concentrations of business services. Because of this, external economies are often referred to as **urbanization economies,** those accruing to producers because of the package of infrastructure, ancillary activities, labor, and markets that is typically associated with urban settings.

Figure 7.14 The haute couture garment district in Paris Specialized districts like this, centered on the Faubourg Saint-Honoré in central Paris, provide good examples of external economies based on a narrow community of interest within which close rivalries breed innovation, while easy contact among producers helps to minimize uncertainty.

Pathways to Development

Patterns of economic development are the product of principles of location and economic interdependence, but they are also historical in origin and cumulative in nature. Even though the fundamental principles of spatial organization hold steady over time, societal and technological conditions change. As a result, economic geographies that were shaped by certain principles of spatial organization during one particular period are inevitably modified, later on, *as the same principles work their way through new technologies and new actors.* Thus we find different pathways of economic development, according to various circumstances of timing and location.

Recognizing this, geographers are interested not only in uncovering the fundamental principles of spatial organization but also in relating them to **geographical path dependence,** the historical relationship between present-day activities in a place and the past experiences of that place. A dynamic relationship exists between past and present geographies; in other words, when spatial structures emerge through the logic of fundamental principles of spatial organization, they do so in ways guided and influenced by preexisting patterns and relationships.

These observations lead to an important principle of economic development, the principle of initial advantage. **Initial advantage** highlights the critical importance of an early start in economic development. It represents a special case of external economies. Other things being equal, new phases of economic development will take hold first in settings that offer external economies: existing labor markets, existing consumer markets, existing frameworks of fixed social capital, and so on. This initial advantage will be consolidated by *localization economies*—cost savings that accrue to particular industries as a result of clustering together at a specific location—and so form the basis for continuing economic growth. This, in turn, provides the preconditions for initial advantage in subsequent phases of economic development.

For places and regions with a substantial initial advantage, therefore, the trajectory of geographical path dependence tends to be one of persistent growth—

Infrastructure (or fixed social capital): the underlying framework of services and amenities needed to facilitate productive activity.

Urbanization economies: external economies that accrue to producers because of the package of infrastructure, ancillary activities, labor, and markets typically associated with urban settings.

Geographical path dependence: the historical relationship between the present activities associated with a place and the past experiences of that place.

Initial advantage: the critical importance of an early start in economic development; a special case of external economies.

reinforcing the core–periphery patterns of economic development found in every part of the world and at every spatial scale. This having been said, geographers recognize that no single pathway exists to development. The consequences of initial advantage for both core and peripheral regions can be—and often are—modified. Old core–periphery relationships can be blurred, and new ones can be initiated.

How Regional Economic Cores Are Created

Regional cores of economic development are created cumulatively, following some initial advantage, through the operation of several of the basic principles of economic geography that we have described. These principles center on external economies, or agglomeration effects, that are associated with various kinds of economic linkages and interdependencies. The trigger for these agglomeration effects can be any kind of economic development—the establishment of a trading port, or the growth of a local industry or any large-scale enterprise. The external economies and economic linkages generated by such developments represent the initial advantage that tends to stimulate a self-propelling process of local economic development.

Given the location of a new economic activity in an area, a number of interrelated effects come into play. *Backward linkages* develop as new firms arrive to provide the growing industry with components, supplies, specialized services, or facilities. *Forward linkages* develop as new firms arrive to take the finished products of the growing industry and use them as inputs to their own processing, assembly, finishing, packaging, or distribution operations. Together with the initial growth, the growth in these linked industries helps to create a threshold of activity large enough to attract *ancillary industries* and activities (maintenance and repair, recycling, security, business services, for example).

The existence of these interrelated activities establishes a pool of specialized labor with the kinds of skills and experience that make the area attractive to still more firms. Meanwhile, the linkages between all these firms help to promote interaction between professional and technical personnel, and allow for the area to support R & D (research and development) facilities, research institutes, and so on, thus increasing the likelihood of local inventions and innovations that might provide further stimulus to local economic development.

Another part of the spiral of local economic growth is a result of the increase in population represented by the families of employees. Their presence creates a demand for housing, utilities, physical infrastructure, retailing, personal services, and so on—all of which generate additional jobs. This expansion, in turn, helps to create populations large enough to attract an even wider variety and more sophisticated kinds of services and amenities. Last—but by no means least—the overall growth in local employment creates a larger local tax base, which allows the local government to provide improved public utilities, roads, schools, health services, recreational amenities, and so on, all of which serve to intensify agglomeration economies and so enhance the competitiveness of the area in being able to attract further rounds of investment.

Swedish economist Gunnar Myrdal was the 1974 Nobel Prize winner who first recognized that any significant initial local advantage tends to be reinforced through geographic principles of agglomeration and localization. He called the process **cumulative causation** (Figure 7.15), referring to the spiral buildup of advantages that occurs in specific geographic settings as a result of the development of external economies, agglomeration effects, and localization economies. Myrdal also pointed out that this spiral of local growth would tend to attract people—enterprising young people, usually—and investment funds from other areas. Because of basic principles of spatial interaction, these flows tend to be

Cumulative causation: a spiral build-up of advantages that occurs in specific geographic settings as a result of the development of external economies, agglomeration effects, and localization economies.

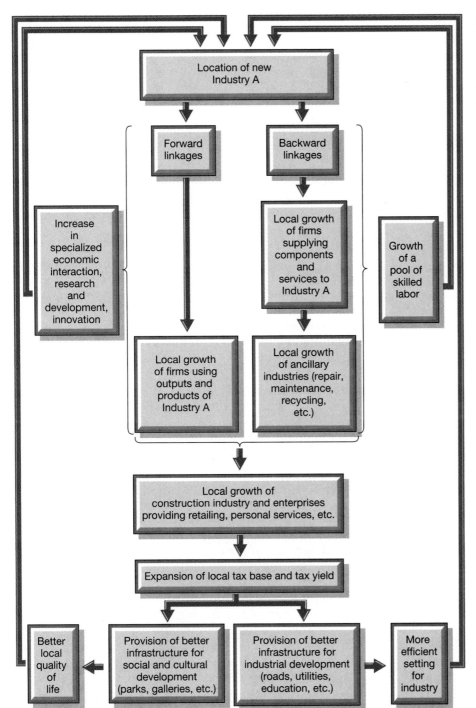

Figure 7.15 Processes of regional economic growth Once a significant amount of new industry comes to be established in an area, it tends to create a self-propelling process of economic growth. As this diagram shows, the initial advantages of industrial growth are reinforced through geographic principles of agglomeration and localization. The overall process is sometimes known as cumulative causation.

strongest from those regions nearby, with the lowest wages, fewest job opportunities, or least attractive investment opportunities.

In some cases, this loss of entrepreneurial talent, labor, and investment capital is sufficient to trigger a cumulative negative spiral of economic disadvantage. With less capital, less innovative energy, and depleted pools of labor, industrial growth in peripheral regions tends to be significantly slower and less innovative than in regions with an initial advantage and an established process of cumulative causation. This, in turn, tends to limit the size of the local tax base, so that local governments find it hard to furnish a competitive infrastructure of roads, schools, and recreational amenities. Myrdal called these disadvantages **backwash effects,**

Backwash effects: the negative impacts on a region (or regions) of the economic growth of some other region.

the negative impacts on a region (or regions) of the economic growth of some other region. These negative impacts take the form, for example, of out-migration, outflows of investment capital, and the shrinkage of local tax bases. Backwash effects are important because they help to explain why regional economic development is so uneven and why core–periphery contrasts in economic development are so common.

How Core–Periphery Patterns Are Modified

Although very important, cumulative causation and backwash effects are not the only processes that affect the geography of economic development. If they were, the world's economic geography would be even more starkly polarized than it is now—little chance would exist for the emergence of new growth regions like Guangdong in Southeast China; and there would be little likelihood of stagnation or decline in once-booming regions like Northern England.

Spread effects: the positive impacts on a region (or regions) of the economic growth of some other region.

Myrdal himself recognized that peripheral regions do sometimes emerge as new growth regions, and he provided a partial explanation of them in what he called spread (or trickle-down) effects. **Spread effects** are the positive impacts on a region (or regions) of the economic growth of some other region, usually a core region. This growth creates levels of demand for food, consumer products, and other manufactures that are so high that local producers cannot satisfy them. This demand provides the opportunity for investors in peripheral regions (or countries) to establish a local capacity to meet the demand. Entrepreneurs who attempt this are also able to exploit the advantages of cheaper land and labor in peripheral regions. If strong enough, these effects can enable peripheral regions to develop their own spiral of cumulative causation, thus changing the interregional geography of economic patterns and flows. The economic growth of South Korea, for example, is partly attributable to the spread effects of Japanese economic prosperity.

Another way in which peripheral regions or countries can develop their own spiral of cumulative causation is through a process of *import substitution* (see Chapter 2). In this process, goods and services previously imported from core regions come to be replaced by locally made goods and locally provided services. Although some things are hard to copy because of the limitations of climate or natural resource endowments, many products and services *can* be copied by local entrepreneurs, thus capturing local capital, increasing local employment opportunities; intensifying the use of local resources, and generating profits for further local investment. The classic example of import substitution is provided by Japan, where import substitution, especially for textiles and heavy engineering, played an important part in the transition from a peripheral economy to a major industrial power in the late nineteenth century. Import substitution also figured prominently in the Japanese "economic miracle" after World War II, this time featuring the automobile industry and consumer electronics. Today, countries like Argentina, Peru, and Ghana are seeking to follow the same sort of strategy, subsidizing domestic industries and protecting them from outside competitors through tariffs and taxes.

Agglomeration diseconomies: the negative economic effects of urbanization and the local concentration of industry.

Core–periphery patterns and relationships can also be modified by changes in the dynamics of core regions—internal changes that can slow or modify the spiral of cumulative causation. The main factor is the development of **agglomeration diseconomies,** the negative economic effects of urbanization and the local concentration of industry. They include the higher prices that must be paid by firms competing for land and labor; the costs of delays resulting from traffic congestion and crowded port and railroad facilities; the increasing unit costs of solid waste disposal; and the burden of higher taxes that eventually have to be levied by local governments in order to support services and amenities previously considered unnecessary—traffic police, city planning, and transit systems, for example.

Diseconomies that are imposed through taxes can often be passed on by firms to consumers in other regions and other countries in the form of higher prices. Charging higher prices, however, decreases the competitiveness of a firm in relation to firms operating elsewhere. Agglomeration diseconomies that cannot be "exported"—noise, air pollution, increased commuting costs, and increased housing costs, for example—require local governments to tax even more of the region's wealth in attempts to compensate for a deteriorating quality of life. In California, this issue has become so pressing that a recent report prepared jointly by the state, the Bank of America, the Greenbelt Alliance, and the Low Income Housing Fund concluded that "California businesses cannot compete globally when they are burdened with the costs of sprawl. . . . An attractive business climate cannot be sustained if the quality of life continues to decline and the cost of financing real estate development escalates."[3] California's response has been to enact some of the nation's toughest traffic and environmental regulations.

Deindustrialization and Creative Destruction The most fundamental cause of change in the relationship between initial advantage and cumulative causation is to be found in the longer-term shifts in technology systems and in the competition between States within the world-system. The innovations associated with successive technology systems generate new industries that are not yet tied down by enormous investments in factories or allied to existing industrial agglomerations. Combined with innovations in transport and communications, this creates *windows of locational opportunity* that can result in new industrial districts, with small towns or cities growing into dominant metropolitan areas through new rounds of cumulative causation.

Equally important as a factor in how core–periphery patterns are modified are the consequent shifts in the profitability of old, established industries in core regions, compared to the profitability of new industries in fast-growing new industrial districts. As soon as the differential is large enough, some disinvestment will take place within core regions. This disinvestment can take place in several ways. Manufacturers can reduce their wage bill by cutting back on production; they can reduce their fixed costs by closing down and selling off some of their factory space and equipment; or they can reduce their spending on research and development for new products. This disinvestment, in turn, leads to deindustrialization in formerly prosperous industrial core regions.

Deindustrialization involves a relative decline (and in extreme cases an absolute decline) in industrial employment in core regions as firms scale back their activities in response to lower levels of profitability (Figure 7.16). This is what happened to the industrial regions of northern England in the early part of the twentieth century; and it is what happened to the Manufacturing Belt (sometimes called the "rustbelt") of the United States in the 1960s and 1970s.

Meanwhile, the capital made available from disinvestment in these core regions becomes available for investment by entrepreneurs in new ventures based on innovative products and innovative production technologies. Old industries— and sometimes entire old industrial regions—have to be "dismantled" (or, at least, neglected) in order to help fund the creation of new centers of profitability and employment. This process is often referred to as **creative destruction**, something that is inherent to the dynamics of capitalism. Creative destruction provides us with a powerful image to understand the entrepreneur's need to withdraw investments from activities (and regions) yielding low rates of profit, to reinvest in new activities (and, often, in new places). In the United States, for example, the

Deindustrialization: a relative decline in industrial employment in core regions.

Creative destruction: the withdrawal of investments from activities (and regions) that yield low rates of profit in order to reinvest in new activities (and new places).

---

[3]*Beyond Sprawl.* Sacramento, CA: State of California: Resources Agency, 1995.

Figure 7.16 Regional economic decline When the locational advantages of manufacturing regions are undermined for one reason or another, profitability declines and manufacturing employment falls. These failures can lead to a downward spiral of economic decline, as experienced by many of the traditional manufacturing regions of Europe during the 1960s, 1970s, and 1980s. In France, Belgium, the Netherlands, Norway, Sweden, and the United Kingdom, manufacturing employment decreased by between one-third and one-half between 1960 and 1990. The most pronounced example of this deindustrialization has been in the United Kingdom, where a sharp decline in manufacturing employment has been accompanied by an equally sharp rise in service employment. (*Source:* Reprinted with permission of Prentice Hall, from P. L. Knox, *Urbanization,* © 1994, p. 55.)

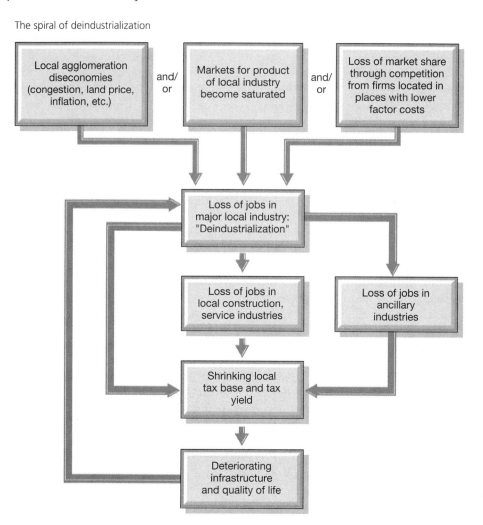

The spiral of deindustrialization

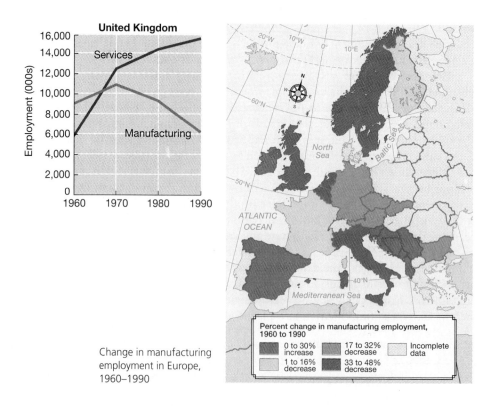

Change in manufacturing employment in Europe, 1960–1990

deindustrialization of the Manufacturing Belt provided the capital and the locational flexibility for firms to invest in the "sunbelt" of the United States and in semiperipheral countries like Mexico and South Korea.

The process does not stop there, however. If the deindustrialization of the old core regions is severe enough, the relative cost of their land, labor, and infrastructure may decline to the point where they once again become attractive to investors. As a result, a see-saw movement of investment capital occurs, which over the long term tends to move from developed to less-developed regions—then back again, once the formerly developed region has experienced a sufficient relative decline. "Has-been" regions can become redeveloped and revitalized, given a new lease on life by the infusion of new capital for new industries. This is what happened, for example, to the Pittsburgh region in the 1980s, resulting in the creation of a postindustrial economy out of a depressed industrial setting. USX, the U.S. Steel group of companies, reduced its work force in the Pittsburgh region from more than 20,000 to less than 5,000 between 1975 and 1995. These losses have been more than made up, however, by new jobs generated in high-tech electronics, specialized engineering, and finance and business services.

Government Intervention In addition to the means of deindustrialization and creative destruction, core–periphery patterns can also be modified by government intervention. National governments realize that regional planning and policy can be an important component of broad economic strategies designed both to stabilize and reorganize their economies, and to maximize their overall competitiveness. Without regional planning and policy, the resources of peripheral regions can remain underutilized, while core regions can become vulnerable to agglomeration diseconomies. For political reasons, too, national governments are often willing to help particular regions to adjust to changing economic circumstances. At the same time, most local governments take responsibility for stimulating economic development within their jurisdiction, if only in order to increase the local tax base.

The nature and extent of government intervention has varied over time and by country. In some countries, special government agencies have been established to promote regional economic development and reduce core–periphery contrasts. Among the best-known examples are the Japanese MITI (Ministry of International Trade and Industry); the Italian Cassa del Mezzogiorno (Southern Development Agency, replaced in 1987 by several smaller agencies); and the U.S. Economic Development Administration. Some governments have sought to help industries in declining regions by undertaking government investment in infrastructure and providing subsidies for private investment; others have sought to devise tax breaks that reduce the cost of labor in peripheral regions. Others still have sought to deal with agglomeration diseconomies in core regions through increased taxes and restrictions on land use.

While each approach has its followers, one of the most widespread governmental approaches to core–periphery patterns involves the exploitation of the principle of cumulative causation through the creation of growth poles. **Growth poles** are places of economic activity deliberately organized around one or more high-growth industries. Economists have noted, however, that not all industries are equal in the extent to which they stimulate economic growth and cumulative causation. The ones that generate the most pronounced effects are known as "propulsive industries," and they have received a great deal of attention from geographers and economists who are interested in helping to shape strategic policies that might promote regional economic development. In the 1920s, shipbuilding was a propulsive industry. In the 1950s and 1960s, automobile manufacturing was a propulsive industry, and today biotechnology is one. The basic idea is for governments to promote regional economic growth by fostering propulsive industries in favorable locations. These are intended to become

Growth poles: economic activities that are deliberately organized around one or more high-growth industries.

Lenowisco, a local development district covering Lee County, the city of Norton, and Wise County in southwestern Virginia, established a growth center in Duffield at the intersection of two of the new highways created by the Appalachian Regional Commission. This photograph shows the industrial park that was created by Lenowisco with the help of federal funding from the commission. While the park has been successful in attracting small factories, which together employ several hundred persons, it has never been large enough to "take off" as a growth center.

Figure 7.17 Growth centers in Appalachia The Appalachian Regional Commission, established by President Kennedy in 1965 in order to promote the economic development of the region, initiated a massive program of road building in order to open up the region for modern economic development. The commission also designated over 200 "growth centers" intended to attract industrialists by furnishing a new physical infrastructure and offering economic incentives in the form of tax breaks. The new industries, it was hoped, would trigger the geographic process of cumulative causation. The growth centers were too small and too numerous, however, with the result that none of them has been able to generate self-sustaining processes of local economic growth. (*Source:* Figure 1 from P. L. Knox and D. E. Short, "Regional Development in Appalachia Since 1965," *Geography,* 62(3), 1977, pp. 271–221.)

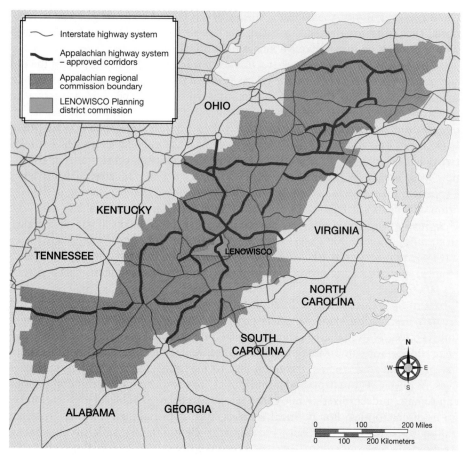

growth poles—places that, given an artificial start, develop a self-sustaining spiral of economic prosperity.

Many countries have used the growth-poles approach as a basis for regional development policies. For example, French governments have designated certain locations as *technopoles*—sites for the establishment of high-tech industries (such as computers and biotechnology)—under the assumption that these leading-edge activities will stimulate further development. In southern Italy, various heavy industries were sited in a number of remote locations in order to stimulate ancillary development. In the United States, a whole series of growth centers were designated within Appalachia to establish a manufacturing base within the region (Figure 7.17). The results of such policies have been mixed, however. The French technopoles have been fairly successful because the French government invested large sums of money in establishing propulsive industries in favorable locations. The Italian and U.S. growth-pole efforts, like many others, however, have been disappointing. In practice, governments often fail to invest in the right industries, and nearly always fail to invest heavily enough to kick-start the process of cumulative causation.

Globalization and Local Economic Development

Now that the world economy is much more globalized, patterns of local and regional economic development are much more open to external influences, and much more interdependent with economic development processes elsewhere. The globalization of the world economy involves a new international division of labor in association with the internationalization of finance; the deployment of a new technology system (using robotics, telematics, biotechnology, and other new technologies); and the homogenization of consumer markets (see Chapter 2).

This new framework for economic geography has already left its mark on the world's economic landscapes, and has also meant that the lives of people in different parts of the world have become increasingly intertwined. You may recall Duong, Hua, Françoise, and Jean-Paul, people mentioned in Chapter 1. The lives of three more people singled out by the World Bank's 1995 *World Development Report* follow:[4]

Joe lives in a small town in southern Texas. His old job as an accounts clerk in a textile firm, where he had worked for many years, was not very secure. He earned $50 a day, but promises of promotion never came through, and the firm eventually went out of business as cheap imports from Mexico forced textile prices down. Joe went back to college to study business administration and was recently hired by one of the new banks in the area. He enjoys a comfortable living even after making the monthly payments on his government-subsidized student loan.

Maria recently moved from her central Mexican village and now works in a U.S-owned firm in Mexico's maquiladora sector. Her husband, Juan, runs a small car upholstery business and sometimes crosses the border during the harvest season to work illegally on farms in California. Maria, Juan, and their son have improved their standard of living since moving out of subsistence agriculture, but Maria's wage has not increased in years: she still earns about $10 a day . . .

Xiao Zhi is an industrial worker in Shenzhen, a Special Economic Zone in China. After three difficult years on the road as part of China's floating population, fleeing the poverty of nearby Sichuan province, he has finally settled with a new firm from Hong Kong that produces garments for the U.S. market. He can now afford more than a bowl of rice for his daily meal. He makes $2 a day and is hopeful for the future.

These examples begin to reveal a complex and fast-changing interdependence that would have been unthinkable just 15 or 20 years ago. Joe lost his job because of competition from poor Mexicans like Maria, and now her wage is held down by cheaper exports from China. Joe now has a better job, however, and the U.S. economy has gained from expanding exports to Mexico. Maria's standard of living has improved, and her son can hope for a better future. Joe's pension fund is earning higher returns through investments in growing enterprises around the world; and Xiao Zhi is looking forward to higher wages and the chance to buy consumer goods. Not everyone has benefited, however, and the new international division of labor has come under attack by some in industrial countries where rising unemployment and wage inequality are making people feel less secure about the future. Some workers in core countries are fearful of losing their jobs because of cheap exports from lower-cost producers. Others worry about companies relocating abroad in search of low wages and lax labor laws.

Most of the world's population now lives in countries that are either integrated into world markets for goods and finance, or rapidly becoming so. As recently as the late 1970s, only a few peripheral countries had opened their borders to flows of trade and investment capital. About a third of the world's labor force lived in countries like the Soviet Union and China with centrally planned economies, and at least another third lived in countries insulated from inter-

---

[4]World Bank, *World Development Report 1995: Workers in An Integrating World*. New York: Oxford University Press, 1995, p. 50.

national markets by prohibitive trade barriers and currency controls. Today, with nearly half the world's labor force among them, three giant population blocs—China, the republics of the former Soviet Union, and India—are entering the global market. Many other countries, from Mexico to Thailand, have already become involved in deep linkages. According to World Bank estimates, fewer than 10 percent of the world's labor force will remain isolated from the global economy by the year 2000. (The World Bank, properly called the International Bank for Reconstruction and Development, is a United Nations affiliate established in 1948 to finance productive projects that further the economic development of member nations.)

In this section, we examine some of the specific impacts of three of the principal components of the global economy: the global assembly line, resulting from the operations of transnational manufacturing corporations; the global office, resulting from the internationalization of banking, finance, and business services; and the pleasure periphery, resulting from the proliferation of international tourism.

The Global Assembly Line

The globalization of the world economy represents the most recent stage in a long process of internationalization. At the heart of this process has been the emergence of private companies that participate not only in international trade but also in production, manufacturing, and/or sales operations in several countries. Many of these transnational corporations have grown so large through a series of mergers and acquisitions that their activities now span a diverse range of economic activities.

Conglomerate corporations: companies that have diversified into a variety of different economic activities, usually through a process of mergers and acquisitions.

When corporations consist of several divisions engaged in quite different activities, they are known as **conglomerate corporations,** those having diversified into a variety of different economic activities, usually through a process of mergers and acquisitions. Philip Morris, for example, primarily known for its tobacco products (Marlboro cigarettes, for example), also controls the single largest group of assets in the beverage industry (including Miller Brewing), and has extensive interests in foods (including Kraft, General Foods, Tobler, Terry's, and Suchard chocolate) as well as real estate (the Mission Viejo company in California), import-export (Duracell do Brasil), and publishing (E.Z. Editions, Zürich). Nestlé, the world's largest packaged food manufacturer, is the largest company in Switzerland, but derives less than 2 percent of its revenue from its home country. Its major U.S. product lines and brand names include beverages (Calistoga, Nescafé, Nestea, Perrier, Quik, Taster's Choice), chocolate and candy (After Eight, Butterfinger, Crunch, KitKat, Raisinets), culinary products (Buitoni, Carnation, Contadina, Libby, Toll House), frozen foods (Lean Cuisine), pet foods (Fancy Feast, Friskies, Mighty Dog), wine (Beringer Brothers), and drugs and cosmetics (Alcon Optical, L'Oreal). In addition to its 438 factories in 63 countries around the world, Nestlé operates more than 40 Stouffer's Hotels.

Transnational corporations first began to appear in the nineteenth century, but until the mid-twentieth century there were only a few, most of them U.S.- or European-based transnationals that were concerned with obtaining raw materials such as oil or minerals for their domestic manufacturing operations. After World War II, an increasing number of large corporations began to invest in overseas production and manufacturing operations as a means of establishing a foothold in foreign consumer markets. Between 1957 and 1967, 20 percent of all new U.S. machinery plants, 25 percent of all new chemical plants, and over 30 percent of all new transport equipment plants were located abroad. By 1970, almost 75 percent of U.S. imports were transactions between the domestic and foreign subsidiaries of transnational conglomerates. By the end of the 1970s, overseas profits accounted for a third or more of the overall profits of the 100 largest transnational corporations. By the early 1980s, 40 percent of all world trade was in the form of intrafirm trade (that is, between different branches and companies of the same transnational conglomerate).

Beginning in the 1970s a sharp increase occurred in the growth of transnational conglomerates, not only U.S.-based but also in Europe, Japan, and even some in semiperipheral countries. By 1995 there were over 35,000 transnational corporations in the world, less than half of which were based in the United States, Japan, and Germany. Of these, the top 300 control approximately one-quarter of the world's productive assets. Many of the largest transnational corporations are now more powerful, in economic terms, than most sovereign nations. Ford's economy is larger than Poland's and Pakistan's; Itochu's (the world's largest corporation) is larger than Saudi Arabia's; and British Petroleum's annual sales exceed Ireland's gross domestic product (Figure 7.18).

The reason for such growth in the number and scale of transnational conglomerate corporations has been that international economic conditions have changed. A recession, triggered by a massive increase in the price of crude oil in 1973, meant that companies everywhere had to re-examine their strategies. At

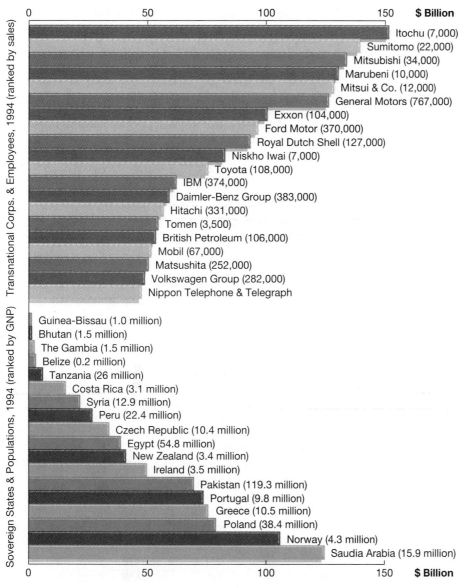

Figure 7.18 Nation-States and transnational corporations compared A comparison of the 20 largest transnational corporations (in terms of annual sales in 1994) and selected nation-states (in terms of their annual GNP). Figures in parentheses indicate the number of employees (for transnational corporations) and total population (for nations). The sheer size of transnational corporations, in economic terms, means that they have considerable influence on the economies, policies, and politics of the countries in which they operate. The world's largest corporation, Itochu, is based in Osaka, Japan. It began as a general trading company but is now a globally integrated corporation with interests in construction, satellite communications, financial services, natural resource development, retailing, cable TV, entertainment broadcasting, and multimedia. With just 7,000 employees, it generated revenues of $166.7 billion in 1994.

around the same time, technological developments in transport and communications provided larger companies with the flexibility and global reach to exploit the steep differentials in labor costs that exist between core countries and peripheral countries. These same developments in transport and communications meanwhile made for intensified international competition, which forced firms to search more intensely for more efficient and profitable global production and marketing strategies. Concurrently, a homogenization of consumer tastes (also facilitated by new developments in communications technologies) has made it possible for companies to cater to global markets.

It was the consequent burst of transnational corporate activity that formed the basis of the recent globalization of the world economy. In effect, the playing field for large-scale businesses of all kinds had been marked out anew. Companies have had to reorganize their operations in a variety of ways, restructuring their activities and redeploying their resources *between different countries, regions, and places* (see Feature 7.2, "Geography Matters—The Changing Geography of the Clothing Industry"). Local patterns of economic development have been recast, and recast yet again as these processes of restructuring, reorganization, and redeployment have been played out.

As these changes have unfolded, local geographies of economic development have become increasingly interdependent, linked together as part of complex transnational commodity chains. *Commodity chains* are networks of labor and production processes whose end result is the delivery of a finished product. They are, effectively, global assembly lines that are geared to produce global products for global markets.

The advantages to manufacturers of a global assembly line are several. First, a standardized global product for a global market allows them to maximize economies of scale. Second, a global assembly line allows production and assembly to take greater advantage of the full range of geographical variations in costs. Basic wages in manufacturing industries, for example, are between 25 and 75 times higher in core countries than in some peripheral countries. With a global assembly line, labor-intensive work can be done where labor is cheap, raw materials can be processed near their source of supply, final assembly can be done close to major markets, and so on. Third, a global assembly line means that a company is no longer dependent on a single source of supply for a specific component, thus reducing its vulnerability to industrial troubles and other disturbances.

The automobile industry was among the first to develop in this way. In 1976, Ford introduced the Fiesta, a vehicle designed to sell in Europe, South America, and the Asian market as well as North America. The Fiesta was assembled in several different locations from components manufactured in an even greater number of locations. The Fiesta became the first of a series of Ford "world cars" that now includes the Escort, the Mondeo, and the Contour. The components of the Ford Escort, for example, are made and assembled in 15 countries across three continents. Ford's international subsidiaries, which used to operate independently of the parent company, are now being functionally integrated, using supercomputers and video teleconferences. Meanwhile, other automobile companies have organized their own global assembly lines for their world cars: the Volkswagen Rabbit/Golf, for example (Figure 7.19), Volkswagen's new Concept 1 vehicle, GM's Corsa, and Fiat's Project 178.

Today, most of the 40 million or so vehicles that roll off production lines each year are made by just 10 global corporations (in order of size: General Motors, Ford, Toyota, Volkswagen, Nissan, Fiat, Peugeot-Citroën, Honda, Mitsubishi, and Renault), all of which not only operate global assembly lines (Figure 7.20) but are also involved in strategic alliances that intensify economic globalization. In 1996, 244 strategic alliances existed between the world's 41 largest auto makers, including parts sharing agreements and joint ventures in research, as well as in manufacturing. Peugeot of France, for example, had 22

Figure 7.19 Volkswagen's global assembly line The map shows the global network of subsidiary manufacturing companies and the flows of parts and finished vehicles between them and the company's main plant in Wolfsburg, Germany. Volkswagen's world car, the Rabbit (marketed as the Golf in much of the world), has been in production since 1975. Volkswagen's global production network allows the company to process raw materials near its sources of supply; to undertake labor-intensive work where labor is cheap; to complete final assembly close to major markets; and to establish multiple sources for components, thus reducing its vulnerability to work stoppages arising from local labor disputes. (*Sources:* B. Marshall (Ed.), *The Real World.* Boston: Houghton Mifflin, 1991, pp. 92–93. Map projection, Buckminster Fuller Institute and Dymaxion Map Design, Santa Barbara, CA. The word Dymaxion and the Fuller Projection Dymaxion™ Map design are trademarks of the Buckminster Fuller Institute, Santa Barbara, California, ©1938, 1967 & 1992. All rights reserved.)

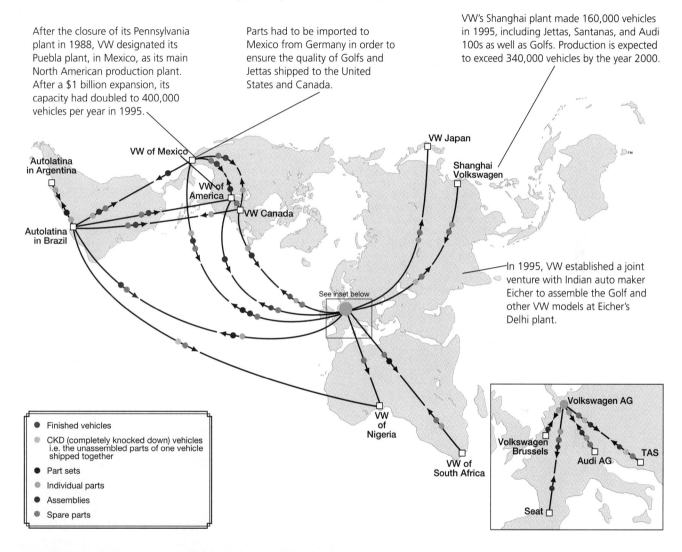

After the closure of its Pennsylvania plant in 1988, VW designated its Puebla plant, in Mexico, as its main North American production plant. After a $1 billion expansion, its capacity had doubled to 400,000 vehicles per year in 1995.

Parts had to be imported to Mexico from Germany in order to ensure the quality of Golfs and Jettas shipped to the United States and Canada.

VW's Shanghai plant made 160,000 vehicles in 1995, including Jettas, Santanas, and Audi 100s as well as Golfs. Production is expected to exceed 340,000 vehicles by the year 2000.

In 1995, VW established a joint venture with Indian auto maker Eicher to assemble the Golf and other VW models at Eicher's Delhi plant.

- ● Finished vehicles
- ● CKD (completely knocked down) vehicles i.e. the unassembled parts of one vehicle shipped together
- ● Part sets
- ● Individual parts
- ● Assemblies
- ● Spare parts

Figure 7.20 Honda's car assembly complex
Honda, like other Japanese auto manufacturers Nissan, Mitsubishi, Isuzu, Toyota, and Fuji Heavy Industries, has developed a global strategy of "localizing" vehicle production. Because of the strength of the yen on international currency exchanges, Japanese-made vehicles are expensive in North American and European markets. Localized production keeps prices down, and also helps avoid trade barriers.

The clothing industry provides a good example of the way that local economic geographies are affected by an industry's response to globalization. In the nineteenth century, the clothing industry developed in the metropolitan areas of core countries, with many small firms using cheap, migrant, or immigrant labor. In the first half of the twentieth century, the clothing industry, like many others, began to modernize. Larger firms emerged, their success based on the exploitation of mass production techniques for mass markets, and on the exploitation of principles of spatial organization within national markets. In the United States, for example, the clothing industry went through a major locational shift as a great deal of production moved out of the workshops of New York to big, new factories in smaller towns in the South, where labor was not only much cheaper but less unionized.

Figure 7.2.1 The changing distribution of clothing manufacturing As the map shows, most of the world's clothing exports come from just a few countries. In detail, however, the geography of clothing manufacturing changes rapidly in response to the changing patterns of costs and opportunities within the world economy.

(*Source:* Map projection, Buckminster Fuller Institute and Dymaxion Map Design, Santa Barbara, CA. The word Dymaxion and the Fuller Projection Dymaxion™ Map design are trademarks of the Buckminster Fuller Institute, Santa Barbara, California, ©1938, 1967 & 1992. All rights reserved.)

Of an estimated 120,000 garment workers in Los Angeles, about 80 percent are Latin American immigrants, mainly from Mexico, Guatemala, and El Salvador.

In 1984, the Guatemalan Congress passed legislation designed to attract investment in garment assembly. By 1993, more than 275 factories were employing more than 50,000 workers (mostly women), who assembled nearly $350 million in garments for export to the United States.

While Levi Strauss & Co is based in San Francisco, it has over 600 different suppliers in more than 50 countries. Since 1992 the company has kept databases to track contractors' work practices and labor relations. Levi withdrew its sourcing from China, Burma, and Peru because of its concern over violations of human rights.

Nike, headquartered in Beaverton, Oregon, is the world's largest seller of sports apparel, with annual revenues of over $2 billion. Most of its products are manufactured in low-wage, underdeveloped countries.

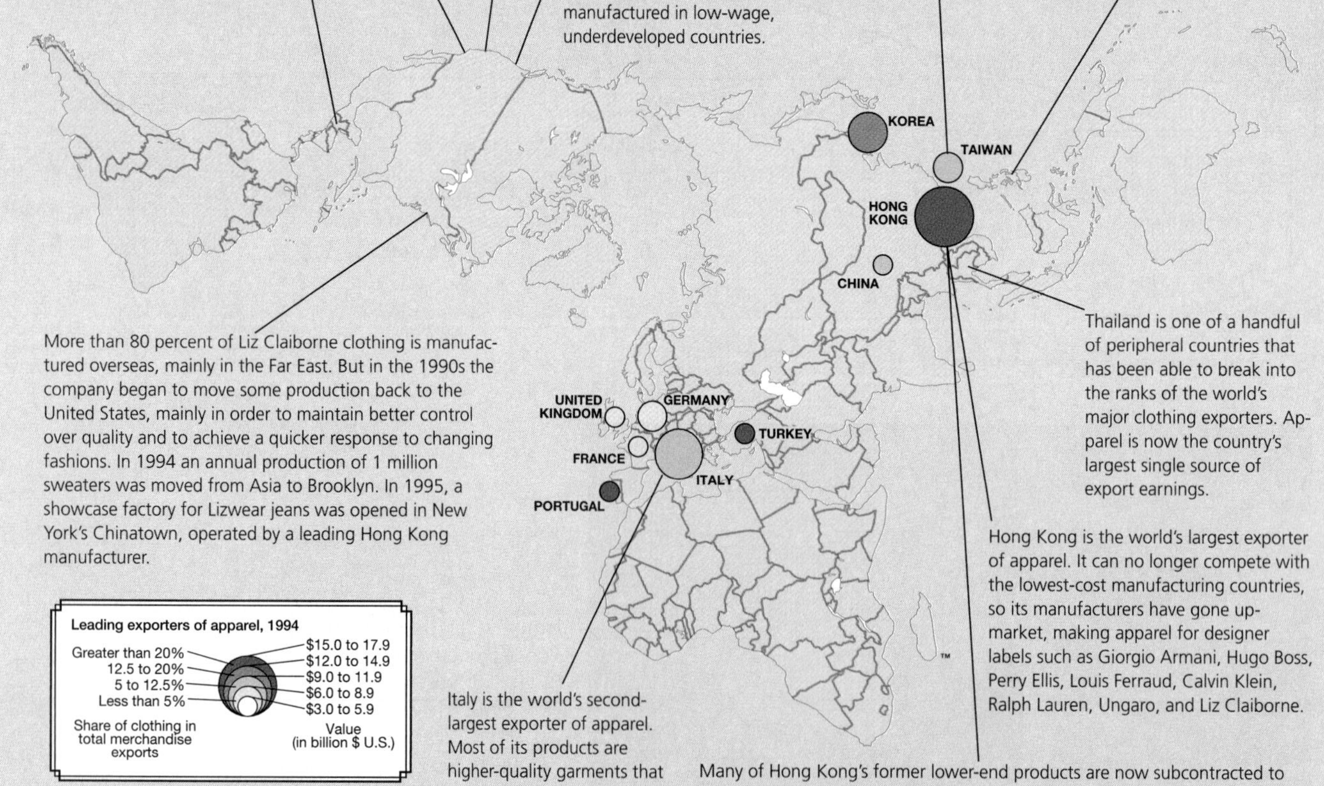

The biggest buyers of Taiwanese clothing include Kmart, Wal-Mart, J. C. Penney, The Limited, and Montgomery Ward. Because of rising wage rates and labor shortages, many Taiwanese firms have established subsidiaries in Southeast Asia and the Caribbean Basin.

The clothing industry employs over 850,000 in the Philippines, 75 percent of whom are women. Orders from apparel companies in core countries are taken by exporting companies in Manila, who then subcontract to provincial manufacturers, who in turn farm out the jobs, all the way down to a rural cottage industry.

More than 80 percent of Liz Claiborne clothing is manufactured overseas, mainly in the Far East. But in the 1990s the company began to move some production back to the United States, mainly in order to maintain better control over quality and to achieve a quicker response to changing fashions. In 1994 an annual production of 1 million sweaters was moved from Asia to Brooklyn. In 1995, a showcase factory for Lizwear jeans was opened in New York's Chinatown, operated by a leading Hong Kong manufacturer.

KOREA

TAIWAN

HONG KONG

CHINA

UNITED KINGDOM

GERMANY

TURKEY

FRANCE

ITALY

PORTUGAL

Thailand is one of a handful of peripheral countries that has been able to break into the ranks of the world's major clothing exporters. Apparel is now the country's largest single source of export earnings.

Hong Kong is the world's largest exporter of apparel. It can no longer compete with the lowest-cost manufacturing countries, so its manufacturers have gone up-market, making apparel for designer labels such as Giorgio Armani, Hugo Boss, Perry Ellis, Louis Ferraud, Calvin Klein, Ralph Lauren, Ungaro, and Liz Claiborne.

Leading exporters of apparel, 1994

Share of clothing in total merchandise exports	Value (in billion $ U.S.)
Greater than 20%	$15.0 to 17.9
12.5 to 20%	$12.0 to 14.9
5 to 12.5%	$9.0 to 11.9
Less than 5%	$6.0 to 8.9
	$3.0 to 5.9

Italy is the world's second-largest exporter of apparel. Most of its products are higher-quality garments that are consumed in other European countries.

Many of Hong Kong's former lower-end products are now subcontracted to manufacturers in China's Guangdong province. China provides the U.S. with 20–25% of its textiles and apparel, worth over $7 billion a year. Another $2 billion of shipments arrive in the U.S. illegally, having been routed through a worldwide network of dealers and relabeled in order to beat U.S. trade quotas.

Then, as the world economy began to globalize, semi-peripheral and peripheral countries became the least-cost locations for mass-produced clothing for "global" markets. In 1960, less than 7 percent of all apparel purchased in the United States had been imported. By 1980, more than half was imported. This globalization of production has resulted in a complex set of commodity chains. Many of the largest clothing companies, such as Liz Claiborne, have most of their products manufactured through arrangements with independent suppliers (over 300 in the case of Claiborne), with no one supplier producing more than 4 or 5 percent of the company's total output. These manufacturers are scattered throughout the world, making the clothing industry one of the most globalized of all manufacturing activities (Figure 7.2.1).

Although cheap leisure wear can be produced most effectively through arrangements with multiple suppliers in peripheral, low-wage regions, higher-end apparel for the global marketplace requires a different geography of production. These products—women's fashion, outerwear, and lingerie, infants' wear, and men's suits—are based on frequent style changes and high-quality finish. This requires short production runs and greater contact between producers and buyers. The most profitable settings for these products are in the metropolitan areas of the core countries—London, Paris, Stuttgart, Milan, New York, and Los Angeles—where, once again, migrant and immigrant labor provides a work force for "designer" clothing that can be shipped in small batches to upscale stores and shopping malls around the world.

The result is that commodity chains in the clothing industry are quite distinctive in terms of the origins of products destined for different segments of the market (Figure 7.2.2). Fashion-oriented retailers in the United States who sell "designer" products to up-market customers obtain most of their goods from manufacturers in a small group of high-value-added countries including France, Italy, Japan, the United Kingdom, and the United States. Department stores that emphasize "private label" products (that is, store brands, such as Nordstrom) and premium national brands obtain most of their goods from established manufacturers in semiperipheral East Asian countries. Mass merchandisers who sell lower-priced brands buy primarily from a third tier of lower-cost, midquality manufacturers, while large-volume discount stores like Wal-Mart import most of their goods from low-cost suppliers in peripheral countries like China, Bangladesh, and the Dominican Republic. Finally, some importers operate on the outer fringes of the international production frontier, seeking out very cheap but low-quality products from new sources in peripheral countries with no significant experience in clothing manufacture.

Figure 7.2.2 Global sourcing by U.S. retailers This diagram shows the geographical division of labor in the clothing industry. The most exclusive imported brands, supplied to the most upscale stores in the United States, are manufactured in Europe and Japan. Brands imported to discount stores, on the other hand, are manufactured in peripheral countries where labor costs are lowest. (*Source:* E. Bonacich et al., *Global Production. The Apparel Industry in the Pacific Rim.* Philadelphia: Temple University Press, 1994, Figure 4.1, p. 65. © 1994 by Temple University Press. Reprinted by permission of Temple University Press.)

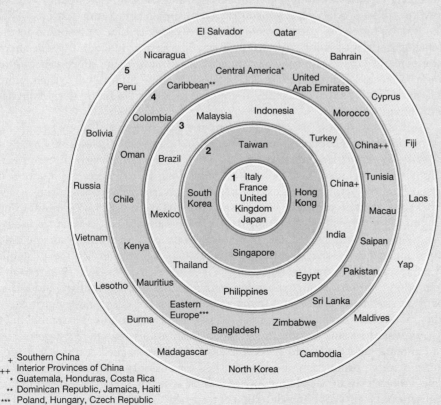

+ Southern China
++ Interior Provinces of China
* Guatemala, Honduras, Costa Rica
** Dominican Republic, Jamaica, Haiti
*** Poland, Hungary, Czech Republic

Retailers and Main Sourcing Areas

Fashion-oriented companies
(Armani, Polo/Ralph Lauren, Donna Karan, Gucci, Hugo Boss, etc.)
- Rings 1,2

Department stores and specialty chains
(Bloomingdale's, Saks Fifth Avenue, Neiman-Marcus, Macy's, Liz Claiborne, The Gap, The Limited, etc.)
- Rings 2, 3, 4 (generally better-quality and higher-priced goods than those sourced by mass merchandisers and discount stores)

Mass merchandisers
(Sears Roebuck, J. C. Penney, Woolworth, Montgomery Ward)
- Rings 2, 3, 4

Discount stores
(Wal-Mart, Kmart, Target)
- Rings 3, 4, 5

Small importers
- Rings 4, 5

agreements with other car companies, including a partnership with Taiwan's Chinese Automobile Co. to build Citroën C15s, and one with Fiat to produce a commercial van. Other products of strategic alliances include the Geo Prizm, a Toyota Corolla that is made in California and marketed in the United States by General Motors; the Geo Metro, made by Suzuki and Isuzu, and marketed by General Motors; the Jaguar, made in England by a wholly owned Ford subsidiary; and the Mazda Navajo, really a Ford Explorer made in Kentucky.

These strategic alliances have become an important aspect of global economic geography in the 1990s, as transnational corporations seek to reduce their costs and to minimize the risks involved in their multimillion dollar projects. Strategic alliances serve several functions:

- allowing transnational companies to link up with local "insiders" in order to tap into new overseas markets
- providing a quick and inexpensive means of swapping information, on a limited basis, about technologies that help to improve their products and their productivity
- reducing the costs of product development
- spreading the costs of market research.

The Nestlé food company, for example, has a strategic alliance with the Swiss Ciba-Geigy AG in the area of microbiology. It cooperates with Calgene in researching the regeneration process of soybean plants and the development of substitutes for cocoa butter and other vegetable oils. Nestlé also cooperates with the Coca-Cola company in exchanging technologies and in marketing. Nestlé will, for example, use Coca-Cola's distribution network for products such as Nescafé instant coffee.

The global assembly line is constantly being reorganized as transnational corporations seek to take advantage of geographical differences between countries and regions. Nike, the athletic footwear marketer, used to have its own manufacturing facilities in both the United States and the United Kingdom. Today, however, all its production is subcontracted to suppliers in South and East Asia. The geography of this subcontracting has evolved over time in response to the changing pattern of labor costs within Asia. The first production of Nike shoes took place in Japan. The company then switched most of its subcontracting to South Korea and Taiwan. Today, as labor costs in these countries are rising, more and more of Nike's subcontracting is going to producers in Indonesia, Malaysia, and China.

Such subcontracting is encouraged by the governments of many peripheral and semiperipheral countries, who see participation in global assembly lines as a pathway to export-led industrialization. They offer incentives to transnational corporations, including tax "holidays" (not having to pay taxes for a specified period) and the setting-up of export-processing zones. **Export-processing zones (EPZs)** are small areas within which especially favorable investment and trading conditions are created by governments in order to attract export-oriented industries. These conditions include minimum levels of bureaucracy surrounding importing and exporting; the absence of foreign exchange controls; the availability of factory space and warehousing at subsidized rents; low tax rates; and exemption from tariffs and export duties. In addition to tax incentives and EPZs, many governments also establish policies that ensure cheap and controllable labor. Sometimes countries are pressured to participate in global assembly lines by core countries and by the transnational institutions that they support. The United States and the World Bank, for example, have backed regimes that support globalized production and have pushed for austerity programs that help to make labor cheap in peripheral countries. Countries pursuing export-led industrialization as an economic development strategy do not plan to remain the

Export-processing zones (EPZs): small areas within which especially favorable investment and trading conditions are created by governments in order to attract export-oriented industries.

providers of cheap labor for foreign-based transnational corporations, however. They hope to shift from labor-intensive manufactures to capital-intensive, high-technology goods, following the path of semiperipheral Asian countries like Singapore and South Korea.

The Global Office

The globalization of production and the growth of transnational corporations have brought about another important change in patterns of local economic development. Banking, finance, and business services are now no longer locally oriented ancillary activities (as depicted in Figure 7.15) but important global industries in their own right. They have developed some specific spatial tendencies of their own—tendencies that have become important shapers of local economic development processes.

The new importance of banking, finance, and business services was initially a result of the globalization of manufacturing; an increase in the volume of world trade; and the emergence of transnational corporate empires. It was helped along by advances in telecommunications and data processing. Satellite communications systems and fiber-optic networks made it possible for firms to operate key financial and business services 24 hours a day, around the globe, handling an enormous volume of transactions. Linked to these communications systems, computers permit the recording and coordination of the data.

As banking, finance, and business services grew into important global activities, however, they were themselves transformed into something quite different from the old, locally oriented ancillary services. The global banking and financial network now handles trillions of dollars every day (estimates in 1996 ranged from three to seven trillion dollars)—no more than 10 percent of which has anything to do with the traditional world economy of trade in goods and services. International movements of money, bonds, securities, and other financial instruments have now become an end in themselves because they are a potential source of high profits from speculation and manipulation. Several factors have supported this development:

- In core countries, the institutionalization of savings (through pension funds and so on) established a large pool of capital managed by professional investors with few local or regional allegiances or ties.

- The quadrupling of crude oil prices in 1973 (organized by OPEC—the Organization of Petroleum Exporting Countries) generated so much capital for oil-rich countries that their banks opened overseas branches in order to find enough borrowers. In many cases, the borrowers were companies and governments in underdeveloped, peripheral countries that had previously been considered poor investment prospects. The internationalization of financial services soon paid off for the big banks. By the mid-1970s, about 70 percent of Citibank's overall earnings came from its international operations, with Brazil alone accounting for 13 percent of the bank's earnings in 1976. For the borrowers, however, the result was a huge increase in international debt, and many peripheral countries became net exporters of capital as they started to make the interest payments on their loans.

- In many countries, governments have lifted restrictions and regulations relating to banking, finance, and business services in the hope of capturing more of their growth.

- A persistent trade deficit of the United States vis-à-vis the rest of the world (a result of the postwar recovery of Europe and Japan) created a growing pool of dollars outside the United States, known as "Eurodollars." This supply in turn created a pool of capital that was beyond the direct control of the U.S. authorities.

- "Hot" money (undeclared business income, proceeds of securities fraud, trade in illegal drugs, and syndicated crime), easily laundered through international electronic transactions, also found its way into the growing pool of Eurodollars. It is estimated that 100 billion U.S. dollars is laundered each year through the global financial system.

- The initial response of many governments (including the U.S. government) to balance-of-payments problems was to print more money—a short-term solution that eventually contributed to a significant surge in inflation in the world economy. This inflation, because it promoted rapid change and international differentials in financial markets, provided a further boost to speculative international financial transactions of all kinds.

Together, these factors have amounted to a change so important that a deep-seated restructuring of the world economy has resulted. The effects of this restructuring have altered economic geography at every scale, affecting the lives of people everywhere. Banks and financial corporations with the size and international reach of Citibank or Nomura or Salomon Brothers are able to influence local patterns and processes of economic development throughout the world, just like the major transnational conglomerates involved in the global assembly line.

We can see the imprint on patterns of economic development in many different ways. A truly global shift has occurred, for example, in the location of major international banks (Figure 7.21). In 1975, three of the four largest banks in the world were headquartered in the United States. By 1995, BankAmerica, the largest in 1975, had slipped to 43rd place. Nine of the top 10 banks in 1995 were Japanese, as were 15 of the top 20. The highest-ranked U.S. bank was Citibank, with total assets of about one-third the size of those of Tokyo's Fuji Bank, the largest bank in the world.

Electronic Offices and Decentralization In more general terms, an important shift has occurred in the economic structure of the world's core economies, with the rapid growth of banking, financial, and business services contributing to the expansion of the quaternary sector. What is particularly important from a geographical perspective is that this growth has been localized, that is, concentrated in relatively small and distinctive settings within major metropolitan centers. This phenomenon is surprising to some observers, who had expected that new communications technologies would allow for the dispersion of "electronic offices" and, with it, the decentralization of an important catalyst for local economic development. A good deal of geographic decentralization of offices has occurred, in fact, but it has mainly involved "back office" functions that have been relocated from metropolitan and business-district locations to small-town and suburban locations.

Back-office functions are record-keeping and analytical functions that do not require frequent personal contact with clients or business associates. The accountants and financial technicians of main street banks, for example, are back-office workers. Developments in computing technologies, database access, electronic data interchanges, and telephone call routing technologies are enabling a larger share of back-office work to be relocated to specialized office space in cheaper settings, freeing space in the high-rent locations occupied by the bank's front office. For example, the U.S. Postal Service is using Optical Character Readers (OCRs) to read addresses on mail, which is then bar-coded and automatically sorted to its appropriate substation. Addresses that the OCRs cannot read are digitally photographed and transmitted to a computer screen, where a person manually types the address into a terminal. In Washington, DC, OCR sorting takes place at the central mail facility, but the manual address entry is done in Greensboro, North Carolina, where wage rates are lower. Workers in Greensboro view images of letters as they are sorted in Washington and enter correct addresses, which are in turn electronically transmitted back to be bar-coded on

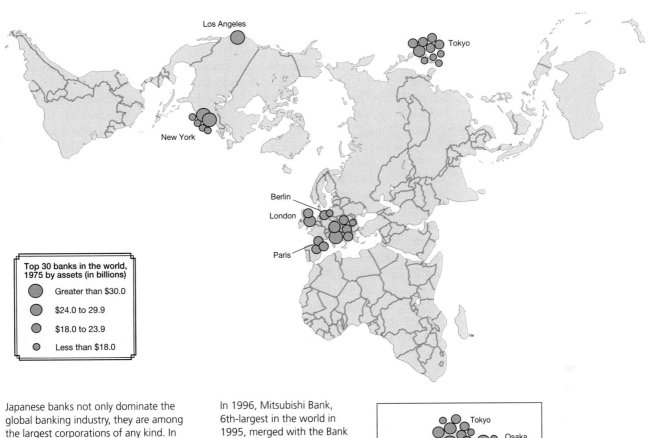

Top 30 banks in the world, 1975 by assets (in billions)

- Greater than $30.0
- $24.0 to 29.9
- $18.0 to 23.9
- Less than $18.0

Japanese banks not only dominate the global banking industry, they are among the largest corporations of any kind. In 1995, Japanese banks constituted 6 of the top 15 largest companies in the world, by market value, and 13 of the top 100.

In 1996, Mitsubishi Bank, 6th-largest in the world in 1995, merged with the Bank of Tokyo (23rd in 1995), creating a "superbank" with 90 overseas branches.

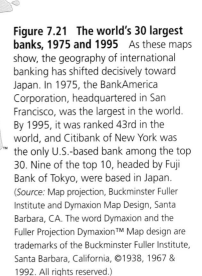

Citibank was chartered in 1812 as the National City Bank, and grew by financing the railroads and major industrial enterprises of the nineteenth century. Its overseas growth began with offices in Shanghai (1900) and Buenos Aires (1914). By the mid-1980s, Citibank was the world's largest, with over 2,000 offices in more than 80 countries. Still the largest American bank, it is now ranked only 30th in the world.

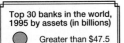

Top 30 banks in the world, 1995 by assets (in billions)

- Greater than $47.5
- $37.5 to 47.5
- $27.5 to 37.5
- Less than $27.5

Crédit Lyonnais, the largest non-Japanese bank in 1995, is controlled by the French government, which had to bail the bank out of financial trouble in 1995.

Many private European banks, like the Deutsche Bank (ranked 13th in the world in 1995), have invested large sums to build global investment banking operations. In 1995, Deutsche Bank acquired Morgan Grenfell. Deutsche Morgan Grenfell is now one of the five largest international investment banks in the world.

Figure 7.21 The world's 30 largest banks, 1975 and 1995 As these maps show, the geography of international banking has shifted decisively toward Japan. In 1975, the BankAmerica Corporation, headquartered in San Francisco, was the largest in the world. By 1995, it was ranked 43rd in the world, and Citibank of New York was the only U.S.-based bank among the top 30. Nine of the top 10, headed by Fuji Bank of Tokyo, were based in Japan. (*Source:* Map projection, Buckminster Fuller Institute and Dymaxion Map Design, Santa Barbara, CA. The word Dymaxion and the Fuller Projection Dymaxion™ Map design are trademarks of the Buckminster Fuller Institute, Santa Barbara, California, ©1938, 1967 & 1992. All rights reserved.)

the piece of mail. Similarly, when residents of London, England, call to inquire about city parking ticket fines, the calls are processed in a small city in northern England.

Among the more prominent examples of back-office decentralization from U.S. metropolitan areas have been the relocation of back-office jobs in American Express from New York to Salt Lake City, Fort Lauderdale, and Phoenix; the relocation of Metropolitan Life's back offices to Greenville, South Carolina, Scranton, Pennsylvania, and Wichita, Kansas; the relocation of Hertz's data-entry division to Oklahoma City, Dean Witter's to Dallas, and Avis's to Tulsa; and the relocation of Citibank's MasterCard and Visa divisions to Tampa and Sioux Falls. Some places have actually become specialized back-office locations as a result of such decentralization. Omaha and San Antonio, for example, are centers for a large number of telemarketing firms, while Roanoke, Virginia, has become something of a mail-order center.

Internationally, this trend has taken the form of offshore back-offices. By decentralizing back-office functions to offshore locations, companies save even more in labor costs. While Zürich-based Swissair, for example, now does its accounting in Mumbai (formerly Bombay), India , several New York–based life-insurance companies have established back-office facilities in Ireland. Situated conveniently near Ireland's main international airport, Shannon Airport, they ship insurance claim documents from New York via Federal Express, process them, and beam the results back to New York via satellite or the TAT-8 transatlantic fiber-optics line (Figure 7.22). Ireland is also becoming a telemarketing center: Call a toll-free number in Europe for Best Western or Gateway 2000 computers and an Irish voice answers the phone.

Clusters of Specialized Offices This decentralization is outweighed, however, by the tendency for a disproportionate share of the new jobs created in banking, finance, and business services to cluster in highly specialized financial districts within major metropolitan areas. The reasons for this localization are to be found in another special case of the geographical agglomeration effects that we discussed earlier in this chapter. Metropolitan areas such as New York City, London, Paris, Tokyo, and Frankfurt have acquired the kind of infrastructure—specialized office space, financial exchanges, teleports (office parks that are equipped with satellite Earth stations and linked to local fiber-optics lines), and communications networks—that is essential for the delivery of services to clients with a national or

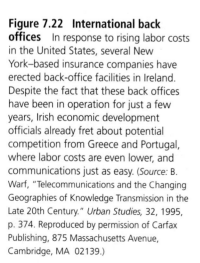

Figure 7.22 International back offices In response to rising labor costs in the United States, several New York–based insurance companies have erected back-office facilities in Ireland. Despite the fact that these back offices have been in operation for just a few years, Irish economic development officials already fret about potential competition from Greece and Portugal, where labor costs are even lower, and communications just as easy. (*Source:* B. Warf, "Telecommunications and the Changing Geographies of Knowledge Transmission in the Late 20th Century." *Urban Studies,* 32, 1995, p. 374. Reproduced by permission of Carfax Publishing, 875 Massachusetts Avenue, Cambridge, MA 02139.)

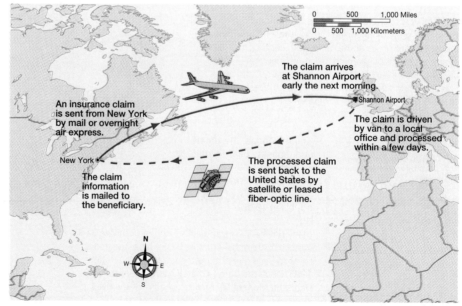

international scope of activity. They have also established a comparative advantage both in the mix of specialized firms and expert professionals that are on hand, and in the high-order cultural amenities that are available (both to high-paid workers and their out-of-town business visitors). Above all, they have established themselves as centers of authority, with a critical mass of people-in-the-know about market conditions, trends, and innovations—people who can gain one another's trust through frequent face-to-face contact, not just in business settings but also in the informal settings of clubs and office bars. They have become **world cities,** places that, in the globalized world economy, are able not only to generate powerful spirals of local economic development but also to act as pivotal points in the reorganization of global space: control centers for the flows of information, cultural products, and finance that, collectively, sustain the economic and cultural globalization of the world (see Chapter 10 for more details).

Offshore Financial Centers The combination of metropolitan concentration and back-office decentralization fulfills most, but not all, of the locational needs of the global financial network. There are some needs—secrecy and shelter from taxation and regulation, in particular—that require a different locational strategy. The result has been the emergence of a series of **offshore financial centers:** islands and micro-States such as the Bahamas, Bahrain, the Cayman Islands, the Cook Islands, Luxembourg, Lichtenstein, and Vanuatu that have become specialized nodes in the geography of worldwide financial flows (Figure 7.23).

World cities: cities in which a disproportionate part of the world's most important business—economic, political, and cultural—is conducted.

Offshore financial centers: islands or micro-States that have become a specialized node in the geography of worldwide financial flows.

Figure 7.23 Offshore financial centers Offshore financial centers offer confidentiality and shelter from the levels of taxation and regulation that prevail throughout most of the world. They have become important nodes in the global flows of finance that are required by the global economy. Some of the most important of these centers are not "offshore" at all in the literal sense (Bahrain, Luxembourg, and Switzerland, for example), but they are "islands" of financial secrecy and independence. (*Source:* Map projection, Buckminster Fuller Institute and Dymaxion Map Design, Santa Barbara, CA. The word Dymaxion and the Fuller Projection Dymaxion™ Map design are trademarks of the Buckminster Fuller Institute, Santa Barbara, California, ©1938, 1967 & 1992. All rights reserved.)

In addition to its role as an offshore tax haven, Costa Rica has become a haven for "pensionados"—retirees from North America who enjoy landed immigrant status, a public health-care system, and no taxes on their pensions, dividends, or trust earnings. In 1995, about 50,000 former Canadian residents and about 120,000 former U.S. residents were living in Costa Rica.

Belize set up shop as an offshore financial center with a 1989 law that authorized "International Business Companies" (IBCs)—anonymous, tax-exempt companies that can be set up with just $100 without the owners ever setting foot in Belize. By 1995, over 1800 IBCs had been established.

Labuan, formerly a penal colony and pirates' lair, has been deliberately established as an International Offshore Financial Center by the Malaysian government in order to attract business from Hong Kong and Japan. By the mid-1990s nearly 30 banks and almost 300 companies had decided to take advantage of Labuan's zero tax on dividends, interest, and royalties, and its 3 percent tax on net profits.

Vanuatu, in the New Hebrides, has no laws against money laundering, no asset-seizure laws, no foreign-exchange controls, and no personal or corporate income taxes. About $50 million passes through its banks every day.

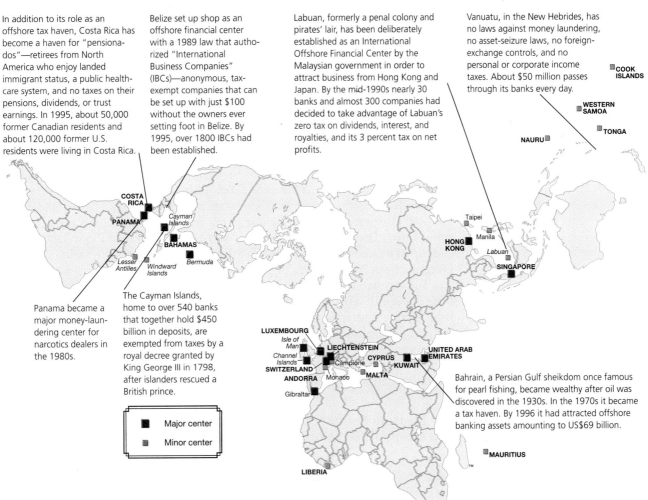

Panama became a major money-laundering center for narcotics dealers in the 1980s.

The Cayman Islands, home to over 540 banks that together hold $450 billion in deposits, are exempted from taxes by a royal decree granted by King George III in 1798, after islanders rescued a British prince.

Bahrain, a Persian Gulf sheikdom once famous for pearl fishing, became wealthy after oil was discovered in the 1930s. In the 1970s it became a tax haven. By 1996 it had attracted offshore banking assets amounting to US$69 billion.

■ Major center
▣ Minor center

Figure 7.24 The Cayman Islands Of more than 550 registered banks in the Cayman Islands, less than 75 maintain a physical presence there, and only half a dozen offer the kind of local services that allow local residents to maintain checking accounts or tourists to cash traveler's checks. Most exist as post-office boxes, as name plates in anonymous office buildings, as fax numbers, or as entries in a computer system. In 1995, over $700 billion passed through the Caymans, making it the most successful of all offshore financial centers.

The chief attraction of these offshore financial centers is simply that they are less regulated than financial centers elsewhere. They provide low-tax or no-tax settings for savings, havens for undeclared income and for hot money. They also provide discreet markets in which to deal currencies, bonds, loans, and other financial instruments without coming to the attention of regulating authorities or competitors. The U.S. Internal Revenue Service estimates that about $300 billion ends up in offshore financial centers each year as a result of tax-evasion schemes. Overall, about 60 percent of all the world's money now resides offshore.

The Cayman Islands (Figure 7.24) provide the classic example of an offshore financial center. This small island State in the Caribbean has transformed itself from a poor, underdeveloped colony to a relatively affluent and modern setting for upscale tourism and offshore finance. With a population of just over 30,000, it has over 24,000 registered companies, including 350 insurance companies and over 550 banks from all around the world. Between them, these banks account for an estimated $250 billion in assets.

The Pleasure Periphery: Tourism and Economic Development

Many parts of the world, including parts of the world's core regions, do not have much of a primary base (that is, in agriculture, fishing, or mineral extraction); are not currently an important part of the global assembly line; and are not closely tied in to the global financial network. For these areas, tourism can offer the otherwise unlikely prospect of economic development. Tourism has, in fact, become enormously important. By the year 2000 the largest single item in world trade will be international tourism, which is already the world's largest nonagricultural employer. One in every 15 workers, worldwide, is occupied in transporting, feeding, housing, guiding, or amusing tourists; and the global stock of lodging, restaurant, and transportation facilities is estimated to be worth about $3 trillion.

While most of the tourists are relatively affluent people from the more-developed parts of the world, tourism is by no means confined to the less-developed ("unspoiled") peripheral regions. Only 10 percent of Americans, for instance, have passports; most U.S. tourist dollars are spent in safe and predictable settings where English is spoken—in national parks, specialized resorts, theme parks, big cities, rural idylls, health spas, and renovated historic towns and districts. The growth of tourism and the economic success of places like Baltimore's renovated Harborplace (see Figure 6.15, page 250), which attracts about 30 million visitors a year, has meant that few localities exist in America—or anywhere else in the developed world, for that matter—that do not encourage tourism as one of the central planks of their economic development strategy. It has been estimated that, because tourism requires only a basic infrastructure, no heavy plant, and little high-tech equipment, the cost of creating one new job in tourism is less than one-fifth the cost of creating a job in manufacturing industry, and less than one-fiftieth of the cost of creating a high-tech engineering job. Consequently, as we saw in Chapter 6, place marketing has become an extremely important aspect of local efforts to promote economic development.

The globalization of the world economy has therefore been paralleled by a globalization of the tourist industry. While Americans have been relatively slow to join in, Europeans and Japanese have been in the forefront of this trend. In aggregate, over 650 million international tourist trips occurred in 1995, compared with just 147 million in 1970. Almost 70 percent of these trips are generated by just 20 of the more affluent countries of the world (Figure 7.25). What is most striking, though, is not so much the growth in the number of international tourists as the increased range of international tourism. Thanks largely to cheap long-distance flights, a significant proportion of tourism is now transcontinental and transoceanic. Visits to peripheral countries in Africa, Asia, and Latin America now account for one-eighth of the industry. This, of course, has made tourism a central component of economic development in countries with sufficiently exotic wildlife (Kenya, for example), scenery (Nepal), beaches (the Seychelle Islands), shopping (Singapore and Hong Kong), culture (China, India, and Indonesia), or sex (Thailand).

Although tourism can provide a basis for economic development in peripheral regions, however, it is often a mixed blessing. While it certainly creates jobs, they are often seasonal. Dependence on tourism also makes for a high degree of economic vulnerability. Tourism, like other high-end aspects of consumption, depends very much on matters of style and fashion. As a result, once-thriving tourist destinations can suddenly find themselves struggling for customers. Some places are sought out by tourists because of their remoteness and their "natural," undeveloped qualities, and it is these that are most vulnerable to shifts of style and fashion. Nepal and New Zealand are recent examples of this phenomenon, but they are now too "obvious" as destinations and are consequently having to work hard to continue to attract sufficient numbers of tourists. Bhutan, Bolivia, Estonia, and Vietnam have been "discovered," and are coping with their first real growth in tourism. It is not only exotic tourist locations that are vulnerable to changing tastes and fashions, however. The Mediterranean beach resort of Rimini, Italy, for example (Figure 7.26), has been all but deserted by its former northwest European, middle-class clientele, who now prefer more distant, more exotic, and more distinctive locations. Consequently, Rimini has had to look to

Figure 7.25 Growth in world tourism
In 1995 about 650 million tourists took trips abroad. By 2010 the number of tourists worldwide is expected to approach one billion. Although the global tourist industry employs nearly 150 million people, the benefits of tourism are not always felt at the local level. In much of the periphery, many of the resorts are owned by transnational corporations, so that a high percentage of the profits leak back to core countries. (*Source:* Map projection, Buckminster Fuller Institute and Dymaxion Map Design, Santa Barbara, CA. The word Dymaxion and the Fuller Projection Dymaxion™ Map design are trademarks of the Buckminster Fuller Institute, Santa Barbara, California, ©1938, 1967 & 1992. All rights reserved.)

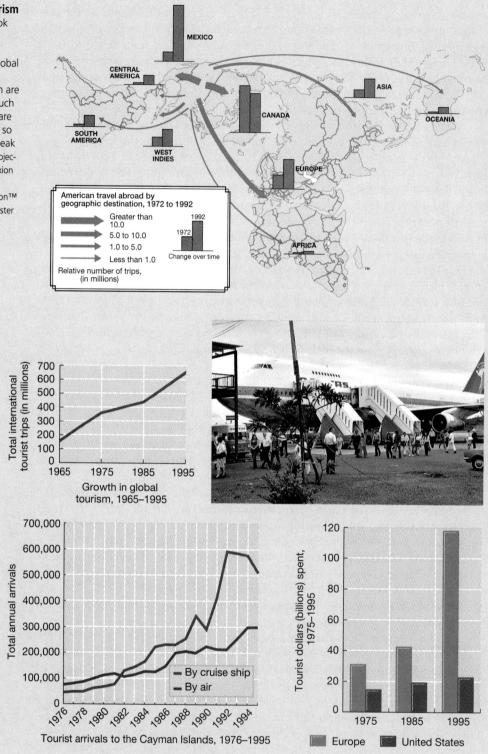

the middle classes of eastern Europe and the former Soviet Union in order to fill its hotel rooms. Unfortunately for Rimini, this is a much less affluent population, which generates far fewer tourist dollars.

Tourism in more exotic tourist destinations, meanwhile, is vulnerable in other ways: to political disturbances, natural disasters, outbreaks of disease or food poisoning, and atypical weather. Fiji's military coup in 1987, for example,

312

resulted in a devastating 70 percent drop in tourism, while an outbreak of the plague in Surat in India in 1994 brought about a virtual halt to the flow of tourists to the whole country. For ski resorts, warm weather represents the equivalent of a harvest failure for an agricultural region. Moreover, although tourism is a multibillion dollar industry, the financial returns for tourist areas are often not as high as might be expected. The greater part of the price of a package vacation, for example, stays with the organizing company and the airline. Typically, only 40 percent is captured by the tourist region itself. If the package involves a foreign-owned hotel, this may fall below 25 percent.

The costs and benefits of tourism are not only economic, of course. On the positive side, tourism can help sustain indigenous lifestyles, regional cultures, arts, and crafts; and provide incentives for wildlife preservation, environmental protection, and the conservation of historic buildings and sites. On the negative side, tourism can adulterate and debase indigenous cultures, and bring unsightly development, pollution, and environmental degradation. In the Caribbean, sewage has poisoned mangrove trees and polluted coastal waters, and boats and divers have damaged coral reefs. In Kenya, the Amboselli National Park has been severely degraded by safari vehicles. In the United States, congestion in Yosemite National Park (with over 4 million visitors per year) has become so acute that Park Service rangers frequently have to turn away as many as 1,000 vehicles per day, and Park administrators are considering the establishment of a strict advance reservation system. In the European Alps, where an incredible 40,000 ski runs attract a winter tourist population 10 times greater than the resident population, forests have been ripped up, pastures obliterated, rivers diverted, and scenic valleys and mountainsides covered with chalets, cabins, and hotels.

Tourism can also involve exploitative relations that debase traditional lifestyles and regional cultural heritages as they become packaged for outsider consumption. In the process, the behaviors and artifacts that are made available to an international market of outsiders can lose much of their original meaning. Traditional ceremonies that formerly had cultural significance for the performers are now enacted only to be watched and photographed. Artifacts like masks and weapons are manufactured not for their original use but as curios, souvenirs, and ornaments. In the process, indigenous cultures are edited, beautified, and altered to suit outsiders' tastes and expectations.

Such problems, coupled with the economic vulnerability occasioned by tourism, have led to the idea of "alternative" tourism as the ideal strategy for economic development in peripheral regions (Figures 7.27 and 7.28). Alternative tourism emphasizes self-determination, authenticity, social harmony, preservation of the existing environment, small-scale development, and greater use of local techniques, materials, and architectural styles. Costa Rica, for example, despite being a poor country, has won high praise from environmentalists for protecting one-quarter of its territory in biosphere and wildlife preserves. The payoff for Costa Rica is the escalating number of tourists who come to visit its active volcanoes, palm-lined beaches, cloud forests, and tropical parks that offer glimpses of toucan and coati. In 1995, the year Costa Rica received over 800,000 tourists, tourism exceeded banana exports as the country's main source of foreign exchange.

Ecuador is another country that has fostered alternative tourism. With six national parks, seven nature reserves, and 20 privately protected areas, Ecuador offers great variety for tourists. In a small area straddling the equator, it has some of the world's oldest rain forest, the world's highest active volcano, Amazon tribes, spectacular wildlife in the Galapagos Islands, a thriving Andean culture, and a well-preserved legacy of Spanish colonialism. Two-thirds of all organized travel to Ecuador is represented by members of the Ecuadorean Ecotourism Association, an organization sponsored by the private sector as well as the government in an effort to ensure sustainable development through environmental

Figure 7.26 Rimini, Italy Rimini, on the Adriatic coast in northeastern Italy, was for many years the destination of package-trip vacationers from the cooler regions of northwestern Europe. The resort grew rapidly in the 1960s and 1970s, but in doing so it became highly dependent on this very specialized clientele. When rising incomes, cheaper air travel, and lower labor costs in more distant locations made other resorts more attractive to vacationers from Britain, Germany, and Scandinavia, Rimini's vulnerability was exposed. Today, the resort is developing a new clientele from the middle classes of eastern Europe and the former Soviet Union.

Figure 7.27 Alternative tourism "Ecotourism" is partly a response to the physical blight, environmental degradation, and cultural intrusion associated with traditional forms of tourist development, and partly a product of the "green" consciousness of affluent vacationers. It is now big business. Magazines like this one offer information and advice on all kinds of alternative and environmentally oriented travel opportunities, from treehouse camping in Brazil to photo safaris in Kenya; from glacier hiking in Iceland to bicycling tours of New Zealand.

Figure 7.28 An award-winning initiative in ecotourism: whale watching at Kaikoura, New Zealand Less than a mile offshore, in water over 2,000 feet deep, the convergence of ocean currents creates ideal conditions for an abundance of marine life. Fishing had been important in Kaikoura for many years, but competition from foreign trawler fleets had depressed the small town's economy. In 1988, a small local company began to offer boat trips to take people to see the area's seabirds, seals, dolphin, and whales. Tourism, which had previously been negligible, is now a mainstay of the small town's economy.

awareness. Other examples include guest-house developments in Papua New Guinea, bungalows in French Polynesia, packaged ecotourism in Belize, and "integrated rural tourism" in Senegal. Such developments must be aimed, however, at tourists who are both wealthy and environmentally conscious—not, perhaps, a large-enough market on which to pin hopes of significant increases in levels of economic development.

Conclusion

The growth of alternative tourism in Costa Rica, like the growth of the Cayman Islands as an offshore financial center, the emergence of Ireland as a center for back-office activities, and the decline of northern England as a manufacturing region, shows that economic development is not simply a sequential process of modernization and increasing affluence. Various pathways to development exist, each involving different ways of achieving increased economic productivity and higher incomes, together with an increased capacity to improve the basic conditions of life. Economic development means not just using the latest technology to generate higher incomes, but also improving the quality of life through better housing, health care, and social welfare systems and enhancing the physical framework, or infrastructure, on which the economy rests.

Local, regional, and international patterns and processes of economic development are of particular importance to geographers. Levels of economic development and local processes of economic change affect many aspects of local well-being, and so contribute to many aspects of human geography. Economic development is an important place-making process that underpins much of the diversity among regions and nations. At the same time, it is a reflection and a product of variations from place to place in natural resources, demographic characteristics, political systems, and social customs.

Economic development is always an uneven geographic phenomenon. Regional patterns of economic development are tied to the geographic distribution of resources, and to the legacy of the past specializations of places and regions within national economies. As we saw with the examples of France and Brazil, a general tendency exists toward the creation of regional cores with dependent peripheries. Nevertheless, such patterns are not fixed or static. Changing economic conditions can lead to the modification or reversal of core–periphery patterns, as in the stagnation of once-booming regions like northern England and the spectacular growth of Guandong province in Southeast China. Over the long term, core–periphery patterns have most often been modified as a result of the changing locational needs and opportunities of successive technology systems. Today, economic globalization has exposed more places and regions than ever to the ups and downs of episodes of creative destruction—episodes played out ever faster, thanks to the way that telematics have shrunk time and space.

At the global scale, the unevenness of economic development takes the form of core–periphery contrasts within the world-system framework. What is most striking about these contrasts today are the dynamism and pace of change involved in economic development. The global assembly line, the global office, and global tourism are all making places much more interdependent, and much faster changing. Parts of Brazil, China, India, Mexico, and South Korea, for example, have developed quickly from rural backwaters into significant industrial regions. The Cayman Islands have been transformed from an insignificant Caribbean colony to an upscale tourist resort and a major offshore financial center. Countries like Ecuador and Costa Rica, with few comparative advantages, suddenly find themselves able to earn significant amounts of foreign exchange through the development of ecotourism. This dynamism has, however, brought with it an expanding gap between rich and poor at every spatial scale: international, regional, and local.

Main Points

- The geography of economic development is the cumulative outcome of decisions guided by fundamental principles of land use, commercial and industrial location, and economic interdependence.

- Geographically, the single most important feature of economic development is that it is highly uneven in nature.

- Successive technology systems have rewritten the geography of economic development as they have shifted the balance of advantages between regions.

- Geographical divisions of labor have evolved with the growth of the world-system of trade and politics, and with the changing locational logic of successive technology systems.

- Regional cores of economic development are created cumulatively, following some initial advantage, through the operation of several of the basic principles of spatial organization.

- Spirals of economic development can be arrested in various ways, including the onset of disinvestment and deindustrialization, which follow major shifts in technology systems and in international geopolitics.

- The globalization of the economy has meant that patterns and processes of local and regional economic development are much more open to external influences than before.

Key Terms

agglomeration diseconomies (p. 292)

agglomeration effects (p. 286)

ancillary activities (p. 287)

backwash effects (p. 291)

conglomerate corporations (p. 298)

creative destruction (p. 293)

cumulative causation (p. 290)

deindustrialization (p. 293)

export-processing zones (EPZs) (p. 304)

external economies (p. 286)

geographical path dependence (p. 289)

Gross domestic product (GDP) (p. 271)

Gross national product (GNP) (p. 271)

growth poles (p. 295)

infrastructure (fixed social capital) (p. 289)

initial advantage (p. 289)

localization economies (p. 287)

offshore financial centers (p. 309)

primary activities (p. 279)

quaternary activities (p. 279)

secondary activities (p. 279)

spread effects (p. 292)

technology systems (p. 278)

tertiary activities (p. 279)

urbanization economies (p. 289)

world cities (p. 309)

Exercises

On the Internet

Economic development is one area that is especially well served by the Internet, since the Internet has become an important medium for economic development agencies and for corporations to publicize themselves. The Internet exercises for this chapter explore the economic relationships between places. After completing them you should have a better understanding of principles of location and of the unevenness of patterns of economic development.

Unplugged

1. While India's per capita income is well below that of the United States (Figure 7.1), India has more people who earn the equivalent of $70,000 a year than the United States does. How can you explain this, and what might be some of the consequences of it, from the point of view of economic geography?

2. Study the diagram of timelines and economic development (Figure 7.7). Can you add examples that relate to the places and regions with which you are familiar?

Chapter 8

Agriculture and Food Production

Orange production in Brazil

In the winter of 1993, Florida orange-growers, after six devastating frosts in the 1980s, were faced with a different and more frightening situation: harvests that were much too bountiful. Mild winter temperatures, the relocation of groves following the frosts, and heavy investments in fertilizers and pesticides had translated into a bumper year. Unfortunately for the orange-growers, the record high volume of fruit picked, packed, and sent to market steadily eroded the overall price paid to them as supply far exceeded demand, and growers were unable to cover their operating costs. Adding to the downward pressure on Florida citrus prices was the fact that Brazil—which in the 1980s became the world's largest orange producer—had also produced a bumper crop. With a corner on the European and Asian markets, Brazil's presence kept Florida from using the international market to absorb its surplus and maintain prices.

As this example illustrates, agriculture, like any other contemporary economic activity, is global. The romanticized view some people hold of farming as a self-sufficient and self-contained livelihood is no longer accurate, as shown by how the ups and downs of weather in Florida can affect the price of oranges in Japan. Furthermore, agriculture itself has been dramatically transformed into the industry of food production, with connections to the manufacturing, distribution, marketing, and finance sectors of the local, national, regional, and global economy. In fact, Florida orange-growers complained that grocery markets had further added to their woes by failing to reduce the retail prices to reflect the wholesale prices and were thus keeping prices artificially high and adding to the glut. Furthermore, Florida orange production is expected to increase in the next few years as the thousands upon thousands of new trees planted after the frosts of the 1980s begin to bear fruit.

In this chapter, we examine the history and geography of agriculture from the global to the household level, exploring the connections and describing the implications of those connections. We begin by looking at traditional agricultural practices. Much of the chapter is devoted to exploring the ways in which geographers have been investigating the dramatic changes that have occurred in agriculture since the middle of the twentieth century.

Traditional Agricultural Geography

The study of agriculture has a long tradition in geography. Because of geographers' interest in the relationships between people and land, it is hardly surprising that agriculture has been of primary concern to us. Unique among research on agriculture is geography's commitment to viewing the physical and human systems as interactively linked. Such an approach combines an understanding of spatial differentiation, the importance of place, and the fact that practices such as agriculture affect and are affected by processes occurring at different scales. It also provides geographers with a powerful perspective for understanding the dynamics of contemporary agriculture.

One of the most widely recognized and appreciated contributions that geographers have made to the study of agriculture is the mapping of the different factors that shape agriculture. They have mapped soil, temperature, and terrain as well as the areal distribution of different types of agriculture and the relationships among and between agriculture and other practices or variables. Figure 8.1 illustrates this point. Such maps have become important bases for shaping policy as well as practice.

The last four decades have been characterized by major changes occurring in agriculture worldwide. One of the most dramatic changes has been the decline in the number of people employed in farming in both the core and the periphery. Meanwhile, farming practices have also been significantly intensified through the use of chemical, mechanical, and biotechnological innovations and applications (Figure 8.2, page 320). Agriculture has also become increasingly integrated into

Figure 8.1 Global distribution of soils, 1995 Knowing the nature of soils in various parts of the world is crucial to understanding the geography of agriculture. Geographers have made important contributions to our understanding of where agricultural activities occur, and why agriculture occurs in those areas and not others. (*Source:* Adapted from U.S. Soil Conservation Service.)

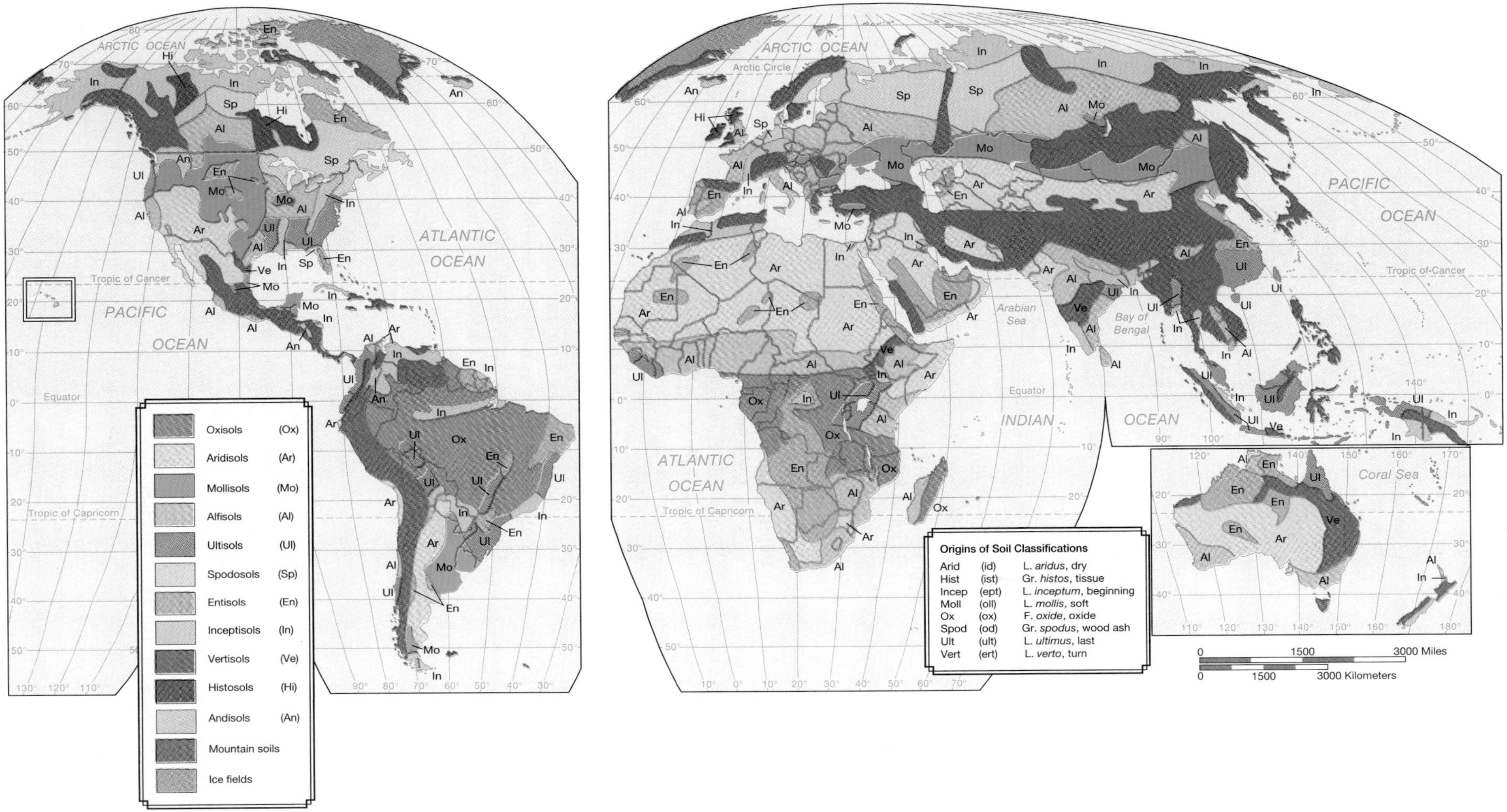

Figure 8.2 Pesticide use In the Cape Verde Islands off the coast of western Africa, an agricultural worker sprays pesticides on irrigated vegetables. Many of the pesticides in use in peripheral countries have been banned in core countries because of human health concerns.

wider regional, national, and global economic *systems* at the same time that it has become more directly linked to other economic *sectors* such as manufacturing and finance. These changes have been profound and have had repercussions from the structure of global finance to the social relations of individual households (Figure 8.3).

Through the examination of agricultural practices, geographers have sought to understand the myriad ways in which humans have learned to adapt and domesticate the natural world around them to sustain themselves, their kin, and ultimately the global community. In addition to understanding agricultural systems, geographers have also been interested in investigating the lifestyle and culture of different agricultural communities. They and other social scientists often use the adjective **agrarian** to describe the way of life that is deeply embedded in the demands of agricultural production. Agrarian not only defines the *culture* of distinctive agricultural communities, but it also refers to the type of *tenure* (or landholding) system that determines who has access to land and what kind of cultivation practices will be employed there.

Agriculture is considered a science, an art, and a business directed at the cultivation of crops and the raising of livestock for sustenance and for profit. The unique and ingenious methods by which humans have learned how to transform the land through agriculture are an important reflection of the two-way relationship between people and their environments (Figure 8.4). Just as geography

Agrarian: referring to the culture of agricultural communities and the type of tenure system that determines access to land and the kind of cultivation practices employed there.

Agriculture: a science, an art, and a business directed at the cultivation of crops and the raising of livestock for sustenance and for profit.

Figure 8.3 Human and machine labor In agriculture worldwide, human labor has been increasingly replaced by machines. Here we see Sudanese women harvesting wheat by hand while modern harvester machines operated by men mow the plants for easier harvesting.

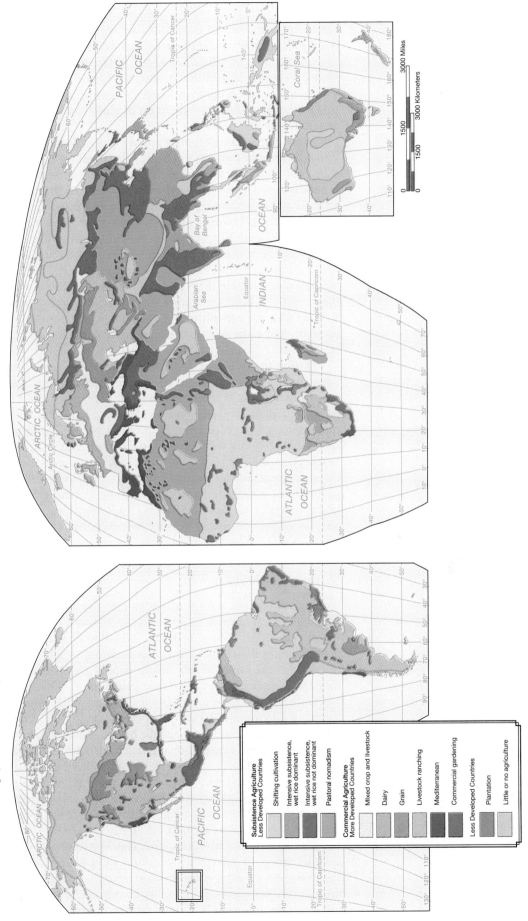

Figure 8.4 Global distribution of agriculture, 1995 The global distribution of agricultural practices is illustrated in this map. Notice the dramatic differences between core and periphery with respect to commercial versus subsistence agriculture. The periphery, though it does contain commercial agriculture, is largely dominated by forms of subsistence, while the core countries contain virtually none. (*Source*: Reprinted with permission of Prentice Hall, from J. M. Rubenstein, *The Cultural Landscape: An Introduction to Human Geography*, 5th Ed., © 1996, pp. 394–395. Reproduced by permission from the *Annals of the Association of American Geographers*, Vol. 26, 1936, p. 241, Fig. 1; D. Whittlesey.)

Subsistence Agriculture
Less Developed Countries
Shifting cultivation
Intensive subsistence, wet rice dominant
Intensive subsistence, wet rice not dominant
Pastoral nomadism

Commercial Agriculture
More Developed Countries
Mixed crop and livestock
Dairy
Grain
Livestock ranching
Mediterranean
Commercial gardening

Less Developed Countries
Plantation
Little or no agriculture

shapes our choices and behaviors, we are able to shape the physical landscape. Most introductory textbooks give considerable attention to tracing the origins of agriculture and the distribution of different agricultural practices across the globe. Although agricultural origins are important, the impact of twentieth-century political and economic changes in agriculture are so transformative that in this textbook we focus on the state of global agriculture at the beginning of the new millennium.

While there is no definitive answer as to where agriculture originated, we do know that before humans discovered the advantages of agriculture they procured their food through hunting (including fishing) and gathering. **Hunting and gathering** characterizes activities whereby people feed themselves through killing wild animals and fish and gathering fruits, roots, nuts, and other edible plants to sustain themselves. Hunting and gathering are considered subsistence activities in that people who practice them procure only what they need to consume. Subsistence agriculture replaced hunting and gathering activities in many parts of the globe when people came to understand that the domestication of plants and animals could enable them to remain settled in one place over time rather than their having to hunt and gather frequently in search of edible wild foods (Figure 8.5). **Subsistence agriculture** is a system in which agriculturalists consume all they produce. While the practice of subsistence agriculture is declining, it is still practiced in many areas of the globe (Figure 8.6, page 324).

During the twentieth century the dominant agricultural system in the core countries has become **commercial agriculture**, a system in which farmers produce crops and animals primarily for sale rather than for direct consumption by themselves and their families. Worldwide, the practice of subsistence agriculture is diminishing as increasing numbers of places are irresistibly incorporated into a globalized economy with a substantial commercial agricultural sector. Still widely practiced in the periphery, however, subsistence activities usually follow one of three dominant forms: shifting cultivation, intensive subsistence agriculture, and pastoralism. Although many people in the global periphery rely on these traditional practices to feed themselves, traditional practices are increasingly being abandoned or modified as peasant farmers convert from a subsistence and barter economy to a cash economy.

Shifting Cultivation

Shifting cultivation, a form of agriculture usually found in tropical forests, is a system in which farmers aim to maintain soil fertility by rotating the *fields* within which cultivation occurs. Shifting cultivation contrasts with another method of maintaining soil fertility called **crop rotation,** in which the fields under cultivation remain the same, but the *crops* planted are changed to balance the types of nutrients withdrawn and delivered to the soil.

Shifting cultivation is globally distributed in the tropics, especially in the rain forests of Central and West Africa, the Amazon in South America, and much of Southeast Asia, including Thailand, Burma, Malaysia, and Indonesia, where climate, rainfall, and vegetation combine to produce soils lacking nutrients (Figure 8.7, page 325). The practices involved in shifting cultivation have changed very little over thousands and thousands of years (Figure 8.8, page 326). As a *land* rotation system, shifting cultivation requires less expenditure of energy than modern forms of farming, though it can only successfully support low population densities (Figure 8.9, page 326).

The typical agrarian system that supports shifting cultivation is one in which a small group of villagers holds land in common tenure. Through collective agreement or a ruling council, sites are distributed among village families and then cleared for planting by family members. As villages increase their populations, tillable sites must be located farther and farther away from the center of the

Hunting and gathering: activities whereby people feed themselves through killing wild animals and fish and gathering fruits, roots, nuts, and other edible plants to sustain themselves.

Subsistence agriculture: farming for direct consumption by the producers, not for sale.

Commercial agriculture: farming primarily for sale, not direct consumption.

Shifting cultivation: a system in which farmers aim to maintain soil fertility by rotating the fields within which cultivation occurs.

Crop rotation: method of maintaining soil fertility where the fields under cultivation remain the same, but the crop being planted is changed.

Figure 8.5 Areas of plant domestication As this map illustrates, plant domestication did not predominate in any one continent but was spread out across the globe. It is important to point out, however, that the origins of plant domestication are not definitively known, and much of what is represented on this map is speculative. Archeological evidence to date supports the distribution shown here and developed in the mid-twentieth century by Carl Sauer. (*Source:* Reprinted with permission of Prentice Hall, from J. M. Rubenstein, *The Cultural Landscape: An Introduction to Human Geography*, 5th Ed., © 1996, p. 392. Adapted from Carl O. Sauer, *Agricultural Origins and Dispersals*, with permission of the American Geographical Society.)

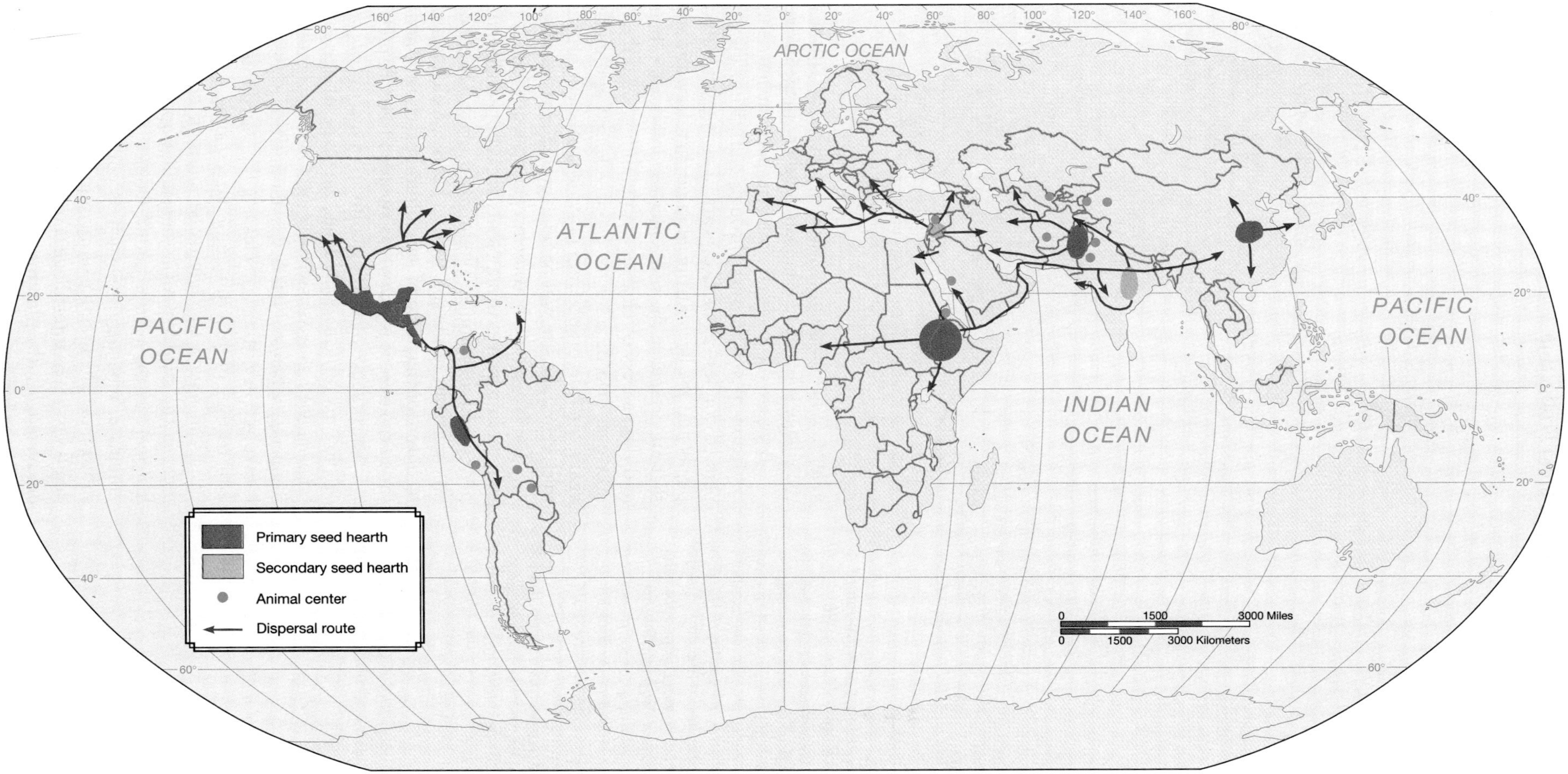

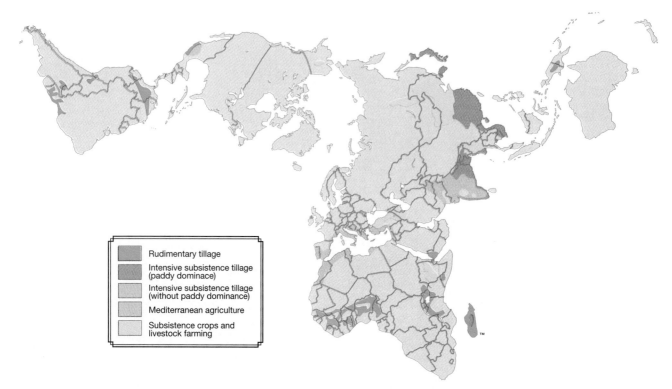

Rudimentary tillage

Intensive subsistence tillage (paddy dominace)

Intensive subsistence tillage (without paddy dominance)

Mediterranean agriculture

Subsistence crops and livestock farming

Figure 8.6 Global distribution of subsistence agriculture, 1980
Subsistence and near-subsistence agriculture is practiced on every continent, though Asia contains the greatest proportion. Subsistence practices are mostly absent from the United States, Japan, and Canada but continue to occur in parts of southern and eastern Europe. (*Sources:* J. Singh, *Agricultural Geography.* New York: McGraw Hill Publishing Company, 1984. Map projection, Buckminster Fuller Institute and Dymaxion Map Design, Santa Barbara, CA. The word Dymaxion and the Fuller Projection Dymaxion™ Map design are trademarks of the Buckminster Fuller Institute, Santa Barbara, California, ©1938, 1967 & 1992. All rights reserved.)

Swidden: land that is cleared through slash-and-burn and is ready for cultivation.

Intertillage: practice of mixing different seeds and seedlings in the same swidden.

village. When population growth reaches a critical stage, several families within the village split off to establish another village in one of the more remote sites.

Because the tropical soils are poor in nutrients, the problem of the rapid depletion of soil fertility through cultivation means that the fields are actively planted for less than five years. The biggest culprits in soil depletion are the plants and the heavy tropical rains that draw off and wash out the few nutrients that are present in the soil. Once the soil nears exhaustion, a new site is identified and the process of clearing and planting, described below, begins again. It may take over two decades for a once-cleared and cultivated site to become tillable again, after decomposition provides sufficient organic material in the soil. When this occurs, the site will be reintegrated into cultivation.

The typical method for preparing a new site is through slash-and-burn agriculture, in which existing plants are cropped close to the ground, left to dry for a period and then ignited (Figure 8.10, page 326). The burning process adds valuable nutrients to the soil such as potash, which is about the only readily available fertilizer for this form of agricultural practice. Once the land is cleared and ready for cultivation, it is known as **swidden** (see Chapter 2).

The practice of shifting cultivation usually occurs without the aid of livestock or plow to turn the soil. Thus, this type of agriculture relies largely on human labor as well as extensive acreage for new plantings as old ones are abandoned frequently when soil fertility is diminished. Although a great deal of human labor is involved in cutting and clearing the vegetation, once the site is planted, there is little tending of the crops until harvest time.

From region to region, the kinds of crops grown and the arrangement of them in the swidden varies depending upon local taste and plant domestication histories. In the warm, humid tropics, tubers—sweet potatoes and yams—predominate, while grains—such as corn or rice—are more widely planted in the subtropics. The practice of mixing different seeds and seedlings in the same swidden is known as **intertillage** (Figure 8.11, page 326). Not only are different plants cultivated, but the planting of different crops is usually staggered so that harvesting of the food supply can continue throughout the year. Such staggered planting

and harvesting are adaptive strategies to reduce the risk of disasters from crop failure and to increase the nutritional balance of the local diet.

Shifting cultivation also involves a gender division of labor that may vary from region to region (Figure 8.12, page 326). For the most part, men are largely responsible for the initial tasks of clearing away vegetation, cutting down trees, and burning the remaining stumps. Women are usually involved with sowing seeds and harvesting the crops. There are places, however, in Asia, Africa, and Latin America, where both men and women are involved equally in planting and harvesting activities. Research on shifting cultivation indicates that the actual division of tasks between men and women (and sometimes children) results from traditional cultural practices as well as the new demands placed upon household by globalization (recall the discussion on Sudanese children in Chapter 6). For instance, many women have found it necessary to complement their subsistence agricultural activities with craft production for local tourist markets. Their absence from routine agricultural activities means those tasks must be taken up by other household members.

Although heralded by many as an ingenious, well-balanced response to the environmental constraints of the tropics and subtropics, shifting cultivation is not without limitations. Its most obvious limitation is the fact that it is a cultivation strategy that can only be effective with small populations. Increasing populations cause cultivation sites to be located farther from villages, with the result that cultivators expend as much energy traveling to sites as they garner energy from the crops they produce. Indeed, at any one time, it is not unusual for land closest to the village to be entirely fallow or unseeded.

Increasingly, population pressures and ill-considered government policies are undermining the practicality of shifting cultivation, resulting in irreparable damage to the environment in many parts of the world. In Central and South America, for example, national governments have used rural resettlement programs to address urban population pressures. In some cases, individuals not familiar with shifting cultivation techniques have employed them improperly. In others, individuals have been relocated to areas unsuitable for such cultivation

Figure 8.7 Global distribution of shifting cultivation and nomadic herding, 1990 Shifting cultivation and nomadic herding practices are on the decline with fewer than 500,000 people still involved in the latter worldwide. These particular forms of subsistence activities are especially labor intensive and occur in relatively marginal environments. (*Sources:* From Fred M. Shelley and Audrey E. Clarke, *Human and Cultural Geography*, Fig . 7.1, p. 185. Data from J. M. Rubenstein, *The Cultural Landscape: An Introduction to Human Geography*, 3rd Ed. New York: Macmillan Publishing, 1993. Copyright © 1994 The McGraw-Hill Companies, Inc. Reprinted by permission. Map projection, Buckminster Fuller Institute and Dymaxion Map Design, Santa Barbara, CA. The word Dymaxion and the Fuller Projection Dymaxion™ Map design are trademarks of the Buckminster Fuller Institute, Santa Barbara, California, ©1938, 1967 & 1992. All rights reserved.)

Figure 8.8 Shifting cultivation Shifting cultivation is usually practiced in tropical forests. It is a system of agriculture that maintains soil fertility by rotating the *fields* within which cultivation occurs.

Figure 8.9 Crop rotation Another method of maintaining soil fertility is crop rotation, where the fields under cultivation remain the same, but the *crops* planted are changed to balance the types of nutrients withdrawn and delivered to the soil. Crop rotation is more likely to be practiced in temperate regions where soil fertility is naturally higher.

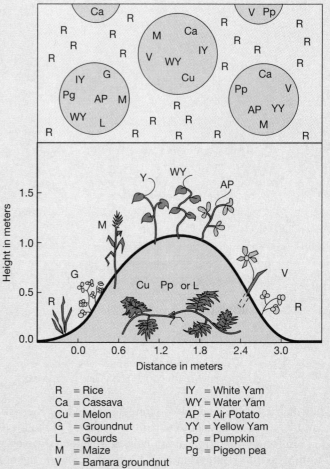

R = Rice	IY = White Yam
Ca = Cassava	WY = Water Yam
Cu = Melon	AP = Air Potato
G = Groundnut	YY = Yellow Yam
L = Gourds	Pp = Pumpkin
M = Maize	Pg = Pigeon pea
V = Bamara groundnut	

Figure 8.11 Intertillage Planting different crops together in the same field has many benefits, not the least of which are the spreading out of food production over the farming season, reduction of disease and pest loss, greater protection from loss of soil moisture, and control of soil erosion.

Figure 8.10 Slash-and-burn agriculture Slash-and-burn is a process of preparing low-fertility soils for planting. In this practice, plants are cleared from a site through cutting, and then the remaining stumps are burned. The burning process helps to add minerals to the soil and thereby improve the overall fertility. Slash-and-burn is a form of agricultural practice that is most effective with low levels of population.

Figure 8.12 Woman and man tending garden In Malawi a woman and her husband work side by side tending the vegetable garden using identical handmade tools. Some tasks such as this one are shared equally within the household division of labor, while others fall exclusively to men or to women.

326

practices. In parts of the Brazilian Amazon, for example, shifting cultivators have acted in concert with cattle grazers, resulting in accelerated environmental degradation.

Despite its negative impacts on the environment, shifting cultivation can be an elegant response to a fragile landscape. The fallow period, which is an essential part of the process, is a passive and perfectly effective way of restoring plant nutrients to the soil. The burning of stumps and other debris makes the soil more workable, and seeding can proceed with a minimum of effort. Intertillage mimics the actual pattern of differing plant heights and types characteristic of the rain forest. It is also a significant factor in protecting the soil from leaching and erosion. Shifting cultivation requires no expensive inputs (except possibly where native seeds are not available) because no manufactured fertilizers, pesticides, herbicides, or heavy equipment—mechanical or otherwise—are necessary. Finally, the characteristically staggered sowing allows for food production throughout the year. Whereas it is likely that shifting cultivation was once practiced by agriculturists throughout the world, it now appears that population growth and the greater need for increased outputs per acre have been responsible for its replacement with more intensive forms of agriculture.

Intensive Subsistence Agriculture

The second dominant form of subsistence activity is **intensive subsistence agriculture,** a practice involving the effective and efficient use of a small parcel of land in order to maximize crop yield; a considerable expenditure of human labor and application of fertilizer is also usually involved. Unlike shifting cultivation, intensive subsistence cultivation often is able to support large rural populations. While shifting cultivation is more characteristic of low agricultural densities, intensive subsistence normally reflects high agricultural density and thus usually occurs in the region of the world with the largest population, namely, Asia, and especially in India, China, and Southeast Asia.

While shifting cultivation involves the application of a relatively limited amount of labor and other resources to cultivation, intensive subsistence agriculture involves fairly constant human labor in order to achieve high productivity from a small amount of land (Figure 8.13). With population pressures fierce and the amount of arable land limited, intensive subsistence agriculture also reflects the inventive ways in which humans confront environmental constraints and reshape the landscape in the process. In fact, the landscape of intensive subsistence agriculture is often a distinctive one including raised fields and hillside farming through terracing.

Intensive subsistence agriculture: practice that involves the effective and efficient use—usually through a considerable expenditure of human labor and application of fertilizer—of a small parcel of land in order to maximize crop yield.

Figure 8.13 Intensive subsistence agriculture Where usable agricultural land is at a premium, agriculturalists have developed ingenious methods for taking advantage of every square inch of usable terrain. Landscapes like this one in Malaysia can be extremely productive when carefully tended, and can feed relatively large rural populations.

It is significant that intensive subsistence agriculture is able to support large rural populations. Unlike shifting cultivation, fields are planted year after year as fertilizers and other soil enhancers are applied to maintain soil nutrients. For the most part, the limitations on the size of plots have more to do with demography than geography. In Bangladesh and southern China, for example, where a significant proportion of the population is engaged in intensive subsistence agriculture, land is passed down from generation to generation—usually from fathers to sons—so that each successive generation receives a smaller and smaller share of the family holdings. Yet each family must produce enough to sustain itself.

Under conditions of a growing population and a decreasing amount of arable land, it is critically important to plant subsistence crops that produce a high yield per hectare. Different crops fulfill this need depending on the regional climate. Generally speaking, the crops that dominate intensive subsistence agriculture are rice and other grains.

Rice production predominates in those areas of Asia—South China, Southeast Asia, Bangladesh, and parts of India—where summer rainfall is abundant. In drier climates and where the winters are too cold for rice production, other sorts of grains—among them wheat, barley, millet, sorghum, corn, and oats—are grown for subsistence. In both situations, the land is intensively used. In fact, it is not uncommon in milder climates for intensive subsistence fields to be planted and harvested more than once a year, a practice known as **double cropping.**

Double cropping: practice in the milder climates where intensive subsistence fields are planted and harvested more than once a year.

Pastoralism

Although not obviously a form of agricultural production, pastoralism is a third, dominant form of subsistence activity associated with a traditional way of life and agricultural practice. **Pastoralism** involves the breeding and herding of animals to satisfy the human needs of food, shelter, and clothing. Usually practiced in the cold and dry climates of deserts, savannas (grasslands), and steppes (lightly wooded, grassy plains) where subsistence agriculture is impracticable, pastoralism can either be sedentary (pastoralists live in settlements and herd animals in nearby pastures) or nomadic (they wander with their herds over long distances, never settling in any one place for very long). Although forms of commercial pastoralism exist—the regularized herding of animals for profitable meat production, as among Basque-Americans in the basin and range regions of Utah and Nevada and the gauchos of the Argentinean grasslands—we are concerned here with pastoralism as a wandering subsistence activity.

Pastoralism: subsistence activity that involves the breeding and herding of animals to satisfy the human needs of food, shelter, and clothing.

Pastoralism is largely confined to parts of North Africa and the savannas of central and southern Africa, the Middle East, and central Asia. Pastoralists generally graze cattle, sheep, goats, and camels, although reindeer are herded in parts of Eurasia. The type of animal herded is related to the culture of the pastoralists as well as the animals' adaptability to the regional topography and foraging conditions (Figure 8.14).

As a subsistence activity, nomadism involves the systematic and continuous movement of groups of herders, their families, and the herds in search of forage. Most pastoralists practice **transhumance,** the movement of herds according to seasonal rhythms: warmer, lowland areas in the winter, and cooler, highland areas in the summer (Figure 8.15). Although the herds are occasionally slaughtered and used directly for food, for shelter, and for clothing, often they are bartered with sedentary farmers for grain and for other commodities. Female and younger members of pastoralist groups may also be involved with cultivation.

Transhumance: movement of herds according to seasonal rhythms: warmer, lowland areas in the winter; cooler, highland areas in the summer.

In such cases, mostly women and children split off from the larger group and plant crops at fixed locations in the spring. They may stay sedentary for the growing season, tending the crops, or they may rejoin the group and return to the fields when the crops are ready for harvesting. The distinguishing characteristic of pastoralists is that they depend on animals, not crops, for their livelihood.

Figure 8.14 Pastoralism Pictured is a Bedouin shepherd and his flock in the Irbid region of Jordan. Bedouins are traditionally nomadic pastoralists. Notice also the dryness of the landscape. Pastoralism usually occurs where agriculture is not feasible.

Figure 8.15 Global distribution of transhumance routes Pictured on this map are the seasonal routes taken by herders as animals are moved from summer to winter pastures. As the map shows, transhumance is a well-established practice characterizing an effective adaptation to temporal rhythms. Just as environmental conditions shape herding practices, so do the herding practices shape the landscape through the emplacement of identifiable trails. (*Source:* "Transhumance" (Figure 7) Map from *The Mediterranean and the Mediterranean World in the Age of Philip II, Volume I,* by Fernand Braudel. Copyright © Librairie Armand Colin 1966. English translation copyright © 1972 by Wm. Collins Sons Ltd. and Harper & Row Publishers, Inc. Reprinted by permission of HarperCollins Publisher, Inc.)

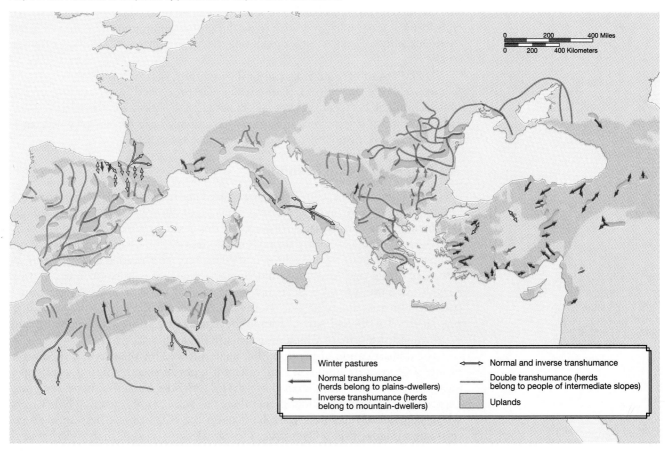

Like the two other traditional forms of agriculture previously mentioned, pastoralism is not simply a subsistence activity but part of a social system as well. Pastoralists consist of groups of families who are governed by a leader or chieftain. Groups of families are divided into units that follow different routes with the herds. The routes taken are well known, with members of the group intimately conversant with the landscape, watering places, and opportunities for contact with sedentary groups. Not surprisingly, pastoralism as a subsistence activity is on the decline as more and more pastoralists have become integrated into a global economy that requires more efficient and regularized forms of production. They have also been forced off the land by competition from other land uses and the need for the State to track citizens for taxation and military reasons.

Agricultural Revolution and Industrialization

For a long time, human geography textbooks treated the differences in agricultural practices worldwide as systems to be described and cataloged as we have done above. In the last 25 to 30 years, however, new conceptual approaches to the agricultural sector have transformed the ways in which we view it. Agriculture has become less a human activity to be *described* through classification and more a complex component of the global economic system to be *explained*. Indeed, while the importance and persistence of traditional agricultural forms are acknowledged, such description must be balanced with an understanding of the ways in which new commercial practices are undermining and otherwise changing these older forms.

Increasingly, geographers and others have come to see world agricultural practices as having proceeded through "revolutionary" phases just as manufacturing did. By seeing agriculture in this new light, it is possible to recognize that, like manufacturing, the changes that have occurred in agricultural practices have transformed geography and society as the global community has moved from predominantly subsistence to predominantly capital-intensive, market-oriented practices.

To understand the new agricultural geography, it is necessary to review the history of world agriculture. This history has proceeded in alternating cycles: long periods of very gradual change punctuated by short explosive periods of radical change. Geographers and others have divided the history of world agriculture into three distinct revolutionary periods (Table 8.1).

The First Agricultural Revolution

The first revolution is commonly recognized as having been founded on the development of seed agriculture and the use of the plow and draft animals. Aspects of this transformation have been discussed in Chapters 2 and 4. The emergence of seed agriculture through the domestication of crops, such as wheat and rice, and animals, such as sheep and goats, replaced hunting and gathering as a way of living and sustaining life. In fact, seed agriculture occurred during roughly the same period in several regions around the world, creating a broad belt of cultivated lands across Southwest Asia from Greece in the west into present-day Turkey and part of Iran in the east as well as in parts of Central and South America, northern China, northeast India, and East Africa.

The domestication of plants and animals allowed for the rise of settled ways of life. Villages were built, creating different types of social, cultural, economic, and political relationships than those that dominated hunter-gatherer societies. Especially important were floodplains along the Tigris, Euphrates, and Nile Rivers where complex civilizations were built upon the fruits of the first agricultural revolution (Figure 8.16). Over time, the knowledge and skill that underlay seed

TABLE 8.1 The Three Agricultural Revolutions		
First Revolution: Beginnings and Spread	**Second Revolution: Subsistence to Market**	**Third Revolution: Industrialization**
Pre-10,000 B.C. to 20th century	c. A.D. 1650 to present	1928 to present
Neolithic, medieval Europe	18th-century England; 19th–20th century in "European" settlement areas	Present day
Europe and Southeast Asia	Western Europe and North America	USSR and eastern Europe, North America, and Western Europe
Domestic food supply and survival; initial selection and domestication of key species	Surplus production and financial return	Lower unit cost of production
Farming replaces hunting and gathering as way of life and basis of rural settlement and society	Critical improvements, mercantilist outlook, and food demands of Industrial Revolution replace subsistence with market	Collective (socialist) and corporate (capitalist) ideologies and common agrotechnology favor integration of agricultural production into total food-industry system
Agrarian societies proliferate and support population growth	Agriculture part of sectoral division of labor: individual family farm becomes "ideal" for way of life and for making a living	Emphasis on productivity and production for profit replace agrarian structure and farm way of life
Subsistence agriculture: labor intensity, low technology, communal tenure	Commercial agriculture develops growing reliance on technological inputs and infrastructure	Collective/corporate production utilizes economies of scale, capital intensity, labor substitution, and specialized production on fewer, larger units

Source: M. J. Troughton, "Farming systems in the modern world." In M. Pacione (Ed.), *Progress in Agricultural Geography.* London: Croom Helm, 1986, p. 98.

Figure 8.16 Present-day agriculture along the Nile
Agriculture along the Nile River dates far back into prehistory. The floodplain of the Nile was one of the important culture hearths for sedentary agriculture and provided the foundation for the growth of complex civilizations in what we now know as Egypt. The Nile floodplain remains a remarkably productive area even today.

agriculture and the domestication of plants diffused outward from these original areas, having a revolutionary impact throughout the globe.

The Second Agricultural Revolution

A great deal of debate exists among historians as to the timing and location of the second agricultural revolution. While agreement exists that it did not occur everywhere at the same time, there is disagreement over which factors were essential to the fundamental transformation of subsistence agriculture, including:

- Dramatic improvements in outputs, like crop and livestock yields
- The importance of innovations, such as the improved yoke for oxen or the replacement of the ox with the horse (Figure 8.17)
- New inputs to agricultural production, such as the application of fertilizers and field drainage systems (Figure 8.18)

It is safe to place the apex of the second agricultural revolution historically and geographically alongside the Industrial Revolution in England and Western Europe. Although many important changes in agriculture preceded the Industrial Revolution, none had more of an impact on everyday life than the rise of an industrialized manufacturing sector, the effects of which spread rapidly to agriculture.

On the eve of the Industrial Revolution—in the middle of the eighteenth century—in Western Europe and England, subsistence peasant agriculture was predominant, though partial integration into a market economy was under way. Many peasants were utilizing a crop rotation system that, in addition to the application of natural and semiprocessed fertilizers, improved soil productivity, leading to both increased crop and livestock yields. Additionally, the feudal landholding system was breaking down and yielding to a new agrarian system based not on service to a lord but on an emerging system of private-property relations. Communal farming practices and common lands were being replaced by enclosed, individually owned land or land worked independently by tenants or renters.

Such a situation was logical in response to the demands for food production that emerged from the dramatic social and economic changes accompanying the Industrial Revolution. Perhaps most important of all these changes was the development—through the creation of an urban industrial work force—of a commercial market for food. Many of the innovations of the Industrial Revolution, such

Figure 8.17 Yoked oxen-drawn plow
In many parts of the world, agriculturalists rely on draught animals to prepare land for cultivation. The introduction of animals to assist in agricultural production was an important element in the first agricultural revolution. The use of animals enabled humans to increase food supplies by expanding the amount of energy applied to production. For many contemporary farmers, draft animals are their most valuable possessions.

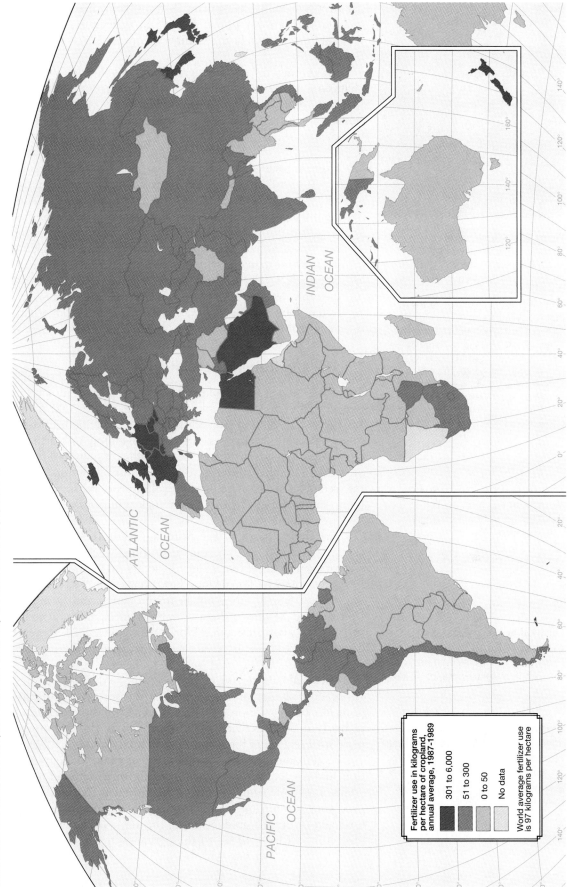

Figure 8.18 Global distribution of fertilizer use, 1991–1992 Western Europe, Egypt, Saudi Arabia, and Japan are among the largest users of fertilizers, with the United States and much of Asia following in their footsteps. While they have rapidly growing populations, many of the the African countries cannot afford expensive agricultural inputs such as fertilizers. The United States and other, core European countries, while they too must feed their populations, export a great deal of food products, and fertilizers enable them to produce exportable surpluses. (*Source:* Reprinted with permission of Prentice Hall, from E. F. Bergman, *Human Geography: Cultures, Connections, and Landscapes*, © 1995, p. 211. Data from World Resources Institute, *World Resources 1994–95*. New York: Oxford University Press, 1994, pp. 29–95.)

Fertilizer use in kilograms per hectare of cropland, annual average, 1987–1989

301 to 6,000

51 to 300

0 to 50

No data

World average fertilizer use is 97 kilograms per hectare

Figure 8.19 Family farm Pictured is an idyllic scene of a prosperous U.S. family farm. Since the third agricultural revolution, the number of family farms in the United States and other core countries has declined dramatically as more corporate forms of farming have emerged. In 1920, about one in every three U.S. citizens lived on farms. By 1978, that number had dropped dramatically to about one in 28. This drastic change in the U.S. farm population has caused some commentators to observe that the family farm in the United States is now just a myth.

Mechanization: term applied to the replacement of human farm labor with machines.

Chemical farming: application of inorganic fertilizers to the soil and herbicides, fungicides, and pesticides to crops in order to enhance yields.

Food manufacturing: adding value to agricultural products through a range of treatments—processing, canning, refining, packing, packaging, etc.—that occur off the farm and before they reach the market.

as improvements in transportation technology, had substantial impacts on agriculture. Innovations applied directly to agricultural practices, such as the new types of horse-drawn farm machinery, improved control over, as well as the quantity of, yields.

By igniting the second agricultural revolution, the Industrial Revolution changed rural life as profoundly as the sedentary requirements of seed agriculture had transformed hunting and gathering society (Figure 8.19). As geographer Ian Bowler writes, this revolution moved rapidly from Europe to other parts of the world.

> From its origins in Western Europe, the new commercialized system of farming was diffused by European colonization during the nineteenth and twentieth centuries to other parts of the world. A dominant agrarian model of commercial capitalist farming was established, based on a structure of numerous, relatively small family farms. From this period can be traced both the dependence of agriculture on manufacturing industry for many farm inputs, and the increasing productivity of farm labor, which released large numbers of workers from the land to swell the ranks of factory workers and city dwellers. Moreover, the production of food surplus to domestic demand enabled international patterns of agricultural trade to be established.[1]

The Third Agricultural Revolution

The third agricultural revolution is a fairly recent one and, unlike the previous two, emanates mostly from the New World, not the Old. Scholars place the third agricultural revolution squarely within the twentieth century, with each of its three important developmental phases originating in North America. Indeed, the globalization trends that frame all of our discussions in this text are the very same ones that have shaped the third agricultural revolution. The difference between the second and third agricultural revolutions is mostly a matter of degree, so that by the late twentieth century technological innovations have virtually industrialized agricultural practices.

The three phases of the third agricultural revolution are mechanization, chemical farming, and food manufacturing. **Mechanization** is the replacement of human farm labor with machines. Tractors, combines, reapers, pickers, and other forms of motorized machines have, since the 1920s, progressively replaced human and animal labor inputs to the agricultural production process in the United States (Figure 8.20). In Europe, mechanization did not become widespread until after the Second World War. Figure 8.21 compares statistics on the declining population involved in agriculture with the increasing use of tractors worldwide.

Chemical farming, the second stage of the third agricultural revolution, is the application of inorganic fertilizers to the soil and herbicides, fungicides, and pesticides to crops in order to enhance yields. Becoming widespread in the 1950s in the United States, chemical farming diffused to Europe in the 1960s and to peripheral regions of the world in the 1970s (Figure 8.22, page 336).

Food manufacturing is the most recent phase of the third agricultural revolution, and it, too, had its origins in North America. **Food manufacturing** involves adding economic value to agricultural products through a range of treatments—processing, canning, refining, packing, packaging, and so on—occurring off the farm and before the products reach the market. The first two phases of the third revolution affected *inputs* to the agricultural production process, whereas the

---

[1]I. Bowler (Ed.), *The Geography of Agriculture in Developed Market Economies.* Harlow, England: Longman Scientific and Technical, 1992, pp. 10–11.

Figure 8.20 Old and new farm machines   In the early part of the twentieth century agriculture in the United States became increasingly linked with mechanization. The black and white photograph shows a 38-horse harvesting machine that also cut, threshed, and sacked the wheat. The color photograph shows contemporary motor-powered harvesting equipment.

final phase affects agricultural *outputs*. While the first two are related to the modernization of farming as an economic practice, the third involves a complication of the relationship of farms to firms in the manufacturing sector, which had increasingly expanded into the area of food early in the 1960s (Figure 8.23, page 336). Considered together, these three developmental phases of the third agricultural revolution constitute the industrialization of agriculture.

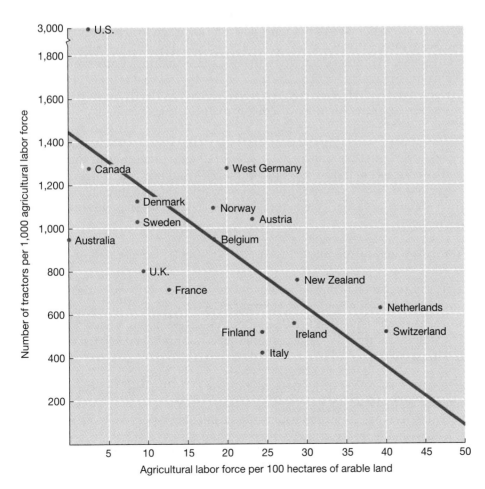

Figure 8.21 Declining agricultural workers relative to the increasing use of tractors   This figure shows the direct impact that mechanization has had on the number of individuals employed in agriculture in core countries. Graphics like these hide the fact that although the number of workers has declined as mechanization has increased, what has remained largely the same is the amount of farmland in cultivation. Mechanization and the increasing size of the average farm has led to higher yields per acre with fewer human inputs. (*Source: Food and Agriculture Organization Production Yearbook, 1981*, Vol. 35, Rome, 1982.)

Figure 8.22 Chemical farming An aircraft sprays fungicide on an orange grove near Ft. Pierce, Florida. The fungicide (trade name Kocide) is mixed with a solution to make it stick to the leaves. Planes spray the fungicide at windless times to minimize the drift of fungicide beyond the groves.

Figure 8.23 Food manufacturing A poultry processing plant in North Carolina illustrates the way in which processing has become as important as growing food. The world broiler (chicken) industry is widespread, with small producers in both the core and the periphery bidding for contracts with large food conglomerates to raise broilers according to strict contract standards.

The Industrialization of Agriculture

Developed over time, the industrialization of agriculture has been largely determined by the changes that have occurred in science and technology, especially by way of mechanical as well as chemical and biological innovations. As with industrialization more generally, the industrialization of agriculture has unfolded as the capitalist economic system has become more advanced and widespread. We regard **agricultural industrialization** as the process whereby the farm has moved from being the *centerpiece* of agricultural production to become but *one part* of an integrated, multileveled (or vertically organized) industrial process including production, storage, processing, distribution, marketing, and retailing. Experts in the study of agriculture have come to see agriculture as clearly linked to industry and the service sector, thus constituting a complex agro-commodity production system.

Geographers have helped demonstrate the changes leading to the transformation of an agricultural product into an industrial food product. This transformation has been accomplished, not only through the indirect and/or direct altering of agricultural outputs such as tomatoes or wheat, but also through changes in rural economic activities. Agricultural industrialization involves three important developments:

- Changes in rural labor activities as machines replace and/or improve human labor
- The introduction of innovative inputs—fertilizers, hybrid seeds, agrochemicals, and biotechnologies—to supplement, alter, or replace biological outputs
- The development of industrial substitutes for agricultural products (Nutrasweet instead of sugar, and thickeners instead of cornstarch or flour, for example)

Recall, however, that the industrialization of agriculture has not occurred simultaneously throughout the globe. Changes in the global economic system affect different places in different ways as different States and social groups respond to and shape these changes. For example, the use of fertilizers and high-yielding seeds occurred much earlier in core-region agriculture than in the periph-

Agricultural industrialization: process whereby the farm has moved from being the centerpiece of agricultural production to become one part of an integrated string of vertically organized industrial processes including production, storage, processing, distribution, marketing, and retailing.

ery, where many places still farm without them. Beginning in the late 1960s, however, core countries exported a technological package of fertilizers and high-yielding seeds to regions of the periphery (largely in Asia and Mexico) in an attempt to boost agricultural production. Known as the **green revolution**, this package also included new machines and institutions, all diffused from the core to the periphery, designed to increase global agricultural productivity as described in Feature 8.1, "Geography Matters—A Look at the Green Revolution."

Green revolution: the export of a technological package of fertilizers and high-yielding seeds, from the core to the periphery, to increase global agricultural productivity.

Global Restructuring of Agricultural Systems

When geographers talk about the globalization of agriculture, they are referring to the incorporation of agriculture into the world economic system of capitalism. A useful way to think about the term **globalized agriculture** is to recognize that as both an economic sector and a geographically distributed activity, modern agriculture is increasingly dependent on an economy and set of regulatory practices that are global in scope and organization.

Globalized agriculture: a system of food production increasingly dependent upon an economy and set of regulatory practices that are global in scope and organization.

Forces of Globalization

Three related processes play a role in the globalization of agriculture:

- The forces—technological, economic, political, and so on—that shape agricultural systems are global in their scope.
- The institutions—trade and finance especially—that most dramatically alter agriculture are organized globally.
- The current form of agriculture reflects integrated, globally organized agro-production systems.

The globalization of agriculture has dramatically changed relationships among, between, and within different agricultural production systems. The result is eventual elimination of some forms of agriculture on the one hand, or the erosion or alteration of systems as they are integrated into the global economy, on the other. Two examples include the current decline of traditional agricultural practices, such as shifting cultivation, and the erosion of a national agricultural system based on family farms.

Not an isolated or independent sector of the regional, national, or global economy, agriculture is one part of a complex and interrelated worldwide economic system. Consequently, it is possible to see that important changes in the wider economy—whether they are technological, social, political, or otherwise—will affect all sectors, including agriculture. Following this logic, national-level problems in agriculture, such as production surpluses, soil erosion, and food price stability among others, affect agriculture as well as other economic sectors globally, nationally, and locally in different ways. The same is also true of crises of a more global proportion—such as the price and availability of oil and other petroleum products critical to commercialized agriculture, the stability of the dollar in the world currency market, and recessions or inflationary tendencies in the economy.

It is because of the systemic impact of problems or crises that the need for integration and coordination of the global economy has become important in order to anticipate or respond to them. In the last several decades, global and international coordination efforts among States has occurred. These efforts include policies advanced by the World Trade Organization (WTO) as well as the formation of supranational economic organizations such as the European Union (EU) or the Association of Southeast Asian Nations (ASEAN). These new forms of cooperation among States indicate that it is becoming increasingly

8.1 Geography Matters
A Look at the Green Revolution

The green revolution constitutes an attempt by agricultural scientists to find ways to feed the world's burgeoning population. The effort began in Mexico in 1943 when the Rockefeller Foundation funded a group of U.S. agricultural scientists to set up a research project in Mexico aimed at increasing that country's wheat production. Scientists distributed the first green revolution wheat seeds only seven years later. The project in Mexico was eventually expanded to include research on maize as well. By 1967, the green revolution scientists were exporting their work to other parts of the world and had added rice to their research agenda. Norman Borlaug, one of the founders of the green revolution, went on to win the Nobel Peace Prize in 1970 for an important component of the project: promoting world peace through the elimination of hunger (Figure 8.1.1).

The initial focus of the green revolution was on the development of seed varieties that would produce higher yields than those traditionally used in the target areas. However, in developing new, higher-yielding seed varieties, agricultural scientists soon discovered that plants were limited in the amount of nitrogen they could absorb and use. The scientists' solution was to increase the nitrogen absorption capacity of plants by delivering nitrogen-based fertilizers in water (this led to the need to build major water and irrigation development projects). Then the scientists discovered that the increased nitrogen and water caused the plants to develop tall stalks. The tall stalks, with heavy heads of seed on top, fell over easily, thus reducing the amount of seed that could be harvested. The scientists went back to the drawing board and came up with dwarf varieties of grains that would support the heavy heads of seeds without falling over. Then another problem arose. The short plants were growing in very moist conditions, which encouraged the growth of diseases and pests. The scientists responded by developing a range of pesticides.

Thus, the green revolution came to constitute a *package* of inputs: new "miracle seeds," water, fertilizers, and pesticides. Farmers had to use all of the inputs—and use them properly—to achieve the yields the scientists produced in their experimental plots (Figure 8.1.2). Green revolution crops, if properly watered, fertilized, and treated for pests, can generate yields that are two to five times larger than the yields of traditional crops. In some countries, yields have been high enough to even engage in export trade, thus generating important sources of foreign exchange. Furthermore, the creation of varieties that produce faster-maturing crops has allowed some farmers to plant two or more crops per year on the same land, thus increasing their individual production—and wealth—considerably.

Thanks to green revolution innovations, rice production in Asia grew 66 percent between 1965 and 1985. India, for example, became largely self-sufficient in rice and wheat by the 1980s. Worldwide, green revolution seeds and agricultural techniques accounted for almost 90 percent of the increase in world grain output in the 1960s and about 70 percent of that in the 1970s. In the late 1980s and 1990s, at least 80 percent of the additional production of grains may be attributed to the use of green revolution techniques. Thus, although hunger and famine surely exist, many argue that conditions might be much worse if the green revolution had never occurred.

The green revolution, however, has not been an unqualified success everywhere in the world (Figure 8.1.3). Even in Mexico, where considerable success was achieved with improved wheat and maize production, green revolution improvements for other dietary items

Figure 8.1.1 Green revolution research

Figure 8.1.2 Green revolution experimental plots

important to poor people, such as legumes, have not been a focus of research, even though one of America's foremost geographers, Carl Sauer, had advocated such research for the original Rockefeller Foundation project. It has had even less spectacular results in other regions of Latin America, and Africa has not seen much improvement at all.

Why is this the case? One important reason is that wheat, rice, and maize are unsuited as crops in many areas of the world, and research on more suitable crops, such as sorghum and millet, has lagged far behind that of the more favored grains. In Africa, poor soils and lack of water make progress even more difficult to achieve. Another important factor is the vulnerability of the new seed strains to pest and disease infestation, often after only a couple of years of planting. Whereas traditional varieties often have a built-in resistance to the pests and diseases characteristic of a particular area, the genetically engineered varieties often lack such resistance.

Another criticism is that the green revolution has been accused of magnifying social inequities because it has allowed more wealth and power to accrue to a small number of agriculturalists, while causing greater poverty and landlessness among poorer segments of the population. In Mexico, a black market developed in green revolution seeds, fertilizers, and pesticides when poorer farmers, who were coerced into using them, accrued high debts that they could not begin to repay. Many ended up losing their lands and becoming migrant laborers, or moved to the cities and joined the ranks of the urban poor. Some critics who have monitored the effects of the green revolution suggest that political and economic conditions may, in fact, be more important than levels of production with regard to a country's food security.

Even in terms of quality the green revolution crops often fall short. The new seed varieties may produce grains that are less nutritious, less palatable, or less flavorful. The chemical fertilizers and pesticides that must be used are derived from fossil fuels—mainly oil—and are thus subject to the vagaries of world oil prices. Furthermore, the use of these chemical inputs, as well as monocropping practices, have produced worrisome levels of environmental contamination and soil erosion and, in many countries, have posed substantial threats to public health, especially among farm workers who are frequently exposed to poisonous (if not lethal) chemicals on

Figure 8.1.3 Global distribution of maize production Increases in the production of grains have been one of the successes of the green revolution. Mexico has been one of the countries that has benefited most from green revolution improvements in maize. (*Source:* Reprinted with permission of Prentice Hall, from E. F. Bergman, *Human Geography: Cultures, Connections, and Landscapes,* © 1995, p. 202.)

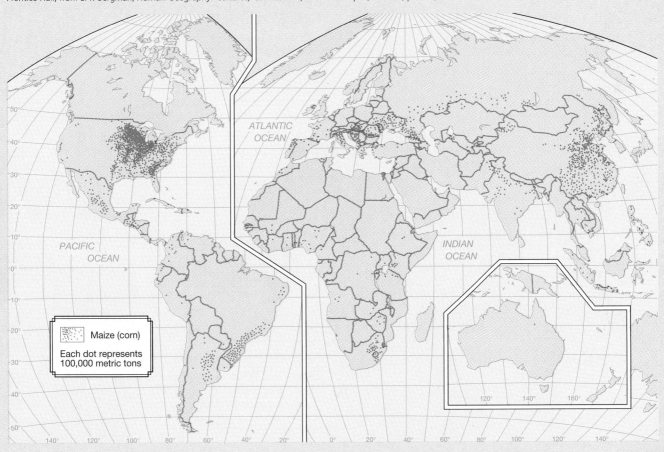

a regular basis. Water developments have benefited some regions, but less well-endowed areas have experienced a deterioration of already existing regional inequities. Worse, pressures to build water projects and to acquire foreign exchange to pay for importation of green revolution inputs has increased pressure on countries to grow even more crops for export, often at the expense of production for local consumption.

In recent years, scientists have been endeavoring to develop seeds with greater pest and disease resistance and more drought tolerance. The new focus is best revealed in Africa. The International Institute of Tropical Agriculture in Ibadan, Nigeria, focuses on foods for the humid and subhumid tropics of Africa, including cassava (interestingly, imported to Africa from South America by the Portuguese in the sixteenth century), yams, sweet potatoes, maize, soybeans, and cowpeas. The International Crops Research Institute for the Semi-Arid Tropics (located in Hyderabad, India, but with a major research center near Niamey, Niger) focuses on researching staples of the Sahel region such as sorghum, millet, pigeonpea,

and groundnut. Research in Africa on new varieties emphasizes testing under very adverse conditions (such as no plowing or fertilizing). New varieties are chosen not just for good yield, but because they will provide stable yields over good and bad years. A focus also exists on developing plants that will increase production of fodder and fuel residues as well as of food, and that give optimal yields when intertilled—a very common practice in Africa. In the Sahel, scientists are working on crops that mature more quickly to compensate for the serious drop in the average length of the rainy season that the region has been experiencing.

Thus, although the green revolution has come under much justified attack over the years, it has focused the world's attention on finding innovative new ways to feed the world's peoples. In the process, the world system has been expanded into hitherto very remote regions, and important knowledge has been gained about how to conduct science and how to understand the role that agriculture plays at all geographical scales of resolution from the global to the local (Figure 8.1.4).

Figure 8.1.4 Effects of the green revolution, 1990 This map illustrates the increases in yields brought about by the green revolution in Asia and Mexico. Burgeoning populations in these regions mean that increased yields are critical. Although yields have indeed improved, the costs of implementing the new practices have driven poorer farmers out of agriculture entirely. It has also been the case that the abandonment of farming by small farmers has meant that arable land has become concentrated in fewer and fewer hands, similar to trends that have occurred in the core. (*Source:* F. Shelley and A. Clarke, *Human and Cultural Geography.* Dubuque, Iowa: William C. Brown, 1994, Fig. 7.8, p. 196.)

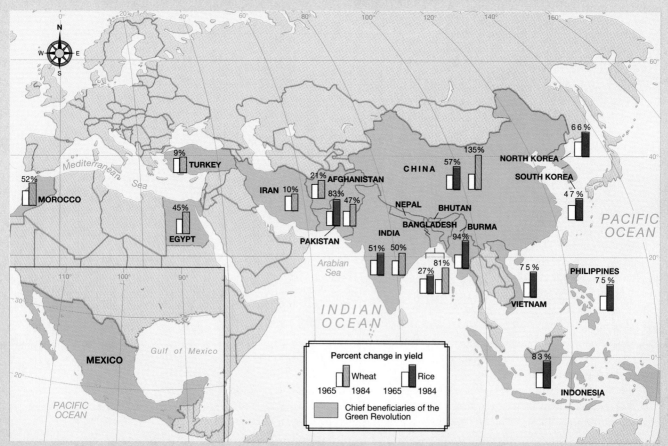

difficult to address economic problems—agricultural and otherwise—strictly at the national level.

At the same time that supranational organizations and coordination efforts have become important to address global-level problems, States continue to be essential to the mediation of crises at national levels. Through changing public policy, the State attempts to regulate agro-industries in order to maintain production, consumption, and corporate profits. One way in which States try to maintain the profitability of the agricultural sector while keeping food prices affordable is through direct and indirect subsidies to agricultural producers. For example, the U.S. government has subsidized agriculture in a number of ways. One of these is by paying farmers not to grow certain crops that are expected to be in excess supply. Another is by buying up surplus supplies and guaranteeing a fixed price for them.

Subsidies, however, while perhaps stabilizing agricultural production in the short term, can lead to problems within the larger national and international agricultural system. For instance, guaranteeing a fixed price for surplus food can act as a disincentive for producers to lower their production, so that the problem of overproduction continues. Once in possession of the surplus, governments must find ways to redistribute it. The U.S. government often will sell or donate its surplus to foreign governments, where the "dumping" of cheap foodstuffs may undermine the local price structure for food as well as reduce economic incentives for farmers to farm. Billions of dollars are paid out each year in agricultural subsidies, the effects of which are complex and global in their impact.

Many reasons exist for State intervention in agriculture. Economic interests, for example, can be both internally and externally driven. At the internal level, governments routinely intervene in one economic sector or another in order, for instance, to correct wider problems of inflation or depression. In the 1920s and 1930s, the U.S. government attempted to address the problem of economic depression through policies intended to reduce overproduction. These policies were initiated over a century ago. Yet, the description of the production of oranges in the United States that began this chapter indicates how complex and enduring a problem overproduction actually is.

States can also intervene in the agricultural sector with respect to consumers' interests. Because of subsidies to farm income, the real cost of food may be quite high, but many States in both the developed and less-developed world also subsidize the price of food in the marketplace. Such policies are meant to keep the work force well fed and healthy as well as to avoid problems of civil unrest should food prices exceed the general population's ability to pay. Nineteenth-century bread riots—a response to the high cost of flour and bread—were a common occurrence in Europe, and similar forms of civil unrest in response to food prices are not uncommon today in many less-developed countries, where States lack the capital to provide adequate subsidies.

In addition to the internal regulation and assistance of agricultural practices within a country, States—especially those of core countries—are also involved directly and indirectly in the agricultural sectors of other countries—especially those of peripheral countries. Food as well as agricultural development aid are widespread and popularly accepted ways in which core States intervene in the agricultural sector of peripheral States. Such intervention is one of the means by which peripheral States are incorporated into the global economy. Table 8.2 (see page 342) lists the major donors and recipients of food aid.

In addition to straightforward food aid, core States are also involved in attempting to improve the capacity of the agricultural sector of peripheral States (Figure 8.24, page 342). Unsuccessful agricultural development projects—whether because of poor design, implementation, or some other reason—as well as successful ones illustrate the many ways in which global forces can and do produce different local consequences and reactions from local people. International

Table 8.2 Major Donors and Recipients of Food Aid, 1988–1989	
Major Donors	**Major Recipients**
United States	Bangladesh
West Germany	China
Australia	Dominican Republic
Japan	Egypt
Italy	El Salvador
Argentina	Ethiopia
Saudi Arabia	Guatemala
Austria	Jamaica
France	Kenya
Denmark	Malawi
Switzerland	Mexico
The Netherlands	Morocco
European Community	Mozambique
Ireland	Pakistan
Belgium and Luxembourg	Peru
Canada	Philippines
Spain	Sri Lanka
Sweden	Sudan
Greece	Tunisia
Norway	Vietnam
Finland	
United Kingdom	

Source: Adapted from I. Bowler (Ed.), *The Geography of Agriculture in Developed Market Economies.* London: Longman Scientific and Technical, 1992, p.178–179.

development organizations and institutions like the World Bank and the Food and Agriculture Organization (FAO) have been involved in agricultural development projects in the periphery for nearly five decades.

All too often, however, international development projects have had a dislocating effect on local populations, such that the "solution" to a particular problem has exacerbated—sometimes with horrifying consequences—a situation rather than relieved it. In Kenya, for example, a succession of World Bank–financed livestock schemes have displaced Masai pastoralists, which has meant the loss of livelihood and way of life for thousands of them.

Besides the World Bank and the FAO, hundreds, if not thousands, of international organizations direct their efforts to agricultural development. Although many of these organizations have created new problems in their attempts to shape national development, others have been more successful in their attempts to alleviate poverty and improve basic needs. Importantly, it seems to be at the local scale, with small projects that are sensitive to local cultural (especially gender), social, economic, and political conditions, that development efforts are most often successful. Among the most favorable have been income-generating projects such as the setting up of small agricultural cooperatives for the production of poultry, market vegetables, or even handicrafts for tourist markets.

Even at the local scale, however, development efforts cannot be successful if the innovations are not diffused efficiently throughout the target area. One scientist who was faced with this problem came up with novel ways of distributing

Figure 8.24 Cotton plantation in Burkina Faso Many development projects such as this one have involved the resettlement of people away from more heavily populated areas to less populated ones. These resettled farmers are picking cotton through a project funded by the World Food Program.

green revolution varieties of cassava in Africa. Sang Ki Kahn, Korean Program Director of the International Institute of Tropical Agriculture, drove through the countryside in his truck, personally distributing cassava cuttings (cassava does not grow from seed) to farmers. Able to reach up to 1,000 farmers a day, he became so popular among them, in fact, that the Yoruba made him an honorary chief with the title Seriki Ageb (King of Farmers).

The Organization of the Agro-Food System

While the history of the changes that have occurred in agriculture worldwide are complex, certain elements help simplify the complexity and serve as important indicators of change. Geographers and other scholars interested in contemporary agriculture have noted three prominent and nested forces that signal a dramatic departure from previous forms of agricultural practice: agribusiness, food chains, and integration of agriculture with the manufacturing, service, finance, and trade sectors.

The concept of agribusiness has received a good deal of attention in the last two decades, and in the popular mind it has come to be associated with large corporations, such as ConAgra or DelMonte. Our definition of agribusiness departs from this popular conceptualization. Although multi- and transnational corporations (TNCs) are certainly involved in agribusiness, the concept is meant to convey more than a corporate form. **Agribusiness** is a *system* rather than a kind of corporate entity. Indeed, it is a set of economic and political relationships that organizes food production from the development of seeds to the retailing and consumption of the agricultural product. Defining agribusiness as a system, however, is not meant to suggest that corporations are not critically important to the food production process. On the contrary, it is certainly the case that, in the core economies, the transnational corporation is *the* dominant player operating at numerous strategically important stages of the food production process. TNCs have become dominant for a number of reasons, but mostly because of their ability to negotiate the complexities of production and distribution in many different geographical locations: a capability requiring special knowledge of national, regional, and local regulations and pricing factors.

Agribusiness: a set of economic and political relationships that organizes agro-food production from the development of seeds to the retailing and consumption of the agricultural product.

A food chain (a special type of commodity chain) is a way to understand the organizational structure of agribusiness as a complex political and economic system of inputs; processing and manufacturing; and outputs. The **food chain** is composed of five central and connected sectors (inputs, production, outputs, distribution, and consumption) with four contextual elements acting as external mediating forces (the State, international trade, the physical environment, and credit and finance). Figure 8.25 (see page 344) illustrates these linkages and relationships including how State farm policies shape inputs, product prices, the structure of the farm, and even the physical environment.

Food chain: five central and connected sectors (inputs, production, outputs, distribution, and consumption) with four contextual elements acting as external mediating forces (the State, international trade, the physical environment, and credit and finance).

The food chain concept illustrates the complex connections that exist among and between producers and consumers *and* regions and places. For example, there are important linkages that connect cattle production in the Amazon and Mexico, the processing of canned beef along the United States–Mexico border, the availability of frozen hamburger patties in core grocery stores, and the construction of McDonald's restaurants in Moscow. Because of complex food chains such as this, it is now common to find that traditional agricultural practices in peripheral regions have been displaced by expensive and capital-intensive practices (Figure 8.26, page 345).

In fact, if we take apart the concept of a food chain, we can begin to see the interdependence of the range of activities that constitute the agricultural production and consumption systems. For example, in Figure 8.27 (see page 346), when comparing grassland beef production in New Zealand with grain-based beef production in the United States and the European Union, we can identify the ways in

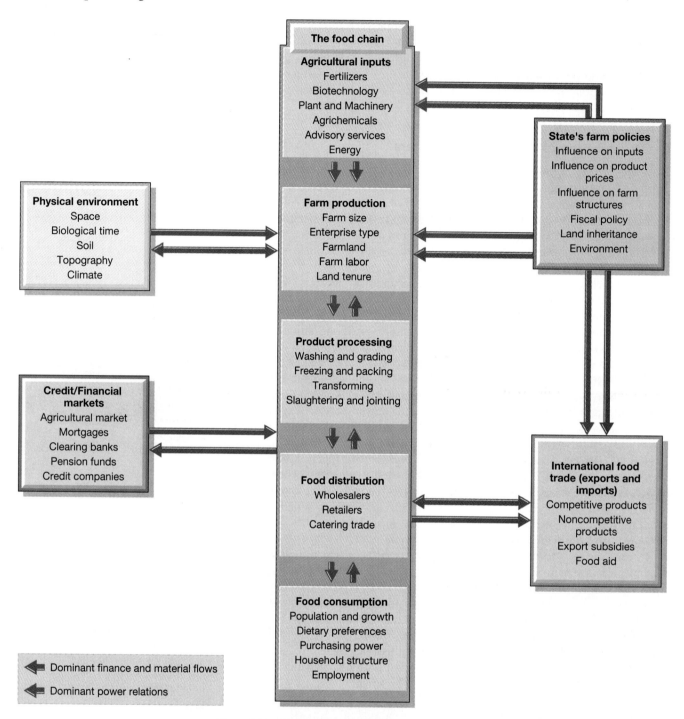

Figure 8.25 The food supply system The production of food has been transformed by industrialization into a complex system that comprises distinctly separate and hierarchically organized sectors. Mediating forces (the State, the structure and processes of international trade, credit and finance arrangements, and the physical environment) influence how the system operates at all scales of social and geographical resolution. (Source: I Bowler (Ed.), *The Geography of Agriculture in Developed Market Economies.* New York: J. Wiley & Sons, 1992, p. 12.)

which the processes differ; the kind of economic competition that exists; the connections that exist among and between political constituencies; and the ways in which geography shapes these processes globally. We can also identify the ways in which the dominant economic sectors—(S)ervice, (I)ndustry, and (A)griculture—interact to constitute a system that is no longer simply about farming, but is instead an integrated, global agro-commodity production system.

That agriculture is not an independent or unique economic activity is not a particularly new realization. Beginning with the second agricultural revolution, agriculture began slowly, but inexorably, to be transformed by industrial practices. What is different about the current state of the food system is the way in which farming has become just one stage of a complex and multidimensional economic process. This process is as much about distribution and marketing—key elements of the service sector—as it is about the growing and processing of agricultural products in the primary sector.

Food Regimes *↙agribusiness.*

A **food regime** is the specific set of links that exists among food production and consumption and capital investment and accumulation opportunities. Like the agricultural revolutions already described, food regimes have developed out of different historical periods where different political and economic forces are in operation. While the *food chain* describes the complex ways in which specific food items are produced, manufactured, and marketed, the concept also indicates the ways in which a particular type of food item is dominant during a specific temporal period. Although hundreds of different food chains may be in operation at any one time, agricultural researchers believe that *only one* food regime dominates a particular period.

The decades surrounding the turn of the nineteenth century were the ones in which an independent system of nation-states emerged and colonization expanded (see Chapters 2 and 9). At the same time, the industrialization of agriculture began. These two forces of political and economic change were critical to the fostering of the first food regime in which colonies became important sources of exportable food stuffs by supplying the industrializing European States with cheap food in the form of wheat and meat. The expansion of the colonial agriculture sectors, however, created a "crisis" in production. The crisis was the result of the higher *cost-efficiency* of colonial food production, which undercut the prices of domestically produced food; put domestic agricultural workers out of work; and forced members of the agricultural sector in Europe to look for new ways to increase cost-efficiency. The response was to industrialize agriculture, which helped to drive down operating costs and restabilize the sector (reducing even more the need for farm workers) while moving toward the integration of agriculture and industry (also known as agro-industrialization).

While a wheat and livestock food regime characterized global agriculture until the 1960s, researchers now believe that a fresh fruit and vegetable regime has emerged. This new pattern of food consumption and production has been called the "postmodern diet" because it represents an important shift away from grains and meats to the more perishable agro-commodities of fresh fruits and vegetables. Integrated networks of food chains, using integrated networks of refrigeration systems, deliver fresh fruits and vegetables from all over the world to the core regions of Western Europe, North America, and Japan. Echoing the former food networks that characterized nineteenth-century imperialism, peripheral production systems supply core consumers with fresh, oftentimes exotic and off-season, produce. Indeed, consumers in the core regions have come to expect the full range of fruits and vegetables to be available year round in their produce sections, and unusual and exotic produce has become increasingly popular.

This emergence of a new food regime based on fresh fruit and vegetables has been helped by retailers who provide symbolic cues and incentives to shoppers to consume the more exotic products. Store managers have introduced them by providing associations between the fruit or vegetable and prevailing ideas about health, class attachment, and epicurean eating. Thus, the transformation of agricultural practices at the global level has enabled the emergence of a new food regime now accompanied by new cultural messages that promote and persuade at

Figure 8.26 Grain-based beef production Cattle at a feedlot in Kenya eat from a trough filled with maize concentrate. Grain-fed beef have become characteristic of beef production not only in the core but in the periphery. These cattle are being raised for domestic consumption by urban, middle-, and upper-class Kenyans.

Food regime: specific set of links that exists among food production and consumption and capital investment and accumulation opportunities.

Grassland-based beef production (e.g., New Zealand)

Grain-based beef production (e.g., U.S., E.C.)

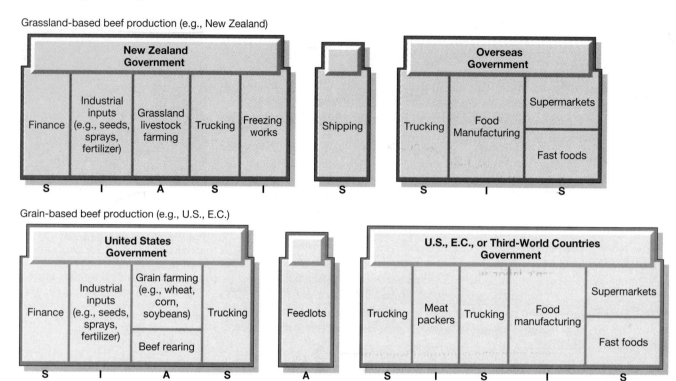

The word system is used to denote the intersection of a range of processes, embracing production to consumption. Designating different spheres of the economy connected with generating "agricultural" outputs are:
A = Agriculture **I** = Industry **S** = Services.

Figure 8.27 Alternative examples of beef chains The progression of beef from pasture to table today constitutes a multidimensional economic process in which distribution and marketing are as important as the traditional activity of growing and processing the product. The beef industry represents a very distinctive example of how a food chain is configured in today's world. (*Source:* R. Le Heron, *Globalized Agriculture: Political Choice 1993.* Tarrytown, NY: Pergamon Press, 1993, p. 46.)

the local level. Furthermore, just as traditional agricultural practices worldwide have been affected by globalization trends, so too are the mainstream eating habits of consumers in core as well as peripheral regions.

Social and Technological Change in Global Agricultural Restructuring

In the previous sections of this chapter, we have tried to show the ways in which the globalization of agriculture has been accomplished through the same kinds of political and economic restructuring that characterized the globalization of industry. Technological change has been of particular importance to agriculture over the last half of the century as mechanical, chemical, and biological revolutions have altered even the most fundamental of agricultural practices. And, just as restructuring in industry has not occurred without innumerable rounds of adjustment and resistance, the same is true of agriculture.

Besides generating economic competition, the newly restructured agro-commodity production system also fosters conflict and competition within socio-cultural systems. For instance, in core locations and peripheral locations, men and women, landowners and peasants, different tribal groups, corporations, and family farmers struggle to establish or to maintain control over production and over ways of life.

Two Examples of Social Change

The impact of a government development scheme to introduce irrigated rice production into the Gambia River Basin illustrates the many ways in which this globalization of agricultural production has affected gender relations among and within households. The Gambian government, with the help of an international development agency, launched a program to grow rice along the banks of the Gambia River, where it had never been traditionally grown.[2] The objective of the project was to develop its own rice-producing sector and thereby decrease its dependence on imported rice. Through local agents employed by the project, the government distributed a (green revolution) package of high-yielding rice varieties, fertilizers, and pesticides with the hope that 2,000 peasant households distributed among 70 villages could attempt a double-cropping rice cultivation program (Figure 8.28). Husbands and wives were involved in the project.

The success of the project required a redistribution of labor as well as the restructuring of land and crop rights. Incorrect assumptions about the availability and cost of women's labor were made, however. These assumptions led to serious problems between spouses when it became apparent that women were not free to work during the season they were most needed to participate in the development scheme. Because of these changes, traditional ways of farming as well as gender relations were significantly challenged. Husbands and wives disagreed to such an extent over who controlled which pieces of land and the crops that were harvested, that the success of the project was compromised.

Lest it seem that only the economies and societies of the periphery are affected by the globalization of agriculture, a case closer to home can be contrasted with the Gambian case. In the 1980s the United States experienced what has become known as the **farm crisis**, the financial failure and eventual foreclosure of thousands of family farms across the Midwest. Caught up in the farm crisis, the small town of Lexington, Nebraska, became the site of community tension and conflict.[3] The agricultural sector of the state of Nebraska, highly specialized in grain production, was particularly hard hit when the international grain market collapsed in 1985. Soaring land prices, a decline in manufacturing employment, a rapid rise in the number of farm bankruptcies, and decreasing revenues from crops sent entrepreneurs looking for ways to overcome the impact of this event on the state's economy. Meatpacking was identified as an alternative generator of economic growth. State-provided tax abatements and other corporate incentives caused IBP, Inc., a giant food conglomerate, to locate a meatpacking plant in Lexington, at the time a small rural town of 6,601 people.

IBP, Inc. opened its plant at the end of 1992 with more than 2,000 workers employed. Most of these workers were immigrants, primarily Mexican and Central American men and women who were actively recruited by the company as a cost-reduction strategy. Such strategies are widespread among large, American meatpacking companies. The arrival of so many new residents to the town created social and economic problems, however, which led to ethnic tensions and resentment among and between the established residents—mostly of European descent—and the newcomers.

Figure 8.28 Gambian Rice Production Pictured in this photograph are Gambian women harvesting rice. Development schemes have created problems within Gambian households, as women want to control their own time and labor to produce rice for family consumption and local markets.

Farm crisis: the financial failure and eventual foreclosure of thousands of family farms across the U.S. Midwest.

---

[2]Adapted from J. Carney, "Converting the Wetlands, Engendering the Environment: The Intersection of Gender with Agrarian Change in The Gambia." *Economic Geography* 69(4), 1993, pp. 329–349.

[3]Adapted from L. Gouveia, "Global Strategies and Local Linkages: The Case of the U.S. Meatpacking Industry," in A. Bonanno et al. (Eds.), *From Columbus to ConAgra: The Globalization of Agriculture and Food.* Lawrence: University of Kansas Press, 1994, pp. 125–148.

After IBP, Inc. arrived in Lexington, the town experienced a severe housing crunch; had to build its first homeless shelter; had to raise new monies to expand the capacities of the school system; and had to build a new and larger jail. A rapid increase in the number of births, especially those not covered by health insurance, also occurred. In 1993, Lexington had the highest crime rate in Nebraska.

Examining the farm crisis is a useful way of demonstrating how core economies—especially farm households—experienced and responded to the changes in American agriculture. While Hollywood movies such as *Country* provided popular dramatizations of the personal impacts of the restructuring of American agriculture, U.S. government statistics provide a less emotionally charged but perhaps a more sobering account. Just as the Gambian government's agricultural development policies and practices transformed gender relations and heightened household tensions, so too did the U.S. government's (See Feature 8.2, "Geography Matters—The U.S. Farm Crisis," page 352).

Biotechnology Techniques in Agriculture

Ever since the nineteenth century, when Austrian botanist Gregor Mendel identified hereditary traits in plants and French chemist Louis Pasteur uncovered fermentation, the manipulation and management of biological organisms has been of central importance to the development of agriculture. The most recent manifestation of the influence of science over agriculture is exemplified by biotechnology. **Biotechnology** is any technique that uses living organisms (or parts of organisms) to make or modify products, to improve plants and animals, or to develop microorganisms for specific uses. Recombinant DNA techniques, tissue culture, cell fusion, enzyme and fermentation technology, and embryo transfer are some of the most talked-about aspects of the use of biotechnology in agriculture (Figure 8.29).

The most common argument for applying biotechnology to agriculture is the belief that it helps reduce agricultural production costs as well as act as a kind of resource-management technique (where certain natural resources are replaced by manufactured ones). Biotechnology has been hailed as a way to address growing concern for the rising costs of cash crop production; surpluses and spoilage; environmental degradation from chemical fertilizers and overuse; soil depletion; and other related sorts of challenges now facing profitable agricultural production.

Indeed, biotechnology has provided impressive responses to these and other challenges. For example, biotechnological research is responsible for the development of *super plants* that produce their own fertilizers and pesticides; can be grown on nutrient-lacking soils; are high-yielding varieties (HYV); and are resistant to disease or the development of microorganisms. Additionally, biotechnologists have been able to *clone,* that is, take the cells of tissues from one plant and use them to form new ones. That tissue culture may be no more than one cubic centimeter in size but it has the potential to produce millions of identical plants. Such a procedure decreases the time needed to grow mature plants ready for reproduction.

While such technological innovations can be seen as miraculous, there is a downside to biotechnological solutions to agricultural problems. For example, cloned plants are more susceptible to disease than natural ones, probably because they have not developed tolerances. This susceptibility leads to an increasing need for chemical treatment. And while industry may reap economic benefits from the development and wide use of tissue cultures, farmers may suffer because they lack the capital or the knowledge to participate in biotechnological applications.

It is no understatement that biotechnology has revolutionized the way in which traditional agriculture has been conducted. Its proponents argue that it

Biotechnology: technique that uses living organisms (or parts of organisms) to make or modify products, to improve plants and animals, or to develop microorganisms for specific uses.

Figure 8.29 Biotechnology
Biotechnology laboratories are typically high-technology greenhouses. Biotechnology possesses both benefits and costs. While the benefits include increased yields and more pest-resistant strains, too often the costs of such technology are too high for the world's neediest populations.

provides a new pathway to the sustainable production of agricultural commodities. By streamlining the growth process with such innovations as tissue cultures, disease and pest-resistant plants, and fertilizer-independent plants, the optimists believe that the biorevolution can maximize global agricultural production to keep up with global requirements of population and demand.

Just as with the green revolution, biotechnology may have deleterious effects for peripheral countries (and for poor laborers and small farmers in core countries). For example, biotechnology has enabled the development of plants that can be grown outside of their natural or currently most suitable environment. Yet cash crops are critical to economic stability for many peripheral nations—such as bananas in Central America and the Caribbean, sugar in Cuba, and coffee from Colombia and Ethiopia (Figure 8.30). These and other export crops are threatened by the development of alternate sites of production. Transformations in agriculture have ripple effects throughout the world system. As an illustration, Table 8.3 (see page 350) compares the different impacts of the biorevolution and the green revolution on various aspects of global agricultural production.

In addition, the availability of technology to these peripheral nations is limited because most advances in biotechnology are the property of private companies. For example, patents protect both the process and the end products of biotechnological techniques. Utilizing biotechnological techniques requires paying fees for permission to use them, and the small farmers of both the core and the periphery are unlikely to be able to purchase or utilize the patented processes. The result of private ownership of biotechnological processes is that control over food production is removed from the farmer and put into the hands of biotechnology firms. Under such circumstances it becomes possible for world food security to be controlled not by publicly accountable governments but by privately held biotechnology firms. Finally, with the refinement and specialization of plant and animal species, those laborers who are currently employed in ancillary activities could face the loss of their jobs. For example, if a grower chooses to plant a bioengineered type of wheat that does not require winnowing (the removal of the chaff, a normally labor-intensive process), then those laborers who once were involved in that activity are no longer needed.

The biorevolution in agriculture is so recent that we are just beginning to understand both its negative and positive impacts. At this point it seems quite clear that these impacts will be distributed unevenly across countries, regions, and locales, and certainly across class, race, and gender lines. It is still too soon to tell what the overall costs and benefits will be.

Figure 8.30 Coffee plantation in Ethiopia For many peripheral countries, the production of cash crops is a way to boost exports and bring in needed income for the national economy. In Ethiopia, coffee has for decades been a cash crop grown for export. Luxury exports such as coffee generate some of the capital needed to import staple foods such as wheat.

TABLE 8.3 Biorevolution Compared with Green Revolution

Characteristics	Green Revolution	Biorevolution
Crops affected	Wheat, rice, maize	Potentially all crops, including vegetables, fruits, agro-export crops, and specialty crops
Other sectors affected	None	Pesticides, animal products, pharmaceuticals, processed food products, energy, mining, warfare
Territories affected	Some developing countries	All areas, all nations, all locations, including marginal lands
Development of technology and dissemination	Largely public or quasipublic sector, International Agricultural Research Centers (IARCs), R & D millions of dollars	Largely private sector, especially transnational corporations, R & D billions of dollars
Proprietary considerations	Plant breeders' rights and patents generally not relevant	Genes, cells, plants, and animals patentable as well as the techniques used to produce them
Capital costs of research	Relatively low	Relatively high for some techniques, relatively low for others
Access to information	Relatively easy, due to public policy of IARCs	Restricted due to privatization and proprietary considerations
Research skills required	Conventional plant breeding and parallel agricultural sciences	Molecular and cell biology expertise plus conventional plant-breeding skills
Crop vulnerability	High-yielding varieties relatively uniform; high vulnerability	Tissue culture crop propagation produces exact genetic copies; even more vulnerability
Side-effects	Increased monoculture and use of farm chemicals, marginalization of small farmer, ecological degradation	Crop substitution replacing Third World exports; herbicide tolerance; increasing use of chemicals; engineered organisms might affect environment; further marginalization of small farmer

Sources: M. Kenney and F. Buttel, "Biotechnology: Prospects and Dilemmas for Third-World Development," *Development and Change* 16, 1995, p. 70, and H. Hobbelink, *Biotechnology and the Future of World Agriculture: The Fourth Resource.* London: Red Books, 1991.

The Environment and Agricultural Industrialization

Agriculture always involves the interaction of both biophysical as well as human systems. In fact, it is this relationship that makes agriculture distinct from other forms of economic activity that do not depend so directly on the environment to function. It is also this interactive relationship that requires attention as to how best to manage the environment in order to enable the continued production of food. Since the relationship between the human system of agriculture and the biophysical system of the environment are highly interactive, it is important to look at the various ways that each shapes the other.

The Impact of the Environment on Agriculture

Management of the environment by farmers has been steadily increasing over the course of the three agricultural revolutions. In fact, the widespread use of fertilizers, irrigation systems, pesticides, herbicides, and industrial greenhouses suggests that agriculture has become an economic practice that can ignore the limitations of the physical environment (Figure 8.31). Yet, it is exactly because agriculture is an economic activity that management of the environment in which it occurs becomes critically important. As geographer Martin Parry writes:

> Soil, terrain, water, weather and pests can be modified and many of the activities through the farming year, such as tillage and spraying, are

Figure 8.31 Modern irrigation system Irrigation is just one way that humans have been able to alter the environment to serve their agricultural needs. In many parts of the core, water prices are heavily subsidized for agricultural users in order to ensure food supplies. For many parts of the periphery, however, access to water is limited to the amount of rain that falls and can be stored behind small dams and in impoundments.

directed toward this. But these activities must be cost-effective; the benefits of growing a particular crop, or increasing its yield by fertilizing, must exceed the costs of doing so. Often such practices are simply not economic, with the result that factors such as soil quality, terrain and climate continue to affect agriculture by limiting the range of crops and animals that can profitably be farmed. In this way the physical environment still effectively limits the *range* of agricultural activities open to the farmer at each location.[4]

While the impact of the environment on agricultural practices that have become heavily industrialized may not at first glance appear obvious, the reverse is more readily observable. In fact, numerous contemporary and historical examples exist of the ways in which agriculture destroys, depletes, or degrades the environmental resources on which its existence and profitability depend.

The Impact of Agriculture on the Environment

One of the earliest treatises on the impact of chemical pesticides on the environment is Rachel Carson's *Silent Spring*, which identified the detrimental impact of synthetic chemical pesticides—especially DDT—on the health of human and animal populations. Although the publication of the book and the environmental awareness that it generated led to a ban on the use of many pesticides in most industrialized nations, chemical companies continued to produce and market them in less-developed countries. While some of these pesticides were effective in combating malaria and other insect-borne diseases, many were applied to crops that were later sold in developed countries' markets. Thus a kind of "circle of poison" was set into motion encompassing the entire global agricultural system (Figure 8.32).

One of the most pressing issues facing agricultural producers today is soil degradation and denudation, which are occurring at rates more than a thousand times the natural erosion rates. Although we in the United States tend to dismiss soil erosion as a historical problem of the 1930s Dust Bowl, in reality the effects of agriculture on worldwide soil resources are dramatic, as Table 8.4 illustrates.

Figure 8.32 Impact of pesticides Pesticides have been shown to have highly damaging impacts on the ecosystem. In addition to fostering pests grown more resistant to chemical suppression, pesticides can also kill and cause genetic damage in larger animals, especially birds. For example, one disorder linked to pesticide use is that the shells of bird eggs, such as the one shown below, are not thick enough to remain intact and protect the embryo through the various stages of maturation. The thin shells crack open prematurely, exposing the embryo before it is viable. In the 1960s, many of the most noxious chemicals, such as DDT, were banned in the United States. Many peripheral countries, however, continue to allow the sale of such chemicals.

---

[4]M. Parry, "Agriculture As a Resource System," in I. Bowler (Ed.), *The Geography of Agriculture in Developed Market Economies*. Harlow, England: Longman Scientific and Technical, 1992, p. 208.

8.2 Geography Matters

The U.S. Farm Crisis

On March 4, 1985, farmers marched on Washington, DC, to demand both higher guaranteed prices for farm products, and stricter control over production. They set up 250 white crosses on the Washington Mall, representing the number of farms they believed were going bankrupt every day. Each cross was inscribed with the name of a victim of the farm crisis (Figure 8.2.1). On May 7, 1985, actresses Jessica Lange, Jane Fonda, and Sissy Spacek spoke before a meeting of Democratic representatives of the U.S. House of Representatives. All three had starred in recent movies about farmers caught up in an economic crisis over which they had little control (Figure 8.2.2). On September 22, 1985, country-and-western singer Willie Nelson, strongly supported by other performers such as John Mellencamp and Neil Young, launched the first of several "Farm Aid" concerts to raise public awareness of the crisis and to raise money to help the farmers weather the bad times. In all of these efforts, the theme was to "save the family farm" as a potent symbol of American history, tradition, and values that many believed was under siege. But what, exactly, was the problem?

To understand the origins and nature of the farm crisis, we must go back to the agricultural boom of the 1970s, when rising land values, low interest rates, and high levels of commodity sales—mainly in exports to other countries—swelled farm incomes. Many farmers went heavily into debt to buy more land, believing the boom would never end. Certain policies of the federal government—allowing for high rates of inflation, encouraging farmers to plant every square inch of their lands, and promoting agricultural commodity exports as a way of dealing with the country's negative international trade balance—also influenced farmers' behavior. By 1980, however, many U.S. farmers were in a very bad position due to soaring interest rates brought on by an abrupt change in Federal Reserve monetary policies in 1979;

declining exports prompted by heavy debt loads in the periphery; increasing competition from the European Economic Community and even from less-developed countries like Argentina; falling land values aggravated by farmers' heavy debt-to-asset ratios; and falling crop prices brought on by over-production. Many farmers saw lenders foreclose on their properties, and scenes of teary-eyed farm families witnessing public auctions of their assets appeared on national television news. Drug and alcohol abuse, child and spousal abuse, and suicide rates all rose at alarming rates.

Figure 8.2.1 Farmers go to Washington Launching a more dramatic and heartfelt objection to the rising number of farm foreclosures, farmers themselves took to the streets to protest. Pictured here is a parade of farm equipment motoring down Pennsylvania Avenue to the White House. Angry protests directed at the U.S. government's changing agricultural policies were not unusual during the mid-1980s as small farmers watched as their farms were foreclosed and bought up by larger-scale producers.

Unfortunately, most forms of agriculture tend to increase natural erosion. Although severe problems of soil erosion exist in the United States—which has a federal agency devoted exclusively to managing soil conservation—more severe problems are occurring in peripheral countries.

The loss of topsoil worldwide is a critical problem because it is a fixed resource that cannot be readily replaced. As it takes, on average, between 100 and 500 years to generate 10 millimeters (one-half an inch) of topsoil, and as it is estimated that nearly 50,000 million metric tons (55,000 million tons) of topsoil are lost each year to erosion, the quantity and quality of soil worldwide is an important determining factor in the quantity and quality of food that can be produced.

Soil erosion due to mismanagement in the semiarid regions of the world has led to desertification where topsoil and vegetation loss have been extensive and

Figure 8.2.2 Star Testimony Following the production of several Hollywood films dealing with the U.S. farm crisis, several Hollywood stars, including Sissy Spacek, Jane Fonda, and Jessica Lange, testified before the U.S. Congress to protest farm foreclosures. Despite the drama of these films and the sincerity of the stars' testimony, the foreclosures continued.

The problem was not unique to the United States. Like American farmers, European farmers were also under stress from falling commodity prices and increased competition. Pressure also came from the dissemination of the green revolution across the globe (discussed in Feature 8.1), which not only enhanced the capability of many less-developed countries to increase production of basic foods such as wheat, rice, and maize, but also enabled some of these countries to become serious competitors on the world food market.

The farm crisis in the United States, however, had some unique characteristics in terms of who was most affected, and what regions were most stressed. The crisis hit hardest at the middle-sized farmers (those whose annual farm sales averaged between $40,000 and $200,000), and the younger farmers (those most willing to take high financial indebtedness risk) more than older farmers. Geographically, the midwestern states were dis-

proportionately affected by the crisis, with Illinois, Iowa, Kansas, Michigan, Missouri, Nebraska, North Dakota, South Dakota, Wisconsin, and Minnesota bearing the brunt of the negative consequences. At the local level, the crisis took its toll on neighboring towns as well as on the farms, for whenever farmers quit the business or reduced their spending and paid less in local taxes, the effects rippled out into adjacent communities, affecting everything from the profitability of local "Main Street" businesses and solvency of local banks to the viability of local schools and churches.

Many commentators argued that the whole federal system of farm policies, price supports, and so on, that had been developed over the previous 50 years needed to be radically revamped—and in many cases eliminated in favor of free market operations—in order to meet the challenges posed by the integration of U.S. agriculture into the world system. In the end, caught between the fiscal conservatism of the Reagan administration and militant demands for support and protection voiced by their farm constituents, members of Congress did little more than enact a series of piecemeal "fixes." These measures may have defused the immediate crisis but did little or nothing to address the future crises certain to occur. In the long run, whether the "family farm" will continue to coexist with a globalized agriculture remains to be seen; many analysts predict that farming will come to be characterized by a relatively few large, productive, and profitable businesses and a number of small operators who earn their main livelihood in nonfarm occupations. Will the family farm remain a cherished image in American iconography even if it disappears from the landscape?

Sources: R. E. Long (Ed.), *The Farm Crisis.* New York: H. W. Wilson, 1987; D. Goodman and M. Redclift (Eds.), *The International Farm Crisis.* New York: St. Martin's Press, 1989.

largely permanent. Desertification is the spread of desert-like conditions in arid or semiarid lands resulting from climatic change or human influences. Desertification not only means the loss of topsoil but can also involve the deterioration of grazing lands and the decimation of forests (Figure 8.33, page 354). In addition to causing soil degradation and denudation problems, agriculture also affects water quality and quantity, through the overwithdrawal of groundwater and the pollution of the same water through agricultural runoff contaminated with herbicides, pesticides, and fertilizers. Deforestation, discussed in Chapter 4, can also result from poor agricultural practices.

Poor land-use practices and the destruction of complex ecosystems through over- or misuse led, in the 1980s, to an innovation called a "debt-for-nature swap." In these swaps a core environmental organization, such as the World

TABLE 8.4 Global Soil Resource Problems (million hectares)

Region	Overgrazing	Deforestation	Agricultural Mismanagement	Other	Total	Degraded Area As Share of Total Vegetated Land
Asia	197	298	204	47	746	20%
Africa	243	67	121	63	494	22%
South America	68	100	64	12	244	14%
Europe	50	84	64	22	220	23%
North & Central America	38	18	91	71	158	8%
Oceania	83	12	8	0	103	13%
World	679	579	552	155	1,965	17%

Source: L. R. Brown et al., *State of the World.* New York: W. W. Norton and Company, 1994, p. 10.

Wildlife Fund, retired some part of the foreign debt of a peripheral country. The debt purchase was made contingent upon the peripheral country agreeing to implement a conservation program to save ecologically sensitive lands from abuse. This usually meant turning the land into a national park or extending the boundaries of an existing park. Funds generated by the swaps were to be used to administer the parks, train personnel, research habitats, and carry out environmental education.

For much of the 1980s, environmental organizations were extremely optimistic about the possible implications of debt-for-nature swaps for both the affected country and worldwide environmental degradation. It turns out, however, that the swaps were not able to address the fundamental causes of environmental degradation in peripheral regions—including extreme poverty, government subsidies for forest clearing, and insecure land tenure. As a result, the debt-for-nature swaps are seen as mere band-aid solutions to extremely complex social, economic, political, and ecological problems.

The nature–society relationship discussed in Chapter 4 is very much at the heart of agricultural practices. Yet as agriculture has industrialized, the impacts of agriculture on the environment have multiplied and in some parts of the globe

Figure 8.33 Desertification Severe, and largely permanent, loss of vegetation and topsoil may result from human activities such as overgrazing or excessive deforestation. The ravaged landscapes of desertification are a compelling testimony to the need for humans to consider the implications of their actions more closely—not always an easy thing to do when ill-informed government policies and grinding hunger and poverty are daily facts of life.

have reached crisis stage. While in some regions the agricultural system leads to overproduction of foodstuffs, in other regions the quantity and quality of water and soil severely limits the ability of a region's people to feed themselves.

Conclusion

Agriculture has become a highly complex, globally integrated system. While traditional forms of agricultural practices such as subsistence farming continue to exist, they have been overshadowed by the global industrialization of agriculture. This industrialization has included not only mechanization and chemical applications but also the linking of the agricultural sector to the manufacturing, service, and finance sectors of the economy. In addition, States have become important players in the regulation and support of agriculture at all levels from the local to the global.

The dramatic changes that have occurred in agriculture have affected different places and different social groups. Households in both the core and the periphery have strained to adjust to these changes, often disrupting existing patterns of authority and access to resources. Just as people have been affected by the transformations in global agriculture, so, too, have the land, air, and water.

The geography of agriculture at the turn of the twentieth century is a far cry from the way it was organized 100 or even 50 years ago. As the globalization of the economy has accelerated in the last several decades of the twentieth century, so too has the globalization of agriculture. The changes in global agriculture do not necessarily mean increased prosperity in the core nor are the impact of these changes simple. As the story that opens this chapter shows, the production of oranges in Florida is directly influenced by the newly emerging Brazilian orange industry. Both industries, in turn, affect the prices of oranges in the marketplaces of Europe and Asia. The U.S. dominance in the citrus market is being directly challenged by an up-and-coming peripheral region, and farmers in Florida will experience this challenge most directly. The long-term effects on the Florida (and the U.S.) economy are as yet indecipherable.

Main Points

- Agriculture has been transformed into a globally integrated system; the changes producing this result have occurred at many scales and have originated from many sources.

- Agriculture has proceeded through three revolutionary phases from the domestication of plants and animals to the latest developments in biotechnology and industrial innovation.

- The State has played an important role in agricultural growth and trade, but the shape of globalized agriculture has been equally influenced by the State's inability to solve the problems of agriculture at either the regional or the national level.

- The introduction of new technologies, political concerns about food security and self-sufficiency, and changing opportunities for investment and employment are among the many forces that have dramatically shaped agriculture as we know it today.

- Although agriculture has been transformed into a globally integrated system, the impact of the new system has not been the same throughout the globe, nor have all people been affected by the transformation in the same way.

- The industrialized agricultural system of today's world has developed from—and has largely displaced—older agricultural practices, including shifting cultivation, subsistence agriculture, and pastoralism.

- The contemporary agro-commodity system is organized around a chain of agribusiness components that begins at the farm and ends at the retail outlet. Different economic sectors, as well as different corporate forms, have been involved in the globalization process.

- Transformations in agriculture have had dramatic impacts on the environment including soil erosion, desertification, deforestation, soil and water pollution, as well as the elimination of some plant and animal species.

Key Terms

agrarian (p. 320)
agribusiness (p. 343)
agricultural industrialization (p. 336)
agriculture (p. 320)
biotechnology (p. 348)
chemical farming (p. 334)
commercial agriculture (p. 322)
crop rotation (p. 322)

double cropping (p. 328)
farm crisis (p. 347)
food chain (p. 343)
food manufacturing (p. 334)
food regime (p. 345)
globalized agriculture (p. 337)
green revolution (p. 337)

hunting and gathering (p. 322)
intensive subsistence agriculture (p. 327)
intertillage (p. 324)
mechanization (p. 334)
pastoralism (p. 328)
shifting cultivation (p. 322)

subsistence agriculture (p. 322)
swidden (p. 324)
transhumance (p. 328)

Exercises

 On the Internet

In this chapter the Internet exercises will explore the relationship between globalization and agriculture. The Internet provides an especially valuable tool for studying this relationship because it contains extensive, worldwide databases on agricultural variables (such as productivity, fertilizer and chemical usage, and distribution of activities). After completing these exercises you should have a better grasp of the geography of agriculture and the ways in which globalization is shaping agricultural practices and patterns.

 Unplugged

1. Your neighborhood grocery store is a perfect location to begin to identify the "global" in the globalization of agriculture. Go to the produce section there and document the source of at least 10 fruits and vegetables you find. You may need to ask the produce manager where they come from, but once you have established that, illustrate those sources on a world map.

2. The Food and Agriculture Organization (FAO) has been publishing a range of yearbooks that provide statistical data on many aspects of global food production since the mid-1950s. Using the *State of Food and Agricultural Production* yearbooks, compare the changes that have occurred in agricultural production between the core and the periphery since mid-century. You may use just two yearbooks for this exercise, or you may want to use several to get a better sense of when and where the most significant changes have occurred. Once you have identified where the changes have been most significant, try to explain why these changes may have occurred.

3. The U.S. Department of Agriculture (USDA) also provides statistics on food and agricultural production, though, of course, limited to the United States. Contained in volumes simply called *Agricultural Statistics*, a range of important variables are included, from what is being grown where, to who is working on farms, and what kinds of subsidies the government is providing. Using the USDA's annual publication of *Agricultural Statistics*, examine the changing patterns of federal subsidies to agriculture over time. Using a map of the United States, show which states since the 1940s (just following the Great Depression) have received subsidies for the decades 1945, 1965, 1985, and the present. Have subsidies increased for some parts of the country and not others? If so or if not, why? Have subsidies increased or decreased overall for the entire country? Which farm sectors and, therefore, which regions have most heavily benefited from federal agricultural subsidies? Why?

Chapter 9

The Politics of Territory and Space

Russian soldiers

Since the end of the cold war, the world political map has changed dramatically. In fact, the entire twentieth century is characterized by dramatic political geographic change. The changing map of Europe, as well as maps of continents such as Africa and Antarctica, are provocative images and telling reminders of the ways in which politics and geography interact. In some cases—for example, the map of Europe following World War II—we can see clearly how the victorious imposed a new political order on the vanquished. In others—for example, the changing shape of Czechoslovakia between 1992 and 1994—we see how the geographic concentration of Czechs in the north and Slovaks in the south led to two new political entities: the Czech Republic and Slovakia (see Figure 9.1).

As this chapter demonstrates, globalization created the world map and has changed it again and again. Exploration, imperialism, colonization, decolonization, and the cold war between East and West are powerful forces that have created national boundaries as well as redrawn them. Much of the civil strife that grips the globe at the turn of the century is a local or regional response to the impacts of globalization of the economy aided by the practices of the State. The complex relationships between politics and geography—both human and physical—are two-way relationships. In addition, political geography is not just about global or international relationships, but is also about the many other geographic scales and political divisions, from the globe to the neighborhood, from large, far-reaching processes to the familiar sites of our everyday lives.

The Development of Political Geography

Political geography is a long-established subfield in the wider discipline of geography. Aristotle is often taken to be the first political geographer because his model of the State is based upon factors such as climate, terrain, and the ratio between population and territory. Since Aristotle, other important political geographers have existed—from Strabo to Montesquieu—who promoted theories of the State that incorporated elements of landscape and the physical environment as well as the population characteristics of regions. From about the fourteenth through the nineteenth century, scholars interested in political geography theorized that the State operated cyclically and organically. What this meant was that States consolidated and fragmented based on complex relationships among and between factors such as population size and composition, agricultural productivity, land area, and the role of the city.

As these factors indicate, political geography at the turn of the century was influenced by two important traditions within the wider discipline of geography: the people–land tradition and environmental determinism. While different theorists placed more or less emphasis on each of these traditions in their own political geographic formulations, the traditions' effects are evident in the factors deemed important to State growth and change. Why these factors were identified as central undoubtedly had much to do with the widespread influence of Charles Darwin on intellectual and social life during this period. Darwin's theory of competition inspired political geographers to conceptualize the State as a kind of biological organism that grew and contracted in response to external factors and forces. It was also during the late nineteenth century that foreign policy as a focus of State activity began to be theorized. This new emphasis came to be called *geopolitics*.

The Geopolitical Model of the State

Geopolitics is the State's power to control space or territory and shape the foreign policy of individual States and international political relations. In Germany, geopolitical theory was influenced by Friedrich Ratzel (1844–1904), a German

Geopolitics: the State's power to control space or territory and shape the foreign policy of individual States and international political relations.

geographer trained in biology and chemistry. Ratzel's greatest influence was Charles Darwin, especially the ideas of *social Darwinism,* which emerged in the mid-nineteenth century. Ratzel employed biological metaphors adopted from Darwin to describe the growth and development of the State as well as seven laws of State growth:

1. The space of the State grows with the expansion of the population having the same culture.

2. Territorial growth follows other aspects of development.

3. A State grows by absorbing smaller units.

4. The frontier is the peripheral organ of the State that reflects the strength and growth of the State; hence it is not permanent.

5. States in the course of their growth seek to absorb politically valuable territory.

6. The impetus for growth comes to a primitive State from a more highly developed civilization.

7. The trend toward territorial growth is contagious and increases in the process of transmission.[1]

Ratzel's model portrays the State as behaving like a biological organism; thus its growth and change are seen as "natural" and inevitable. Although Ratzel advanced his model of the State at the turn of the nineteenth century, his views have continued to influence State theorizing. What has been most enduring about Ratzel's conceptualization is the conviction that geopolitics stems from the interactions of power and territory.

Although it has evolved since Ratzel first introduced the concept, geopolitics has become one of the cornerstones of twentieth-century political geography and State foreign policy more generally. Although adherence to an organic view of the State has been abandoned, the twin features of power and territory still lie at the heart of political geography. In fact, the changes that have occurred in Europe and the former Soviet Union suggest that Ratzel's most important insights about geopolitics are still being played out.

Figure 9.1 (see page 360) portrays Ratzel's conceptualization of the interaction of power and territory through the changing map of Europe following the end of the First World War to the present. In it we see how the fluidity of maps reflects the instability between power and territory, especially some States' failure to achieve stability. The most recent map of Europe is a reflection of the precariousness of nation-state boundaries in the post–cold war period. In fact, the 1996 map of Europe has more in common with the 1924 map than with 1989's. Estonia, Latvia, and Lithuania have returned to their sovereign status. Czechoslovakia has dissolved into the Czech Republic and Slovakia. The former Soviet Union is now a Commonwealth of Independent States, with Russia the largest and most powerful. Yugoslavia has dissolved into four States but not without much civil strife and loss of life. Indeed, the difference between Europe in 1990 and Europe now is, in fact, far more dramatic than any other two maps from the previous 50 years. The maps also illustrate the centrality of territorial boundaries to the operations of the State.

---

[1]Adapted from Martin I. Glassner and Harm Deblij, *Systematic Political Geography,* 3rd ed. New York: J. Wiley & Sons, 1980, p. 164.

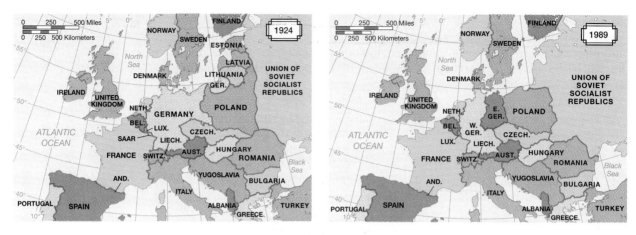

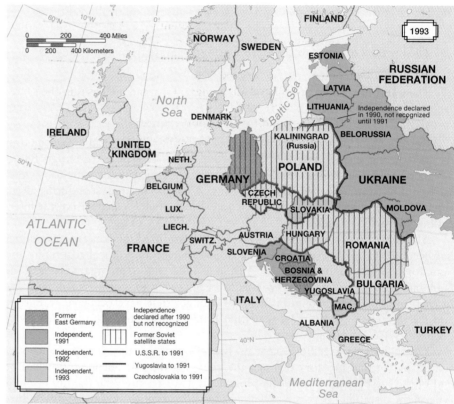

Figure 9.1 The changing map of Europe: 1924, 1989, 1996 The boundaries of the European States have undergone dramatic changes since World War I. The changing map of Europe illustrates the instability of international politics and the resultant dynamism in the geography of the nation-state system. (Source: Reprinted with permission from Prentice Hall, from J. M. Rubenstein, *The Cultural Landscape: An Introduction to Human Geography,* 5th Ed., © 1996, p. 338.)

Boundaries and Frontiers

Boundaries are important phenomena because they allow territoriality to be defined and enforced, and because they allow conflict and competition to be managed and channeled. The creation of boundaries is, therefore, an important element in place making. It follows from the concept of territoriality that boundaries are normally inclusionary (Figure 9.2). That is, they are constructed in order to regulate and control specific sets of people and resources. Encompassed within a clearly defined territory, all sorts of activity can be controlled and regulated—everything, in fact, from birth to death. The delimited area over which a State

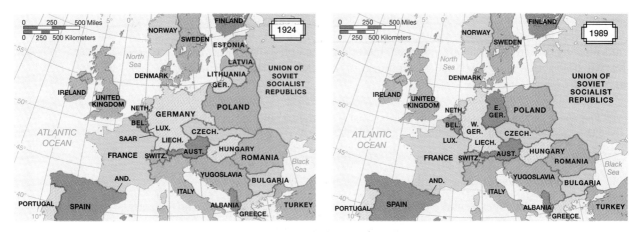

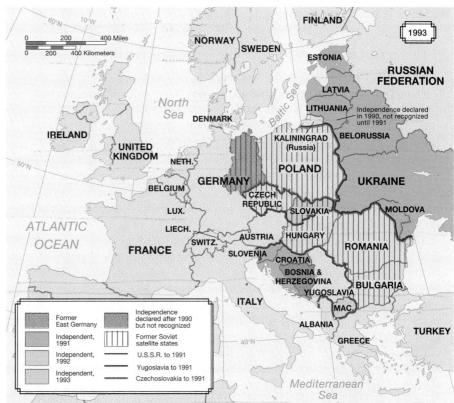

Figure 9.1 The changing map of Europe: 1924, 1989, 1996 The boundaries of the European States have undergone dramatic changes since World War I. The changing map of Europe illustrates the instability of international politics and the resultant dynamism in the geography of the nation-state system. (Source: Reprinted with permission from Prentice Hall, from J. M. Rubenstein, T*he Cultural Landscape: An Introduction to Human Geography,* 5th Ed., © 1996, p. 338.)

Boundaries and Frontiers

Boundaries are important phenomena because they allow territoriality to be defined and enforced, and because they allow conflict and competition to be managed and channeled. The creation of boundaries is, therefore, an important element in place making. It follows from the concept of territoriality that boundaries are normally inclusionary (Figure 9.2). That is, they are constructed in order to regulate and control specific sets of people and resources. Encompassed within a clearly defined territory, all sorts of activity can be controlled and regulated—everything, in fact, from birth to death. The delimited area over which a State

geographer trained in biology and chemistry. Ratzel's greatest influence was Charles Darwin, especially the ideas of *social Darwinism*, which emerged in the mid-nineteenth century. Ratzel employed biological metaphors adopted from Darwin to describe the growth and development of the State as well as seven laws of State growth:

1. The space of the State grows with the expansion of the population having the same culture.

2. Territorial growth follows other aspects of development.

3. A State grows by absorbing smaller units.

4. The frontier is the peripheral organ of the State that reflects the strength and growth of the State; hence it is not permanent.

5. States in the course of their growth seek to absorb politically valuable territory.

6. The impetus for growth comes to a primitive State from a more highly developed civilization.

7. The trend toward territorial growth is contagious and increases in the process of transmission.[1]

Ratzel's model portrays the State as behaving like a biological organism; thus its growth and change are seen as "natural" and inevitable. Although Ratzel advanced his model of the State at the turn of the nineteenth century, his views have continued to influence State theorizing. What has been most enduring about Ratzel's conceptualization is the conviction that geopolitics stems from the interactions of power and territory.

Although it has evolved since Ratzel first introduced the concept, geopolitics has become one of the cornerstones of twentieth-century political geography and State foreign policy more generally. Although adherence to an organic view of the State has been abandoned, the twin features of power and territory still lie at the heart of political geography. In fact, the changes that have occurred in Europe and the former Soviet Union suggest that Ratzel's most important insights about geopolitics are still being played out.

Figure 9.1 (see page 360) portrays Ratzel's conceptualization of the interaction of power and territory through the changing map of Europe following the end of the First World War to the present. In it we see how the fluidity of maps reflects the instability between power and territory, especially some States' failure to achieve stability. The most recent map of Europe is a reflection of the precariousness of nation-state boundaries in the post–cold war period. In fact, the 1996 map of Europe has more in common with the 1924 map than with 1989's. Estonia, Latvia, and Lithuania have returned to their sovereign status. Czechoslovakia has dissolved into the Czech Republic and Slovakia. The former Soviet Union is now a Commonwealth of Independent States, with Russia the largest and most powerful. Yugoslavia has dissolved into four States but not without much civil strife and loss of life. Indeed, the difference between Europe in 1990 and Europe now is, in fact, far more dramatic than any other two maps from the previous 50 years. The maps also illustrate the centrality of territorial boundaries to the operations of the State.

---

[1]Adapted from Martin I. Glassner and Harm Deblij, *Systematic Political Geography,* 3rd ed. New York: J. Wiley & Sons, 1980, p. 164.

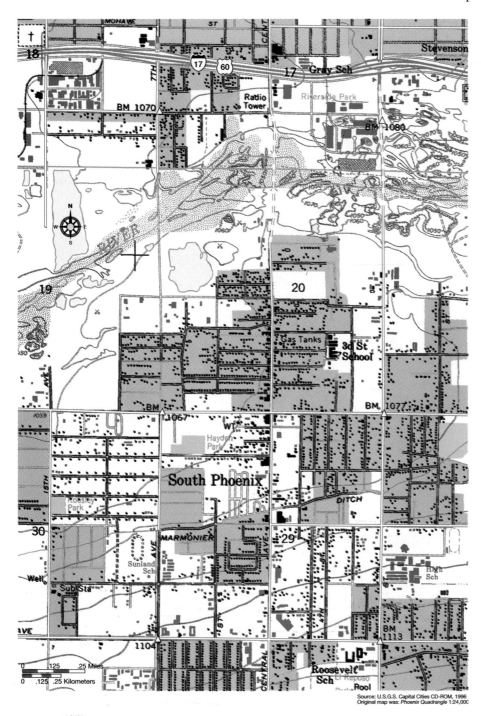

Figure 9.8 Township-and-range system The checkerboard landscapes of the midwestern and western United States can be traced to the U.S. Land Ordinance of 1785 and the Northwest Territories Act of 1803, which divided much of the country into "townships" in order to facilitate the surveying of federal land. Most townships were laid out as squares, oriented North–South and East–West, with sides 8.9 kilometers (six miles) long. Some were given irregular shapes that allowed for natural features such as rivers and mountains. The square-shaped townships were each divided into 36 "sections," each one being one square mile, or 640 acres (259 hectares). Sections, in turn, were each divided into four "quarter-sections." The Homestead Act of 1863, designed to encourage the settlement of the Great Plains, stipulated that the quarter-section would be the basic unit of land ownership, giving each settler 160 acres of land. With this recti-linear pattern of land ownership, it also made sense for the boundaries of admin-istrative areas—counties and states—to be rectilinear.

Source: U.S.G.S. Capital Cities CD-ROM, 1996
Original map was: *Phoenix Quadrangle* 1:24,000

evolved from a loose patchwork of territories (with few formally defined or delimited boundaries) to nested hierarchies (Figure 9.9, page 366) and overlapping systems of de jure territories.

These de jure territories are often used as the basic units of analysis in human geography, largely because they are both convenient and significant units of analysis. They are often, in fact, the only areal units for which reliable data are available. They are also important units of analysis in their own right, because of their importance as units of governance or administration. A lot of regional analysis and nearly all attempts at regionalization, therefore, are based on a framework of de jure spaces.

Figure 9.9 Nested hierarchy of de jure territories De jure territories are constructed at various spatial scales, depending on their origin and function. Administrative and governmental territories are often "nested," with one set of territories fitting within the larger framework of another, as in this example of states, districts, and municipalities in India.

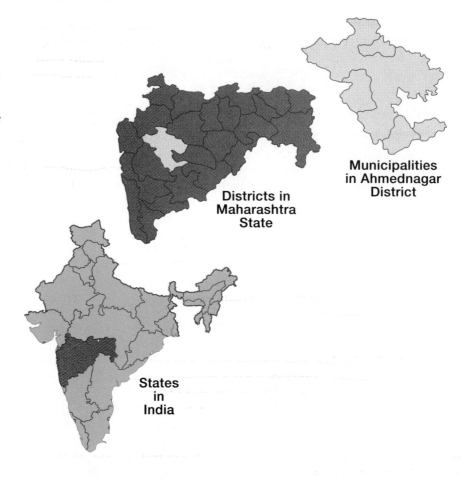

Municipalities in Ahmednagar District

Districts in Maharashtra State

States in India

Geopolitics and the World Order

There is, arguably, no other concept to which political geographers devote more of their attention than the State. The State is one of, if not *the*, most powerful institutions cultivating the process of globalization. The State effectively regulates, supports, and legitimates the globalization of the economy.

States and Nations

As described in Chapter 2, the State is an independent political unit with recognized boundaries, even if some of these boundaries are in dispute. In contrast to a State, a **nation** is a group of people often sharing common elements of culture such as religion or language, or a history or political identity. Members of a nation recognize a common identity, but they need not reside within a common geographical area. For example, the Jewish nation refers to members of the Jewish culture and faith throughout the world regardless of their place of origin. The term **nation-state** is an ideal form consisting of a homogeneous group of people governed by their own State. In a true nation-state, no significant group exists that is not part of the nation. Furthermore, **sovereignty** is the exercise of State power over people and territory, recognized by other States and codified by international law.

In fact, few pure or true nation-states exist today. Rather, multinational States exist—States composed of more than one regional or ethnic group. Spain is such a multinational State (composed of Catalans, Basques, Gallegos, and

Nation: a group of people often sharing common elements of culture such as religion or language, or a history or political identity.

Nation-state: an ideal form consisting of a homogeneous group of people governed by their own State.

Sovereignty: the exercise of State power over people and territory, recognized by other States and codified by international law.

Castilians), as is France, Kenya, the United States, and Bolivia. Multinational States are far more typical than homogeneous nation-states. Since World War I, it has become increasingly common for groups of people sharing an identity different from the majority, yet living within the same political unit, to agitate to form their own State separate from the existing one. This has been the case with the Quebeçois in Canada and the Basques in Spain. It is out of this desire for autonomy that the term "nationalism" emerges. **Nationalism** is the feeling of belonging to a nation as well as the belief that a nation has a natural right to determine its own affairs. Nationalism can accommodate itself to very different social and cultural movements, from the white supremacy of the "Aryan nation" movements in the United States and Europe to the movements for independence in Estonia, Latvia, and Lithuania. The impact of minority nationalism on the world map has been pronounced during the twentieth century (see Feature 9.1, "Geography Matters—Nationalism Around the Globe").

The history and the present status of the former Soviet Union also provides a clear illustration of the tensions among and between States, nations, and nationalism. Demonstrating how enduring are both nationalism and the desire for sovereignty is the history of the Russian Empire: the overthrow of the czar, the subsequent establishment of the Union of Soviet Socialist Republics, and the recent events that have resulted in the establishment of the Commonwealth of Independent States.

The Russian Empire, like other colonial empires in Europe such as Spain and Britain, has a long history. Although it has a medieval origin, most relevant to issues of nationalism is the territorial expansion of the Muscovite State beginning in the fifteenth century. In 1462, Muscovy was a principality of approximately 5,790 square kilometers (15,000 square miles) centered on the present-day city of Moscow. Over a 400-year period, the Muscovite State expanded at a rate of about 20 square kilometers (50 square miles) per day so that by 1914, on the eve of the Russian Revolution, the Empire occupied more than 3.25 million square kilometers (roughly 8.5 million square miles), or one-seventh of the land surface of Earth. The expansion of the Muscovite State was mostly eastward but also to the west and south. This remarkable expansion occurred in discernible stages (Figure 9.10, page 368).

By the end of the fifteenth century, the Muscovite State had annexed most of the other Russian principalities of present-day European Russia. During the sixteenth century, Moscow conquered the non-Russian Tartar States. Desirous of more forest resources—especially furs—Moscow expanded into Siberia. By the mid-seventeenth century Russia had wrested the eastern and central parts of Ukraine from Poland. In the eighteenth century under Catherine the Great, Russia secured the territory of what would eventually become southern Latvia, Lithuania, Belarus, and western Ukraine. The remaining expansion of the Russian State occurred in the late eighteenth and nineteenth century with the acquisition of territory in the Transcaucasus, what is now Kazakhstan, and desert lands to the south bordering on China and Afghanistan.

Russia's imperial expansion followed the same impulses as other European empires. The factors behind expansion were the drive for more territorial resources (especially a warm water port) and additional subjects. Different for Russia, however, was that vast stretches of adjacent land on the Eurasian continent were annexed, whereas other empires established new territories overseas. Strikingly similar was that different nations had to be incorporated into one State—and *not* as colonies. All the problems attendant to that fact came to challenge Russia's consolidation of its extensive empire.

To meet the challenge of different nationalities under one State, Russia needed to apply binding policies and practices. **Centripetal forces** are those that strengthen and unify the State. Centripetal forces are put into place to counter **centrifugal forces**, which are those that divide or tend to pull the State apart.

Nationalism: has been coined to describe the feeling of belonging to a nation as well as the belief that a nation has a natural right to determine its own affairs.

Centripetal forces: forces that strengthen and unify the State.

Centrifugal forces: forces that divide or tend to pull the State apart.

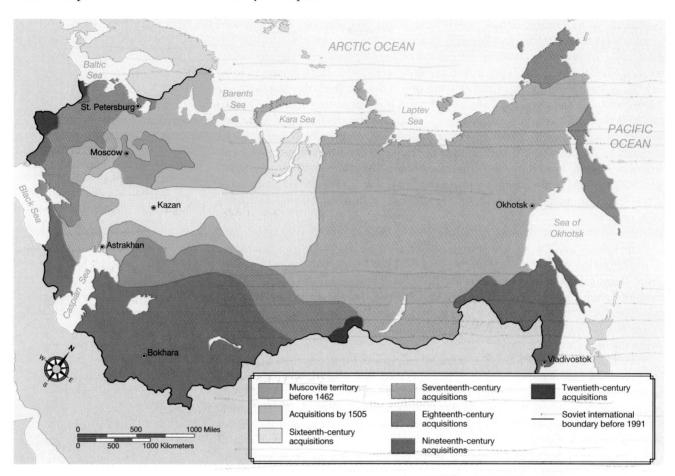

Figure 9.10   Territorial growth of the Muscovite/Russian state   The Muscovite Empire was vast and was conquered over the same period (fifteenth century to the present) that corresponds with the globalization of the world economy. What makes the Russian case different is the fact that the lands conquered were adjacent ones and not overseas. When the Bolsheviks came to power at the beginning of the twentieth century, some of the territory was lost. Eventually, however, the Bolsheviks were able to control most of the territories formerly ruled by the czars, and it was upon this that they also built the Soviet State. (*Source:* D. J. B. Shaw, Ed., *The Post Soviet Republics: A Systematic Geography.* New York: John Wiley & Sons, 1995, p. 164.)

Russia's centripetal strategies to bind together the 100-plus nationalities (non-Russian ethnic peoples) into a unified Russian State were oftentimes punitive and not at all successful. Non-Russian nations were simply expected to conform to Russian cultural norms. Those which did not were more or less persecuted. The result was opposition and, among many if not most of the nationalities, sometimes rebellion and refusal overall to bow to Russian cultural dominance.

 Such was the legacy that Lenin and the Bolsheviks inherited from the Russian Empire following the overthrow of the aristocracy in 1917. The solution to the "national problem" orchestrated by Lenin was *recognition* of the many nationalities through the newly formed Union of Soviet Socialist Republics (USSR). Lenin believed that a *federal system,* with *federal units* delimited according to the geographic extent of ethnonational communities, would ensure political equality among at least the major nations in the new State. This political arrangement recognized the different nationalities and provided them a measure of independence. Federation was also a way of bringing reluctant areas of the former Russian Empire into the Soviet fold. A **federal State** allocates power to units of local government within the country. The United States is a federal State with its system of state, county, and city/town government. A federal State can be contrasted with a

Federal State: a form of government in which power is allocated to units of local government within the country.

9.1 Geography Matters

Nationalism Around the Globe

Europe

1. **Bosnia and Herzegovina**—Serbian forces have captured about 70 percent of the country and carried out an ethnic cleansing campaign that has expelled and killed Muslims and Croats.
2. **Croatia**—Serbian separatists control about a third of Croatia's territory, having killed an estimated 25,000 since Croatia declared independence in 1991.
3. **Spain**—Basque nationalists seek an independent state on the border of France and Spain.
4. **Britain**—The Protestant majority in Northern Ireland seeks continued union with Britain, while the Catholic minority wants to join the rest of Ireland. Scotland seeks independence from Britain.
5. **Germany**—The influx of 650,000 foreigners seeking asylum from Bulgaria, Romania, the Balkans, Turkey, and other areas has stirred up right-wing neo-Nazi attacks.
6. **Romania**—Ethnic Hungarians in Transylvania want greater autonomy; sporadic attacks continue on Gypsies.
7. **Russia**—Chechnya and Ingushetia have broken apart and seek greater autonomy within Russia.
8. **Moldova**—Moldova's mainly Romanian population seeks economic, political, and cultural ties with Romania.
9. **Georgia**—Abkhazia, dominated by Muslims, seeks independence or union with Russia. Southern Ossetia, also dominated by Muslims, seeks union with Northern Ossetia, an autonomous republic in Russia.

Middle East and North Africa

10. **Azerbaijan**—Troops from Muslim Azerbaijan aided by Russian forces, are fighting to end a rebellion by Nagorno-Karabakh, an enclave within Azerbaijan populated largely by Christian Muslims, who favor independence or affiliation with Armenia.
11. **Turkey**—Kurdish separatists represented by the Marxist Kurdish Workers Party have sought a separate Kurdish State.
12. **Iraq**—In the north, two major Kurdish parties rule an enclave protected by the United States and its allies. In the south, Shiite Muslims are under attack from the Sunni-dominated Baghdad government.
13. **Israel**—The intifada, a popular uprising of Palestinians against Israeli occupation of the West Bank of the Jordan River and the Gaza Strip, has been ongoing since 1987.
14. **Algeria**—A revolt by Islamic militants has led to deaths and clashes with government forces.
15. **Egypt**—Clashes and attacks between Islamic militants and government security forces have occurred as well as attacks by militants on foreigners and Coptic Christians.
16. **Sudan**—The government, dominated by Arab Muslims from the north, is fighting a long-standing insurgency by black Christians and animists in the south.

Africa South of the Sahara

17. **Mauritania**—Government security forces under the Arab-dominated regime have clashed with opposition groups who are angry over expulsions and oppression of the black majority.
18. **Mali**—A demand for sovereignty by ethnic Tuaregs, living in both Mali and neighboring Niger, has led to fighting in both places.
19. **Chad**—Clashes continue to occur between the Zakawa and the Gourane tribes.
20. **Somalia**—Clan fighting escalated into full-scale civil war in which hundreds of thousands have died and millions have been made homeless.
21. **Senegal**—In Casamance, a coastal region mostly populated by the Diola tribe, opposition to Muslim domination in the government has surfaced.
22. **Liberia**—Civil war has occurred between the Gio and Mano ethnic groups against the Krahn group.
23. **Togo**—Clashes have occurred between the Kabiye tribe and the Ewe tribe.
24. **Nigeria**—Conflict has flared among many of Nigeria's 200 ethnic groups, but mostly between the Hausas in the north and the Yorubas in the south.
25. **Uganda**—The army, composed mainly of members of the Baganda and Banyarwanda tribes, continues to wage war with northern rebels, mainly from the Acholi and Langi tribes.
26. **Rwanda**—Civil war has broken out between the government-dominated Hutu tribe and the minority Tutsi tribe.
27. **Burundi**—Ethnic clashes persist between the majority ethnic group, the Hutus, and the minority Tutsis.
28. **Kenya**—Clashes among and between tribes have resulted in deaths and displacement.
29. **Democratic Republic of Congo**—Civil war with ethnic overtones is under way due to persistent competition between ethnic groups or tribes.
30. **Angola**—Fighting between government and guerrilla forces has taken on ethnic overtones due to the strong support of guerrillas from disenfranchised ethnic groups.
31. **South Africa**—Despite the "end of apartheid," violence continues between blacks and whites as well as between Zulus and other rival black groups.

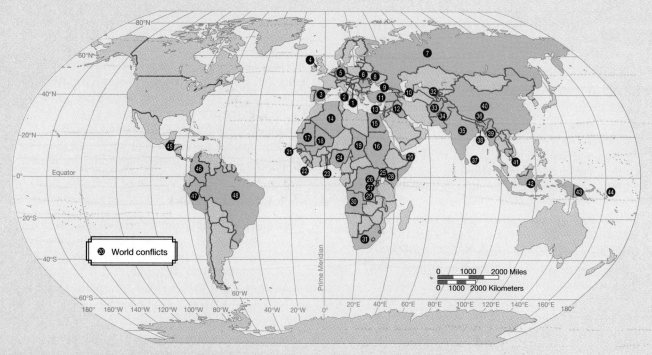

Figure 9.1.1 Global conflicts The map depicts the locations on the globe where nationalism has provoked serious disagreement or outright conflict. While native peoples in North America have agitated for independence at various times (though most continue to defend tribal sovereignty) and although appreciable support for the separation of Quebec from Canada persists, at present the continent of North America is the only one where outright conflict is not presently in evidence.

Asia

32. **Tajikistan**—Tens of thousands of Tajik Muslims have been driven from their land by resurgent communist armies seeking to suppress Islamic political power.

33. **Afghanistan**—Since the withdrawal of Soviet troops in 1992, the country has collapsed into civil war among competing ethnic factions. The Hazars control central and western areas near Iran; the Pathans are largely in control in the east; and the Tajiks largely control the north.

34. **Pakistan**—Conflicts persist between government forces and groups of secessionists and dissidents in Sindh and the Northwest Frontier Province; Rioting in Karachi among Muslims continues to flare up.

35. **India**—Conflict between Hindus and Muslims has beset most of the country. In Punjab, conflicts continue to occur between Hindus and Sikhs. In Nagaland, insurgent Bodos have been fighting for a separate State.

36. **Bhutan**—A revolt by ethnic Nepalese against the government has occurred as well as reprisals by the government.

37. **Sri Lanka**—An insurgency by mostly Hindu Tamils in the north and east has been carried out against the government, which is mostly dominated by Buddhist Sinhalese.

38. **Bangladesh**—A migration by members of the country's Muslim majority into the thinly populated Chittagong Hill Tracts region in the south has led to an insurgency by the area's Chakmas, a mainly Buddhist people.

39. **Burma (Myanmar)**—Clashes between Burmese soldiers and separatist Karen rebels along the Thai/Burmese border have led to death and displacement.

40. **China**—Tibetans rebelled against Chinese rule with tens of thousands killed. China has also recently suppressed a rebellion among Muslims of Turkic descent.

41. **Cambodia**—Khmer Rouge soldiers, who blame Vietnam for many of Cambodia's problems, continue both to carry out attacks on Vietnamese living in Cambodia and to vie for control of the country.

42. **Indonesia**—After a civil war broke out in 1975 in East Timor, Indonesia crushed the pro-independence rebellion among mostly Roman Catholic East Timorese. A separatist movement, which Indonesian forces are attempting to crush, also exists in northern Sumatra.

43. **Papua New Guinea**—After rebels on the island of Bougainvillea declared independence in 1990, the Papua New Guinea government subdued the rebellion.

370

44. **Fiji**—Violence erupted after the Indian-dominated government was elected in 1987. The government was overthrown and the current government consolidates ethnic Fijian dominance.

Latin America

45. **Guatemala**—An essentially political conflict between the government and leftist guerrillas has had ethnic overtones because of the long history of repression of native peoples in Guatemala.
46. **Mexico**—Insurgency in Chiapas State against government forces by native peoples reflects the long history of Mexico's repression of them.
47. **Colombia**—A group representing the rights of native peoples, Quintin Lamee, suspended an armed rebellion in 1991, although other Marxist peasant groups are continuing guerrilla attacks on the government.
48. **Peru**—Since 1980, a Maoist guerrilla group known as the Shining Path has waged war and won control of about a third of Peruvian territory. It has drawn support from largely native or mixed-race populations who are resisting control from the mostly Hispanic elite in Lima.
49. **Brazil**—Native tribes in the Amazon region are pressing the government in Brasilia to recognize their traditional homelands. In the north, the federal government is attempting to expel gold miners from the lands of the Yanomamo tribes.

unitary State in which power is concentrated in the central government. The Russian State under the czar had been a unitary State.

Lenin was optimistic that once international inequalities were diminished, and once the many nationalities became united as one Soviet people, the federated State would no longer be needed. Nationalism would be replaced by communism. Lenin's vision was short-lived and, following his death in 1924, a true federal system in the USSR also declined. Stalin came to power and enforced a new nationality policy, the aim of which was to construct a unified Soviet people whose interests transcended nationality. Although the federal system remained in place, nations increasingly lost their independence and by the 1930s were punished for displays of nationalism. Figure 9.11 (see page 372) shows the administrative units and nationalities that were part of the USSR during Stalin's tenure as premier. Figure 9.11 also shows how, during and immediately after World War II, Stalin expanded the power of the Soviet State eastward to include Albania, Bulgaria, Czechoslovakia, the German Democratic Republic, Hungary, Poland, Romania, and Yugoslavia.

While a federal system remained in place through the administration of Mikhail Gorbachev, the USSR actually operated as a unitary State with power concentrated in Moscow. When Gorbachev came to lead the USSR in 1985, he assumed that the diverse nationalities of the USSR had been "Sovietized," that an international worker orientation had replaced nationalism (see Figure 9.12, page 373). This was not the case, however, and his economic and political policies of *perestroiyka* and *glasnost* created the conditions that encouraged the re-emergence of nationalism in the USSR (Figure 9.13, page 374).

Gorbachev's goal was a massive restructuring of the Soviet economy through radical economic and governmental reforms (*perestroiyka*) and the direct democratic participation of the Union republics in shaping these reforms through open discussions, freer dissemination of information, and independent elections (*glasnost*). Effectively, Gorbachev lifted the restrictions that had been placed on the legal formation of national identity. By 1988, grassroots national movements were emerging, first in the Baltic republics and later in Transcaucasia, Ukraine, and Central Asia. By 1991, the breakup of the Soviet Union was under way, and new nation-states had emerged to claim their independence. In fact, the federated structure that existed under the USSR enabled the relatively peaceful breakup of the Soviet Union. Figure 9.14 (see page 374) is a recent map of the former USSR, including all the newly independent States, renamed the Commonwealth of Independent States (CIS). The CIS is a **confederation**, a group of States united for

Unitary State: a form of government in which power is concentrated in the central government.

Confederation: a group of States united for a common purpose.

Figure 9.11 Soviet State expansionism, 1940s and 1950s World War II gave the Soviet State the opportunity to move eastward for additional territories. Insisting that these countries would never again be used as a base for aggression against the USSR, Stalin retained control over Poland, East Germany, Czechoslovakia, Hungary, Romania, Bulgaria, Albania, and western Austria. In 1945 Stalin promised democratic elections in these territories. After 1946, however, Soviet control over Eastern and Central Europe became complete as noncommunist parties were dissolved and Stalinist governments installed. (Source: *Atlas of Twentieth Century World History.* New York: HarperCollins Cartographic, 1991, p. 86–87.)

a common purpose. These newly independent States have confederated mostly for economic (and to a lesser extent for military) purposes. A similar case is that of the Confederate States of America, the 11 southern states that seceded from the United States between 1860–1861 for economic and political solidarity. This secession led ultimately to the Civil War.

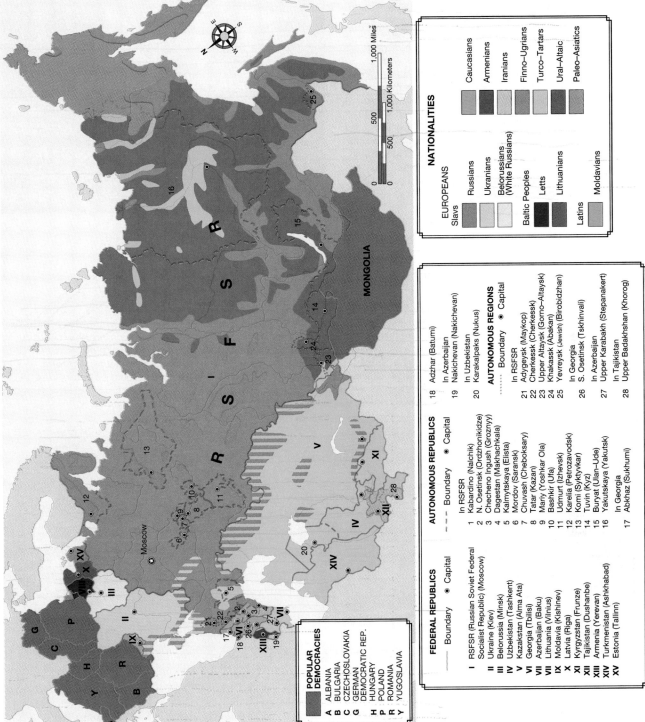

Figure 9.12 Major nationalities of the USSR, 1989

The Soviet State incorporated a vast expanse of territory as well as numerous nationalities. In 1991, just before the fall of the Soviet Union, 120 officially recognized nationalities existed. In order to administer the different regions, the territory was divided into a hierarchical system of federal republics with a nominal degree of autonomy, autonomous republics, and regions that were subunits of the Russian Federation of Soviet States, the largest administrative unit in the USSR. (*Source:* Harper Atlas of World History. New York: HarperCollins, 1992, p. 166–167.)

POPULAR DEMOCRACIES

A ALBANIA
B BULGARIA
C CZECHOSLOVAKIA
G GERMAN DEMOCRATIC REP.
H HUNGARY
P POLAND
R ROMANIA
Y YUGOSLAVIA

FEDERAL REPUBLICS

— Boundary   ● Capital

I RSFSR (Russian Soviet Federal Socialist Republic) (Moscow)
II Ukraine (Kiev)
III Belorussia (Minsk)
IV Uzbekistan (Tashkent)
V Kazakstan (Alma Ata)
VI Georgia (Tbilisi)
VII Azerbaijan (Baku)
VIII Lithuania (Vilnius)
IX Moldavia (Kishinev)
X Latvia (Riga)
XI Kyrgyzstan (Frunze)
XII Tajikistan (Dushanbe)
XIII Armenia (Yerevan)
XIV Turkmenistan (Ashkhabad)
XV Estonia (Tallinn)

AUTONOMOUS REPUBLICS

--- Boundary   ● Capital

In RSFSR
1 Kabardino (Nalchik)
2 N. Osetinsk (Ordzhonikidze)
3 Checheno Ingush (Groznyy)
4 Dagestan (Makhachkala)
5 Kalmytskaya (Elista)
6 Mordov (Saransk)
7 Chuvash (Cheboksary)
8 Tatar (Kazan)
9 Mariy (Yoshkar Ola)
10 Bashkir (Ufa)
11 Udmurt (Izhevsk)
12 Karelia (Petrozavodsk)
13 Komi (Syktyvkar)
14 Tuvin (Kyz)
15 Buryat (Ulan-Ude)
16 Yakutskaya (Yakutsk)

In Georgia
17 Abkhaz (Sukhumi)

18 Adzhar (Batumi)

In Azerbaijan
19 Nakichevan (Nakichevan)

In Uzbekistan
20 Karakalpaks (Nukus)

AUTONOMOUS REGIONS

········· Boundary   ● Capital

In RSFSR
21 Adygeysk (Maykop)
22 Cherkessk (Cherkessk)
23 Upper Altaysk (Gorno-Altaysk)
24 Khakassk (Abakan)
25 Yevreysk (Jewish) (Birobidzhan)

In Georgia
26 S. Osetinsk (Tskhinvali)

In Azerbaijan
27 Upper Karabakh (Stepanakert)

In Tajikistan
28 Upper Badakhshan (Khorog)

NATIONALITIES

EUROPEANS
Slavs
  Russians
  Ukranians
  Belorussians (White Russians)

Baltic Peoples
  Letts
  Lithuanians

Latins
  Moldavians

Caucasians
Armenians
Iranians
Finno-Ugrians
Turco-Tartars
Ural-Altaic
Paleo-Asiatics

MONGOLIA

Moscow

Figure 9.13 Nationalism in Latvia The Baltic Republics (Latvia, Estonia, and Lithuania) were the first to push for re-establishing their sovereignty in the wake of *perestroiyka* introduced by Gorbachev. All three have a history of independence movements that began with the early expansion of the Russian Empire. One obvious reason for their resistance to Russian and later Soviet State incorporation is the fact that all were largely populated by non-Russian people until their forced incorporation into the Soviet Union in 1940. These three republics were independent nation-states from 1918 to 1940. With a well-developed economic base and higher standards of living than most of the rest of the USSR, Latvia, Estonia, and Lithuania were quick to move from asserting their nationalism to demanding separate nation-state status when the former Soviet Union crumbled in 1991.

Figure 9.14 Newly independent states of the former Soviet Union, 1995 For the most part, the administrative structure of the USSR has remained in place since the events of 1991. The differences are that autonomous regions and republics now have more than nominal local control, and the former federal republics have become independent States as have the popular democracies of Eastern and Central Europe. Despite, or perhaps because of, recent democratic reforms, nationalist movements continue to plague the consolidation of the State in autonomous republics like Chechnya. (Source: Reprinted with permission from Prentice Hall, from J. M. Rubenstein, *The Cultural Landscape: An Introduction to Human Geography*, © 1996, p. 318.)

Theories and Practices of States

The definition of the State provided in the previous section is a static one. In fact, the State, through its institutions—such as the military or the educational system—can act to protect national territory and harmonize the interests of its people. We can also say, therefore, that the State is a *set of institutions* for the protection and maintenance of society. Thus it is not only a place—a bounded territory—it is also an active entity that operates through the rules and regulations of its various institutions from governing bodies to social service agencies to the courts. Recognizing that the State has the power to act through its institutions, it is not difficult to see why political geographers and related scholars have advanced theories and models to explain State actions. For political geographers, theories and models of geopolitics have been their most prominent contributions to understanding the role and behavior of the State.

As mentioned previously, perhaps the most important influence on geopolitics was the nineteenth-century German geographer Friedrich Ratzel. Ratzel was a scholar whose organic theory of the State used biological constructs to describe its actions. He believed that States, like living organisms, progress through stages of youth, maturity, old age, and even a possible return to youth. He also believed that one could determine the general well-being of the State by regarding its size as measured in terms of its geographic expansion or contraction over time. Ratzel, though certainly an environmental determinist, did not believe that the State *is* an organism, but only that it *acts* like one, and that States grow—increase their territory—as their populations grow and require more territory.

Ratzel wrote during a period of tremendous change in Europe when the system of States that largely persists to this day was being solidified. It was also a period when Europe, Japan, and the United States were maintaining or initiating imperialist practices. The States' efforts to expand were seen as a response to population pressures for more territory and resources. These pressures also led to the need for new markets for manufactured goods and increased colonization of less economically developed places. Eventually, all-out war within Europe occurred as States fought each other over territories.

Geopolitics, as a theory of the State, was eventually taken up and distorted by German political geographers who transformed it into *"geopolitik."* In Germany, geopolitik became a particularly aggressive, antidemocratic tool used to justify Nazism—racist nationalism and national expansion—and Hitler's extermination of tens of millions of European Jews, Gypsies, Catholics, homosexuals, and other minorities. Geopolitics, through its German adaptation geopolitik, came to be regarded as synonymous with Nazism and as a result was rejected as a useful explanatory concept following the end of World War II. It was not revived until the advent of the cold war, this time under the guise of domino theory discussed later in this chapter.

Imperialism, Colonialism, and the North/South Divide Geopolitics may involve extension of power by one group over another. Two ways in which it may occur are imperialism and colonialism. As was discussed in Chapter 1, imperialism is the extension of State authority over the political and economic life of other territories. As Chapter 2 describes, over the last 500 years, imperialism has resulted in the political or economic domination of strong core States over the weaker States of the periphery. Imperialism does not necessarily imply formal governmental control over the dominated area. It can also involve a process by which some countries pressure the independent governments of other countries to behave in certain ways. This pressure may take many forms such as military threat, economic sanctions, or cultural domination (described in Chapter 5). Imperialism involves some form of *authoritative control* of one state over another.

Figure 9.15 Principal elements in the process of exploration This diagram illustrates the main elements in the process of exploration beginning with a need in the home country, which prompts a desire to look outward to satisfy that need. Geographers have figured prominently in the process of exploration by identifying areas to be explored as well as actually traveling to these far-flung places and providing catalogs of resources and people. Nineteenth-century geography textbooks are records of these explorations and the way in which geographers conceptualized the worlds they encountered. Exploration is one aspect of the process of imperialism. Colonization is another. (*Source:* J. D. Overton, "A theory of exploration," *Journal of Historical Geography* 7, 1981, p. 57.)

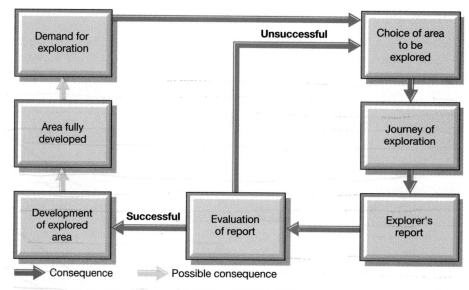

Figure 9.15 provides a theorization of the process of imperialism beginning with exploration and culminating in development via either colonization or the exploitation of people and resources or both. At the beginning of the process, a State perceives a need for exploration. This need is often the result of a scarcity or lack of a critical natural resource.

Generally speaking, in the first phases of imperialism, the core exploits the periphery for raw materials. Later as the periphery becomes developed, colonization may occur, and cash economies are introduced where none have previously existed. The periphery may also become a market for the manufactured goods of the core. Eventually, though not always, the periphery, because of the availability of cheap labor, land, and other inputs to production, can become a new arena for large-scale capital investment. In some cases, it is possible for peripheral countries to improve their status to become semiperipheral or even core countries. Figure 9.16 illustrates European imperialism in Africa.

Colonialism differs from imperialism in that it involves formal establishment and maintenance of rule by a sovereign power over a foreign population through the establishment of settlements. The colony does not have any independent standing within the world system, but it is considered an adjunct of the colonizing power. From the fifteenth to the early twentieth century, colonization constituted an important component of core expansion. Between 1500 and 1900, the *primary* colonizing States were Britain, Portugal, Spain, the Netherlands, and France. Figure 9.17 (see page 378) illustrates the colonization of South America, largely by Spain and Portugal and to a very limited extent by Italy.

Other important States more recently involved in both colonization and imperialist wars include the United States and Japan. Although it is often the case that colonial penetration results in political dominance by the colonizer, such is not always the case. For example, Britain may have succeeded in setting up British colonial communities in China, but it never succeeded in imposing British administrative or legal structures in any widespread way. And, at the end of the colonial era, a few colonies, such as the United States, Canada, and Australia, eventually became core States themselves. Others such as Rwanda, Bolivia, and Cambodia remain firmly within the periphery. Some former colonies, such as Mexico and Brazil, have come close, but have not fully attained core status, and therefore are categorized as being within the semiperiphery. Two examples of the colonization process are the extension of British rule to India and French rule to Algeria.

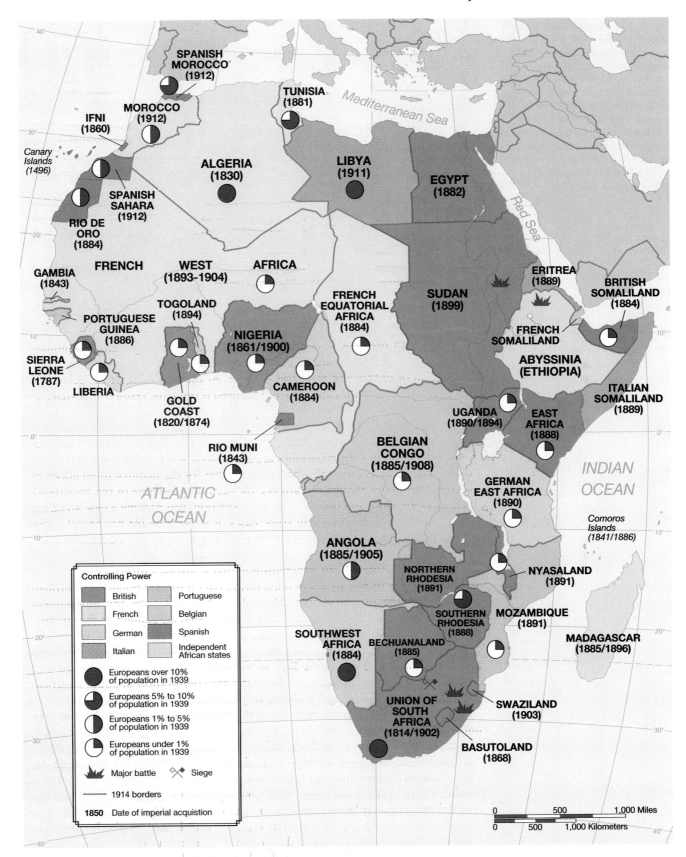

Figure 9.16 European imperialism in Africa, 1418–1912 The partitioning of the African continent by the colonial powers created a crazy quilt that cross-cut preexisting affiliations and alliances among the African peoples. Lying directly within easy reach of Europe, Africa was the most likely continent for early European expansion. The Belgian, Italian, French, German, and Portuguese States all laid claim to various parts of Africa and in some cases went to war to protect those claims. (*Source: Harper Atlas of World History.* New York: HarperCollins, 1992, p. 139.)

Figure 9.17   Colonization in South America, 1496–1667   The Spanish and Portuguese dominated the colonization and settlement of South America, with the Dutch, French, Italians, and English maintaining only a minor and largely tentative presence. While African colonization focused upon new subjects and the simple acquisition of additional territories, South American colonization yielded rich commodity and mineral returns. (*Source: Rand McNally Atlas of World History.* Skokie, IL: Rand McNally, 1992, p. 85.)

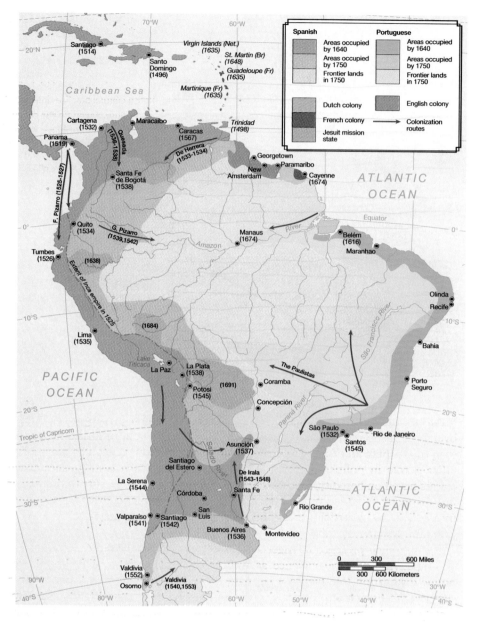

The substantial British presence in India began with the establishment of the East India Trading Company in the mid-eighteenth century. The British government gave the company the power to establish forts and settlements as well as to maintain an army. The company soon established settlements—including factories—in Bombay (now Mumbai), Madras (now Chennai), and Calcutta. What began as a small trading and manufacturing operation over time burgeoned into a major military, administrative, and economic presence by the British government, which did not end until Indian independence in 1947 (Figure 9.18). During that 200-year period the Indian population was brutalized and killed, and their society transformed by British influence. That influence permeated nearly every institution and practice of daily life—from language and judicial procedure to railroad construction and cultural identity.

The postcolonial history of the Indian subcontinent has included partition and repartition as well as the eruption of regional and ethnic conflicts. Regional conflicts include radical Sikh movements for independence in the states of Kashmir and Punjab. Ethnic conflicts include decades of physical violence

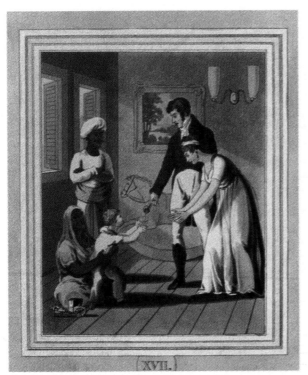

Figure 9.18 British colonialism in India The British presence in India affected culture, politics, the economy, and the layout of cities as well as numerous other aspects of everyday life. This painting illustrates the way in which Indian and British cultural practices intermingled. Importantly, Indian society absorbed and remolded many British political and cultural practices so that contemporary Indian government, for example, is a hybrid of British and Indian ideals and practices.

between Muslims and Hindus over religious beliefs and the privileging of Hindus over Muslims in the national culture and economy. It would be misleading, however, to attribute all of India's current strife to colonialism. Caste also plays a significant role in political conflict. The caste system, which distinguishes social classes based on heredity, preceded British colonization and persists to this day. It is a distinctly Hindu institution that perpetuates racism and related discriminatory beliefs and practices. In Hindu society there are four major social classes: Brahmin, Kshatriya, Vaisya, and Sudra.

The French presence in Algeria is a story of 132 years of colonization with accompanying physical violence as well as cultural, social, political, and economic dislocation (see Figure 9.16, page 377). Over the course of more than a century, the French appropriated many of the best agricultural lands and completely transformed Algiers, the capital, into a Westernized city. They also imposed a veneer of Western religious and secular practice over the native and deeply rooted Islamic culture.

In the cases of both India and Algeria, achieving independence has been a painful and sometimes bloody process for both colonizer and colonized. In Algeria, for example, in the aftermath of the protracted war of independence, there occurred a major exodus of French settlers, many of whose families had lived in Algeria for several generations. With few Westerners in positions of power, important aspects of Islamic society and culture were recuperated. At the same time, Western-oriented politicians, bureaucrats, and citizens have been threatened and killed by radical Islamists who seek to return Algeria to a State governed by strict religious tenets (see Chapter 5 on cultural nationalism).

Since the turn of the nineteenth century, the effects of colonialism continue to be felt as peoples all over the globe struggle for political and economic independence. The civil war in Rwanda in 1994 is a sobering example of the ill effects of colonialism. As occurred in India, where an estimated one million Hindus and Muslims died in civil war when the British pulled out, the exit of Belgium from Rwanda left colonially created tribal rivalries unresolved and seething (Figure 9.19, page 380). Although the Germans were the first to colonize Rwanda, the

Figure 9.19 Refugees returning to Rwanda
Fleeing civil unrest in their own country, Rwandans from the Hutu tribe increasingly sought refuge in the Democratic Republic of Congo (formerly Zaire) when the Tutsi-led government assumed power in 1994. Two and a half years later, over half a million Rwandan refugees in the Democratic Republic of Congo occupied some of the largest refugee camps in the world. In late 1996, refugees began streaming back into Rwanda. Tens of thousands of Rwandans moving on foot jammed the road between eastern Democratic Republic of Congo and Rwanda for over three days. The Tutsi-led government urged the return of the refugees so they might help in the efforts to rebuild the country. Faced with two difficult alternatives—extremely difficult conditions in the camps or a possible return to violence in Rwanda—the refugees chose to go home.

Belgians, who arrived after World War I, set up what was to become a highly problematic political hierarchy. They established political dominance among the Tutsi by allowing them special access to education and the bureaucracy.

Previously a symbiotic relationship had existed between Tutsi, who were cattle herders, and Hutu, who were agriculturists. The Belgians changed that by establishing a stratified society with the Tutsis on top. In effect, colonialism introduced difference into an existing political and social structure that had operated more or less peacefully for centuries. In 1959, the Hutus rebelled, and the Belgians abandoned their Tutsi favorites to side with the Hutus. In 1962, the Belgians ceded independence to Rwanda, leaving behind a volatile political situation that has erupted periodically ever since, most tragically in 1994's civil war, which continues in reduced form at present. Each side in the conflict has practiced extreme violence upon the other such that mutual genocide has been the outcome.

Civil wars following independence have not been unusual in formerly colonized countries. Figure 9.20 shows the civil wars that followed independence in Africa. Yet another legacy of colonialism has been a deep and sometimes bitter divide between the colonizer and the colonized.

North/South divide: the differentiation made between the colonizing States of the Northern Hemisphere and the formerly colonized States of the Southern Hemisphere.

The **North/South divide** is the differentiation made between the colonizing States of the Northern Hemisphere and the formerly colonized States of the Southern Hemisphere. The colonization of Africa, South America, parts of the Pacific, Asia, and smaller territories scattered throughout the Southern Hemisphere resulted in a political geographic division of the world into North and South. In the North—roughly the Northern Hemisphere—were the imperialist States of Europe, the United States, Russia, and Japan. In the South—roughly the southern Hemisphere—were the colonized. Though the equator has been used as a dividing line, it is clear that some so-called Southern territories actually are part of the North, in an economic sense, such as Australia and New Zealand.

The crucial point is that a relation of dependence was set up between countries in the South, or periphery, on those in the North, or the core, that began with colonization and persists even today. Very few peripheral countries of the South have become prosperous and economically competitive since achieving political autonomy. Political independence is markedly different from economic independence, and the South, even to this day, is very much oriented to the economic demands of the North. An example of this one-way orientation from South to North is the transformation of agricultural practices in Mexico as increasing amounts of production have become directed not toward subsistence

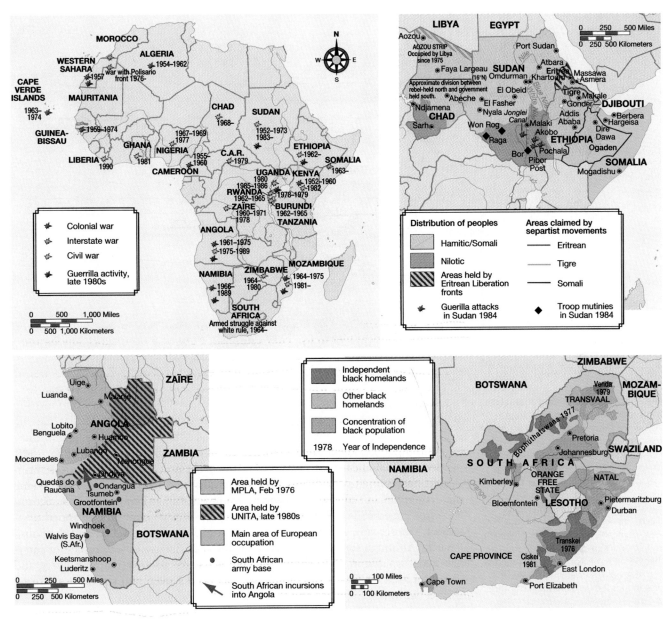

for the local peasant populations but toward consumption in American markets. Mexico, for example, has been described as the "salad bowl" of North America.

Twentieth-Century Decolonization   Decolonization relates to the acquisition, by colonized peoples, of control over their own territory. In many cases, however, sovereign Statehood has only been achievable through armed conflict. From the Revolutionary War in the United States to the twentieth-century decolonization of Africa, the world map created by the colonizing powers has repeatedly been redrawn. Today this map comprises an almost universal mosaic of sovereign States.

Many of the former colonies achieved their independence only after the First World War, however. Deeply desirous of averting future wars like the one that had just ended, the victors (excluding the United States, which entered a period of isolationism following World War I) established the League of Nations. One of the first international organizations formed, the League of Nations had a goal of international peace and security. Figure 9.21 (see page 382) shows the member countries of the short-lived League of Nations.

Figure 9.20   Civil Wars in Africa following independence, 1952–1989
The withdrawal in the 1960s of European colonial influence from Africa and ensuing independence for African States culminated in civil wars in various parts of the continent. The wars were a direct or indirect result of colonial occupation and the imposition of largely arbitrary, contested boundaries that do not correspond to regional ethnic distributions and loyalties. Over 30 years after the decolonization movement in Africa, civil wars continue to rage in countries such as Sudan, Chad, Somalia, and Rwanda. (*Source: Harper Atlas of World History.* New York: HarperCollins, 1992, p. 285.)

Decolonization: the acquisition, by colonized peoples, of control over their own territory.

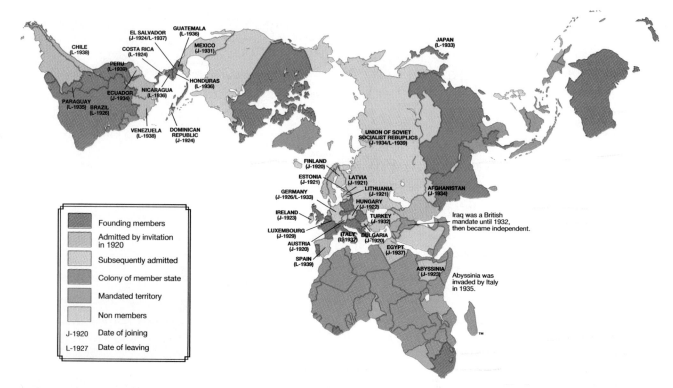

League of Nations map legend:
- Founding members
- Admitted by invitation in 1920
- Subsequently admitted
- Colony of member state
- Mandated territory
- Non members
- J-1920 Date of joining
- L-1927 Date of leaving

Figure 9.21 Participating countries in the League of Nations Although Woodrow Wilson was a central figure in the creation of the League of Nations, his inability to convince the people of the United States to join the League was a major blow to the effectiveness of this, the first international organization of the twentieth century. Britain and France played important roles in the League but were never able to secure arms limitations and security agreements among the membership. Perhaps its greatest success before it was dissolved in 1946 was pressing for the decolonization of Africa. (*Source:* Map projection, Buckminster Fuller Institute and Dymaxion Map Design, Santa Barbara, CA. The word Dymaxion and the Fuller Projection Dymaxion™ Map design are trademarks of the Buckminster Fuller Institute, Santa Barbara, California, ©1938, 1967 & 1992. All rights reserved.)

Within the League a system was designed to assess the possibilities for independence of colonies and to assure that the process occurred in an orderly fashion. Known as the *colonial mandate system,* it did have successes in overseeing the dismantling of numerous colonial administrations. Figures 9.22, 9.23, and 9.24 (see pages 383–386) illustrate the decolonization during the twentieth century of, respectively, Africa; South and Central America; and Asia and the South Pacific. (Although the League of Nations proved effective in settling minor international disputes, it was unable to prevent aggression by major powers and dissolved itself in 1946. It did, however, serve as the model for the more enduring United Nations.)

Decolonization does not necessarily mean an end to domination within the world system, however. Even though a former colony may exhibit all the manifestations of independence, including its own national flag, governmental structure, currency, educational system, and so on, its economy and social structures may continue to be dramatically shaped, in a variety of ways, by core States. Participation in foreign aid, trade, and investment arrangements originating from core countries subjects the periphery to relations that are little different from those they experienced as colonial subjects.

For example, core countries' provision of foreign aid monies, development expertise, and educational opportunities to selected individuals in Kenya has created a class of native civil servants. In many ways this class of individuals is more strongly connected to core processes and networks than those operating within Kenya. This relatively small group of men and women, often foreign-educated, now comprises the first capitalist middle class in Kenyan history. Their Swahili name indicates their strong ties to core globalization processes. They are called *wabenzi:* people (*wa*) who drive Mercedes-Benzes (*benzi*).

Commercial relations also enable core countries to exert important influence over peripheral, formerly colonized, countries. For example, *contract farming* has become the main vehicle around which agricultural production in the periphery is

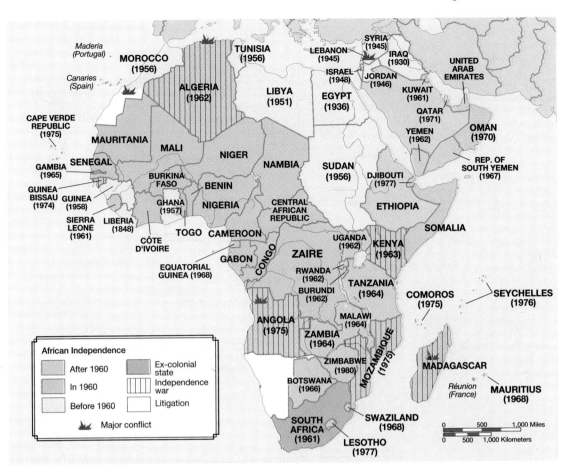

Figure 9.22 Decolonization of Africa, before and after 1960 Britain, France, and Belgium—the dominant European presences in African colonization—were also the first to divest themselves of their colonies. Britain began the process in 1957 when Ghana was granted its independence. France granted independence to Equatorial and West Africa soon after Britain made the first move. In the French-speaking former colonies, the transition to independence occurred largely without civil strife. Belgium's withdrawal as well as the withdrawal of Britain from the remainder of its colonial holdings did not go at all smoothly, with civil wars breaking out. Portugal did not relinquish its possession of Guinea Bissau, Mozambique, or Angola until 1974. (*Source: The Harper Atlas of World History, Revised Edition,* Librairie Hachette, p. 285. Copyright © 1992 by HarperCollins Publishers, Inc. Reprinted by permission of HarperCollins Publishers, Inc.)

organized for core consumption. Dictated by core countries to growers in the periphery are the conditions of production of specified agricultural commodities. Whether the commodity is tea, processed vegetables, fresh flowers, rice, or something else, a contract sets the standards that are involved in producing the commodity.

For example, Japanese firms issue contracts that set the conditions of production for the Thai broiler (chicken) industry; the United Fruit Company, an American firm, issues contracts for Honduran banana production; Lebanese merchants set contracts in Senegal for the production of fresh fruit and vegetables. As the contracting example demonstrates, it is also possible for a core country to invoke a new form of colonialism in places it never formally colonized. Known as neocolonialism, this new form is the domination of peripheral States by core States, not by direct political intervention (as in colonialism) but by economic and cultural influence and control.

The spread of a capitalist world order has had the undeniable impact of "modernizing" traditional societies through education, health care, and other factors. It is also the case that this world order, based on imperialism and colonialism, has been financed with a great deal of bloodshed and human lives. An example is the imperial historical geography of South Africa, a country still laboring under the burdens of colonization as the twentieth century draws to a close (see Feature 9.2, "Geography Matters—Imperialism, Colonization, and the Dismantling of Apartheid in South Africa," page 387). South Africa, as well as other formerly colonized locations around the globe such as Egypt and Indonesia, provides an example of the way in which the global process of imperialism and colonialism unfolded locally.

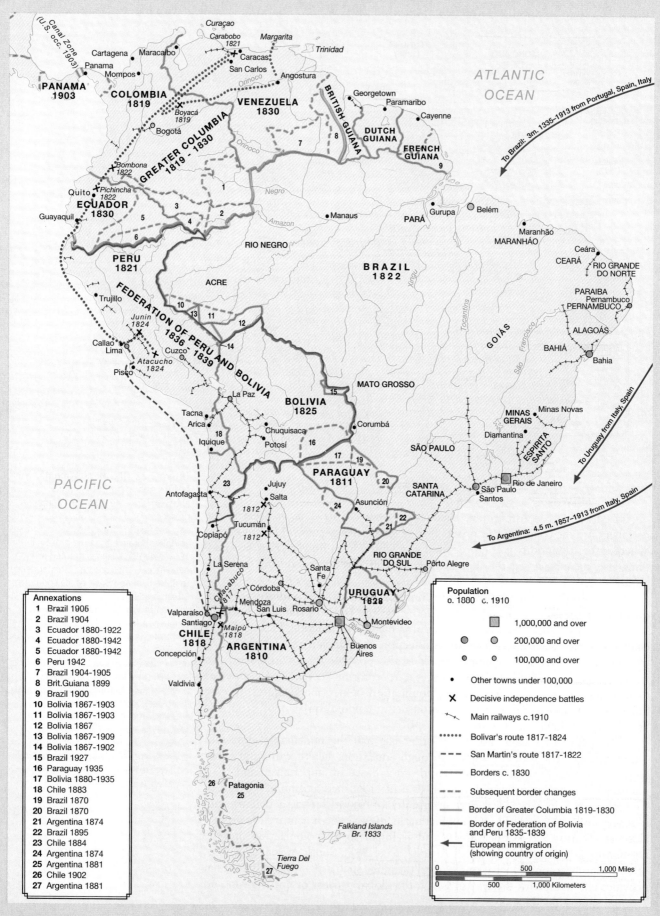

Annexations
1 Brazil 1906
2 Brazil 1904
3 Ecuador 1880-1922
4 Ecuador 1880-1942
5 Ecuador 1880-1942
6 Peru 1942
7 Brazil 1904-1905
8 Brit.Guiana 1899
9 Brazil 1900
10 Bolivia 1867-1903
11 Bolivia 1867-1903
12 Bolivia 1867
13 Bolivia 1867-1909
14 Bolivia 1867-1902
15 Brazil 1927
16 Paraguay 1935
17 Bolivia 1880-1935
18 Chile 1883
19 Brazil 1870
20 Brazil 1870
21 Argentina 1874
22 Brazil 1895
23 Chile 1884
24 Argentina 1874
25 Argentina 1881
26 Chile 1902
27 Argentina 1881

Population
o. 1800   c. 1910

☐ 1,000,000 and over
● ○ 200,000 and over
● ○ 100,000 and over
• Other towns under 100,000
✗ Decisive independence battles
⊬ Main railways c.1910
··· Bolivar's route 1817-1824
--- San Martin's route 1817-1822
— Borders c. 1830
--- Subsequent border changes
— Border of Greater Columbia 1819-1830
— Border of Federation of Bolivia and Peru 1835-1839
← European immigration (showing country of origin)

0   500   1,000 Miles
0   500   1,000 Kilometers

PANAMA 1903
COLOMBIA 1819
VENEZUELA 1830
GREATER COLUMBIA 1819 - 1830
ECUADOR 1830
PERU 1821
FEDERATION OF PERU AND BOLIVIA 1836 1839
BOLIVIA 1825
PARAGUAY 1811
CHILE 1818
ARGENTINA 1810
URUGUAY 1828
BRAZIL 1822
BRITISH GUIANA
DUTCH GUIANA
FRENCH GUIANA
RIO NEGRO
ACRE
MATO GROSSO
GOIAS
MINAS GERAIS
ESPIRITA SANTO
PARÁ
MARANHÁO
CEARÁ
RIO GRANDE DO NORTE
PARAIBA
PERNAMBUCO
ALAGOÁS
BAHIÁ
SÃO PAULO
SANTA CATARINA
RIO GRANDE DO SUL
PATAGONIA
TIERRA DEL FUEGO

ATLANTIC OCEAN
PACIFIC OCEAN

To Brazil: 3m. 1335-1913 from Portugal, Spain, Italy
To Uruguay from Italy, Spain
To Argentina: 4.5 m. 1857-1913 from Italy, Spain

Falkland Islands Br. 1833

Figure 9.23 Independent South and Central America, nineteenth and twentieth centuries   In comparison to Africa, independence came much earlier to Latin America (it had also been colonized much earlier than Africa). What was most influential in the independence movements in Latin America was the presence of local Spanish and Portuguese elites. This colonial ruling class became frustrated with edicts and tax demands from the home country and eventually waged wars—not unlike the U.S. Revolutionary War—against Spain and Portugal for independence. (*Source:* Page 384, *Rand McNally Atlas of World History.* Skokie, IL: Rand McNally, 1992, p. 113. Above, *The Harper Atlas of World History, Revised Edition,* Librairie Hachette, p. 209. Copyright © 1992 by HarperCollins Publishers, Inc. Reprinted by permission of HarperCollins Publishers, Inc.)

Figure 9.24 Independence in Asia and the South Pacific, before and after 1960

Decolonization and resultant independence are not uniform phenomena. Different factors influence the shape that independence takes. The form of colonial domination that was imposed is as much a factor as the composition and level of political organization that existed in an area before colonization occurred. Some former colonies gained independence without wars of liberation; in Asia these include India and Australia. In other places, the former colonizers were only prepared to surrender their colonies after wars of liberation were waged. This was the case in Indochina where domino theory influenced the French colonizers, and later the United States, to go to war against the Vietcong. The war for independence in Vietnam lasted from 1954 until 1976, exacting huge costs for all involved. Mostly, decolonization and political independence forced societies into the mold of a nation-state for which they had little, if any, preparation. It is little wonder then, that few former colonies have succeeded in competing effectively in a world economy, given the complex legacy of colonialism and an inadequate preparation for independence. (Source: *The Harper Atlas of World History, Revised Edition,* Librairie Hachette, p. 283. Copyright © 1992 by HarperCollins Publishers, Inc. Reprinted by permission of HarperCollins Publishers, Inc.)

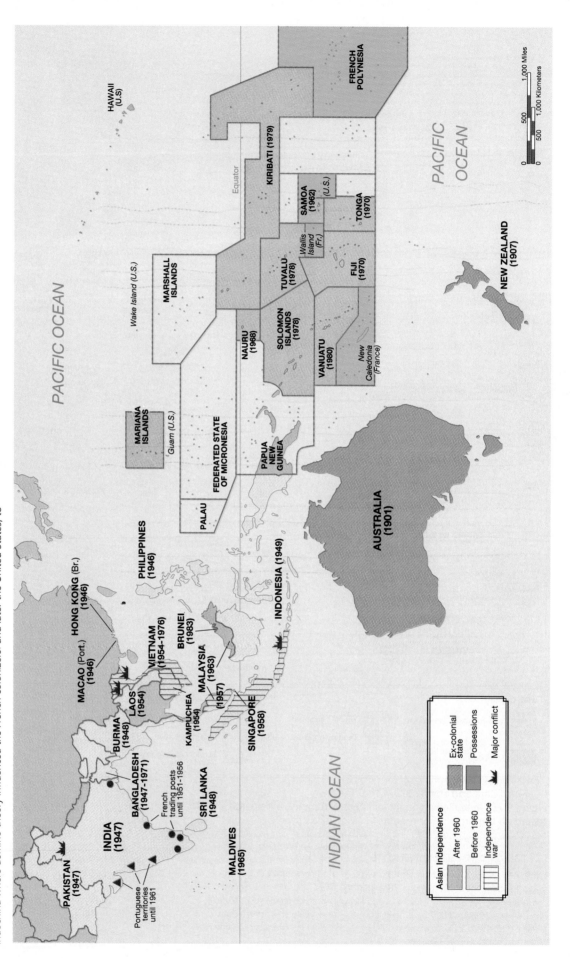

9.2 Geography Matters

Imperialism, Colonization, and the Dismantling of Apartheid in South Africa

Colonial Conquest

The history of imperialism and colonization in South Africa is a long one, dating back to the establishment of a supply station by the Dutch East India Company in Cape Town in 1652. The Dutch, whose settlement developed slowly at first, were segregationists and, from the first, attempted to prevent contact between whites and aboriginal peoples. The early settlers included French Huguenots and Germans as well as Dutch free burghers, who were mostly farmers. Over time, as white settlers moved into the dry lands to the east of Cape Town, they pushed into areas already occupied by native peoples. In this push for land in the western Cape area, the native people were decimated in frontier skirmishes as well as by diseases contracted from the white settlers. During these conflicts between the Boers and the native peoples, an Afrikaner (South African Dutch) identity was constructed. As the native peoples were pushed from the land around Cape Town, some survived by migrating north into the Karoo Desert, while others were incorporated as servants into the emerging Boer economy.

By 1806, Britain had established political control over the Cape. Like the Dutch, the British set about expropriating land and setting up defendable boundary lines between the European immigrant settlements and the largely Bantu-speaking Nguni and Sotho people. Conflict between native peoples and whites persisted. Whites claimed increasing control over land and water as aboriginal peoples resisted and then retreated south and westward.

An important component of the mid-nineteenth-century colonization of South Africa was the intermittent waging of the "Kaffir Wars" between 1835 and 1879. In contrast to previous conflicts, these wars were not aimed at securing additional lands and resources, but at securing a labor supply. Missionaries and traders were especially important in convincing the defeated blacks to work as wage labor on white farms or in the white urban areas. The previously Boer-established policy of strict racial segregation between blacks and Afrikaners was directly confronted by a British policy of racial co-mingling where blacks were intentionally exposed to white value systems and institutions.

As British influence came increasingly to undermine Dutch control of native peoples, Boer farmers—in the Great Trek of 1836—moved northward to areas beyond the influence of British colonial influence. Known as the *Voortrekkers,* the Boers abandoned the well-established European settlements in and around Cape Town. They did this in order to preserve their own value system and to protest the abolition of slavery and the repeal of the pass laws designed to control the movements of black laborers.

In their migration east and northward, the Voortrekkers displaced, through direct conflict and expropriation of tribal land, the Sotho and Zulu people, and founded republics in Natal, the Orange Free State, and in the Transvaal. In these areas, the Boers established native reserves while continuing to wrest land from the independent Zulu peoples. Their efforts not only displaced the native peoples, but also forced them to participate in a cash economy in order to obtain money to pay taxes and to purchase particular types of clothes, which they were required to wear when working or traveling in white urban areas.

Although the British annexed the Republic of Natal as the colony of Natal soon after its establishment by the Voortrekkers, in the other two republics native peoples were incorporated into the economy as servants, squatter tenants, or semifeudal serfs. Ultimately, the "Fundamental Law," established in 1852, legally enshrined the inequality between blacks and whites. It had taken about two centuries for white people to colonize and extend their control throughout what was to become the Union of South Africa.

Exploitation of Resources

No history of South Africa would be complete without a discussion of diamond-, gold-, and coal-mining activities, which began with the discovery of diamonds near Kimberley in 1867 and linked white South Africa to the world economy. This discovery spurred investment of British capital in the mines as well as the construction of the railway network to connect the mines to the ports. Skilled labor, machinery, technology, and capital as well as the dividends and profits garnered from mining were the important linkages between South Africa and the world economy. Importantly, this connection to the world economy and the struggle between the British and Dutch colonizers to control it led eventually to the Anglo-Boer War of 1899 to 1902.

The mining of resources was also responsible for the tremendous growth in population in the mining centers and in the ports. By 1911, the major diamond- and gold-mining centers (including Kimberley, Pretoria-Witwatersrand, and Johannesburg) comprised 37 percent of the total urban population, with the four ports (Cape Town, Durban, Port Elizabeth, and East London) accounting for an additional 23 percent. This population increase was a result of increased European migration to

South Africa, as well as a result of the temporary migration of black males to the mines and urban centers in search of work.

The Era of Territorial Segregation

The first half of the twentieth century witnessed the strengthening and extension of the Boer principles of *racial* segregation through *territorial* segregation. Black ownership of land was restricted, as was black settlement activity. In addition, the permanent residence of blacks in white urban areas was prohibited. The Natives (Urban Areas) Act codified this latter restriction, defining blacks as temporary urban residents who were to be repatriated to the tribal reserves if not employed. It also established that blacks, while within urban areas, were to be physically, socially, and economically separated from the white population.

The intention was for the black reserves to operate as independent economies supporting the black population and separate from the operations of the white economy. Unfortunately, low wages for black laborers as well as high rates of landlessness among blacks living on the reserves undermined the viability of an independent subsistence economy. The reserves were unable to support the black migrant labor system so necessary to the success of the white economy. In addition, blacks increasingly flowed into the urban areas for work, creating a growing and permanent black population in the white cities. By 1946, blacks were the largest racial group in the urban areas, a direct result of the demand for black labor in the growing urban manufacturing sector. Clearly, territorial segregation was becoming increasingly ineffective as a method of separating the races. By midcentury, the policies of segregation were abandoned, and new policies of apartheid were introduced. With the British effectively controlling the politics and economy of South Africa until the mid-1940s, the separation and unequal treatment of races were ubiquitous practices with a loose set of laws, practices, and procedures to uphold them. When the British lost control of national political power in 1946, segregationist practices became more solidly codified. In the wake of their victory, the Afrikaners imposed strict racial separation policies transforming apartheid from practice to rule.

FIGURE 9.2.1 Homelands in South Africa (*Source:* D. M. Smith (Ed.), *Living under Apartheid.* London: Allen & Unwin, 1982, Figs. 2.1, 2.3, and 2.5, pp. 26, 34, and 40.)

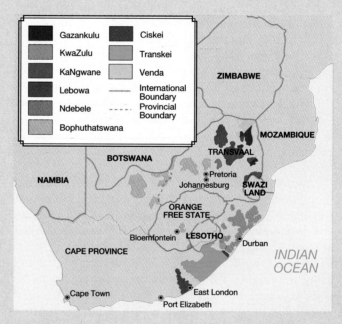

The system of South African "homelands"

Geographers have historically played very central roles in the imperialist efforts of European States. As Figure 9.15 (see page 376) shows, imperialism usually begins with exploration. Most, if not all, of the early geographical expeditions undertaken by Europeans were intended to evaluate the possibilities for resource extraction, colonization, and the expansion of empire. In fact, organiza-

The Era of Apartheid

By 1960, whites were a minority in every South African city. The government introduced the apartheid system to allay the fears of whites who were being crowded out. Apartheid was a system of control of the movement, employment, and residences of blacks. Its main vehicle was the creation of "homelands," a new version of the tribal reserves (Figure 9.2.1). The pass system was revived as well, which further restricted the movement of blacks in white urban areas.

For nearly 40 years, apartheid was the method of control of a white minority over a black majority. Through containment of urban blacks, regional decentralization of employment, and the suppression of dissent, Afrikaners attempted to maintain white supremacy while they continued to exploit black labor to fuel a burgeoning economy. Leaving the homeland areas as well as entering the "proclaimed areas"—all urban areas of the country—was strictly controlled by a permission and pass system. Legislation to remove blacks from urban areas was also enacted. Industrial decentralization, though encouraged, was not a successful strategy and instead fostered the settlement of blacks in homeland townships close to white urban areas, such as Soweto near Johannesburg. Protests against apartheid were also quickly and ruthlessly repressed, with African National Congress leaders like Nelson Mandela being jailed or killed.

A New South Africa?

The late 1980s saw the beginning of the end of apartheid in South Africa—Nelson Mandela was freed from jail, and President P. W. Botha agreed to the sharing of political power between blacks and whites. In 1994, South Africa held the first election in its history in which blacks were allowed to vote. Nelson Mandela was elected the first black president there. The months that preceded and followed his election have been difficult ones as the entire world watches the dismantling of over 300 years of racism by European colonizers.

Source: J. Browett, "The Evolution of Unequal Development within South Africa: An Overview," In *Living Under Apartheid*, Ed. D. M. Smith. London, England: George Allen and Unwin, 1982, pp. 10–23.

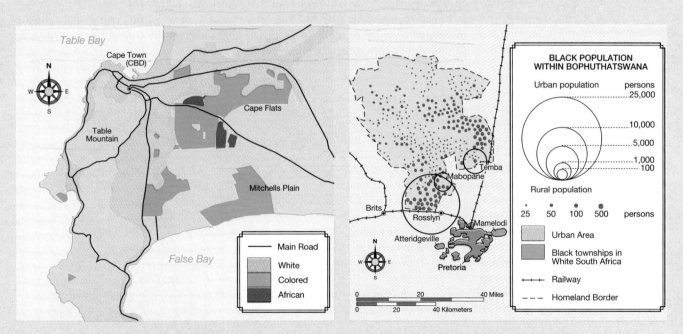

Major residential areas designated for different race groups in the Cape Town Metropolitan Area (before the dissolution of apartheid in 1994)

Wherever possible, the white South African government drew the boundaries of homelands in ways that would maximize the benefits and minimize the costs to the white population. Bophuthatswana, for example, was drawn up as an "independent" homeland, responsible for its own services and infrastructure, but was near enough to the city of Pretoria to provide a large pool of cheap labor for the city.

tions like the royal geographical societies in England and Scotland were explicitly formed to aid in the expansionary efforts of their home countries.

Exploration and colonization did not cease at the midpoint of the twentieth century, however. In fact, exploration, and to a lesser extent, colonization are still occurring in Antarctica (Figure 9.25, page 390). This iced land mass is a somewhat exceptional example of twentieth-century imperialism, where strong States have exerted power in an area of land where no people and, therefore, no indigenous State power have existed. At present, while no one country exclusively "owns" the continent, 15 countries lay claim to territory and/or have established research stations in Antarctica: Argentina, Australia, Belgium, Brazil, Chile, China, France, Germany, India, Japan, New Zealand, Norway, United Kingdom, United States, and Uruguay.

Heartland Theory Because imperialism and colonialism are important forces that shaped the world political map, it is helpful to understand one of the theories that shaped their deployment. By the end of the nineteenth century, numerous formal empires were well established and imperialist ideologies were at their peak. To justify the strategic value of colonization and explain the dynamic processes and possibilities behind the new world map that imperialism had created, Halford Mackinder (1861–1947) developed a theory. Mackinder was a professor of geography at Oxford University and Director of the London School of Economics. He later went on to serve as a member of Parliament from 1910 to 1922, and Chairman of the Imperial Shipping Committee from 1920 to 1945. With such a background in geography, economics, and government, it is not surprising that his theory highlighted the importance of geography to world political and economic stability and conflict.

Mackinder believed that Eurasia was the most likely base from which a successful campaign for world conquest could be launched. He considered Eurasia to be the "geographical pivot" and "heartland," the location central to establishing global control. Mackinder premised his model on the conviction that the age of maritime exploration, beginning with Columbus, was drawing to a close. He

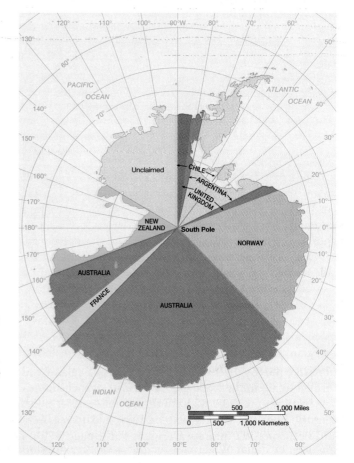

Figure 9.25 Territorial divisions of Antarctica, 1990 Even the uninhabitable terrain of Antarctica has become a site for competition among States. Note the radial lines delineating the various claims: These lines bear no relationship to the physical geography of Antarctica; rather, they are cartographic devices designed to formalize and legitimate colonial designs on the region. (Source: Reprinted with permission from Prentice Hall, from J. M. Rubenstein, *The Cultural Landscape: An Introduction to Human Geography,* 5th Ed., © 1996, p. 294.)

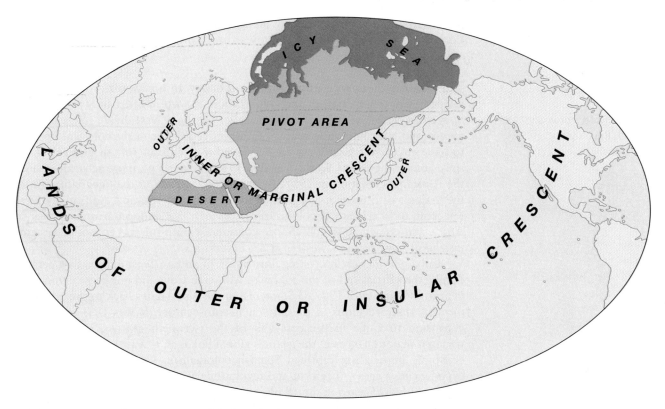

Figure 9.26 Inner and outer crescent of the heartland A quintessential geographical conceptualization of world politics, Mackinder's heartland theory has formed the basis for important geopolitical strategies throughout the decades since its inception. While the pivot area of Eurasia is wholly continental, the outer crescent is wholly oceanic and the inner crescent part continental and part oceanic. It is interesting to compare the Mercator map projection, which Mackinder used to promote his geopolitical theory, with the Dymaxion projection used in this text. This is a classic example of how maps can be used for ideological purposes. The Mercator projection decreases the importance of the northern and southern oceans which are vast and significant natural barriers. The spatial distortions inherent in the Mercator Projection overemphasize the importance of Asia. And the splitting of North and South America so that they appear on both sides of the map adds even more exaggerated emphasis to the centrality of Asia. The Dymaxion projection, as a northern polar representation, de-emphasizes the centrality of any one land mass (see Chapter 1). Mackinder's world view map provides a good example of how cartographic representations can be employed to support ideological arguments. (*Source:* M I. Glassner and H. DeBrig, *Systematic Political Geography,* 3rd Ed., New York: J. Wiley & Sons, 1980, p. 291.)

theorized that land transportation technology, especially the railways, would reinstate land-based power, rather than sea prowess, as essential to political dominance. Eurasia, which had been politically powerful in earlier centuries, would rise again. The reason was because it was adjacent to the borders of so many important countries; was not accessible to sea power; and was strategically buttressed by an inner and outer crescent of land masses. The inner crescent was made up of somewhat marginal States like Egypt and India. The outer crescent was composed of the high-profile sea powers, including the United States, Britain, and Japan (Figure 9.26).

When Mackinder presented his geopolitical theory in 1904, Russia controlled a large portion of the Eurasian land mass protected from British sea power. In an address to colleagues of the British Royal Geographical Society, Mackinder urged that Russia be prevented from expanding its sphere of influence and thereby achieving world domination. Colonial outposts like Ethiopia and India were to be supported militarily as were France, Italy, and Egypt in order to arrest the further expansion of Russia. As many authors have pointed out, Mackinder's theory was providing a strategic rationale for British imperialism, especially in India. When considering why Britain adopted this rationale, it is important to keep in mind that antagonism was increasing among the core European States. That antagonism toward each other led to the First World War a decade later.

The East/West Divide and Domino Theory In addition to a North/South divide based on imperialism and colonization, the world order of States can also be seen to cluster along an East/West split. The **East/West divide** refers to communist and noncommunist countries, respectively. Though the cold war appears to have ended, the East/West divide played a significant role in global politics since at least the end of World War II in 1945 and perhaps, more accurately, since the Russian Revolution in 1917. By the second decade of the twentieth century the major world powers had backed away from colonization. Still, many were reluctant to accelerate decolonization for fear that independent countries in Africa and elsewhere would choose communist political and economic systems instead of some form of Western-style capitalism.

Cuba provides an interesting illustration of an East/West tension that persists despite the official end of the cold war. Although Cuba did not become independent from Spain until 1902, U.S. interest in the island dates back to the establishment of trade relations in the late eighteenth century. It was U.S. commercial expansion to Cuba in the first half of the twentieth century, however, that wrought major changes in the island's global linkages. It was then that American economic imperialism replaced Spanish colonialism. Cuba, at that time, was experiencing a series of reform and revolutionary movements that culminated in the rise of Fidel Castro in the 1950s.

Following the end of the Second World War, anticommunist sentiment was at its peak, and American fears about Cuba "going communist" began to accelerate (Figure 9.27). With Castro in power by 1959, the United States began training Cuban exiles in Central America for an attack on Cuba. This training proceeded despite the fact that in a 1959 visit to the United States, Castro publicly declared his alliance with the West in the cold war. By 1961, the Bay of Pigs, an unsuccessful attempt by United States–trained Cuban exiles to invade Cuba and overthrow Castro had undermined Cuba's relations with the United States and paved the way for improving Cuban–Soviet relations.

Since that time, America's method of dealing with communism in Cuba has been economic, not military. Having failed to overthrow Castro militarily, the United States fell back upon a 1960 embargo designed to destabilize the Cuban economy. The intent of the embargo—which is still in place—was to create such popular dissatisfaction with economic conditions that Cubans themselves would eventually overthrow Castro.

East/West divide: communist and noncommunist countries, respectively.

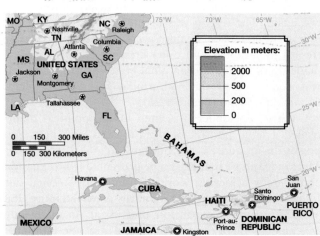

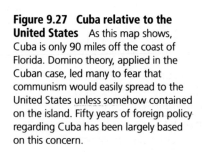

Figure 9.27 Cuba relative to the United States As this map shows, Cuba is only 90 miles off the coast of Florida. Domino theory, applied in the Cuban case, led many to fear that communism would easily spread to the United States unless somehow contained on the island. Fifty years of foreign policy regarding Cuba has been largely based on this concern.

TABLE 9.1 The Domino Theory and U.S. Foreign Policy

Date	Speaker	Dominoes
1947	Acheson (Undersecretary of State)	Greece ⟶ Turkey ⟶ Europe
1948	General Clay (Army Commander in Chief)	Berlin ⟶ West Germany ⟶ Europe
1954	Eisenhower (President)	Vietnam ⟶ Southeast Asia
1955	Eisenhower (President)	Quemoy/Matsu ⟶ Formosa ⟶ West Pacific
1961	Kennedy (President)	Berlin ⟶ Europe
1964	Rostow (Presidential Adviser)	South Vietnam ⟶ Thailand ⟶ Southeast Asia
1964	Johnson (President)	South Vietnam ⟶ Hawaii ⟶ San Francisco
1980	Carter (President)	Afghanistan ⟶ Iran ⟶ Gulf States
1990	Reagan (President)	Kuwait ⟶ Saudi Arabia

Source: J. R. Short, *Introduction to Political Geography.* London: Routledge, 1982, p. 61.

The embargo forbids economic trade between the United States and Cuba across a wide range of activities and products. While the embargo has not been successful—Castro remains in power—the country has become increasingly embattled as the collapse of the Soviet Union has left it without powerful and economically generous allies. Yet the standoff between Cuba and the United States persists, most recently aggravated by Castro's tacit approval of the release of tens of thousands of refugees—many from Cuban prison populations or suffering from HIV infection—to the United States.

The end of World War II marked the rise of the United States to a dominant position among countries of the core. The tension that arose, following the war, between East and West translated into an American foreign policy that pitched it against the former Soviet Union. Domino theory was the source of the foreign policy that included economic, political, and military objectives directed at undermining the possibility for Soviet world domination. The **domino theory** held that if one country in a region chose or was forced to accept a communist political and economic system, then neighboring countries would be irresistibly susceptible to falling to communism as well. The concept behind domino theory was that one falling domino in a line of dominos causes all the others in its path to fall. The antidote to preventing the domino-like spread of communism was often military aggression.

Domino theory: if one country in a region chose or was forced to accept a communist political and economic system, then neighboring countries would be irresistibly susceptible to falling to communism.

Table 9.1 illustrates the way in which domino theory influenced more than 40 years of American foreign policy around the globe. Adherence to the theory began in 1947, when postwar America feared communism would spread from Greece to Turkey to Europe. It culminated in the more recent events of U.S. wars in Korea, Vietnam, Nicaragua, El Salvador, and the Persian Gulf. Yet preventing the domino effect was not just based on military aggression. Cooperation was also emphasized, such as in the 1949 establishment of international organizations like NATO (North Atlantic Treaty Organization), which had the stated purpose of *safeguarding* the West against Soviet aggression. Following the end of the Second World War, the core countries set up a range of foreign aid, trade, and banking organizations. All were intended to open up foreign markets and bring peripheral countries into the global capitalist economic system. The strategy not only improved productivity in the core countries, but was also seen as a way of strengthening the position of the West in its cold war confrontation with the East.

International and Supranational Organizations

Just as States are seen as key players in political geography, so, too, have international and supranational organizations, in the last century, become important participants in the world system. These organizations have become increasingly important as a way of dealing with situations in which international boundaries stand in the way of specific goals. These goals include, among other things, the freer flow of goods and information and more cooperative management of shared resources such as water.

International organization: group that includes two or more States seeking political and/or economic cooperation with each other.

An **international organization** is one that includes two or more States seeking political and/or economic cooperation with each other. One well-known example of an international organization operating today is the United Nations (UN) (Figure 9.28). Other examples of international organizations include the Organization for Economic Cooperation and Development (OECD); the Organization of Petroleum Exporting Countries (OPEC); the Association of South-East Asian Nations (ASEAN); and the now-disbanded Council for Mutual Economic Assistance (COMECON). Though these organizations were formed to accomplish very different ends, they all aim to achieve cooperation while maintaining full sovereignty of the individual States. The countries involved in these organizations are shown in Figure 9.29.

Figure 9.28 United Nations member countries, 1990 Following World War II and the demise of the League of Nations, renewed effort was made to establish an international organization aimed at instituting a system of international peace and security. The United Nations Charter was approved by the U.S. Senate in July 1945, raising hopes for a more long-lived organization than the ineffective League of Nations. Located in New York City, the United Nations is composed of a Security Council, which includes the permanent members of the United States, Britain, China, France, and Russia; as well as a General Assembly, which includes all those countries identified on the map. At the same time that the United Nations was set up, the United States lobbied for the creation of the International Monetary Fund (IMF) and the World Bank. The U.S. government had believed that World War II had erupted due to the collapse of world trade and financial dislocation caused by the Great Depression. The task of the IMF and the World Bank was to provide loans to stabilize currencies and enhance economic growth and trade. (*Source:* Map projection, Buckminster Fuller Institute and Dymaxion Map Design, Santa Barbara, CA. The word Dymaxion and the Fuller Projection Dymaxion™ Map design are trademarks of the Buckminster Fuller Institute, Santa Barbara, California, ©1938, 1967 & 1992. All rights reserved.)

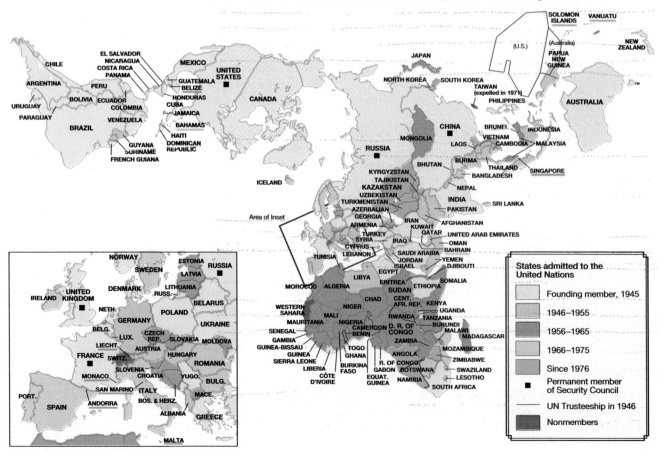

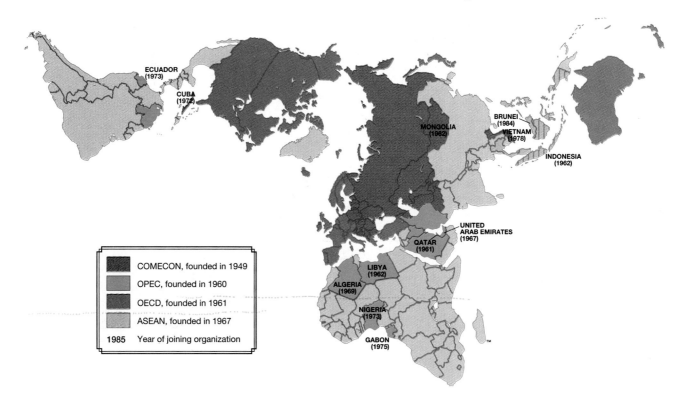

Figure 9.29 International economic groups, 1990 OPEC (Organization of Petroleum Exporting Countries) States joined together to foster cooperation in the setting of world oil prices in 1960. ASEAN (Association of Southeast Asian Nations), founded in 1967, exists to further economic development in Southeast Asia. The OECD (Organization for Economic Cooperation and Development), founded by the United States, Canada, and 18 European States in 1961, aims to increase world trade through the provision of financial security. Groupings such as these suggest that the independent State has lost effectiveness in promoting its own and other States' economic progress and stability. Founded in 1949, COMECON (the Council for Mutual Economic Assistance) was the grouping organized to promote trade among former communist and socialist States. It is no longer functioning. (*Source:* Map projection, Buckminster Fuller Institute and Dymaxion Map Design, Santa Barbara, CA. The word Dymaxion and the Fuller Projection Dymaxion™ Map design are trademarks of the Buckminster Fuller Institute, Santa Barbara, California, ©1938, 1967 & 1992. All rights reserved.)

The postwar period has seen the rise and growth not only of large international organizations, but also of new regional arrangements. These arrangements vary from highly specific, such as the Swiss–French cooperative management of Basel-Mulhouse airport, to the more general, such as the North American Free Trade Agreement (NAFTA), which joins Canada, the United States, and Mexico into a single trade region. Regional organizations and arrangements now exist to address a wide array of issues, including the management of international watersheds and river basins (such as the Great Lakes of North America and the Danube and Rhine Rivers in Europe). They also exist to oversee the maintenance of acceptable levels in health and sanitation, coordinated regional planning, and tourism management. Such regional arrangements seek to overcome the barriers to the rational solution of shared problems posed by international boundaries. They also provide larger arenas for the pursuit of political, economic, social, and cultural objectives.

Unlike international organizations, **supranational organizations** *reduce* the centrality of individual States. Through organizing and regulating designated operations of the individual member States, these organizations diminish, to some extent, individual State sovereignty in favor of the collective interests of the large membership. The European Union (EU) is perhaps the best example of a supranational organization.

Dating back to the end of the Second World War, European leaders realized that Europe's fragmented State system was insufficient to the demands and levels of competition coalescing within the world political and economic system. They endeavored to create an entity that would preserve important features of State sovereignty and identity. They have also intended to create a more efficient intra-European marketing system and a more competitive entity in global transactions. Figure 9.30 (see page 396) shows the progression of integration of European countries into the European Union since 1957.

The EU holds elections, has its own parliament and court system, and decides whether and when to allow new members to join. Generally speaking, the EU

Supranational organizations: collections of individual States with a common goal that may be economic and/or political in nature and which diminish, to some extent, individual State sovereignty in favor of the group interests of the membership.

Figure 9.30 Map of membership in EEC to EFTA The European Economic Community, also known as the Common Market, was established in 1957. The aim of the original member countries was to abolish internal tariffs and institute a common external tariff within 15 years. Britain declined to join out of concern that a common external tariff would destroy Commonwealth trade. In 1959 Britain formed the European Free Trade Association (EFTA) initially with Norway, Sweden, Denmark, Switzerland, Austria, and Portugal. The goal of EFTA was to abolish internal tariffs only on industrial goods with no attempt to establish a common external tariff. During the 1960s the EEC proved more successful than EFTA as the British economy declined. Britain sought admission to the EEC but was twice refused by France. In 1973, Britain, along with Denmark and Ireland, joined the group. Portugal and Greece joined in the 1980s. Finland, Norway, Sweden, Austria, and the former East Germany were all admitted in the 1990s. (Source: *Time Atlas of European History*. London: Times Books, 1994, p. 190.)

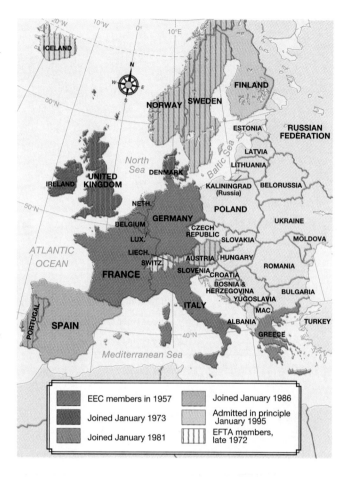

EEC members in 1957	Joined January 1986
Joined January 1973	Admitted in principle January 1995
Joined January 1981	EFTA members, late 1972

aims to create a common geographical space within Europe in which goods, services, people, and information move freely and, eventually, in which a single monetary currency will prevail. Whether an EU foreign policy or a single common currency will ever be accomplished remains to be seen, however, for nationalism within the individual member countries remains strong. Ironically, just as the European system of States is on the threshold of dissolving into the larger EU organizational form, national and regional movements (discussed in Feature 9.1 and in Chapter 5) have become potent forces operating against full integration.

The Two-Way Street of Politics and Geography

Political geography can be seen to proceed according to two contrasting orientations. The first orientation sees political geography as being about the *politics of geography*. This perspective emphasizes that *geography*—or the areal distribution/differentiation of people and objects in space—has a very real and measurable impact on politics. Regionalism and sectionalism, discussed below, provide examples of how geography shapes politics. The politics-of-geography orientation is also a reminder that politics occurs at all levels of the human experience from the international order down to the scale of the neighborhood and household.

The second orientation sees political geography as being about the *geography of politics*. In contrast to the first orientation, this approach analyzes how *politics*—the tactics or operations of the State—shapes geography. Mackinder's heartland theory and the domino theory are examples of how the geography of politics works at the international level. In heartland theory the State expands into new territory in order to relieve population pressures. In domino theory, as communism seeks new members, it must expand geographically to incorporate

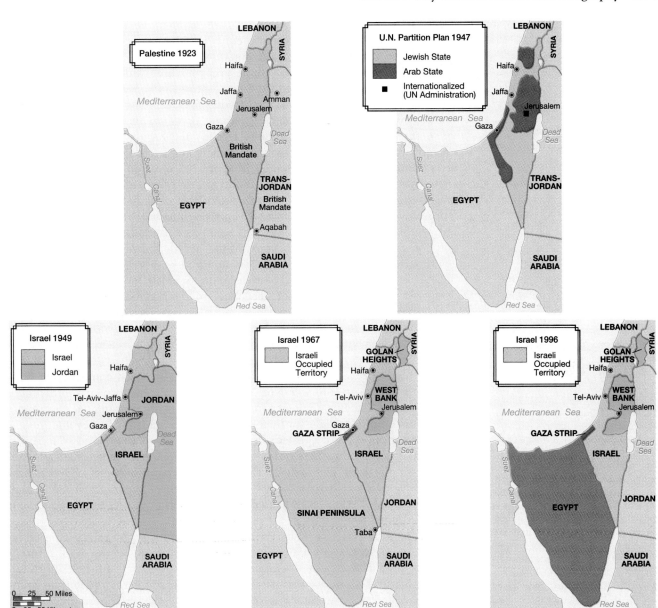

Figure 9.31 Changing geography of Israel/Palestine, 1923–1996   Since the creation of Israel out of the former Palestine in 1947, the geography of the region has undergone significant modifications. A series of wars between Israelis and Arabs and a number of political decisions regarding how to cope with both resident Palestinians and large volumes of Jewish people immigrating to Israel from around the world have produced the changing geographies we see here. (Source: Reprinted with permission from Prentice Hall, from J. M. Rubenstein, *The Cultural Landscape: An Introduction to Human Geography*, 5th Ed., © 1996, p. 233.)

new territories. Likewise, examination of a series of maps of Palestine/Israel since 1947 reveals how the changing geography of this area is a response to changing international, national, regional, and local politics (Figure 9.31).

The Politics of Geography

Territory is often regarded as a space to which a particular group attaches its identity. Related to this concept of territory is the notion of **self-determination,** which refers to the right of a group with a distinctive politico-territorial identity to determine its own destiny, at least in part, through the control of its own territory.

Regionalism and Sectionalism It is sometimes the case that different groups with different identities—religious or ethnic—coexist within the same State boundaries. At times, discordance between legal and political boundaries and the distribution of populations with distinct identities becomes manifested as a movement to claim particular territories. These movements, whether conflictual or

Self-determination: the right of a group with a distinctive politico-territorial identity to determine its own destiny, at least in part, through the control of its own territory.

Regionalism: a feeling of collective identity based on a population's politico-territorial identification within a State or across State boundaries.

Sectionalism: extreme devotion to local interests and customs.

Figure 9.32 Basque independence poster This independence poster is plastered over the door of a shop in Donostia (San Sebastian) in one of the Basque provinces of Spain. The poster is a sign of the passionate opposition the Basques have adopted toward the central government in Madrid. Acts of terrorism continue to occur throughout Spain as the Basques maintain their desire for independence. What is most interesting about the sign is that it is not written in either Castillian Spanish (the national language) or Euskadi (the Basque language). It appears as if this and declarations like it are directed at the tourists who have made Donostia a popular destination.

peaceful in their claim to territory, are known as regional movements. **Regionalism** is a feeling of collective identity based on a population's politico-territorial identification within a State or across State boundaries.

Regionalism often involves ethnic groups whose aims include autonomy from an interventionist State and the development of their own political power. For example, in spring 1993, several leading Basque guerrillas were arrested in France, raising hopes for an end to Basque terrorism in Europe. For over 25 years, the French, Spanish, and more recently the Basque regional police have attempted to undermine the Basque Homeland and Freedom movement through arrests and imprisonments. "Basquism" represents a regional movement that has roots back to industrialization and modernization beginning at the turn of the nineteenth century. The Basque people feared that *cultural forces* accompanying industrialization would undermine Basque preindustrial traditions. Because of this, the Basque provinces of northern Spain and southern France have sought autonomy from those States for most of the twentieth century. Since the 1950s, agitation for political independence has occurred—especially for the Basques in Spain—through terrorist acts. Not even the Spanish move to parliamentary democracy and the granting of autonomy to the Basque provinces, however, could squelch the thirst for self-determination among the Basques in Spain (Figure 9.32). Meanwhile, on the French side of the Pyrenees, although a Basque separatist movement does exist, it is neither as violent nor as active as the movement in Spain.

We need only look at the long list of territorially based conflicts that have emerged in the post–cold war world to realize the extent to which territorially based ethnicity remains a potent force in the politics of geography. For example, the Kurds continue to fight for their own State separate from Turkey and Iraq. A significant proportion of Quebec's French-speaking population, already accorded substantial autonomy, persists in advocating complete independence from Canada. In the 1996 plebiscite on the issue, the separatists were only very narrowly defeated. Consider the former Yugoslavia, whose geography has fractured along the lines of ethnicity. Regionalism also underlies efforts to sever Scotland from the United Kingdom.

Regionalism need not be based, however, on ethnicity. An example based on economics is the case of a nonbinding referendum put before the California voters in 1993. Drawing on a sentiment over a century old, California's northern and mostly rural counties voted to separate from the southern, more urban counties. The desire for separation was based on the belief that political representation in the state legislature economically and politically advantaged the south over the north. Many citizens expressed the opinion that the state has grown unmanageably large and that government has become unresponsive to the people. Although many steps would have to be taken beyond support for the advisory referendum—such as state senate approval, approval by the governor and by the voters, and finally the United States Congress—the California separatist movement does represent an interesting example of *economic* regionalism.

Not to be confused with regionalism is the concept of **sectionalism,** an extreme devotion to local interests and customs. Sectionalism has been identified as an overarching explanation for the United States Civil War. It was an attachment to the institution of slavery and to the political and economic way of life slavery enabled that prompted the southern states to secede from the Union. The Civil War was fought to ensure that sectional interests would not take priority over the unity of the whole; that is, that state's rights would not undermine the power of the federal government. Although the Civil War was waged around the real issue of permitting or prohibiting slavery, it was also fought at another level, a level that dealt with issues of the power of the State. As Figure 9.33 shows, the election of Abraham Lincoln to the Presidency in 1860 reflected the sectionalism that dominated the country: He received no support from slave states.

Suburbs Versus Cities and Rural Versus Urban Today, sectionalism persists, but in different forms. In the United States, one of the most apparent manifestations of sectionalism may be found in the politics of differentiation between suburbs and cities. Two examples are the "taxpayer revolts" such as Proposition 13 in California and Proposition 2 1/2 in Massachusetts and the growth-control movements. The former were voter-approved mandates to cut property taxes and limit government spending, while the latter were government-enforced caps on population growth approved by residents to limit the density and extent of growth. Both actions were dominated by suburban U.S. voters.

Proposition 13 in California was an initiative that had tremendous negative fiscal impacts on the municipalities, especially on public education and social programs. What began as a localized movement of angry southern California taxpayers in South Bay and San Gabriel Valley grew into a statewide, suburban, antitax protest directed at lowering county government spending on social programs. Cutting across class lines and unified by the subculture of homeownership, Howard Jarvis's California Taxpayers League collected 1.5 million signatures to put Proposition 13 on the ballot in 1978. Suburban homeowners became a powerful political force all over California and elsewhere, a force that continues to exert itself in local elections throughout the United States as well as many European countries (Figure 9.34, page 400).

A child of the taxpayer revolts, the growth-control movements of the 1980s also drew their support from suburban homeowning constituencies. Arguing for the maintenance of open space, low density, large lots, and community control over land use planning and development, the growth control movements have been attempts to keep the city out of the suburbs by retaining a "country feel." Critics of growth control complain that it discriminates against newcomers to the community. Related to growth-control movements is *NIMBYism,* which stands for "Not In My Backyard" actions. NIMBYism is action by neighborhood residents against the introduction of unwanted land uses. These unwanted land uses can range from group homes for developmentally disabled adults or battered women's shelters to low-income housing and other residential forms. NIMBY supporters object to these land uses because they are perceived as a threat to the composition and quality of the neighborhood and the market value of single-family residences.

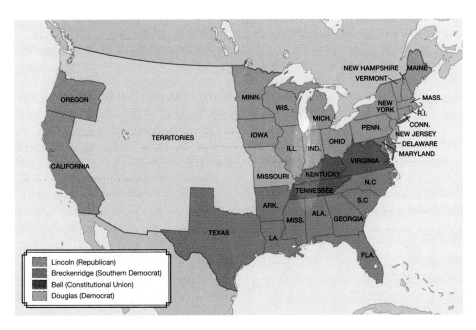

Legend:
- Lincoln (Republican)
- Breckenridge (Southern Democrat)
- Bell (Constitutional Union)
- Douglas (Democrat)

Figure 9.33 The 1860 presidential election The U.S. presidential election of 1860 graphically illustrates the role of sectionalism in determining who gets votes from which geographical regions. Here, in a four-way race, we see that Abraham Lincoln failed to win the support of any of the slave states. (*Source: Presidential Elections Since 1789,* 4th Ed., Congressional Quarterly Inc., 1987.)

Figure 9.34 Proposition 13 campaign poster
Suburbanites in California strongly favored Proposition 13, which appeared on the ballot in 1978. Fed up with the voracious appetites of local jurisdictions—especially cities—for tax dollars to support schools and welfare programs, voters mandated severe restrictions on the amount that property taxes could be raised. The capping of property taxes under Proposition 13 had a distorting effect on the housing market: the removal of an important limiting factor to uncontrolled inflation. As housing prices rose over the years following the implementation of Proposition 13, newer houses were taxed at a much higher rate than older ones, which continued to be occupied by pre-Proposition 13 owners. Thus, on the same street, two houses next to each other of roughly the same quality and same market value could be (and often are) taxed at dramatically different rates. By reducing the flow of tax dollars into the government coffers, Proposition 13 also affected the provision of social services such as police and fire protection as well as libraries, parks, and museums.

The politics of geography, in terms of regionalism, also finds strong focus today in rural versus urban politics. In France, for example, attitudes about birth control (and birthrates themselves) are significantly different between the urbanized north of the country and the more rural south. Throughout the EU, farmers have fought against removal of farm subsidies and tariff arrangements, advocated by urban-based policymakers, that have long protected agricultural productivity. Here, the dispute pits the politics of local farmers against an international organization.

In Mexico, the contest is between local and national levels. Chiapan peasants have forced the federal government in Mexico City to address profound problems of rural poverty. These rural problems are seen to have been aggravated by the government's largely urban (and industrial agriculture) orientation. Likewise, rural-to-urban migration throughout the periphery (grossly inflating the populations of cities such as Lima, Peru; Nairobi, Kenya; and Djakarta, Indonesia) has generated enormous social and political pressures and poses overwhelming challenges. Policymakers must ask themselves difficult questions: How much of the country's scarce resources should be devoted to slowing (or reversing) rural out-migration through development projects? What level of resources should be devoted to accommodating the throngs of new urban-dwellers, most of whom have worse living conditions in the city than they did in the countryside?

Competition also exists among and between cities as well as among and between states. The most ubiquitous form of this competition revolves around the desire by local and state authorities to attract corporate investment to their city or state (as discussed in Chapter 7). Oftentimes, it is the corporations that play the jurisdictions against each other in attempts to obtain the most attractive investment packages. At other times it is government facilities that cities and states vie to induce to locate within their jurisdictions. The location of military bases during and after the Second World War illustrates this point. Cities and states were especially keen to attract the military because it meant increased employment opportunities and all of the economic growth attached to employment generation.

The Geography of Politics

An obvious way to show how politics shapes geography is to show how systems of political representation are geographically anchored. For instance, the United States has a political system in which democratic rule and territorial organization are linked together under the concept of territorial representation.

Geographical Systems of Representation Democratic rule is a system in which public policies and officials are directly chosen by popular vote. **Territorial organization** is a system of government formally structured by *area,* not by social groups. Thus, voters vote for officials and policies that will represent them and affect them *where they live.* The territorial bases of the United States system of representation are illustrated in Figure 9.35.

The United States is a federation of 50 states. The states are themselves subdivided into counties or parishes, of which there are over 3,000. Counties and parishes are further broken down into municipalities, townships, and special districts, which include school districts, water districts, library districts, and others.

The electoral divisions established for choosing elected officials in the United States range from precincts and wards, to congressional districts and states. State power is applied within geographical units and State representatives are chosen from geographical units. The bottom line is that in the United States—as in many other representative democracies—politics *is* geography. This means that people and their interests gain representation in government through the location of their interests in particular places and through their relative ability to capture political control of *geographically based* political units.

For example, election of the president involves a popular vote, carried out at the precinct level but totaled at the state level. Thus, even though particular precincts, cities, or counties may give a majority to one candidate, if the majority of votes at the state level of aggregation supports the opposing candidate, then that person is declared the winner in that state. This arrangement is particularly important because the president is not elected by the popular vote but by the electoral college. The electoral college is composed of a specified number of delegates allocated to each state based on that state's population as of the most recent official census. Thus, it is the state-level voting tally that drives the process.

Occasionally, a candidate may win the countrywide popular vote but lose in the electoral college if that candidate failed to win enough states to acquire the required majority of the electoral votes. Conversely, as in the 1992 election of Bill Clinton, a candidate may win with considerably less than 50 percent of the popular vote if a third-party candidate (like H. Ross Perot) siphons off enough of the popular vote to prevent the opposing candidate (in this case, George Bush) from winning a sufficient number of electoral votes. The geographical implications of the U.S. presidential voting arrangement are crucial to candidates' campaign strategies. To be a winner requires concentrating enormous time and energy to capture a majority of votes in the nation's most populous states.

The American tradition of territorial representation promotes stronger acceptance of the legitimacy of local interests. Such an arrangement also makes

Democratic rule: a system in which public policies and officials are directly chosen by popular vote.

Territorial organization: a system of government formally structured by area, not by social groups.

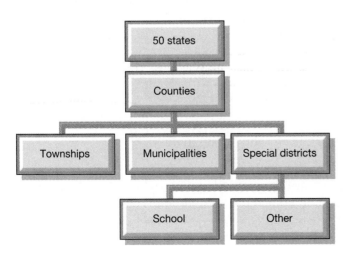

Figure 9.35 U.S. geographical basis of representation This diagram gives a breakdown in the number of voter districts at each level of political representation. Each of these districts is territorially defined, creating a complicated and overlapping pattern of political units.

it difficult for third parties to succeed. The reason for this is that unless third parties are geographically concentrated, they can never win enough states—as in the case of presidential elections, for instance—to capture office. The failed presidential campaign of H. Ross Perot in 1992 is an excellent example of this. Figure 9.36 illustrates this point. Although Perot received 19 percent of the vote, he did not win a single state.

Reapportionment and Redistricting The U.S. Constitution requires the allocation of legislators among states, guaranteeing that each state will have a representative system of government. For U.S. presidential and senatorial elections, candidates are elected at-large, not on the basis of electoral districts. U.S. representatives, however, are elected from congressional districts of roughly equal population size. This is also the case for state senators, representatives, and, often, other elected officials from city council members to school board members. It is the responsibility of each state's legislature to create the districts that will elect most federal and state representatives. Other levels of government— from counties to special districts—also establish their own electoral districts. The result is that representatives are elected at any number of levels of government in a collection of districts that is complicated, extensive, and by no means systematic.

Problems of the "fit" between political representation and territory emerge when population changes. Since most forms of representation are based on population, it often becomes necessary to change electoral district boundaries to distribute the total population more evenly among districts. For example, the number of congressional representatives in the United States as a whole is fixed at 435. These 435 seats must be reapportioned in accordance with population change every 10 years. (Recall from Chapter 3 that the federal government is required to count the American population every 10 years. One of the chief reasons for this is to maintain the proper match between population and representation). **Reapportionment** is the process of allocating electoral seats to geographical

Reapportionment: the process of allocating electoral seats to geographical areas.

Figure 9.36 The vote for president, 1992 The total popular vote, aggregated at the state level, determines the number of electoral college votes for each candidate. Voter preferences shows a clear geographical pattern, with George Bush running particularly strongly in the middle of the country and in some southern states, as well as Alaska. H. Ross Perot, despite the fact that he won 19 percent of the popular vote, was unable to translate those votes into any electoral votes.

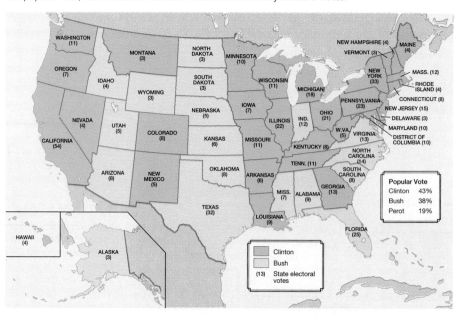

areas. **Redistricting** is the defining and redefining of territorial district boundaries. Both are political, geographical, and statistical exercises.

Redistricting: the defining and redefining of territorial district boundaries.

The process is *political* in that the design and approval of systems of districts is usually done by bodies of elected representatives; the balance of power between groups and areas is often involved; identification of citizens with a traditional electoral territory is altered; and the incumbency of individuals is usually at stake. . . . The process is *geographic* in that areas must be allocated to districts and boundaries drawn (or territories partitioned into districts); communities of interest which may have arisen in part from pre-existing systems of districts, may well be affected; restructuring of basic electoral geography is altered; and accessibility of voters to their representatives or centers of decision-making may be changed. . . . Redistricting is also *statistical* or mathematical in that there is a requirement for reasonably current and accurate data on population and its characteristics and, sometimes of property and its valuation.[2]

For example, following the results of the 1990 census, legislative leaders in Arizona redrew the state's political boundaries. Statistics show that growth in Maricopa County—where Phoenix, the largest city, is located—exceeded growth in Pima County—where Tucson, the second largest city, is located. Because Phoenix experienced relatively more growth, Tucson is likely to lose one state senator and two members of the House to Phoenix. Legislators also have had to draw a new Sixth Congressional District (federal level) because the state has grown so rapidly since 1980 (Figure 9.37).

Gerrymandering The intent of redistricting is to insure the equal probability of representation among all groups. The practice of redistricting for partisan purposes, however, is known as **gerrymandering.** In this abuse, boundaries of dis-

Gerrymandering: the practice of redistricting for partisan purposes.

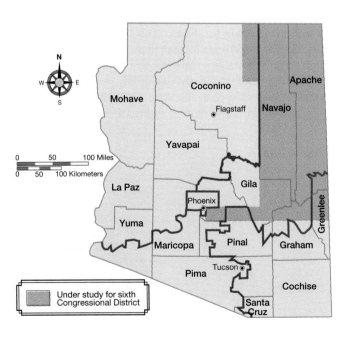

Figure 9.37 Arizona's proposed Congressional District, 1991 The 1990 census revealed rapid growth in Arizona, necessitating a redrawing of the state's boundaries for electoral districts. Among the changes brought on by the high growth rate was creation of a new Sixth Congressional District. Arizona's districts vary wildly in size and have strange shapes, reflecting the very uneven distribution of population in the state as well as the effects of political maneuvering for votes between Democrats and Republicans. (Source: *Arizona Daily Star,* 1991.)

---

[2]R. Morrill, *Political Redistricting and Geographic Theory.* Washington, DC: Association of American Geographers, 1981, p. 1.

Figure 9.38 Gerrymandering salamander, 1812 This strange beast was the result of political shenanigans that used the geography of electoral district boundaries to concentrate Federalist votes within a single "sacrifice" district in order to avert the possibility of Federalist supporters influencing elections in the other districts. Today, laws attempt to prevent such blatant manipulations of the geography of voting. (*Source: Boston Gazette*, March 26, 1812.)

Figure 9.39 North Carolina's proposed Twelfth Congressional District The drawing of electoral district boundaries remains a politically volatile exercise. A 1996 case before the Supreme Court was the redrawn Twelfth Congressional District of North Carolina, whose shape is as contorted as the original dragon. Accusations of gerrymandering circulated around the drawing of the district, which was intended to consolidate African American voting strength in the district. (*Source:* Reprinted with permission from Prentice Hall, from E. F. Bergman, *Human Geography: Cultures, Connections, and Landscapes,* © 1995, Fig. 10.28.)

tricts have been redrawn to advantage a particular political party or candidate, or to prevent any loss of power to a particular subpopulation (like African Americans). The term immortalized Governor Elbridge Gerry of Massachusetts who signed into law a bill designed to maximize the election of Republican-Democrats over Federalists in the election of 1812. Although Federalists won the most votes overall, Republican-Democrats took 29 of 40 seats because they won the most districts. An electoral district north of Boston, illustrated in the cartoon in Figure 9.38, was redrawn so that most of the Federalist votes were contained there. With the majority of Federalist votes restricted to a small number of districts, their voting power was diluted.

Gerrymandering persists as an issue today. Although the Federal Voting Rights Act intended that redistricting enhance the chances of minority representation in Congress, a fine line exists between "enhancement" and creating a district solely to insure that a minority person be elected. Despite overturning some districts, the Supreme Court has yet to provide a clear mandate on the appropriate parameters for the use of race or minority status in drawing district boundaries. Figure 9.39 is a map of the state of North Carolina's proposed Twelfth District, which was later invalidated by the Supreme Court. Compare its tortured boundaries to the classic gerrymandering salamander in Figure 9.38.

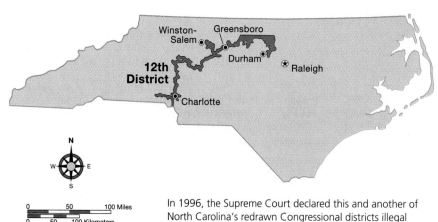

At the same time that the Supreme Court declared these boundaries illegal, they did not set a time limit for when the boundaries were to be redrawn. As a result, the 1996 district elections proceeded on the basis of these boundaries.

In 1996, the Supreme Court declared this and another of North Carolina's redrawn Congressional districts illegal because race was the predominant factor in drawing its irregular boundary.

Conclusion

The globalization of the economy has been largely facilitated by the actions of States extending their spheres of influence and paving the way for the smooth functioning of markets and industries. Political geography is as much about what happens at the global level as it is about what happens at other levels of spatial resolution, from the region to the neighborhood to the household and the individual.

Theories of the State have been one of geography's most important contributions to understanding politics. Ratzel's emphasis on the relationship between power and territory and Mackinder's model of the geographical pivot remind us that space and territory shape the actions of States in both dramatic and mundane ways. Time as well as space shape politics, and events distant in time and space—such as colonialism—continue to have impacts long after decolonization. The civil war in Northern Ireland, instigated by English colonial practices now centuries old, has only recently shown credible signs of ceasing. The impacts of English colonization have been felt in countries throughout the Northern Hemisphere, as well as by neighbors living unhappily for several generations side-by-side in cities like Belfast and Boston.

Continuing strife is also the case with the enduring North/South divide that pits core countries against peripheral, mostly formerly colonial, countries. Perhaps the most surprising political geographical transformation of this century has been the near dissolution of the East/West divide. Although it is too soon to tell whether communism has truly been superseded by capitalism, it is certainly the case that the distinctions between them are more blurred than clear.

The pairing of the terms "politics" and "geography" serves to remind us that politics is clearly geographical at the same time that geography is unavoidably political. The simple divisions of area into states, counties, cities, towns, and special districts means that where we live shapes our politics and vice versa. Geography is politics just as politics is geography, and geographical systems of representation as well as identity politics based on regional histories confirm these interactive relationships.

Main Points

- A subfield of the discipline of geography, political geography examines complex relationships between politics and geography (both human and physical).

- The "where" of politics and the nature of spatial interactions are crucial to understanding relations among people and government.

- Political geographers recognize that the relationship between politics and geography is two-way: politics $\longleftrightarrow$ geography. Thus, political geography can be seen both as the geography of politics and the politics of geography.

- The relations between politics and geography are often driven by particular theories and practices of the world's States. Understanding imperialism, colonialism, heartland theory, and domino theory is key to comprehending how, within the context of the world-system, geography has influenced politics and how politics has influenced geography.

- Political geography deals with the phenomena occurring at all scales of resolution from the global to the household. Important East/West and North/South divisions dominate international politics, whereas regionalism, sectionalism, and similar divisions dominate intra-State politics.

Key Terms

centrifugal forces
   (p. 367)
centripetal forces (p. 367)
confederation (p. 371)

decolonization (p. 381)
democratic rule (p. 401)
domino theory (p. 393)
East/West divide (p. 392)

federal State (p. 368)
geopolitics (p. 358)
gerrymandering
   (p. 403)

international organization
   (p. 394)
nation (p. 366)
nation-state (p. 366)

Exercises

On the Internet

In this chapter, the exercises will explore the relationship between politics and geography. The Internet provides an especially valuable tool for studying this relationship because it provides current information on political events at the international, national, regional, and local level. After completing these exercises you will have a clearer sense of how politics shapes geography and geography shapes politics, especially with respect to how events in one part of the globe affect people and places everywhere.

Unplugged

1. International boundaries are a prominent feature of the political geography of the contemporary world. In this exercise you are asked to explore the impact of a boundary on nationalist attitudes and behaviors. You will need to use the *New York Times Index* to complete this assignment. Using the United States/Mexico border as your key word, describe the range of issues that derive from this juxtaposition of two very different nations. You should concentrate on a five-year period and show which issues grew in importance, which issues declined, and which issues continued to have a consistent news profile throughout the period.

2. National elections usually tell a story about the ways in which regional ideas and attitudes shape the national political agenda. In the 1996 U.S. presidential election, pollsters considered gender an important issue with so-called soccer moms playing a crucial role in re-electing Bill Clinton. Using national election result data available through the Government Documents Division of your college or university library, describe the political geography of soccer moms. The demographics of soccer moms include suburban women between 32 and 50 who have children under 18 and work at least part-time. Did soccer moms vote for Clinton in all regions of the country? If not, which ones did not, and what might explain the regional distribution of this powerful voting bloc?

3. Not all countries elect their leaders on the basis of territorial representation. Many other countries use the system of proportional representation, in which voters cast their ballots for parties rather than for individual candidates. Italy and Israel, for example, employ a proportional representation electoral system. Using international newspaper coverage of the 1996 national elections in Israel, see if you can identify any other ways in which proportional representation affects political party practices and regional interests.

Chapter 10

Urbanization

Port au Prince market

A 1995 report by the United Nations International Children's Fund (UNICEF)[1] blamed "uncontrollable urbanization" in less-developed countries for the widespread creation of "danger zones" in which increasing numbers of children are forced to become beggars, prostitutes, and laborers before they reach their teens. Pointing out that urban populations are growing at twice the general population rate, the report concluded that too many people are being squeezed into cities that have no jobs, no shelter, and no schools to accommodate them. As a consequence, the family and community structures that support children are being destroyed, with the result that increasing numbers of children have to work. For hundreds of thousands of street kids in less-developed countries, "work" means anything that contributes to survival: shining shoes; guiding cars into parking spaces; chasing other street kids away from patrons at an outdoor cafe; working as domestic help; making fireworks; selling drugs. In Abidjan, in the Ivory Coast, 15-year-old Jean-Pierre Godia, who cannot read or write, spends about six hours every day trying to sell 10-roll packets of toilet paper to motorists at a busy intersection. He buys the packets for about $1.20 and sells them for $2. Some days, he doesn't sell any. In the same city, seven-year-old Giulio guides cars into parking spaces outside a chic pastry shop. Since he was five, he has been doing this to help his mother and four siblings who beg on a nearby corner.

Urbanization is one of the most important geographic phenomena in today's world. Another 1995 report by the United Nations[2] concluded that the growth of cities and the urbanization of rural areas are now irreversible due to the global shift to technological, industrial, and service-based economies. The proportion of the world's population living in urban settlements is growing at a rapid rate, and the world's economic, social, cultural, and political processes are increasingly being played out within and between the world's systems of towns and cities. The UN report concluded that few countries are able to handle the urban population crush, which is causing problems on an unprecedented scale with everything from clean water to disease prevention. Already, 10 million people are dying annually in densely populated urban areas from conditions produced by substandard housing and poor sanitation. About 500 million people worldwide are either homeless or living in unfit housing that is life-threatening. This chapter describes the extent and pattern of urbanization across the world, and explains its causes and the changes wrought in people and places as a result of it.

Urban Geography and Urbanization

From small, market towns and fishing ports to megacities of millions of people, the urban areas of the world are the linchpins of human geographies. They have always been a crucial element in spatial organization and the evolution of societies, but today they are more important than ever. Between 1960 and 1995, the number of city-dwellers worldwide rose by 1.5 billion. Cities now account for almost half the world's population. Much of the developed world has become almost completely urbanized (Figure 10.1), while in many peripheral and semiperipheral regions the current *rate* of urbanization is without precedent (Figure 10.2). In 1996 the United Nations held a major conference in Istanbul on human settlement. Statistics prepared for the conference report that some 5 billion urban dwellers will exist by the year 2025, of which 80 percent will be in peripheral and semiperipheral countries. Urbanization on this scale is a remarkable geographical phenomenon—one of the most important sets of processes shaping the world's landscapes.

---

[1]*The Progress of Nations.* New York: United Nations International Children's Fund (UNICEF), 1995.
[2]*Global Report on Human Settlements: An Urbanizing World.* New York: United Nations Center for Human Settlements, 1995.

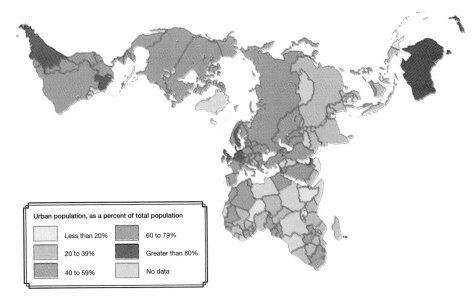

Figure 10.1 The percentage of each country's population living in urban settlements, 1995 The lowest levels of urbanization are found in the African countries of Rwanda and Burundi, where only 7 percent of the population lived in urban settlements in 1995. Most of the core countries are highly urbanized, with between 65 and 95 percent of their populations living in urban settlements. (*Source:* Map projection, Buckminster Fuller Institute and Dymaxion Map Design, Santa Barbara, CA. The word Dymaxion and the Fuller Projection Dymaxion™ Map design are trademarks of the Buckminster Fuller Institute, Santa Barbara, California, ©1938, 1967 & 1992. All rights reserved.)

Towns and cities are engines of economic development and centers of cultural innovation, social transformation, and political change. While they often pose social and environmental problems, they are essential elements in human economic and social organization. Experts on urbanization point to four fundamental aspects of the role of towns and cities in human economic and social organization:

- The *mobilizing function* of urban settlement. Urban settings, with their physical infrastructure and their large and diverse populations, are places where entrepreneurs can get things done. Cities, in other words, provide efficient and effective environments for organizing labor, capital, and raw materials, and for distributing finished products. In developing countries, urban areas produce as much as 60 percent of total gross domestic product with just one-third of the population.

- The *decision-making capacity* of urban settlement. Because urban settings bring together the decision-making machinery of public and private institu-

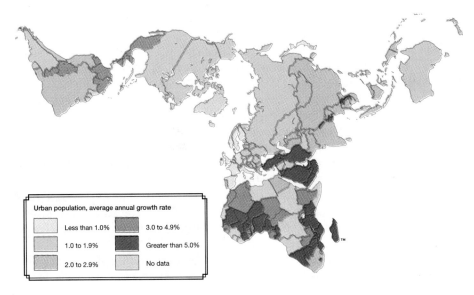

Figure 10.2 Rates of growth in urbanization This map shows the annual average growth rate between 1980 and 1995 in the proportion of people in each country living in urban settlements. Core countries, already highly urbanized, grew quite slowly. The urban populations of peripheral countries such as Burkina Faso, Kenya, Lesotho, Mauritania, and Tanzania, on the other hand, grew by more than six percent each year, creating tremendous pressure on cities' capacities to provide jobs, housing, and public services. (*Source:* Map projection, Buckminster Fuller Institute and Dymaxion Map Design, Santa Barbara, CA. The word Dymaxion and the Fuller Projection Dymaxion™ Map design are trademarks of the Buckminster Fuller Institute, Santa Barbara, California, ©1938, 1967 & 1992. All rights reserved.)

tions and organizations, they come to be concentrations of political and economic power.

- The *generative functions* of urban settlement. The concentration of people in urban settings makes for much greater interaction and competition, which facilitates the generation of innovation, knowledge, and information.
- The *transformative capacity* of urban settlement. The size, density, and variety of urban populations tends to have a liberating effect on people, allowing them to escape the rigidities of traditional, rural society and to participate in a variety of lifestyles and behaviors.

The study of urban geography is concerned with the development of towns and cities around the world, with particular reference to the similarities and differences both *among* and *within* urban places. For urban geographers, some of the most important questions include: What attributes make towns and cities distinctive? How did these distinctive identities evolve? What are the relationships and interdependencies between particular sets of towns and cities? What are the relationships between cities and their surrounding territories? Do significant regularities exist in the spatial organization of land use within cities, in the patterning of neighborhood populations, or in the layout and landscapes of particular kinds of cities?

Urban geographers also want to know about the causes of the patterns and regularities they find. How, for example, do specialized urban subdistricts evolve, why did urban growth occur in a particular region at a particular time, and why did urban growth exhibit a distinctive physical form during a certain period? In pursuing such questions, urban geographers have learned that the answers are ultimately to be found in the wider context of economic, social, cultural, and political life. In other words, towns and cities must be viewed as part of the economies and societies that maintain them.

Urbanization, therefore, is not simply a process of the demographic growth of towns and cities. It also involves many other changes, both quantitative and qualitative. From the geographer's perspective, these changes can be conceptualized in several different ways. One of the most important of these is by examining the attributes and dynamics of urban systems. An **urban system,** or city-system, is any interdependent set of urban settlements within a given region. Thus, for example, we can speak of the French urban system, the African urban system, or even the global urban system. As urbanization takes place, the attributes of urban systems will, of course, reflect the fact that increasing numbers of people are living in ever-larger towns and cities. They will also reflect other important changes, such as changes in the relative size of cities; changes in their functional relationships with one another; and changes in their employment base and population composition.

Urban system: an interdependent set of urban settlements within a specified region.

Another important aspect of change associated with urbanization processes concerns urban form. **Urban form** refers to the physical structure and organization of cities in terms of their land use, layout, and built environment. As urbanization takes place, not only do towns and cities get bigger physically, extending upward and outward, but they also get reorganized, redeveloped, and redesigned in response to changing circumstances.

Urban form: the physical structure and organization of cities.

These changes, in turn, are closely related to a third aspect of change: transformations in patterns of urban ecology. **Urban ecology** is the social and demographic composition of city districts and neighborhoods. Urbanization not only brings more people to cities, it also brings a greater variety of people. As different social, economic, demographic, and racial subgroups become sorted into different territories, so distinctive urban ecologies emerge. As new subgroups arrive or old ones leave, so these ecologies change.

Urban ecology: the social and demographic composition of city districts and neighborhoods.

A fourth aspect of change associated with urbanization concerns people's attitudes and behavior. New forms of social interaction and new ways of life are brought about by the liberating and transformative effects of urban environments. These changes have given rise to the concept of urbanism, which refers to the distinctive nature of social and cultural organization in particular urban settings. **Urbanism** describes the way of life fostered by urban settings, in which the number, physical density, and variety of people often results in distinctive attitudes, values, and patterns of behavior. Geographers are interested in urbanism because of the ways in which it varies both within and between cities.

Urbanism: the way of life, attitudes, values, and patterns of behavior fostered by urban settings.

Urban Origins

It is important to put the geographic study of towns and cities in historical context. After all, many of the world's cities are the product of a long period of development. We can only understand a city, old or young, if we know something about the reasons behind its growth; about the rate at which it has grown; and about the processes that have contributed to its growth.

In broad terms, the earliest urbanization developed independently in the various hearth areas of the first agricultural revolution. The very first region of independent urbanism was in the Middle East, in the valleys of the Tigris and Euphrates (in Mesopotamia) and in the Nile Valley from around 3500 B.C. (see Chapter 4). Together, these intensively cultivated river valleys formed the so-called Fertile Crescent. By 2500 B.C. cities had appeared in the Indus Valley, and by 1800 B.C. they were established in northern China. Other areas of independent urbanism include Mesoamerica (from around 100 B.C.) and Andean America (from around A.D. 800). Meanwhile, the original Middle Eastern urban hearth continued to produce successive generations of urbanized world-empires, including those of Greece (Figure 10.3, page 412), Rome, and Byzantium.

Experts differ in their explanations of these first transitions from subsistence minisystems to city-based world-empires. The classical archeological interpretation emphasizes the availability of an agricultural surplus large enough to allow the emergence of specialized, nonagricultural workers. Some urbanization, however, seems to have been the result of the pressure of population growth. This pressure, it is thought, disturbed the balance between population and resources, causing some people to move to marginal areas. Finding themselves in a region where agricultural conditions were unfavorable, these people either had to devise new techniques of food production and storage or had to establish a new form of economy based on services such as trade, religion, or defense. Any such economy would have required concentrations of people in urban settlements.

Most experts agree that changes in social organization were an important precondition for urbanization. Specifically, urbanization required the emergence of groups who were able to exact tributes, impose taxes, and control labor power, usually through some form of religious persuasion or military coercion. Once established, this elite provided the stimulus for urban development by using its wealth to build palaces, arenas, and monuments in order to show off its power and status. This activity not only created the basis for the physical core of ancient cities (Figure 10.4, page 413) but also required an increased degree of specialization in nonagricultural activities—construction, crafts, administration, the priesthood, soldiery, and so on—which could only be organized effectively in an urban setting.

The urbanized economies of world-empires were a precarious phenomenon, however, and many of them lapsed into ruralism before being revived or recolonized. In a number of cases, the decline of world-empires was a result of demographic setbacks associated with wars or epidemics. Such disasters left too few people to maintain the social and economic infrastructure necessary for urbanization. This lack of labor power seems to have been a major contributing factor to the eventual collapse of the Mesopotamian empire and may also have contributed to the abandonment of much of the Mayan empire more than 500 years

Corinth began to develop as a commercial center in the eighth century B.C., though its site had been inhabited since before 3000 B.C. This photograph shows the ruins of the Temple of Apollo.

Figure 10.3 Greek colonization The Greeks traded and colonized throughout the Aegean from the eighth century B.C. onward. Later, they extended their activities to the central Mediterranean and Black Sea, where their settlements formed important nuclei for subsequent urbanization. The shaded area on the map shows the extent of the "polis," the Greek ideal of a democratic city-state. (*Source:* Map, J. Rich and A. Wallace-Hadrill (Eds.), *City and Country in the Ancient World.* London: Routledge, 1991, Fig. 1b.)

Ancient Athens was at its peak in the fifth century B.C., when, as the largest and wealthiest *polis* (city-state), it became the cultural and intellectual center of the Classical Greek world. The Parthenon, built on top of the Acropolis, dates from this time.

before the arrival of the Spanish. Similarly, the population of the Roman empire began to decline in the second century A.D., giving rise to labor shortages, abandoned fields, and depopulated towns, and allowing the infiltration of "barbarian" settlers and tribes from the German lands of east-central Europe.

The Roots of European Urban Expansion

In Europe, the urban system introduced by the Greeks and re-established by the Romans almost collapsed during the Dark Ages of Early Medieval Europe (A.D. 476–1000). During this period, feudalism gave rise to a fragmented landscape of inflexible and inward-looking world-empires. Feudalism was a rigid, rurally oriented form of economic and social organization based on the communal chiefdoms of Germanic tribes that had invaded the disintegrating Roman Empire. From this unlikely beginning, however, an elaborate urban system developed, its largest centers eventually growing into what would become the nodal centers of a global world-system.

Early Medieval Europe, divided into a patchwork of feudal kingdoms and estates, was mostly rural. Each feudal estate was more or less self-sufficient in terms of foodstuffs, and each kingdom or principality was more or less self-sufficient in terms of the raw materials needed to craft simple products. Most

Figure 10.4 The ancient Middle Eastern city Erbil, in present-day Iraq, has been continuously occupied for over 6,000 years. The old core of the city shown in this photograph is typical of ancient Middle Eastern cities, with a large defensive wall around a round site, and a central cluster of temples and public buildings from which major streets radiated outward, through a close-knit, cellular pattern of alleys and buildings.

regions, however, did support at least a few small towns (Figure 10.5, page 414). The existence of these towns depended mainly on their role:

- **Ecclesiastical or university centers**—Examples include St. Andrews in Scotland; Canterbury, Cambridge, and Coventry in England; Rheims and Chartres in France; Liège in Belgium; Bremen in Germany; Trondheim in Norway; and Lund in Sweden.
- **Defensive strongholds**—Examples include the hilltop towns of central Italy such as Urbino, Foligno, and Montecompatri; and the bastide, or fortress, towns of southwestern France, such as Aigues Mortes and Montauban.
- **Administrative centers for the upper tiers of the feudal hierarchy**—Examples include Köln, Mainz, and Magdeburg in Germany; Falkland in Scotland; Winchester in England; and Toulouse in France.

From the eleventh century onward, however, the feudal system faltered and disintegrated in the face of successive demographic, economic, and political crises that were caused by steady population growth in conjunction with only modest technological improvements and limited amounts of cultivable land. In order to bolster their incomes and raise armies against one another, the feudal nobility began to levy increasingly higher taxes. Their rural peasants were consequently obliged to sell more of their produce for cash on the market. As a result, a more extensive money economy developed, along with the beginnings of a pattern of trade in basic agricultural produce and craft manufactures. Some long-distance trade even began in luxury goods such as spices, furs, silks, fruit, and wine. Towns began to increase in size and vitality on the basis of this trade.

The regional specializations and trading patterns that emerged provided the foundations for a new phase of urbanization based on merchant capitalism (Figure 10.6, page 415). Beginning with networks established by the merchants

Figure 10.5 Some towns of early medieval Europe

▶ In Winchester, England, the influence of the gridiron structure of Roman Winchester (called *Venta Bulgarum*) resulted in an uncharacteristically regular medieval street pattern.

▶ St. Andrews, Scotland, was an important ecclesiastical center. The cathedral was built in the twelfth century, the castle (an episcopal residence), c. 1200. The university was founded in 1410.

▶ Lund, Sweden, was founded in 1020 by the Danish King Canute, and became the seat of the archbishop of all Scandinavia in 1103.

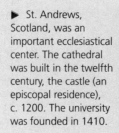

▶ York, England, had been captured by the Danes in the ninth century and was retaken by the English King Athelstan in 926. The Minster (cathedral) was built between about 1250 and 1450 on an old Roman site, about half a kilometer (0.31 miles) from the Danish royal palace.

▶ From the eighth century, Liège, Belgium, was the seat of a dynasty of independent prince-bishops, under whom the town became an important intellectual center with a strong citizens' charter and influential guilds.

▶ Urbino, Italy, on a classic hilltop defensive site, had been settled by Etruscans and Romans before coming under church rule in the ninth century. It then became the center of the dukedom of the Montefeltro family and by the late medieval period had become a hub of artistic and literary activity.

▶ Aigues-Mortes, France, is a classic example of a planned medieval town, laid out on a grid plan, with rectilinear walls. These planned towns, or *bastides,* were established as military garrisons in the outlying regions of French and English kingdoms in the thirteenth century.

▶ Magdeburg, Germany, was a tiny trading settlement until Otto I founded the Benedictine Abbey of Saints Peter, Maurice, and Innocent in the tenth century.

414

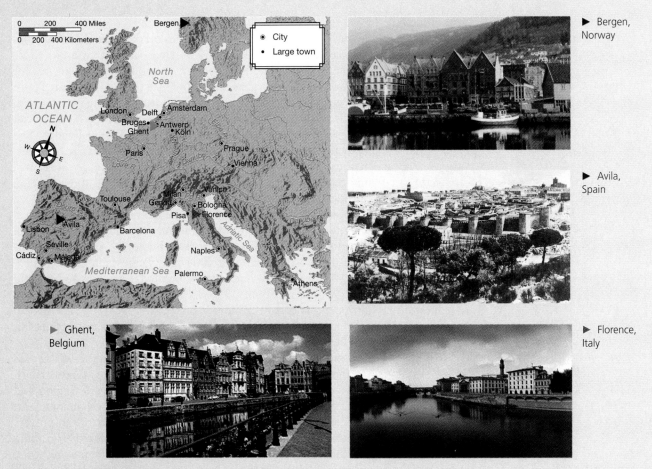

Figure 10.6 The towns and cities of Europe, c. 1350 Cities with more than 10,000 residents were uncommon in Medieval Europe except in northern Italy and Flanders, where the spread of cloth production and the growth of trade permitted relatively intense urbanization. Elsewhere, large size was associated with a complex of administrative, religious, educational, and economic functions. By 1350, many of the bigger towns (for example, Barcelona, Köln, Prague) supported universities as well as a variety of religious institutions. Most urban systems, reflecting the economic and political realities of the time, were relatively small in extent. (*Source:* Map, P. M. Hohenberg and L. H. Lees, *The Making of Urban Europe 1000–1950.* Cambridge, MA: Harvard University Press, 1985, Fig. 2.1.)

of Venice, Pisa, Genoa, and Florence (in northern Italy) and the trading partners of the Hanseatic League (a federation of city-states around the North Sea and Baltic coasts), a trading system of immense complexity soon came to span Europe from Bergen to Athens and from Lisbon to Vienna. By 1400, long-distance trading was well established, based not on the luxury goods of the pioneer merchants but on bulky staples such as grains, wine, salt, wool, cloth, and metals. Milan, Genoa, Venice, and Bruges had all grown to 100,000 or more. Paris was the dominant European city, with a population of about 275,000. This was the Europe that stood poised to extend its grasp to a global scale.

Between the fifteenth and seventeenth centuries, a series of changes occurred that transformed not only the cities and city-systems of Europe but also the entire world economy. Merchant capitalism increased in scale and sophistication; economic and social reorganization was stimulated by the Protestant Reformation and the Scientific Revolution. Meanwhile, aggressive overseas colonization made Europeans the leaders, persuaders, and shapers of the rest of the world's economies and societies. Spanish and Portuguese colonists were the first to extend the European urban system into the world's peripheral regions. They established the basis of a Latin American urban system in just 60 years, between 1520 and 1580. Spanish colonists founded their cities on the sites of Indian cities (as, for example in Oaxaca and Mexico City, Mexico; Cajamarca and Cuzco, Peru; and Quito, Ecuador) or in regions of dense indigenous populations (as, for example, in Puebla and Guadalajara, Mexico; and Arequipa and Lima, Peru). These colonial towns were established mainly as administrative and military centers from which the Spanish Crown could occupy and exploit the New World. Portuguese colonists, in contrast, situated their cities—Recife, Salvador, São Paulo, and Rio de Janeiro—with commercial rather than administrative considerations in mind. They, too, were motivated by exploitation, but their strategy was to establish colonial towns in locations best suited to organizing the collection and export of the products of their mines and plantations.

In Europe, Renaissance reorganization saw the centralization of political power and the formation of national States; the beginnings of industrialization; and the funneling of plunder and produce from distant colonies. In this new context, the port cities of the North Sea and Atlantic coasts enjoyed a decisive locational advantage. By 1700, London had grown to 500,000, while Lisbon and Amsterdam had each grown to about 175,000. The cities of continental and Mediterranean Europe expanded at a more modest rate. By 1700 Venice had added only 30,000 to its 1400 population of 110,000, while Milan's population had not grown at all between 1400 and 1700.

The most important aspect of urbanization during this period, however, was the establishment of gateway cities around the rest of the world (Figure 10.7). **Gateway cities** are those that serve as a link between one country or region and others because of their physical situation. They are control centers that command entrance to, and exit from, their particular country or region. European powers founded or developed literally thousands of towns as they extended their trading networks and established their colonies. The great majority of them were ports. Protected by fortifications and European naval power, they began as trading posts and colonial administrative centers. Before long, they developed manufacturing of their own to supply the pioneers' needs, along with more extensive commercial and financial services.

Gateway city: city that serves as a link between one country or region and others because of its physical situation.

As colonies were developed and trading networks expanded, some of these ports grew rapidly, acting as gateways for colonial expansion into continental interiors. Into their harbors came waves of European settlers; through their docks were funneled the produce of continental interiors. Rio de Janeiro (Brazil) grew on the basis of gold mining; Accra (Ghana) grew on the basis of cocoa; Buenos Aires (Argentina) on mutton, wool, and cereals; Calcutta (India) on jute, cotton, and textiles; São Paulo (Brazil) on the basis of coffee; and so on. As they grew into major population centers, they became important markets for imported European manufactures, adding even more to their functions as gateways for international transport and trade.

Industrialization and Urbanization

It was not until the late eighteenth century, however, that urbanization came to be an important dimension of the world-system in its own right. In 1800, less than 5 percent of the world's 980 million people lived in towns and cities. By 1950, however, 16 percent of the world's population was urban, and there were more than 900 cities of 100,000 or more around the world. The Industrial Revolution and European imperialism had created unprecedented concentrations of humanity that were intimately linked in networks and hierarchies of interdependence.

Cities were synonymous with industrialization. Industrial economies could only be organized with the large pools of labor; the transportation networks; the physical infrastructure of factories, warehouses, stores, and offices; and the consumer markets provided by cities. As industrialization spread throughout Europe

Figure 10.7 Gateway cities in the world-system periphery Many of the world's most important cities grew to prominence as gateway cities because they commanded routeways into and out of developing colonies. Gateway cities are control centers that command entrance to, and exit from, their particular country or region. (Source: Map projection, Buckminster Fuller Institute and Dymaxion Map Design, Santa Barbara, CA. The word Dymaxion and the Fuller Projection Dymaxion™ Map design are trademarks of the Buckminster Fuller Institute, Santa Barbara, California, ©1938, 1967 & 1992. All rights reserved.)

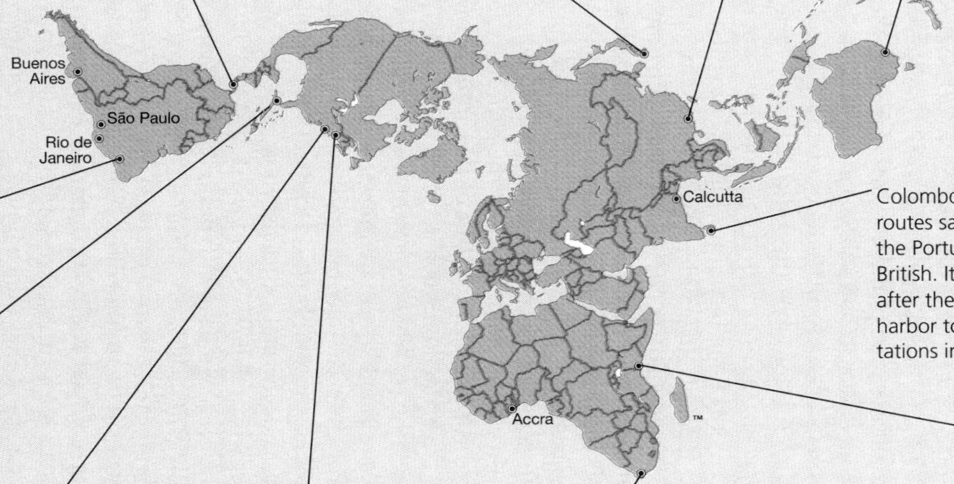

Sydney, Australia, was not settled until the late eighteenth century, and even then many of the settlers were convicts who had been forcibly transported from Britain. It soon became the gateway for agricultural and mineral exports (mostly to Britain) and for imports of manufactured goods and European immigrants.

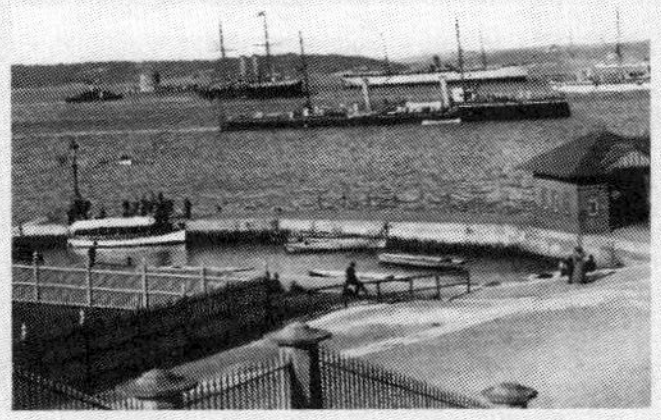

Guangzhou was the first Chinese port to be in regular contact with European traders—first Portuguese in the sixteenth century and then British in the seventeenth century.

Colombo's strategic situation on trade routes saw it occupied successively by the Portuguese, the Dutch, and the British. It became an important gateway after the British constructed an artificial harbor to handle exports from tea plantations in Ceylon (now Sri Lanka).

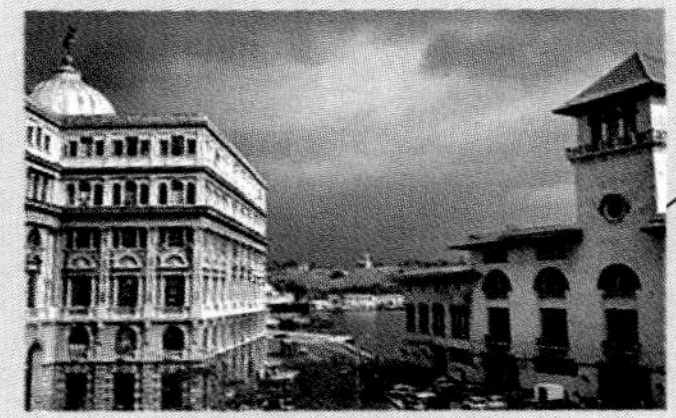

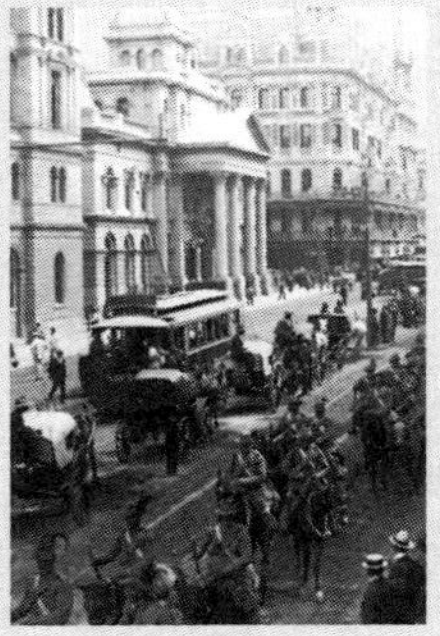

Mombasa (in present-day Kenya) was already a significant Arab trading port when Vasco da Gama visited it in 1498 on his first voyage to India. The Portuguese used it as a trading station until it was recaptured by the Arabs in 1698. It did not become an important gateway port until it fell under British imperial rule in the nineteenth century, when railroad development opened up the interior of Kenya, along with Rwanda, Uganda, and northern Tanzania.

Nagasaki was the only port that feudal Japanese leaders allowed open to European traders, and for more than 200 years Dutch merchants held a monopoly of the import-export business through the city.

Panama City, founded by the Spanish in 1519, became the gateway for gold and silver on its way by galleon to Spain.

Cape Town was founded in 1652 as a provisioning station for ships of the Dutch East India Company. Later, under British rule, it developed into an import-export gateway for South Africa.

Boston first flourished as the principal colony of the Massachusetts Bay Company, exporting furs and fish and importing slaves from West Africa, hardwoods from central America, molasses from the Caribbean, manufactured goods from Europe, and tea (via Europe) from South Asia.

Salvador, Brazil, was the landfall of the Portuguese in 1500. They established plantations that were worked by slave labor from West Africa. Salvador became the gateway for most of the 3.5 million slaves who were shipped to Brazil between 1526 and 1870.

New York, at first a modest Dutch fur-trading port, became the gateway for millions of European immigrants and for a large volume of U.S. agricultural and manufacturing exports.

Havana was founded and developed by the Spanish in 1515 because of its excellent harbor. It was used as the assembly point for annual convoys returning to Spain.

Calcutta

Accra

São Paulo

Buenos Aires

Rio de Janeiro

417

in the first half of the nineteenth century and then to other parts of the world, so urbanization increased at a faster pace. The higher wages and greater variety of opportunities in urban labor markets attracted migrants from surrounding areas. The countryside began to empty. In Europe, the *demographic transition* caused a rapid growth in population as death rates dropped dramatically (see Chapter 3). This growth in population provided a massive increase in the labor supply throughout the nineteenth century, further boosting the rate of urbanization, not only within Europe itself but also in Australia, Canada, New Zealand, South Africa, and the United States, as emigration spread industrialization and urbanization to the frontiers of the world-system.

The shock city of nineteenth-century European industrialization was Manchester, England, which grew from a small town of 15,000 in 1750 to a city of 70,000 in 1801; a metropolis of 500,000 in 1861; and a world city of 2.3 million by 1911. A **shock city** is one that is seen at the time as the embodiment of surprising and disturbing changes in economic, social, and cultural life. As industrialization took hold in North America, the shock city was Chicago, which grew from under 30,000 in 1850 to 500,000 in 1880, 1.7 million in 1900, and 3.3 million in 1930. Both Manchester and Chicago were archetypal forms of an entirely new kind of city—the *industrial city*—whose fundamental reason for existence was not, as in earlier generations of cities, to fulfill military, political, ecclesiastical, or trading functions. Rather, it existed simply to assemble raw materials and to fabricate, assemble, and distribute manufactured goods. Both Manchester and Chicago had to cope, however, with unprecedented rates of growth and the unprecedented economic, social, and political problems that were a consequence of their growth (see Feature 10.1, "Visualizing Geography—Shock Cities: Manchester and Chicago"). Both were also *world cities* in which a disproportionate part of the world's most important business—economic, political, and cultural—is conducted. At the top of a global urban system, they experienced growth largely as a result of their role as key nodes in the world economy.

Throughout the nineteenth century, European imperialism gave a significant impetus to urbanization in the world's peripheral regions. New gateway cities were founded and, as Europeans raced to establish economic and political control over continental interiors, colonial cities were established as centers of administration, political control, and commerce. **Colonial cities** are those that were deliberately established or developed as administrative or commercial centers by colonial or imperial powers. In fact, geographers often distinguish between two distinct types of colonial city. One, the pure colonial city, was usually established, or "planted," by colonial administrations in a location where no significant urban settlement had previously existed. Such cities were laid out expressly to fulfill colonial functions, with ceremonial spaces, offices, and depots for colonial traders, plantation representatives, and government officials; barracks for a garrison of soldiers; and housing for colonists. Subsequently, as these cities grew, they added housing and commercial land uses for local peoples drawn by the opportunity to obtain jobs such as servants, clerks, or porters. Examples of such "pure" colonial cities are Mumbai (formerly Bombay), Calcutta, Ho Chi Minh City (Saigon), Hong Kong, Jakarta, Manila, and Nairobi.

In the other type of colonial city, colonial functions were grafted onto an existing settlement, taking advantage of a good site and a ready supply of labor. Examples include Delhi, Mexico City, Shanghai, and Tunis. In these cities, the colonial imprint is most visible in and around the city center in the formal squares and public spaces; the layout of avenues; and the presence of colonial architecture and monuments. This architecture includes churches, city halls, and railway stations (Figure 10.8); the palaces of governors and archbishops; and the houses of wealthy traders, colonial administrators, and landowners.

The colonial legacy can also be read in the building and planning regulations of many of these cities. Often, colonial planning regulations were copied from

Shock city: city that is seen as the embodiment of surprising and disturbing changes in economic, social, and cultural life.

Colonial cities: cities that were deliberately established or developed as administrative or commercial centers by colonial or imperial powers.

Statue of Queen Victoria, Mumbai
(formerly Bombay), India

Victoria Station, Mumbai, India

**Figure 10.8 Colonial architecture and urban
design** Cities in the periphery of the world-system
have grown very rapidly since the colonial era, but the
legacy of the colonial period can still be seen in the
architecture, monuments, and urban design of the
period.

those that had been established in the colonizing country. Because these regulations were based on Western concepts, many of them turned out to be inappropriate to colonial settings. Most colonial building codes, for example, are based on Western models of family and work, with a small family living in a residential area that is some distance from the adults' places of work. This is at odds with the needs of large, extended families whose members are involved with a busy domestic economy and with family businesses that are traditionally integrated with the residential setting. Colonial planning, with its gridiron street layouts; zoning regulations that do not allow for a mixture of land uses; and building codes designed for European climates, ignored the specific needs of local communities and misunderstood their cultural preferences.

World Urbanization Today

It is difficult to say just how urbanized the world has become. In many parts of the world, urban growth is taking place at such a pace and under such chaotic conditions that it is impossible even for experts to do more than provide informed estimates. The most comprehensive source of statistics is the United Nations, whose data suggest that almost half of the world's population is now urban. These data incorporate the very different definitions of "urban" used by different countries. Some countries (Australia and Canada, for example) count any settlement of 1,000 people or more as urban, while others (including Italy and Jordan) use 10,000 as the minimum for an urban settlement; and Japan uses 50,000 as the cut-off. This, by the way, tells us something about the nature of urbanization itself: It is a *relative* phenomenon. In countries like Peru, where population is thin and scattered, a settlement of 2,000 represents a significant center. In countries like Japan, however, with greater numbers, higher densities, and a tradition of centralized rather than scattered agricultural settlement, a much larger concentration of people is required to count as "urban."

Taking 100,000 as an arbitrary minimum size for an "urban" place, just over one-third of the world's population is now urbanized. As Table 10.1 (page 422) shows, North America is the most urbanized continent in the world, with almost 80 percent of its population living in cities of 100,000 or more. In contrast, Africa is only 20 percent urban.

"One day I walked with one of these middle class gentlemen into Manchester. I spoke to him about the disgraceful unhealthy slums and drew his attention to the disgusting condition of that part of the town in which the factory worker lived. I declared that I had never seen so badly built a town in my life. He listened patiently and at the corner of the street he remarked: 'And yet there is a great deal of money to be made here. Good morning, Sir!'"

Friedrich Engels, *The Condition of the Working Class in England in 1844*

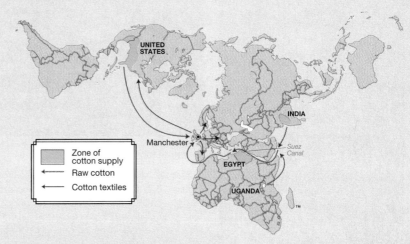

Zone of cotton supply
Raw cotton
Cotton textiles

The opening of the Suez Canal in 1869 halved the traveling time between Britain and India. It ruined the Indian domestic cotton textile industry, but it allowed India to export its raw cotton to Manchester. Around the same time, British colonialists established cotton plantations in Egypt and Uganda, providing another source of supply. (*Source:* Map projection, Buckminster Fuller Institute and Dymaxion Map Design, Santa Barbara, CA. The word Dymaxion and the Fuller Projection Dymaxion™ Map design are trademarks of the Buckminster Fuller Institute, Santa Barbara, California, ©1938, 1967 & 1992. All rights reserved.)

In the mid-nineteenth century, the United States produced over 80 percent of the world's raw cotton, much of it from plantations like this one in Georgia. Manchester was the chief consumer of this cotton; and it in turn became the world's chief exporter of cotton textiles.

The Manchester Ship Canal (1894), a joint undertaking of the municipality and private enterprise, connected Manchester with the Irish Sea and world markets beyond. The canal revived the city's trade and made possible the development of a concentration of heavy industry in nearby Trafford Park Estate.

Manchester's first cotton mill was built in the early 1780s, and by 1830 there were 99 cotton-spinning mills. As the city grew, it spilled out into the surrounding countryside, bringing its characteristic landscape of red-brick terrace housing and "Dark Satanic Mills" with their tall brick chimneys.

Manchester City Hall, a classic example of Victorian Gothic architecture, was built to show the world that the city had arrived. Manchester in the nineteenth century was a city of enormous vitality not only in its economic life but also in its political, cultural, and intellectual life.

In the late nineteenth century, working-class housing was built to conform with local building codes—but only just. Much of it has now been replaced, but a good deal still remains.

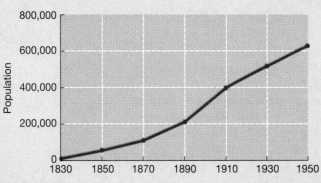

Migrants from Ireland and northern England contributed to Manchester's rapid growth from the mid-nineteenth century.

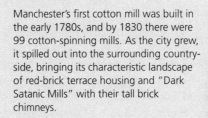

When Chicago was first incorporated as a city in 1837, its population was only 4,200. Its growth followed the arrival of the railroads, which made the city a major transportation hub. By the 1860s, lake vessels were carrying iron ore from the Upper Michigan ranges to the city's blast furnaces, and railroads were hauling cattle, hogs, and sheep to the city for slaughtering and packing. The city's prime geographic situation also made it the nation's major lumber-distributing center by the 1880s.

Chicago's immigrant and African American neighborhoods were an entirely new urban phenomenon—highly segregated, and with very distinctive social and cultural attributes. The 1880 and 1890 censuses showed that more than three-fourths of Chicago's population was made up of foreign-born immigrants and their children. These photographs of ethnic neighborhoods in Chicago's Southside were taken in 1941.

Chicago announced its prosperity through elaborate skyscrapers and towers. This picture is of the Tribune Tower, built in Gothic Revival style, based on the winning entry in an international design competition organized by the *Chicago Tribune* in 1922. The city has regarded itself ever since as a sponsor of landmark architecture.

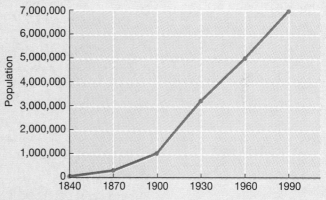

Immigrants from Europe fueled Chicago's phenomenal early growth.

Hog-butcher for the World,
Tool Maker, Stacker of Wheat
Player with Railroads and the Nation's
Freight Handler;
Stormy, husky, brawling,
City of the Big Shoulders.

Carl Sandburg, 1916

1870

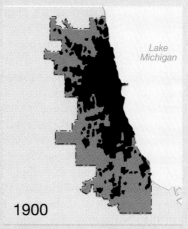

1900

1930

In 1870, when Manchester was already a thriving metropolis, Chicago was at the beginning of a period of explosive growth. A year later, nine square kilometers (4 sq. mi.) of the city, including the business district, were destroyed by fire. It was rebuilt rapidly, with prosperous industrialists taking the opportunity to build impressive new structures in the downtown area. The city's economic and social elite colonized the Lake Michigan shore, while heavy industry, warehouses, and railyards crowded the banks of the Chicago River, stretching northwestward from the city center. To the south of the city center were the Union Stockyards and a pocket of heavy industry where the Calumet River met Lake Michigan. All around were the homes of working families, in neighborhoods that spread rapidly outward as wave after wave of immigrants arrived in the city.

421

TABLE 10.1 Urbanization by Major World Regions, 1990		
	Percentage of Total Population		
	In Cities of 1 Million or More	In Cities of 100,000 to 999,999	In All Cities of 100,000 or More
Africa	11.2	7.9	19.1
Asia	15.5	7.7	23.2
Latin America	32.7	14.8	47.5
North America	52.6	25.2	77.8
Europe	24.2	23.7	47.9
Oceania	44.5	17.7	62.2
World	**19.5**	**12.0**	**31.5**

Source: United Nations, *World Urbanization Prospects.* New York: U.N. Department of Economic and Social Information, 1993.

To put these figures in perspective, only 16 percent of the world's population was urbanized in 1950, using the same cut-off of 100,000 inhabitants. In that year there were only 71 metropolitan areas of a million or more, and only six of five million or more existed; today the statistics are approximately 350 and 40, respectively. Looking ahead, population projections for 2010 suggest that more than 35 percent of the world's population will be living in cities of 100,000 or more; and there will be around 500 cities with a population of a million or more, including at least 60 of five million or more.

Regional Trends and Projections

The single most important aspect of world urbanization, from a geographical perspective, is the striking difference in trends and projections between the core regions and the semiperipheral and peripheral regions. In 1950, 21 of the world's largest 30 metropolitan areas were located in core countries—11 of them in Europe and 6 in North America. By 1980 the situation was completely reversed, with 19 of the largest 30 located in peripheral and semiperipheral regions. By 2010, all but 5 of the 30 largest metropolitan areas are expected to be located in peripheral and semiperipheral regions (Table 10.2; see also Figure 10.9, page 424).

Asia provides some of the most dramatic examples of this trend. From a region of villages, Asia is fast becoming a region of cities and towns. Between 1950 and 1985, for example, its urban population rose nearly fourfold to 480 million people. By 2020, half of Asia's population will be living in urban areas. Nowhere is the trend toward rapid urbanization more pronounced than in China, where for decades the communist government imposed strict controls on where people were allowed to live, fearing the transformative and liberating effects of cities. By tying people's jobs, school admission, and even the right to buy food to the places where people were registered to live, the government made it almost impossible for rural residents to migrate to towns or cities. As a result, more than 70 percent of China's one billion population still lived in the countryside in 1985. Now, however, China is rapidly making up for lost time. The Chinese government, having decided that towns and cities can be engines of economic growth within a communist system, has not only relaxed residency laws but also drawn up plans to establish over 430 new cities. Between 1982 and 1994 the number of people living in cities in China more than doubled, from 152 million to 350 million.

In the world's core countries, levels of urbanization are high, and have been high for some time. According to their own national definitions, the populations of Belgium, the Netherlands, and the United Kingdom are more than 90 percent urbanized, while those of Australia, Canada, Denmark, France, Germany, Japan, New Zealand, Spain, Sweden, and the United States are all more than 75 percent urbanized. In these core countries, however, *rates* of urbanization are relatively low, just as their overall rate of population growth is slow (see Chapter 3).

Levels of urbanization are also very high in many of the world's semiperipheral countries. Brazil, Hong Kong, Mexico, Taiwan, Singapore, and South Korea, for example, are all at least 75 percent urbanized. Unlike the core countries, however, their rate of urban growth has been high. In peripheral countries, the contrast is even greater.

Whatever the current level of urbanization in peripheral countries, almost all are experiencing high rates of urbanization, with growth forecast of unprecedented speed and unmatched size. Karachi, Pakistan, a metropolis of 1.03 million in 1950, had reached 10.77 million in 1995, and is expected to be 20.6 million by 2015. Likewise, Cairo, Egypt, grew from 2.41 million to 10.65 million between

TABLE 10.2 The World's 30 Largest Metro Areas, Ranked by Population Size, 1950, 1980, and 2010 (in millions)

1950	Population	1980	Population	2010	Population
New York	12.3	Tokyo	21.9	Tokyo	28.9
London	8.7	New York	15.6	São Paulo	25.0
Tokyo	6.9	Mexico City	13.9	Mumbai (Bombay)	24.4
Paris	5.4	São Paulo	12.1	Shanghai	21.7
Moscow	5.4	Shanghai	11.7	Lagos	21.1
Shanghai	5.3	Osaka	10.0	Mexico City	18.0
Essen	5.3	Buenos Aires	9.9	Beijing	18.0
Buenos Aires	5.0	Los Angeles	9.5	Dacca	17.6
Chicago	4.9	Calcutta	9.0	New York	17.2
Calcutta	4.4	Beijing	9.0	Jakarta	17.2
Osaka	4.1	Paris	8.7	Karachi	17.0
Los Angeles	4.0	Rio de Janeiro	8.7	Metro Manila	16.1
Beijing	3.9	Seoul	8.3	Tianjin	15.7
Milan	3.6	Moscow	8.2	Calcutta	15.7
Berlin	3.3	Bombay (Mumbai)	8.0	Delhi	15.6
Mexico City	3.1	London	7.8	Los Angeles	13.9
Philadelphia	2.9	Tianjin	7.7	Seoul	13.8
St. Petersburg	2.9	Cairo	6.9	Buenos Aires	13.7
Bombay (Mumbai)	2.9	Chicago	6.8	Cairo	13.4
Rio de Janeiro	2.9	Essen	6.7	Rio de Janeiro	13.3
Detroit	2.8	Jakarta	6.4	Bangkok	12.7
Naples	2.8	Metro Manila	6.0	Teheran	11.9
Manchester	2.5	Delhi	5.5	Istanbul	11.8
São Paulo	2.4	Milan	5.4	Osaka	10.6
Cairo	2.4	Teheran	5.4	Moscow	10.4
Tianjin	2.4	Karachi	5.0	Lima	10.1
Birmingham	2.3	Bangkok	4.8	Paris	9.6
Frankfurt	2.3	St. Petersburg	4.7	Hyderabad	9.4
Boston	2.2	Hong Kong	4.5	Lahore	8.8
Hamburg	2.2	Lima	4.4	Madras (Chenai)	8.4

Source: United Nations, *World Urbanization Prospects.* New York: U.N. Department of Economic and Social Information, 1993.

Figure 10.9 Population growth in the world's largest metropolitan areas, 1950–2000 While the metropolitan areas of the world's core countries have continued to grow, most of them have been overtaken by the startling growth of the "unintended' metropolises of peripheral and semiperipheral countries. (*Source:* United Nations Development Program, *Human Development Report.* New York: Oxford University Press, 1990.)

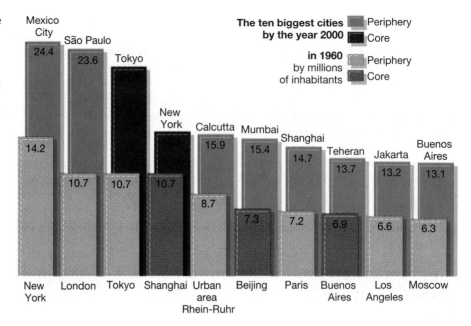

1950 and 1995, and is expected to reach 16 million by 2015. Mumbai, formerly Bombay (India), Jakarta (Indonesia), Lagos (Nigeria), São Paulo (Brazil), and Shanghai (China) are all projected to have populations in excess of 20 million by 2015. The reasons for this urban growth vary. Wars in Liberia and Sierra Leone have pushed hundreds of thousands of people into their capitals, Monrovia and Freetown. In Mauritania, Niger, and other countries bordering the Sahara, deforestation and overgrazing have allowed the desert to expand and swallow up villages, forcing people toward cities. For the most part, though, urban growth in peripheral countries is a consequence of the onset of the demographic transition (see Chapter 3), which has produced fast-growing rural populations in regions that face increasing problems of agricultural development (see Chapter 8). As a response, many people in these regions migrate to urban areas seeking a better life.

Many of the largest cities in the periphery are growing at annual rates of between 4 and 7 percent, which means that they have a doubling time between 10 and 17 years. The doubling time of a city's population is the time needed for it to double in size, at current growth rates. To put the situation in numerical terms, metropolitan areas like Mexico City and São Paulo are adding half a million persons to their population each year: nearly 10,000 every week, even after making up for deaths and out-migrants. It took London 190 years to grow from half a million to 10 million; it took New York 140 years. By contrast, Mexico City, São Paulo, Buenos Aires, Calcutta, Rio de Janeiro, Seoul, and Mumbai all took less than 75 years to grow from half a million to 10 million inhabitants.

Urban Systems

Every town and city is part of one of the interlocking urban systems that link regional-, national-, and international-scale human geographies in a complex web of interdependence. These urban systems organize space through hierarchies of cities of different sizes and functions. Many of these hierarchical urban systems exhibit certain common attributes and features, particularly in terms of the relative size and spacing of individual towns and cities.

Central Places Geographers have long recognized the tendency for the functions of towns and cities as market centers to result in a hierarchical system of central places. A **central place** is a settlement in which certain types of products and services are available to consumers. The tendency for central places to be organized in hierarchical systems was first explored by Walter Christaller, a

Central place: a settlement in which certain products and services are available to consumers.

German geographer, in the 1930s. Christaller's ideas gave rise to central place theory. **Central place theory** seeks to explain the relative size and geographic spacing of towns and cities as a function of people's shopping behavior. Christaller had noticed that, in southern Germany, quite numerous were the smaller places, each offering a limited assortment of stores, services, and amenities for their residents and the residents of nearby areas. He noticed that they also tended to be located at relatively short and consistent distances from one another. Large towns and cities, on the other hand, were fewer and farther between, but offered a much greater variety of stores, services, and amenities, many of them catering to customers and clients from quite distant towns and intermediate rural areas. In between were intermediate-sized places serving intermediate-sized markets with middling collections of stores, services, and amenities.

Christaller termed these collections of stores, services, and amenities "central place functions." Towns and cities, as different-sized central places, serve different-sized territories and populations. In order to explain the observed tendency for a hierarchy of central places, Christaller drew on principles concerning the range and threshold of central place functions. The **range** of a product or service is the maximum distance that consumers will normally travel to obtain it. The area that will be served by a particular central place function from a particular central place will be a roughly circular area whose radius is equal to the range of the function. "High-order" goods and services are those that are relatively costly and generally required infrequently (specialized equipment, professional sports, and specialized medical care, for example). They have the greatest range—100 or more kilometers (62 miles) is not unusual. At the other extreme, "low order" goods and services are those that are relatively inexpensive and required at frequent intervals (bakery and dairy products, and groceries, for example). They have a very short range, perhaps as low as 500 meters (about one-third of a mile).

The **threshold** of a good or service can be thought of in terms of the minimum market area with enough potential buyers to make the enterprise profitable. High-order services such as hospitals have thresholds in the tens of thousands of people. Low-order services, such as small grocery stores, can have thresholds of between 200 and 300 people.

It follows that in any given region a need will exist for only a limited number of large central places in which all of the higher-order goods and services are provided. It is also logical that these places will also provide the entire spectrum of central-place functions. The number and spacing of other, smaller central places will depend on the conjunction of different-sized ranges and thresholds. Using a series of different starting assumptions, Christaller was able to demonstrate that, under ideal circumstances (on flat plains, with good transportation in every direction), towns and cities tend to be arranged in clear hierarchies, with hexagonal-shaped market areas of different sizes arranged around different-sized places (Figure 10.10). Although such circumstances are never approached in real life, geographers did subsequently find many instances where hierarchies of central places existed, with a high degree of regularity in their size and spacing. Today, in fact, relatively few regions exist where the functions of most towns and cities are still dominated by local markets and shopping. Nevertheless, the urban systems of most regions do exhibit a clear hierarchical structure. This is partly a legacy of past eras, when towns and cities did function mainly as market centers for surrounding agricultural areas. Figure 10.11 (see page 426) shows a typical example: the Spanish urban system, with smaller towns and cities functioning interdependently with successively larger ones, the whole system dominated by one or two metropolitan areas whose linkages are national in scope.

Urban systems also exhibit clear *functional* differences within such hierarchies, yet another reflection of the interdependence of places. The geographical division of labor resulting from such processes of economic development (Chapter 7) means that many medium- and larger-sized cities perform quite specialized economic functions and so acquire quite distinctive characters. Thus, for

Central place theory: a theory that seeks to explain the relative size and spacing of towns and cities as a function of people's shopping behavior.

Range: the maximum distance that consumers will normally travel to obtain a particular product or service.

Threshold: the minimum market size required to make the sale of a particular product or service profitable.

Figure 10.10 Central places and locational hierarchies This diagram illustrates Walter Christaller's basic concept of a hierarchy of settlements (central places) of different sizes, with successively larger settlements offering a greater variety of goods and services, thereby commanding a broader market territory. The hexagonal market areas were hypothetical, allowing Christaller to avoid gaps or overlapping market areas.

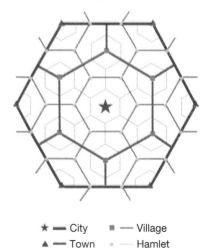

★ — City   ■ — Village
▲ — Town   • — Hamlet

Figure 10.11 The Spanish urban system   (1) National metropolises, (2) regional metropolises, (3) middle-order cities, (4) small cities, (5) principal intercity linkages, (6) high-density regions, (7) regions of high-income growth, and (8) regions of high density and high income growth. Note how the smaller cities tend to be linked to middle-order cities, while these, in turn, are linked to regional metropolises, which are linked to the national metropolises, Madrid and Barcelona. These linkages represent the major flows of capital, information, and goods within the Spanish urban system. (*Source:* L. Bourne, R. Sinclair, M. Ferrer, and A. d'Entremont (Eds.), *The Changing Geography of Urban Systems.* Department of Human Geography, Universidad de Navarra, Navarra, Spain, 1989, Fig. 2, p. 46.)

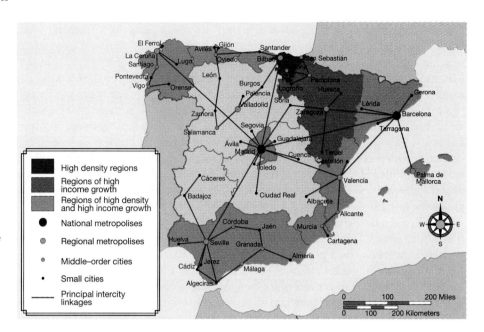

example, there are steel towns (for example, Pittsburgh, Pennsylvania; Sheffield, England), textile towns (for example, Lowell, Massachusetts; Manchester, England), and auto-manufacturing towns (for example, Detroit; Oxford, England; Turin, Italy; Toyota City, Japan; Togliattigrad, Russia). Some towns and cities, of course, do evolve as general-purpose urban centers, providing an evenly balanced range of functions for their own particular sphere of influence. Figure 10.12 shows the example of the urban system in the United States, where the top

Figure 10.12 Functional specialization within an urban system   Different cities tend to specialize in particular kinds of economic activities, while some provide a much greater range of functions than others. This map shows a functional classification of the urban system in the United States. The top tier of the system consists of world cities that provide high-order functions to a global marketplace. Among these, New York stands alone as the dominant metropolis. The next tier consists of cities with diverse functions but only regional importance (for example, Atlanta, Minneapolis). The third tier consists of more specialized centers of business, government, and producer services (for example, Austin, Texas; Albany, New York; and Hartford, Connecticut). The fourth tier consists of still more specialized cities of various kinds: manufacturing centers (for example, Buffalo, New York; Chattanooga, Tennessee), mining/industrial centers (Charleston, West Virginia; Duluth, Minnesota), industrial/military centers (Newport News, Virginia; San Diego, California), and resort/retirement centers (Las Vegas, Nevada; Orlando, Florida). (*Source:* Reprinted with permission of Prentice Hall, from P. L. Knox, *Urbanization,* © 1994, p. 64.)

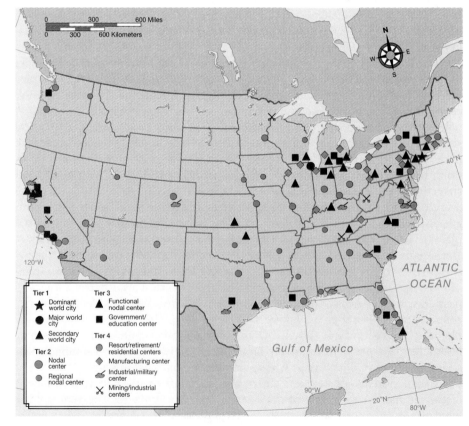

tier of cities consists of centers of global importance (including Chicago, New York, and Los Angeles) that provide high-order functions to an international marketplace. The second tier consists of general-purpose cities with diverse functions but only regional importance (including Atlanta, Miami, and Boston), and the third and fourth tiers consist of more specialized centers of subregional and local importance.

City-Size Distributions, Primacy, and Centrality The functional interdependency between places within urban systems tends to result in a distinctive relationship between the population size of cities and their rank within the overall hierarchy. This relationship is known as the **rank-size rule,** which describes a certain statistical regularity in the city-size distributions of countries and regions. The relationship is such that the *n*th largest city in a country or region is 1/*n* the size of the largest city in that country or region. Thus, if the largest city in a particular system has a population of one million, the fifth-largest city should have a population one-fifth as big (that is, 200,000); the hundredth-ranked city should have a population one-hundredth as big (that is, 10,000), and so on. Plotting this relationship on a graph with a logarithmic scale for population sizes would produce a perfectly straight line. The actual rank-size relationship for the U.S. urban system has always come close to this (Figure 10.13). Over time, the slope has moved to the right on the graph, reflecting the growth of towns and cities at every level in the urban hierarchy.

In some urban systems, the top of the rank-size distribution is distorted as a result of the disproportionate size of the largest (and sometimes also the second-largest) city (Figure 10.14, page 428). In Argentina, for example, Buenos Aires is more than 10 times the size of Rosario, the second-largest city. In the United Kingdom, London is more than nine times the size of Birmingham. In France, Paris is more than eight times the size of Marseilles. In Brazil, both Rio de Janeiro and São Paulo are five times the size of Belo Horizonte, the third-largest city. Geographers call this condition **primacy,** occurring when the population of the largest city in an urban system is disproportionately large in relation to the second- and third-largest cities in that system. Cities like London and Buenos Aires are termed "primate cities."

Rank-size rule: a statistical regularity in city-size distributions of cities and regions.

Primacy: condition in which the population of the largest city in an urban system is disproportionately large in relation to the second- and third-largest cities in that system.

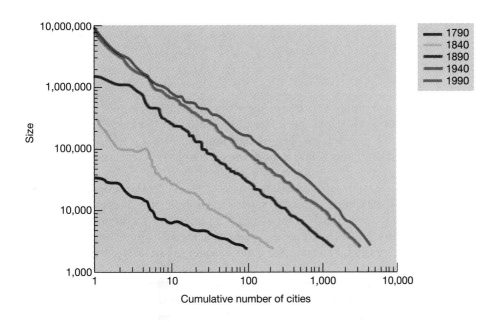

Figure 10.13 The rank-size distribution of cities in the U.S. urban system, 1790–1990 This graph shows that the U.S. urban system has conformed fairly consistently to the rank-size rule. As urbanization brought increased populations to cities at every level in the urban hierarchy, the rank-size graph has moved to the right. Meanwhile, the growth of some cities (for example, San Diego, California) has sent them from the lower end of the hierarchy to the very top, while other cities (Savannah, Georgia, for example) have declined, at least in relative terms, causing them to fall in the hierarchy. (*Source:* Reprinted with permission of Prentice Hall, from P. L. Knox, *Urbanization,* © 1994, p. 32.)

Figure 10.14 Rank-size distributions of city systems Some urban systems are dominated by primate cities whose populations are much larger than would be expected according to the rank-size rule. These rank-size graphs show four examples, each for 1950, 1975, and 2000. In the United Kingdom, London has always been disproportionately large, as has Jakarta within Indonesia. Japan and Brazil both have two primate cities: Tokyo and Osaka in Japan, and São Paulo and Rio de Janeiro in Brazil. Note how the rank-size curves for Indonesia and Brazil are spaced further apart, reflecting rapid population growth throughout their urban systems. (*Source:* After S. Brunn and J. F. Williams, *Cities of the World.* New York: HarperCollins, 2nd Ed., 1993, Fig. 1.11, p. 24. Reprinted by permission of Addison-Wesley Educational Publishers Inc.)

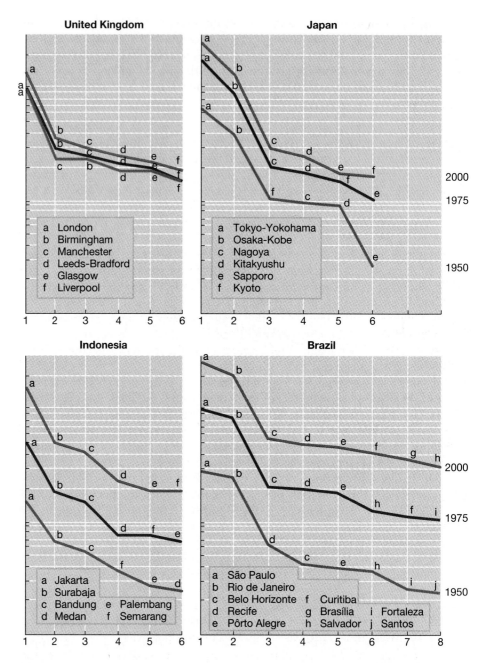

Primacy is not simply a matter of sheer size. Some of the largest metropolitan areas in the world—Karachi, New York, and Mumbai (formerly Bombay), for example—are not primate. Further, as the examples in Figure 10.14 show, primacy is a condition that is found in both the core and the periphery of the world-system. This suggests that primacy is a result of the roles played by particular cities within their own national urban systems. A relationship does exist to the world-economy, however. Primacy in peripheral countries is usually a consequence of primate cities' early roles as gateway cities. In core countries it is usually a consequence of primate cities' roles as imperial capitals and centers of administration, politics, and trade for a much wider urban system than their own domestic system.

When cities' economic, political, and cultural functions are disproportionate to their population, the condition is known as **centrality,** or the functional dominance of cities within an urban system. Cities that account for a disproportion-

Centrality: the functional dominance of cities within an urban system.

ately high share of economic, political, and cultural activity have a high degree of centrality within their urban system. Very often, primate cities exhibit this characteristic, but cities do not necessarily have to be primate in order to be functionally dominant within their urban system. Figure 10.15 (see page 430) shows some examples of centrality, revealing the overwhelming dominance of some cities within the world-system periphery. Bangkok, for instance, with around 10 percent of the Thai population, accounts for approximately 38 percent of the country's overall gross domestic product (GDP), over 85 percent of the country's GDP in banking, insurance, and real estate, and 75 percent of its manufacturing.

World Cities Ever since the evolution of a world-system in the sixteenth century, certain cities known as *world cities* (sometimes referred to as "global cities"), have played key roles in organizing space beyond their own national boundaries. In the first stages of world-system growth, these key roles involved the organization of trade and the execution of colonial, imperial, and geopolitical strategies. The world-cities of the seventeenth century were London, Amsterdam, Antwerp, Genoa, Lisbon, and Venice. In the eighteenth century, Paris, Rome, and Vienna also became world cities, while Antwerp and Genoa became less influential. In the nineteenth century, Berlin, Chicago, Manchester, New York, and St. Petersburg became world cities, while Venice became less influential.

Today, with the globalization of the economy, the key roles of world cities are concerned less with the deployment of imperial power and the orchestration of trade and more with transnational corporate organization, international banking and finance, supranational government, and the work of international agencies. World cities have become the control centers for the flows of information, cultural products, and finance that collectively sustain the economic and cultural globalization of the world.

World cities also provide an interface between the global and the local. They contain the economic, cultural, and institutional apparatus that channels national and provincial resources into the global economy and that transmits the impulses of globalization back to national and provincial centers. As such, world cities possess several functional characteristics:

- They are the sites of most of the leading global markets for commodities, commodity futures, investment capital, foreign exchange, equities, and bonds.
- They are the sites of clusters of specialized, high-order business services, especially those that are international in scope and that are attached to finance, accounting, advertising, property development, and law.
- They are the sites of concentrations of corporate headquarters—not just of transnational corporations but also of major national firms and large foreign firms.
- They are the sites of concentrations of national and international headquarters of trade and professional associations.
- They are the sites of most of the leading nongovernmental organizations (NGOs) and intergovernmental organizations (IGOs) that are international in scope (for example, the World Health Organization, United Nations Educational, Scientific, and Cultural Organization (UNESCO), the International Labor Organization, and the International Federation of Agricultural Producers).
- They are the sites of the most powerful and internationally influential media organizations (including newspapers, magazines, book publishing, satellite television); news and information services (including newswires and on-line information services); and culture industries (including art and design, fashion, film, and television).

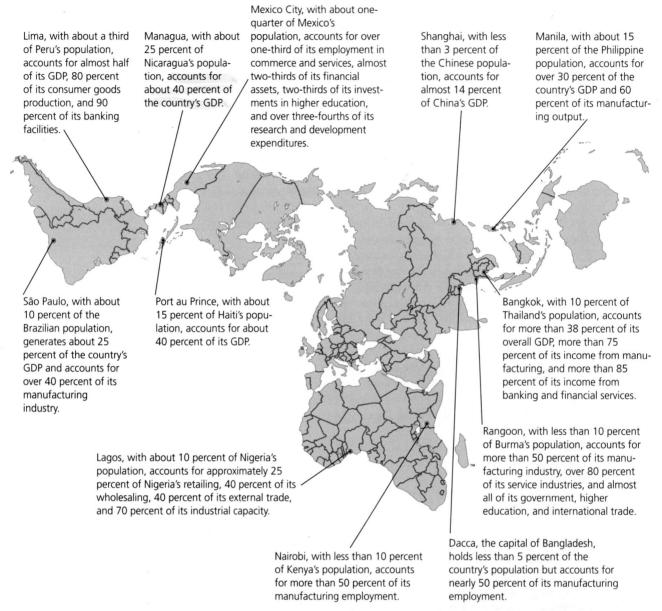

Lima, with about a third of Peru's population, accounts for almost half of its GDP, 80 percent of its consumer goods production, and 90 percent of its banking facilities.

Managua, with about 25 percent of Nicaragua's population, accounts for about 40 percent of the country's GDP.

Mexico City, with about one-quarter of Mexico's population, accounts for over one-third of its employment in commerce and services, almost two-thirds of its financial assets, two-thirds of its investments in higher education, and over three-fourths of its research and development expenditures.

Shanghai, with less than 3 percent of the Chinese population, accounts for almost 14 percent of China's GDP.

Manila, with about 15 percent of the Philippine population, accounts for over 30 percent of the country's GDP and 60 percent of its manufacturing output.

São Paulo, with about 10 percent of the Brazilian population, generates about 25 percent of the country's GDP and accounts for over 40 percent of its manufacturing industry.

Port au Prince, with about 15 percent of Haiti's population, accounts for about 40 percent of its GDP.

Bangkok, with 10 percent of Thailand's population, accounts for more than 38 percent of its overall GDP, more than 75 percent of its income from manufacturing, and more than 85 percent of its income from banking and financial services.

Rangoon, with less than 10 percent of Burma's population, accounts for more than 50 percent of its manufacturing industry, over 80 percent of its service industries, and almost all of its government, higher education, and international trade.

Lagos, with about 10 percent of Nigeria's population, accounts for approximately 25 percent of Nigeria's retailing, 40 percent of its wholesaling, 40 percent of its external trade, and 70 percent of its industrial capacity.

Nairobi, with less than 10 percent of Kenya's population, accounts for more than 50 percent of its manufacturing employment.

Dacca, the capital of Bangladesh, holds less than 5 percent of the country's population but accounts for nearly 50 percent of its manufacturing employment.

Figure 10.15 Examples of urban centrality The economic, political, and cultural importance of some cities is disproportionate to their population size, making them central to their economies. This is a reflection of core–periphery differentials within countries and often becomes a political issue because of the economic disparities. The centrality of these cities also leads to localized problems of congestion, land price inflation, and pollution. (*Source:* Map projection, Buckminster Fuller Institute and Dymaxion Map Design, Santa Barbara, CA. The word Dymaxion and the Fuller Projection Dymaxion™ Map design are trademarks of the Buckminster Fuller Institute, Santa Barbara, California, ©1938, 1967 & 1992. All rights reserved.)

A great deal of synergy exists among these various functional components. A city like New York, for example, attracts transnational corporations because it is a center of culture and communications. It attracts specialized business services because it is a center of corporate headquarters and of global markets, and so on. These interdependencies represent a special case of the geographical *agglomeration effects* that we discussed in Chapter 7. Agglomeration is the clustering of functionally related activities. In the case of New York City, corporate headquarters and specialized legal, financial, and business services cluster together because of the mutual cost savings and advantages of being close to one another. At the same time, different world cities fulfill different roles within the world-system, making for different emphases and combinations (that is, differences in the nature of their world-city functions) as well as for differences in the absolute and relative localization of particular world-city functions (that is, differences in their degree of importance as world cities).

Today, the global urban system is dominated by three world cities whose influence is truly global: London, New York, and Tokyo (Figure 10.16). The second tier of the system consists of world cities with influence over large regions

Top-tier World Cities
London
New York
Tokyo

2nd-tier World Cities
Brussels
Chicago
Frankfurt
Los Angeles
Paris
São Paulo
Singapore
Washington, DC
Zürich

3rd-tier World Cities
Amsterdam
Bangkok
Berlin
Mumbai (Bombay)
Buenos Aires
Hong Kong
Houston
Johannesburg

Madrid
Manila
Mexico City
Miami
Milan
Osaka
Rio de Janeiro
San Francisco

Seoul
Sydney
Taipei
Toronto
Vancouver

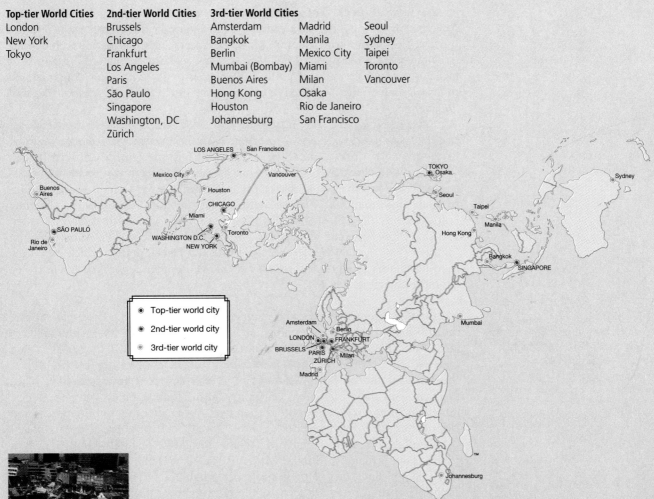

Top-tier world city

2nd-tier world city

3rd-tier world city

Brussels qualifies as a world city because it is the administrative center of the European Union and because it has attracted a large number of nongovernmental organizations that are transnational in scope.

London, New York, and Tokyo are major world cities because of their major financial markets, their transnational corporate headquarters, and their concentrations of financial services. Because of their geographical locations in different time zones, office hours in each city overlap just enough to allow 24-hour trading and decision making.

Milan has global status in terms of cultural influence (especially fashion and design) and is an important regional financial center, but it is relatively dependent in terms of corporate control and information-processing activities.

Figure 10.16 World cities   World cities are not simply the world's largest and economically most important cities. Rather, they are the control centers of the world economy, places where critical decision making and interaction take place with regard to global economic, cultural, and political issues. (*Source:* Map projection, Buckminster Fuller Institute and Dymaxion Map Design, Santa Barbara, CA. The word Dymaxion and the Fuller Projection Dymaxion™ Map design are trademarks of the Buckminster Fuller Institute, Santa Barbara, California, ©1938, 1967 & 1992. All rights reserved.)

of the world-system. These include, for example, Brussels, Frankfurt, Los Angeles, Paris, Singapore, and Zürich. A third tier consists of important international cities with more limited or more specialized international functions (including Amsterdam, Madrid, Miami, Mexico City, Seoul, and Sydney). A fourth tier exists of cities of national importance and with some transnational functions (including Barcelona, Dallas, Manchester, Munich, Melbourne, and Philadelphia).

Some geographers, recognizing the increasing interdependence of the global and the local, would add a fifth tier that includes such cities as Atlanta, Georgia; Rochester, New York; Columbus, Ohio; and Charlotte, North Carolina: places where an imaginative and aggressive local leadership has sought to carve out distinctive niches in the global marketplace. Columbus, for example—with a substantial "informational" infrastructure that includes CompuServe, Sterling Software/Ordernet, Chemical Abstracts, the Online Computer Library Center, and the Ohio Supercomputer Center—managed to get itself designated as an "Infoport" by the UN Conference on Trade and Development, which is seeking to foster international trade through the use of computer networks and electronic data interchange.

Megacities Megacities are not necessarily world cities, as described above, though some of them are (London, and Tokyo, for example). **Megacities** are very large cities characterized by both primacy and a high degree of centrality within their national economy. Their most important common denominator is their sheer size—most of them number 10 million or more in population. This, together with their functional centrality, means that in many ways they have more in common with one another than with the smaller metropolitan areas and cities within their own countries.

Examples of such megacities include Bangkok, Beijing, Cairo, Calcutta, Dacca, Jakarta, Lagos, Manila, Mexico City, New Delhi, São Paulo, Shanghai, and Teheran. Each one of these has more inhabitants than 100 of the member countries of the United Nations. While most of them do not function as world cities, they do provide important intermediate roles between the upper tiers of the system of world cities and the provincial towns and villages of large regions of the world. They not only link local and provincial economies with the global economy, but also provide a point of contact between the traditional and the modern, and between formal and informal economic sectors. The **informal sector** of an economy involves a wide variety of economic activities whose common feature is that they take place beyond official record and are not subject to formalized systems of regulation or remuneration.

Urban Growth Processes

The large-scale urbanization triggered in the world's core countries by the evolution of a world-system and, later, reinforced by the Industrial Revolution, was based on growth processes that were self-sustaining. Cities themselves were the engines of economic growth, and this growth, in turn, attracted the migrants, settlers, and immigrants that made for rapid population growth. In this urbanization process, a close and positive relationship existed between rural and urban development (Figure 10.17). The appropriation of new land for agriculture, together with mechanization and the innovative techniques that urbanization allowed, resulted in increased agricultural productivity. This extra productivity released rural labor to work in the growing manufacturing sector in towns and cities. At the same time, it provided the additional produce needed to feed growing urban populations. The whole process was further reinforced by the capacity of urban labor forces to produce agricultural tools, machinery, fertilizer, and other products that made for still greater increases in agricultural productivity.

Megacity: very large city characterized by both primacy and high centrality within its national economy.

Informal sector: economic activities that take place beyond official record, not subject to formalized systems of regulation or remuneration.

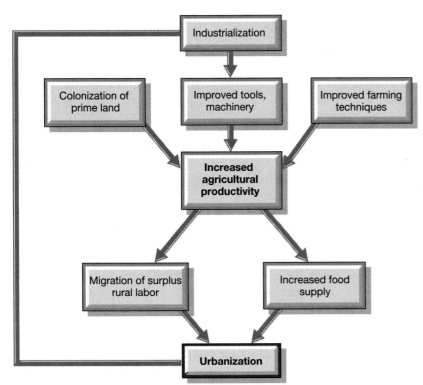

Figure 10.17 The urbanization process in the world's core regions Urbanization was stimulated by advances in farm productivity that (1) provided the extra food to support the increased numbers of townspeople, and (2) made many farmers and farm laborers redundant, prompting them to migrate to cities. Labor displaced in this way ended up consuming food rather than producing it, but this was more than compensated for by the increases in agricultural productivity and by the increased capacity of enlarged urban labor forces to produce agricultural tools, machinery, fertilizers, and so on that contributed further to agricultural productivity.

Urbanization and Economic Development

In this self-sustaining process of urbanization, the actual rate and amount of a city's growth depends on the size of its economic base. A city's **economic base** consists of those economic functions that involve the manufacture, processing, or trading of goods or the provision of services for markets beyond the city itself. Activities that provide income-generating "exports" for a city are termed **basic functions**. In contrast, **nonbasic functions** are those that cater to the city's own population and so do not generate profit from "outside" customers. Examples of nonbasic activities include local newspapers, local bakeries and restaurants, schools, and local government.

The fundamental determinant of cities' growth in population, employment, and income in the world's core countries is the percentage of their economies that is devoted to basic activities. The prosperity generated by basic economic activities leads to increased employment in nonbasic activities in order to satisfy the demand for housing, utilities, retailing, personal services, and other services. The incomes generated by the combination of basic and nonbasic economic activities allow for higher potential tax yields, which can be used to improve public utilities, roads, schools, health services, recreational amenities, and other infrastructure improvements. These activities are also nonbasic, but they all serve to improve the efficiency and attractiveness of the city for further rounds of investment in basic economic activities. The whole process is one of cumulative causation (Chapter 7), in which a spiral buildup of advantages is enjoyed by a particular place as a result of the development of external economies, agglomeration effects, and localization economies.

Economic base: set of manufacturing, processing, trading or services activities that serve markets beyond the city.

Basic functions: economic activities that provide income from sales to customers beyond city limits.

Nonbasic functions: economic activities that serve a city's own population.

Decentralization and Counterurbanization

The logic of economic development does not always work uniformly in the direction of population concentration and urban growth. The forces of cumulative causation are refocused from time to time as new technologies, new resources, and new opportunities alter the balance of comparative advantages enjoyed by particular places within the world's core and semiperipheral countries. New rounds of urbanization are initiated in the places most suited to the new circumstances, while those least suited are likely to suffer a spiral of *deindustrialization* and urban decline. Deindustrialization involves a decline in industrial employment in core regions as firms scale back their activities in response to lower levels of profitability (see Chapter 7). Such adversity particularly affected cities such as Pittsburgh and Cleveland, Sheffield and Liverpool (United Kingdom), Lille (France), and Liège (Belgium)—places where heavy manufacturing constituted a key economic sector. Cities like these suffered substantial reductions in employment throughout the 1970s and 1980s. During the same period, better and more flexible transport and communications networks allowed many industries to choose from a broader range of potential locations. The result has been a *decentralization* of jobs and people from larger to smaller cities within the urban systems of core countries. In some cases, routine production activities relocated to smaller metropolitan areas or to rural areas with low labor costs and more hospitable business climates. In other cases, these activities moved overseas—as part of the new international division of labor (see Chapter 2)—or were eliminated entirely.

These trends toward deindustrialization and decentralization have been intensified by the dampening effects of *agglomeration diseconomies* (Chapter 7) on the growth of larger metropolitan areas. Agglomeration diseconomies are the negative effects of urban size and density and include noise, air pollution, increased crime, commuting costs, the costs of inflated land and housing prices, traffic congestion, crowded port and railroad facilities, and higher taxes levied to rebuild decaying infrastructure and to support services and amenities previously considered unnecessary—traffic police, city planners, and shelters for the homeless, for example.

The combination of deindustrialization in core manufacturing regions, agglomeration diseconomies in major metropolitan areas, and the improved accessibility of smaller towns and rural areas can give rise to the phenomenon of counterurbanization. **Counterurbanization** occurs when cities experience a net loss of population to smaller towns and rural areas. This process results in the deconcentration of population within an urban system. This is, in fact, what happened in the United States, Britain, Japan, and many other developed countries in the 1970s and early 1980s. Metropolitan growth slowed dramatically, while the growth rates of small- and medium-sized towns and of some rural areas increased. In the United States, population increases in nonmetropolitan localities in the South, Southwest, and West were of stunning proportions in the 1970s. Counties that for decades had recorded stable populations grew by 15 or 20 percent. Some of the strongest gains were registered in those counties that were within commuting range of metropolitan areas, but some remote counties also registered big population increases. Meanwhile, at the top of the urban hierarchy, eight metropolitan areas of over one million population actually lost population. All of these were in the Northeast and Midwest: Buffalo, Cleveland, Detroit, Milwaukee, New York, Philadelphia, Pittsburgh, and St. Louis. Metropolitan areas of similar size in the South and Southwest did not decline, but their rate of growth slowed dramatically, and many of their inner-city cores did actually lose population. In contrast, small- and medium-sized metropolitan areas in the South and West grew vigorously—faster, in fact, than the surrounding nonmetropolitan areas.

Counterurbanization was a major reversal of long-standing trends, but it seems to have been a temporary adjustment rather than a permanent change. The globalization of the economy and the growth of postindustrial activities in revamped and expanded metropolitan settings has restored the trend toward the concentration of population within urban systems. Cities that were declining fast

Counterurbanization: net loss of population from cities to smaller towns and rural areas.

in the 1970s are now either recovering (New York, London) or bottoming out (Paris, Chicago). Those once growing slowly (Tokyo, Barcelona) are now expanding more quickly.

The Unintended Metropolis

Urban growth processes in the world's peripheral regions have been entirely different from those in core regions. In contrast to the self-sustaining urban growth of the world's core regions, the urbanization of peripheral regions has been a consequence of demographic growth that has preceded economic development. Although the demographic transition is a fairly recent phenomenon in the peripheral regions of the world (see Chapter 3), it generated large increases in population well in advance of any significant levels of industrialization or of rural economic development.

The result, for the mainly rural populations of peripheral countries, was more and more of worse and worse. Problems of agricultural development (see Chapter 8) meant that fast-growing rural populations faced an apparently hopeless future of drudgery and poverty. Emigration provided one potential safety valve, but as the frontiers of the world-system closed out, the more affluent core countries put up barriers to immigration. The only option for the growing numbers of impoverished rural residents was—and still is—to move to the larger towns and cities, where at least there is the hope of employment and the prospect of access to schools, health clinics, piped water, and the kinds of public facilities and services that are often unavailable in rural regions. Cities also have the lure of modernization and the appeal of consumer goods—attractions that are now directly beamed into rural areas through satellite TV. Overall, the metropolises of the periphery have absorbed four out of five of the 1.2 billion city dwellers added to the world's population since 1970.

Rural migrants have poured into cities out of desperation and hope, rather than being drawn by jobs and opportunities. Because these migration streams have been composed disproportionately of teenagers and young adults, an important additional component of urban growth has followed—exceptionally high rates of natural population increase. In most peripheral countries, the rate of natural increase of the population in cities exceeds that of net in-migration. On average, about 60 percent of urban population growth in peripheral countries is attributable to natural increase.

The consequence of all this urban population growth has been described as **overurbanization**, which occurs when cities grow more rapidly than the jobs and housing they can sustain. In such circumstances, urban growth produces instant slums—shacks set on unpaved streets, often with open sewers and no basic utilities. The shacks are constructed out of any material that comes to hand, such as planks, cardboard, tarpaper, thatch, mud, and corrugated iron. Such is the pressure of in-migration that many of these instant slums are squatter settlements, built illegally by families who are desperate for shelter. **Squatter settlements** are residential developments that take place on land that is neither owned nor rented by its occupants. Squatter settlements are not always slums, but many of them are. In Chile, squatter settlements are called *callampas*, meaning "mushroom cities"; in Turkey, they are called *gecekindu*, meaning that they were built after dusk and before dawn. In India, they are called *bustees*; in Tunisia, *gourbevilles*; in Brazil, *favelas*; and in Argentina, simply *villas miserias*. They typically account for well over one-third and sometimes up to three-quarters of the population of major cities (Figure 10.18, page 436). A 1996 United Nations conference report on human settlements (known as *Habitat II*) estimated that 600 million people worldwide live under health- and life-threatening situations in cities, and some 300 million live in extreme poverty. The same report showed that in some peripheral countries, including Bangladesh, El Salvador, Gambia, Guatemala, Haiti, and

Overurbanization: condition in which cities grow more rapidly than the jobs and housing they can sustain.

Squatter settlements: residential developments that take place on land that is neither owned nor rented by its occupants.

436

Figure 10.18 Slum housing in peripheral cities Throughout much of the world, the scale and speed of urbanization, combined with the scarcity of formal employment, have resulted in very high proportions of slum housing, much of it erected by squatters. The world's worst city for housing, according to United Nations statistics, is Addis Ababa, the capital of Ethiopia where, in 1995, 79 percent of the population was homeless or living in unfit accommodations. (*Source: Map projection, Buckminster Fuller Institute and Dymaxion Map Design, Santa Barbara, CA. The word Dymaxion and the Fuller Projection Dymaxion™ Map design are trademarks of the Buckminster Fuller Institute, Santa Barbara, California, ©1938, 1967 & 1992. All rights reserved.*)

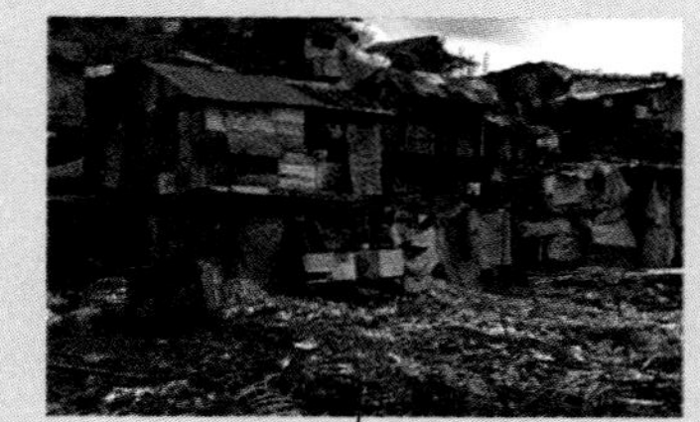

Manila's squatter settlements vary in size from the very large Tondo Foreshore area, with 30,000 families, to tiny groups of squatter homes on small sites. Less than 15 percent of Manila's population can afford to buy or rent a legal house or flat on the open market.

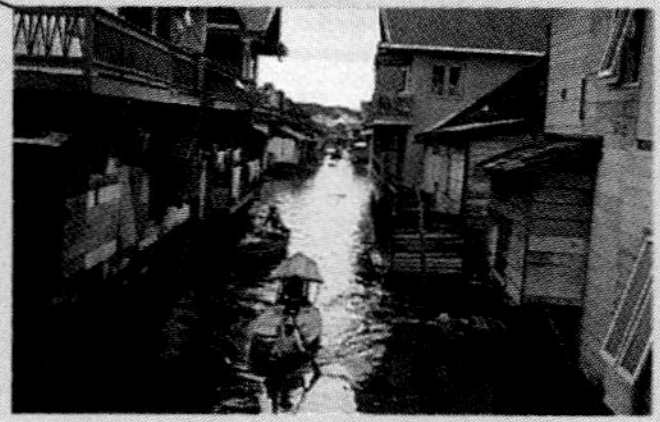

In Bandjarmasin, Indonesia, shanty housing spills over to waterlogged sites on the River Barito.

More than a quarter of Bangkok's 7 million population live in slums and unauthorized settlements, many of which are on swampy land that is prone to flooding.

More than 70 percent of Delhi's residents are unable to afford new legal housing of any kind. Over 35 percent of them live in housing that is officially substandard, and 17 percent live in housing that is officially illegal. Thousands have to live on the streets.

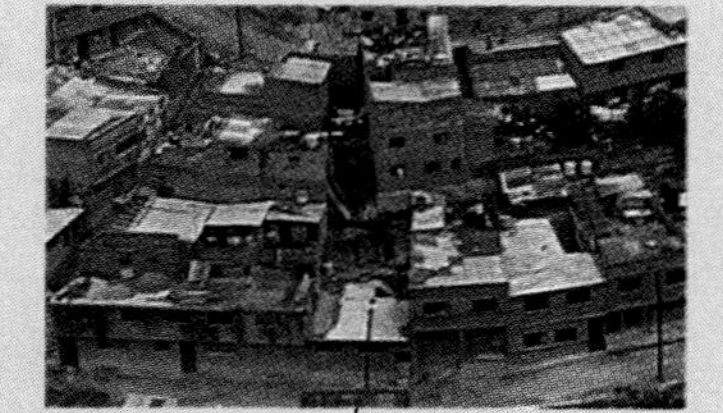

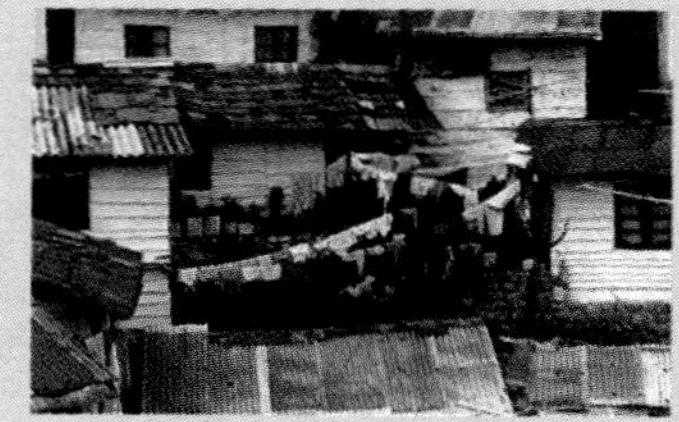

In Bogotá, Colombia, pressure on housing has been intensified by refugees from a three-decade-old war between the army and Marxist rebels, together with families uprooted by land-hungry drug traffickers.

In Mexico City, more than 7 million people live in some sort of uncontrolled or unauthorized settlement.

Noukachott, Mauritania, grew from 6,000 inhabitants in 1965 to 500,000 in 1995. Sixty-four percent of its residents live in self-built housing on land that is occupied illegally.

In Nairobi, Kenya, only one-third of the households can afford the rents charged for low-income public housing. The Mathare Valley squatter area grew from 4,000 inhabitants in 1964 to 90,000 in 1979 and to 250,000 in 1995. Overall, 40 percent of Nairobi's population live in unauthorized settlements.

Sixty percent of the population of Guyaquil, Ecuador's largest city, live in settlements built by their inhabitants over tidal swamplands. Some homes are a 40-minute walk along rickety boardwalks to dry land.

In Lima, Peru, 80 percent of the residents are unable to afford the least expensive new housing available in the formal sector. Over 40 percent live in housing that is officially substandard, of which three-quarters is shanty housing.

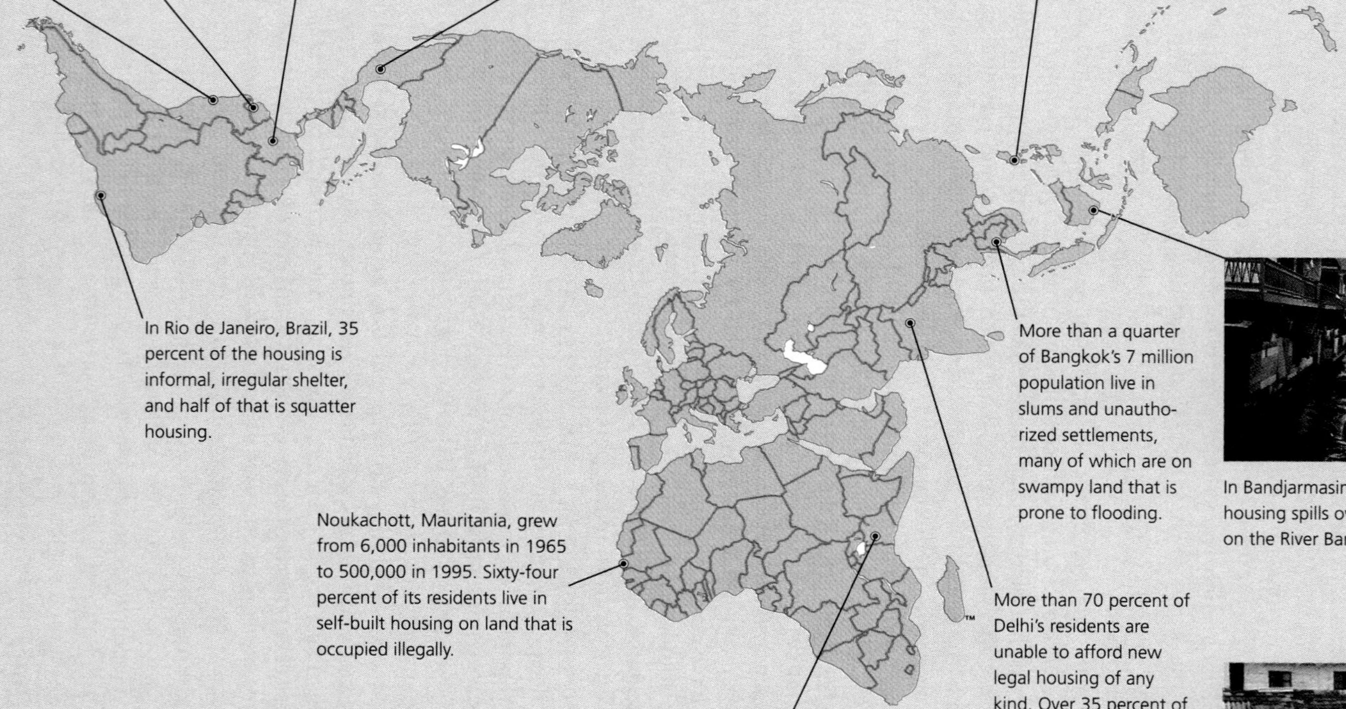

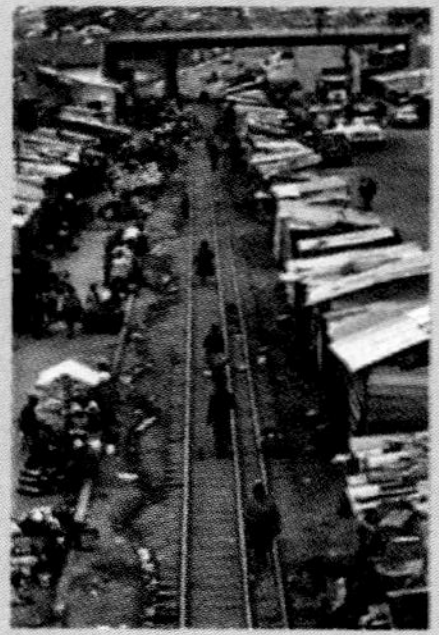

In Rio de Janeiro, Brazil, 35 percent of the housing is informal, irregular shelter, and half of that is squatter housing.

Honduras, more than 50 percent of the urban population live below their respective national poverty line.

Collectively, it is these slums and squatter settlements that have to absorb the unprecedented rates of urbanization in the megacities of the periphery. As we shall see in the next chapter, this often leads to severe problems of social disorganization and environmental degradation. Nevertheless, many neighborhoods are able to develop self-help networks and organizations that form the basis of community amid dauntingly poor and crowded cities.

Frontier Urbanization

Urban growth in the periphery is not all channeled into megacities and older urban centers. In some parts of the world, urbanization is a consequence of the commercial exploitation of regions that are only just becoming incorporated, selectively, into the world-system. The best example is the urbanization of the Amazon rain forest, where global economic forces have created frontier towns and cities in order to organize the local exploitation of gold, hardwood, rubber, and raw land. Here is an extract from the logbook of John Browder, an American researcher, who gives a graphic impression of this frontier urbanization:[3]

> No map in my possession showed a town called Rolim de Moura in the Federal Territory of Rondônia. Yet I was assured . . . by several prominent mahogany exporters in Curitiba, São Paulo and Rio de Janeiro that, although *meia precária* (primitive), this town did indeed exist. . . . [This was] the last 65-kilometer leg of a 3,000 kilometer journey from Belém, at the mouth of the Amazon River, to the frontier town of Rolim de Moura, *capital mundial de mogno* ("world capital of mahogany"). . . .
>
> The bus descended into a sprawling bowl of single-story wood frame business establishments, buzzing with activity in a continuous whirlwind of dust. Beyond the choking confusion of downtown, glimpses of a larger landscape passed in and out of view between great pylons of smoke and dust. A landscape almost randomly strewn with shacks and patches of remnant jungle. . . .
>
> The municipal prefect, Adegildo Aristides Ferreira, immediately agreed to meet the first norteamericano to set foot in this town in the Brazilian outback. . . . [He] wasted no time reciting a long list of hardships endured by his fellow citizens, who numbered, he believed, somewhere between 10,000 and 20,000. Rolim de Moura suffered from no running water or plumbing, no electricity, malaria and dysentery of epidemic proportions, terrible communications, daily homicides, prostitution, runaway inflation, and choking dust.

Since 1950, the population of northern Brazil has increased from under 2 million to nearly 10 million. Over the same period, the urban population of the region has increased from just over 30 percent of the total population to just under 60 percent. In 1960, the region had only 22 towns of 5,000 inhabitants or more, only two of which were larger than 100,000. By 1991, 133 towns existed of 5,000 or more inhabitants, including eight of at least 100,000 and two of one million or more (Figure 10.19, page 438). The growth rate of some of these towns has been phenomenal. Porto Velho, for example, the capital of the land-rush state of Rondônia, grew from just 48,839 in 1970 to 229,410 in 1991, an annual growth rate of over 7.5 percent. Remote Boa Vista, capital of gold-rich Roraima, grew by almost 10 percent a year.

---

[3]John Browder and Brian J. Godfrey, *Rainforest Cities*. New York: Columbia University Press, 1997.

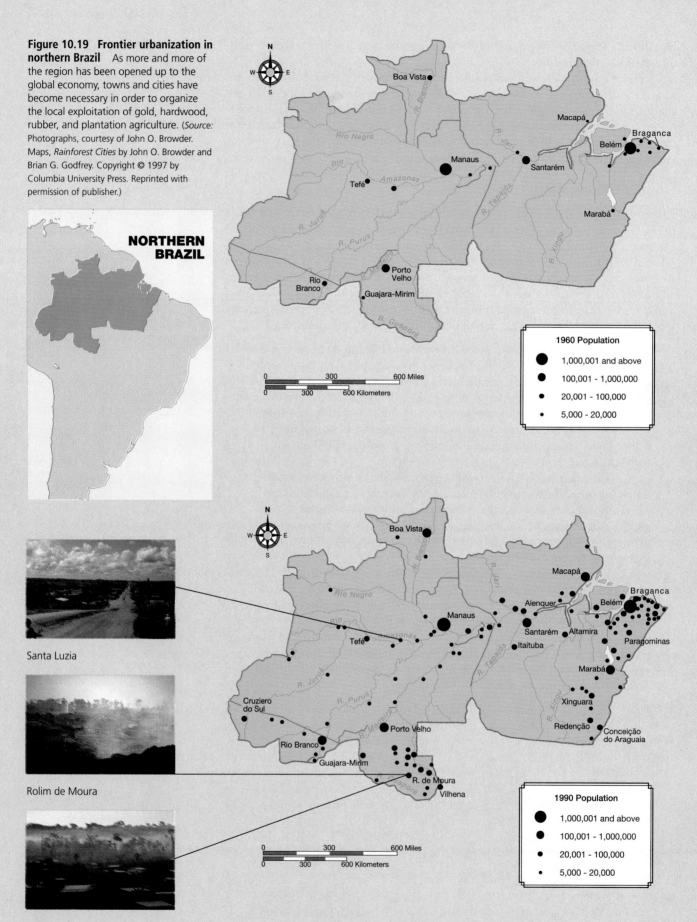

Figure 10.19 Frontier urbanization in northern Brazil As more and more of the region has been opened up to the global economy, towns and cities have become necessary in order to organize the local exploitation of gold, hardwood, rubber, and plantation agriculture. (*Source:* Photographs, courtesy of John O. Browder. Maps, *Rainforest Cities* by John O. Browder and Brian G. Godfrey. Copyright © 1997 by Columbia University Press. Reprinted with permission of publisher.)

NORTHERN BRAZIL

1960 Population

- 1,000,001 and above
- 100,001 – 1,000,000
- 20,001 – 100,000
- 5,000 – 20,000

Boa Vista
Macapá
Bragança
Belém
Manaus
Santarém
Tefé
Marabá
Rio Branco
Porto Velho
Guajara-Mirim

0   300   600 Miles
0   300   600 Kilometers

Santa Luzia

Rolim de Moura

Rolim de Moura

1990 Population

- 1,000,001 and above
- 100,001 – 1,000,000
- 20,001 – 100,000
- 5,000 – 20,000

Boa Vista
Macapá
Alenquer
Bragança
Belém
Manaus
Santarém
Altamira
Paragominas
Tefé
Itaituba
Marabá
Cruziero do Sul
Xinguara
Rio Branco
Redenção
Conceição do Araguaia
Guajara-Mirim
Porto Velho
R. de Moura
Vilhena

0   300   600 Miles
0   300   600 Kilometers

The main function of this region of Brazil has always been to provide raw materials and new markets for national, and sometimes global, economic expansion. The Brazilian government has encouraged frontier urbanization through the construction of major infrastructure projects, such as the Trans-Amazon highway, and through the creation of free-trade zones in frontier cities like Manaus. More recently, State corporations have created planned company towns as part of their resource extraction megadevelopments for which the Brazilian Amazon is famous. Other towns have appeared spontaneously, as smaller businesses based on extractive forest industries, crop processing, regional commerce, auto repair, banking, and other services have been established throughout the frontier region.

As with the unintended metropolises of the periphery, the population growth of these frontier towns is fueled mainly by poor migrants. Drawn to frontier towns and cities because of the prospects of economic opportunities, they number so many, however, that up to one-third or one-half wind up having to survive through informal-sector activities. As a result, frontier urbanization is characterized by the same problems of shanty and squatter housing as elsewhere in the periphery. Official statistics for Belém, for example, show that about 15 percent of the city's dwellings in 1991 consisted of squatter housing. Unofficial statistics, compiled by researchers and journalists, suggest that about 60 percent of the housing in the city, accommodating about 450,000 people, is a product of spontaneous, unregulated settlement in low-lying shantytowns that are subject to frequent flooding. Not surprisingly, frontier urbanization is also characterized by acute but widespread problems of disease.

Conclusion

Urbanization is one of the most important geographic phenomena. Cities can be seedbeds of economic development and cultural innovation. Cities and groups of cities also organize space—not just the territory immediately around them but, in some cases, national and even international space. The causes and consequences of urbanization, however, are very different in different parts of the world. The urban experience of the world's peripheral regions stands in sharp contrast to that of the developed core regions, for example. This contrast is a reflection of some of the demographic, economic, and political factors that we have explored in previous chapters.

Much of the developed world has become almost completely urbanized, with highly organized systems of cities. Today, levels of urbanization are high throughout the world's core countries, while rates of urbanization are relatively low. At the top of the urban hierarchies of the world's core regions are world cities such as London, New York, Tokyo, Paris, and Zürich that have become control centers for the flows of information, cultural products, and finance that, collectively, sustain the economic and cultural globalization of the world. In doing so, they help to consolidate the hegemony of the world's core regions.

Few of the metropolises of the periphery, on the other hand, are world cities that occupy key roles in the organization of global economics and culture. Rather, they operate as connecting links between provincial towns and villages and the world economy. They have innumerable economic, social, and cultural linkages to their provinces on one side and to major world cities on the other. Almost all peripheral countries, meanwhile, are experiencing high rates of urbanization, with forecast growth of unprecedented speed and unmatched size. In many peripheral and semiperipheral regions, current rates of urbanization have given rise to unintended metropolises and fears of "uncontrollable urbanization" with urban "danger zones" where "work" means anything that contributes to survival. The result, as we shall see in Chapter 11, is that these unintended metropolises are quite different from the cities of the core as places in which to live and work.

Main Points

- The urban areas of the world are the linchpins of human geographies at the local, regional, and global scales.

- Towns and cities are engines of economic development and centers of cultural innovation, social transformation, and political change. They now account for almost half the world's population.

- The earliest urbanization developed independently in the various hearth areas of the first agricultural revolution.

- The expansion of trade around the world, associated with colonialism and imperialism, established numerous gateway cities.

- The Industrial Revolution generated new kinds of cities, and many more of them.

- Today, the single most important aspect of world urbanization, from a geographical perspective, is the striking difference in trends and projections between the core regions on the one hand, and the semiperipheral and peripheral regions on the other.

- A small number of "world cities," most of them located within the core regions of the world-system, occupy key roles in the organization of global economics and culture.

- Many of the largest cities in the periphery have a doubling time of only 10 to 15 years.

- Many of the megacities of the periphery are primate and exhibit a high degree of centrality within their urban systems.

Key Terms

basic functions (p. 433)
central place (p. 424)
central place theory (p. 425)
centrality (p. 428)
colonial cities (p. 418)
counterurbanization (p. 434)

economic base (p. 433)
gateway city (p. 416)
informal sector (p. 432)
megacity (p. 432)
nonbasic functions (p. 433)
overurbanization (p. 435)
primacy (p. 427)

range (p. 425)
rank-size rule (p. 427)
shock city (p. 418)
squatter settlements (p. 435)
threshold (p. 425)
urban ecology (p. 410)
urban form (p. 410)

urban system (p. 410)
urbanism (p. 411)

Exercises

 ### On the Internet

The Internet exercises for this chapter explore the relationships among globalization, economic development, and urbanization. The Internet itself is a rich source of information on cities: By the end of 1996 more than 1,700 cities around the world had established home pages on the Internet. After completing the Internet exercises for this chapter, you should have a better understanding of the different processes behind urbanization in different parts of the world.

 ### Unplugged

1. From census volumes in your library, find out the population of the town or city you know best. Do the same for every census year, going back from 1990 to 1980, 1970, and so on, all the way back to 1860. Then plot these populations on a simple graph. What

explanations can you offer for the pattern that the graph reveals? Now draw a larger version of the same graph, annotating it to show the landmark events that might have influenced the city's growth (or decline).

2. The following cities all have populations in excess of 2 million. How many of them could you locate on a world map? Their size reflects a certain degree of importance, at least within their regional economy. What can you find out about each? Compile for each a 50-word description of each that explains its chief industries and a little of its history.

Kanpur	Poona	Xian
Medan	Surabaja	Recife
Bangalore	Kinshasa	Turin
Tashkent	Ankara	Ibadan

Chapter 11

City Spaces: Urban Structure

Paris: Chaillot Palace

The neighborhood shopping street is at the end of the block. It is a narrow lane, barely wide enough for one car to pass, and is lined on both sides with small shops whose fronts open widely to the street . . . and invite customers in. There are more and more boutiques and other new arrivals on the street, including an extremely busy supermarket, but there are still quite a few of the older establishments left as well: fishmongers, rice sellers, a noodle maker, a cracker bakery, a cubbyhole that sells only buttons, a glazier's shop, and countless other, small places for the local market. Tucked away to the side is the neighborhood's Buddhist temple. It is a new building but designed in a traditional style, and has a welcome open space for community fairs and other gatherings in front, and a lovely Japanese garden at the back. The garden is such a contrast to the harsh lines and bustling activity of the surrounding city that at times it seems to me to be the most secluded and contemplative place in the world.[1]

We can recognize in this description of a Tokyo neighborhood several elements that are fairly common in central cities throughout the world's core regions: the mixture of old stores, new boutiques, and local supermarket, for example. On the other hand, there are some that are unique: the noodle maker and the Buddhist temple with its Japanese garden. In this chapter we turn our attention to the internal dynamics of cities, looking at the ways in which patterns and processes tend to vary according to the type of city and its history. In many ways, the most striking contrasts are to be found between the cities of the core regions and those of the periphery. The evolution of the unintended metropolis of the periphery has been very different from the evolution of metropolitan areas in the world's core regions. Similarly, the problems they face are very different.

Urban Structure and Land Use

The internal organization of cities reflects the way that they function, both to bring people and activities together and to sort them out into neighborhoods and functional subareas. The simple stereotype of the U.S. city, for example, is based on several main elements of land use. Traditionally, the very center of the city has been the principal hub of shops and offices, together with some of the major institutional land uses such as the city hall, libraries, and museums. This center, known as the **central business district,** or **CBD** (Figure 11.1), is a city's nucleus of commercial land uses. It always contains the densest concentration of shops, offices, and warehouses and the tallest group of nonresidential buildings in the city. It usually developed at the nodal point of transportation routes, so that it also contains bus stations, railway termini, and hotels. The CBD typically is surrounded by a zone of mixed land uses: warehouses, small factories and workshops, specialized stores, apartment buildings, and older residential neighborhoods (Figure 11.2). This zone is often referred to as the **zone in transition,** because of its mixture of growth, change, and decline. Beyond this zone are residential neighborhoods, suburbs of various ages and different social and ethnic composition.

As cities have grown larger and become more complex, so this simple stereotype has had to accommodate additional elements. *Secondary business districts* and *commercial strips* have emerged in the suburbs to cater to neighborhood

Central business district (CBD): central nucleus of commercial land uses in a city.

Zone in transition: area of mixed commercial and residential land uses surrounding the CBD.

---

[1]Roman Cybriwsky, *Tokyo*. London: Belhaven Press, 1991, p. 3.

Figure 11.1   The central business district (CBD)   This photograph of Chicago's central business district, known in Chicago as "the Loop," shows the concentration of high-rise office buildings typical of CBDs in U.S. metropolitan areas. In Chicago, as in other major cities, the CBD originally grew up around the point of maximum accessibility—near railway stations and at the intersection of the city's principal transit lines.

shopping and service needs. *Industrial districts* have developed around large factories and around airports; and, in larger metropolitan areas, edge cities have emerged as new suburban hubs of shops and offices that overshadow the old CBD. **Edge cities** are nodal concentrations of shopping and office space that are situated on the outer fringes of metropolitan areas, typically near major highway intersections. Meanwhile, other changes have occurred in more central locations as older buildings and neighborhoods have been restructured to meet new needs. One of the most striking of these changes has been the gentrification of older,

Edge cities: nodal concentrations of shopping and office space that are situated on the outer fringes of metropolitan areas, typically near major highway intersections.

Figure 11.2   The zone in transition   This photograph shows part of the zone in transition around the central business district in Manhattan, New York City. In most U.S. cities the CBD is surrounded by such a zone, which consists of older neighborhoods with mixed land uses, parts of which are in long-term decline, and parts of which have been redeveloped.

Gentrification: the invasion of older, centrally located working-class neighborhoods by higher-income households seeking the character and convenience of less expensive and well-located residences.

Congregation: the territorial and residential clustering of specific groups or sub-groups of people.

Minority groups: population subgroups that are seen—or that see themselves—as somehow different from the general population.

centrally located, working-class neighborhoods (Figure 11.3). **Gentrification** involves the invasion of such neighborhoods by higher-income households seeking the character and convenience of less expensive and centrally located residences. These invasions result in the physical renovation and upgrading of housing, but they also result in the displacement of many of the original occupants.

Territoriality, Congregation, and Segregation

In cities, as at other geographic scales, territoriality provides a means of establishing and preserving group membership and identity. The first step in forming group identity is to define "others" in an exclusionary and stereotypical way. **Congregation**—the territorial and residential clustering of specific groups or subgroups of people—enables group identity to be consolidated in relation to people and places outside the group. Congregation is thus a place-making activity and an important basis for urban structure and land use (Figure 11.4). It is particularly important in situations where there is one or more distinctive minority group. Defined in relation to a general population or host community, **minority groups** are population subgroups that are seen—or that see themselves—as somehow different from the general population. Their defining characteristics can be based on race, language, religion, nationality, caste, sexual orientation, or lifestyle.

Several specific advantages of congregation exist for minority groups:

- Congregation provides a means of cultural preservation. It allows religious and cultural practices to be maintained, and strengthens group identity through daily involvement in particular routines and ways of life. Particularly important in this regard is the way that clustering fosters within-group marriage and kinship networks.

- Congregation helps minimize conflict and provides defense against "outsiders."

- Congregation provides a place where mutual support can be established through minority institutions, businesses, social networks, and welfare organizations.

Figure 11.3   Gentrification This photograph shows part of Georgetown, in Washington, DC, where a lengthy process of gentrification has established an elite area among neighborhoods of older town houses, most of which had previously been occupied by lower- and middle-income households.

Figure 11.4 Congregation and segregation In most larger cities a patchwork of distinctive neighborhoods results from processes of congregation and segregation. Most distinctive of all are neighborhoods of ethnic minorities, such as the Chinatowns, Little Italys, and Koreatowns of major American cities.

Chinatown, San Francisco

Asian neighborhood, Bradford, England

Vietnamese neighborhood, San Francisco

Protestant neighborhood, Belfast, Northern Ireland

One of the most widely used methods of calculating the degree to which a particular minority group is residentially segregated is the "index of dissimilarity," an index that produces a theoretical range of values from 0 (no segregation—the minority group is distributed across the city evenly, in proportion to its overall numbers) to 100 (complete segregation). This table shows the index for the black population of major cities in 1990.

Index of Dissimilarity, 1990	
Baltimore	76.1
Boston	76.0
Chicago	85.9
Cleveland	87.6
Los Angeles	73.9
Milwaukee	73.7
Oklahoma City	59.2
Philadelphia	82.3
Phoenix	57.3
San Francisco	57.5

Congregation is an important place-making activity. It enables group identity to be established and preserved in relation to "other" people and places. For minority groups, it also brings certain advantages in terms of cultural preservation and mutual support. For these reasons, congregation is often voluntary. In many cases, however, it is, to some degree involuntary, driven by discrimination. The combined result of congregation and discrimination is segregation, the spatial separation of specific subgroups within a wider population.

- Congregation helps establish a power base in relation to the host society. This power base can be democratic, organized through local elections, or it can take the form of a territorial heartland for insurrectionary groups.

Congregation is not always voluntary, of course. Host populations are also impelled by territoriality, and they may respond to social and cultural differences by *discrimination* against minority groups. Discrimination can also have a strong territorial basis, the objective being to restrict the territory of minority groups and to resist their assimilation into the host society. This resistance can take a variety of forms. Social hostility and the voicing of "keep out" attitudes is probably the most widespread, although other forms of discrimination can have more pronounced spatial effects. These effects include exclusion and prejudice in local labor markets; the manipulation of private land and housing markets; the steering of capital investment away from minority areas; and the institutionalization of discrimination through the practices and spatial policies of public agencies.

Segregation: the spatial separation of specific population subgroups within a wider population.

The combined result of congregation and discrimination is **segregation,** the spatial separation of specific subgroups within a wider population. Segregation varies a great deal, both in intensity and form, depending on the relative degree and combination of congregation and discrimination. Geographers and demographers have developed indexes of segregation that measure the relative spatial concentration of population subgroups. Comparisons are often problematic, however, because of the influence of spatial scale on the computation and construction of such indexes. In terms of form, geographers have identified three principal situations:

- **Enclaves,** in which tendencies toward congregation and discrimination are long-standing, but dominated by internal cohesion and identity. The Jewish districts of many of today's cities in Europe and the eastern United States are examples of enclaves.
- **Ghettos,** which are also long-standing, but which are more the product of discrimination than congregation. Examples are the segregation of African Americans and Hispanics in American cities.
- **Colonies,** which may result from congregation, discrimination, or both, but in relatively weak and short-lasting ways. Their persistence over time therefore depends on the continuing arrival of new minority-group members. For example, U.S. cities in the early twentieth century contained distinctive colonies of German, Scandinavian, Irish, and Italian immigrants that have now all but disappeared. Today, colonies of Greek and Yugoslav neighborhoods exist in Australian cities, and colonies of Koreans in Japanese cities.

Competing for Space in American Cities

The overall structure of American cities is shaped primarily by competition for territory and location. Individual households and population groups compete for the most socially desirable residences and neighborhoods, while all categories of land users—commercial and industrial, as well as residential—compete for the most convenient and accessible locations within the city. Geographers draw on several different perspectives in looking at these aspects of competition between urban land users. Four are particularly useful: an economic perspective that emphasizes accessibility; an economic perspective that emphasizes functional linkages between land uses; a sociocultural perspective that emphasizes ethnic congregation and segregation; and a historical perspective that emphasizes the influence of the growth of transportation corridors.

Accessibility and Land Use Most urban land users want to maximize the *utility* that they derive from a particular location. The utility of a specific place or location refers to its usefulness to particular persons or groups. The price they are prepared to pay for different locations will be a reflection of this utility. In general, utility will be a function of *accessibility*. Commercial land users want to be accessible to one another, to markets, and to workers; private residents want to be accessible to jobs, amenities, and friends; public institutions want to be accessible to their clients. In an idealized city built on an isotropic surface, the point of

Isotropic surface: a hypothetical, uniform plain: flat, and with no variations in its physical attributes.

maximum accessibility is the city center. An **isotropic surface** is a hypothetical, uniform plain: flat, and with no variations in its physical attributes. Under these conditions, accessibility falls off steadily with distance from the city center. Likewise, utility falls off, but at different rates for different land users. The result is a tendency for concentric zones of different mixes of land use (Figure 11.5).

One counterintuitive implication of this model is that it will be the poorest households that end up occupying the periphery of the city. While this is true in some parts of the world, we know that in the core countries the farthest suburbs are the territory of wealthier households, while the poor occupy more accessible locations nearer to city centers. Some modification of the assumptions is clearly required. In this case, we must assume that wealthier households trade off the convenience of accessibility for the greater utility of being able to consume larger amounts of (relatively cheap) suburban space. Poorer households, unable to afford the recurrent costs of transportation, must trade off living space for accessibility to jobs, so that they end up in high-density areas, at expensive locations near their low-wage jobs. Because of the presumed trade-off between accessibility and living space, this urban land use model is often referred to as a *trade-off model*.

Functional Clustering: Multiple Nuclei
The multiple-nuclei model of urban land use is based on the observation that some activities attract one another, while others repel one another. Without denying the concentric patterns that result from principles of distance, accessibility, and utility, geographers recognize that certain categories of land use are drawn together into functional clusters, or nuclei, while others tend to repel one another. At the broadest scale, economic relationships draw manufacturing, transportation, and warehousing together. These activities need to be in proximity to one another so that each can function as effectively and efficiently as possible. Similarly, functional relationships exist between these land uses and blue-collar housing, which tends to result in their mutual attraction. On the other hand, upper-middle-class housing is repelled by industrial and working-class districts.

We can see the same principles operating at a more detailed scale. Within the sphere of commercial land use in downtown areas, for example, fashion clothing and shoe stores are drawn together near larger department stores, all seeking to exploit the same pool of potential shoppers. All these high-end retail land uses are compatible with the large commercial hotels and professional office buildings that also seek busy and accessible locations. On the other hand, they all tend to be repelled by certain other commercial land uses, such as meat and vegetable markets, budget stores, and X-rated businesses. The result is that urban land use becomes spatially segregated, with nodes or nuclei of different groupings of land users (Figure 11.6).

Social and Ethnic Clustering: Social Ecology
Just as different categories of land use attract and repel one another, so do different social and ethnic groups. The third model of land use is based on an ecological perspective developed by sociologists to explain this phenomenon with special reference to cities in the United States whose rapid growth has been fueled by streams of migrants and immigrants with very different backgrounds. It is based on the idea of city neighborhoods being structured by the "invasion" of successive waves of migrants and immigrants.

When immigrants first arrive in the city looking for work and a place to live, they will have little choice but to cluster in the cheapest accommodation, typically to be found in the zone in transition around the CBD. The classic example is provided by Chicago in the 1920s and 1930s. Immigrants from Scandinavia, Germany, Italy, Ireland, Poland, Bohemia (now part of the Czech Republic), and Lithuania established themselves in Chicago's low-rent areas, the only places they

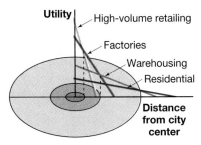

Figure 11.5 Accessibility, bid-rent, and urban structure Competition for accessible sites near the city center is an important determinant of land use patterns. Different land users are prepared to pay different amounts for locations at various distances from the city center. The result is a tendency for a concentric pattern of land uses. (*Source:* Reprinted with permission of Prentice Hall, from P. Knox, *Urbanization,* @ 1994, p. 99.)

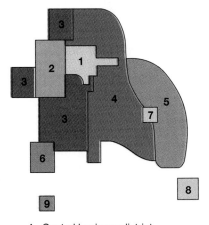

1. Central business district
2. Wholesale light manufacturing
3. Low-income residential
4. Medium-income residential
5. High-income residential
6. Heavy manufacturing
7. Outlying business district
8. Residential suburb
9. Industrial suburb

Figure 11.6 Multiple-nuclei model of urban land use When cities reach a certain size, the traditional downtown (1) is no longer sufficient to serve the commercial needs of the whole city, and so additional nodes of shops and offices emerge in outlying districts (7). Functional groupings of related activities—manufacturing (2, 6), wholesaling (2), and so on—also tend to develop, creating multiple nuclei of economic activities around which the city is organized. (Source: C. D. Harris and E. L. Ullman, "The Nature of Cities," *Annals of the American Academy of Political and Social Science,* CCXLII, 1945, Fig. 5.)

could afford. By congregating together in these areas, however, immigrants accomplished several things—they were able to establish a sense of security; to continue speaking their native language; to have familiar churches or synagogues, restaurants, bakeries, butcher shops, and taverns; and to support their own community newspapers and clubs. These immigrants were joined in the city's zone-in-transition by African American migrants from the South, who also established their own neighborhoods and communities. In Chicago, as in other U.S. cities of the period, the various ethnic groups formed a patchwork or mosaic of communities encircling the CBD.

These ethnic communities lasted from one to three generations, after which they started to break up. Many of the younger, city-born individuals did not feel the need for the security and familiarity of ethnic neighborhoods. Gradually, increasing numbers of them were able to establish themselves in better jobs and move out into newer, better housing. As the original immigrants and their families moved out, their place in the zone-in-transition was taken by a new wave of migrants and immigrants. In this way, Chicago became structured into a series of *concentric zones* of neighborhoods of different ethnicity and status (Figure 11.7). The same problem can occur in other cities where rapid growth is fueled by streams of migrants and immigrants from very different backgrounds.

Invasion and succession: a process of neighborhood change whereby one social or ethnic group succeeds another.

Throughout this process of invasion and succession, people of the same background tend to stick together—partly because of the advantages of residential clustering and partly because of discrimination. **Invasion and succession** is a process of neighborhood change whereby one social or ethnic group succeeds another in a residential area. The displaced group, in turn, invades other areas, creating over time a rippling process of change throughout the city. The result is that within each concentric zone, there exists a mosaic of distinctive neighborhoods. Classic examples include the Chinatowns, Little Italys, Koreatowns, and African American ghettos of big American cities. Such neighborhoods can be thought of as *ecological niches* within the overall metropolis—settings where a particular mix of people have come to dominate a particular territory and a particular physical environment, or habitat.

Corridors and Sectors The fourth model of land use is a historical one. In cities such as Richmond, Virginia; Kansas City, Missouri; and Memphis, Tennessee—where growth has been less dominated by successive waves of different immigrant ethnic groups—neighborhood patterns are often structured around the development of two different types of district: industrial districts and high-class residential districts. Over time, both of these tend to grow outward from the center of the city, but for different reasons, and in different directions. Industry tends to follow transportation corridors along low-lying, flat land where space exists for large factories, warehouses, and railway marshaling yards. Working-class residential areas grow up around these corridors, following them out in sectoral-shaped neighborhoods as they grow (Figure 11.8).

High-status residential districts, on the other hand, tend to grow outward from a different side of town, often following a ridge of high ground (free from flooding and with panoramic views). This outward growth creates another sectoral component of urban structure. The social status of this sector attracts middle-class housing, which in turn creates additional sectors of growth, thus completing the city's overall structure.

Comparative Urban Structure

Putting together the essential components of the four models of urban structure, we can derive a composite model of cities in the United States (Figure 11.9, page 450) in which a web of sectors and zones stretches around the CBD and various outlying nodal centers. More than anything else, the radial sectors reflect patterns

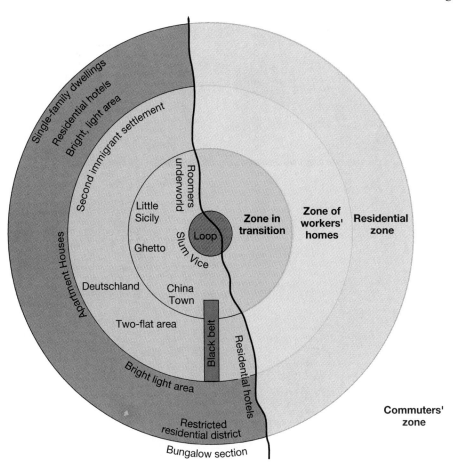

Figure 11.7 The ecological model of urban land use: Chicago in the 1920s Competition between members of different migrant and immigrant groups for residential space in the city often results in distinctive neighborhoods that have their own social "ecology." The classic example, shown in this diagram, is Chicago of the 1920s, which had developed a series of concentric zones of distinctive neighborhoods formed as successive waves of immigrants established themselves. Over time, most immigrant groups made their way from low-rent, inner-city districts surrounding the CBD (known in Chicago as the Loop) to more attractive and expensive districts farther out. (*Source:* R. E. Park, E. W. Burgess , and R. D. McKenzie, *The City,* Chicago: University of Chicago Press, 1925, p. 53.)

Figure 11.8 Corridors and sectors
Cities that have not been dominated by successive waves of migrant or immigrant ethnic groups tend to be organized around the linear development of two main features that grow outward from the CBD (1)—corridors of industrial development (2), and sectors of high-status residential development (5). Sectors of working-class residential districts (3) surround the industrial corridors, while sectors of middle-class residential districts (4) surround the high-status developments. (*Source:* C. D. Harris and E. L. Ullman, "The Nature of Cities," *Annals of the American Academy of Political and Social Science* CCXLII, 1945, Fig. 5.)

of socioeconomic status, while the concentric zones reflect demographic patterns, and the nodes reflect commercial patterns. Because they are the product of some fundamental forces—economic competition for space and accessibility, social and ethnic discrimination and congregation, functional agglomeration, and residential search behavior—they can be traced in many of the world's cities, particularly in affluent core regions where economic, social, and cultural forces are broadly similar. Nevertheless, urban structure varies considerably because of the influence of history, culture, and the different roles that cities have played within the world-system.

European Cities European cities are typically the product of several major epochs of urban development. As we saw in Chapter 2, because many of today's most important cities were founded in the Roman period, it is not uncommon for the outlines of Roman and medieval urban development to be preserved in their street plans (Figure 11.10, page 451). Many of the distinctive features of European cities derive from their long history. Following is a list of several important features, which are illustrated in Figure 11.11 (see page 452).

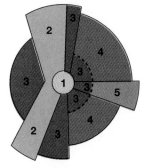

1. Central business district
2. Industrial area
3. Working-class residential district
4. Middle-class development
5. High-income residential district

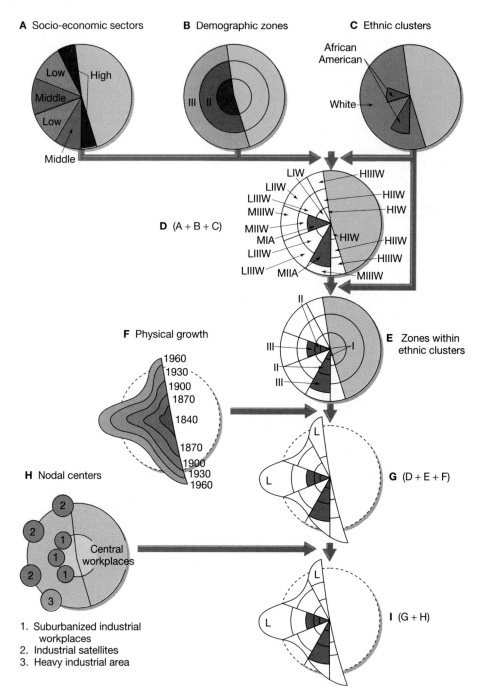

A Socio-economic sectors

B Demographic zones

C Ethnic clusters

D (A + B + C)

E Zones within ethnic clusters

F Physical growth

G (D + E + F)

H Nodal centers

1. Suburbanized industrial workplaces
2. Industrial satellites
3. Heavy industrial area

I (G + H)

Figure 11.9 Composite model of urban structure This diagram shows the concentric zones, functional nodal nuclei, and corridors of development that are common features of urban structure in many of the world's core regions. (*Source:* After B. J. L. Berry and P. Rees, *American Journal of Sociology,* 74, 1969, Fig. 13.)

- **Complex street patterns**—In the historic cores of some older cities, the layout of streets reflects ancient patterns of rural settlement and field boundaries. Beyond these historic cores, narrow, complex streets are the product of the long, slow growth of cities in the pre-automobile era, when hand-pushed and horse-drawn carts were the principal means of transportation, and urban development was piecemeal and small-scale (Figure 11.11a, page 452).

Figure 11.10 Towns of medieval Europe
The towns of medieval Europe were typically small, with compact outlines defined by defensive walls. In addition to their defensive function, these walls were status symbols, signs of wealth and power. At their gates, tolls were collected and goods checked. They enclosed a separate administrative area in which residents were free and usually self-governing. Within the walls, towns were typically organized around a fortress, or cité, though the foci of activity and sources of growth were more likely to be religious institutions, market squares, and merchants' quarters. The example depicted here, Arras, in northeastern France, shows the "organic" pattern of irregular-shaped blocks with narrow, winding streets typical of many of the medieval towns of Europe. Its growth was focused successively on three elements: (a) the cité, or fortress, (b) the monastery, and (c) two market squares.
(*Source:* P. M. Hohlenberg and L. H. Lees, *The Making of Urban Europe 1000–1500.* Cambridge, MA: Harvard University Press, 1985, p. 32.)

a. Cité
b. Monastery
c. Market squares

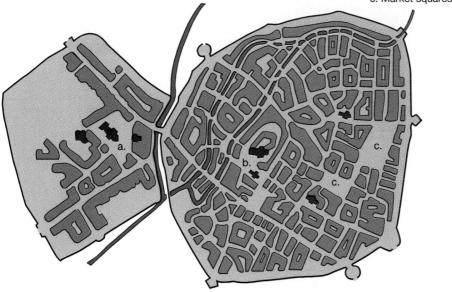

- **Plazas and squares**—Greek, Roman, and medieval cities were all characterized by plazas, central squares, and marketplaces; they are still important nodes of urban activity (Figure 11.11b).
- **High density and compact form**—High levels of urbanization, a long history of urban development preceding the automobile, and the constraints of peripheral defensive walls all made urban land expensive and encouraged a tradition of high-density living in tenements and apartment houses. More recently, strong city planning regulations have allowed planners to place strict limits on the sprawl of cities (Figure 11.11c).

Figure 11.11 Distinctive characteristics of European cities

(a) York, England—The narrow, winding streets that lead to York Minster (cathedral) date from medieval times. They now provide an ideal retail setting for visitors to the cathedral.

(b) Beaune, France—Life in Beaune, an ancient market town in the Burgundy wine district, is centered on the market square.

(c) Dundee, Scotland—This aerial photograph shows the city's compact form, with high-density urban development contained within a boundary that is strictly enforced through land use regulations.

(d) Milan, Italy—Milan's skyline is dominated by its cathedral, whose spires overlook a low-key cityscape that is punctuated by just a few high-rise office towers and apartment blocks.

(e) Munich, Germany—The streets in the city's central business district still contain a lot of the city's best cafes and stores. As in many other European cities, people use these streets for evening and weekend strolls, window-shopping, and people-watching.

- **Low skylines**—Although the larger European cities do have a fair number of high-rise apartment buildings and a sprinkling of office skyscrapers, they all offer a predominantly low skyline. This is partly because much of their growth came before the invention of the elevator and the development of steel-reinforced, concrete building techniques, and partly because of master plans and building codes (some written as long ago as the sixteenth century) seeking to preserve the dominance of monumental buildings like palaces and cathedrals (Figure 11.11d).

- **Lively downtowns**—The CBDs of European cities have been able to retain their focal position in terms of shopping and social life because of the relatively late arrival of the suburbanizing influence of the automobile and because of strong planning controls that have been directed against urban sprawl (Figure 11.11e).

- **Neighborhood stability**—Europeans change residence, on average, about half as often as Americans. In addition, the physical life cycle of city neighborhoods tends to be longer because of the past use of durable construction materials such as brick and stone. As a result, European cities provide relatively stable socioeconomic environments (Figure 11.11f).

(f) Newport-on-Tay, Scotland—Older neighborhoods like this, solidly built for upper-middle-income groups toward the end of the nineteenth century, tend to change very slowly in both their physical appearance and their social composition.

(g) Siklós, Hungary—As late as the eighteenth century, the development of towns and cities across continental Europe was strongly influenced by the need for defensible sites. At Siklós, a small town in southwestern Hungary, the castle has been continuously occupied since the Middle Ages. The eighteenth-century castle building shown here is now used as a hotel.

(h) Sacré Coeur, Paris—Churches and monuments are an important part of the rich symbolic content of European cities. The Sacré Coeur was built between 1879 and 1919, sponsored by conservative Catholics and monarchists as a symbolic counterpoint to French liberal republicanism.

(i) Municipal Housing, Sheffield, England—Public housing is a very important element in many European cities, often housing more than 1 in every 5 households and, in some cities, 3 or 4 in every 5 households.

- **The scars of war**—Europe's history of international conflict is etched on its cities in a number of ways. The legacy of defensive hilltop sites and city walls has limited and shaped the growth of modern cities (Figure 11.11g), while in more recent times the bombings and shellings of World War II destroyed many buildings in a large number of cities.
- **Symbolism**—The legacy of a long and varied history includes a rich variety of symbolism in the built environment. Europeans are reminded of their past not only by large numbers of statues and memorials, but also by cathedrals, churches, and monasteries; by guildhalls and city walls; by the palaces of royalty and the mansions of aristocracy; and by city halls and the libraries, museums, sports stadia, and galleries that are monuments to civic achievement (Figure 11.11h).
- **Municipal socialism**—European welfare states provide a distinctively broad range of municipal services and amenities, from clinics to public transit systems. Perhaps the most important in terms of urban structure is social housing (public housing), which accounts for between 20 and 40 percent of all housing in most of the larger English, French, and German cities (Figure 11.11i).

Putting all this together, we can produce a composite model of the western European city (Figure 11.12). We should note, however, that the richness of European history and the diversity of its geography mean that several variations on this theme do exist. One of the most interesting is the eastern European city, in which the legacy of an interlude of 44 years of socialism (1945–1989) was grafted onto cities that had already developed mature patterns of land use and social differentiation. Major examples include Belgrade, Budapest, Katowice, Kraków, Leipzig, Prague, and Warsaw. State control of land and housing meant that huge public housing estates and industrial zones were created in outlying districts (Figure 11.13). The structure of the older city was little altered, however, apart from the addition of socialist monuments and the renaming of streets.

Colonial Cities Colonial cities are those that were deliberately established or developed as administrative or commercial centers by colonial or imperial powers. Examples are Accra (Ghana), Hanoi (Vietnam), Macau (China), Nairobi (Kenya), and New Delhi (India). The stereotypical colonial city reflects a

CBD, Vienna

Alt-Erlaa Complex of Social Housing, Vienna

Low-density Villa Belt, Vienna

Figure 11.12 Urban structure in continental Western Europe In the typical continental Western European city, the CBD has a more dominant role in commercial and social life than in North American cities. The density of residential development is also high, with large amounts of nineteenth-century housing—including stable, high-status neighborhoods close to the city center—and significant amounts of social housing. (*Source:* P. White, *The West European City.* London: Longman, 1984, Fig. 7.6, p. 188. Reprinted by permission of Addison Wesley Longman Ltd.)

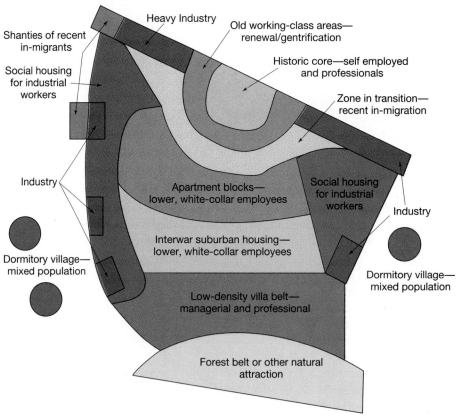

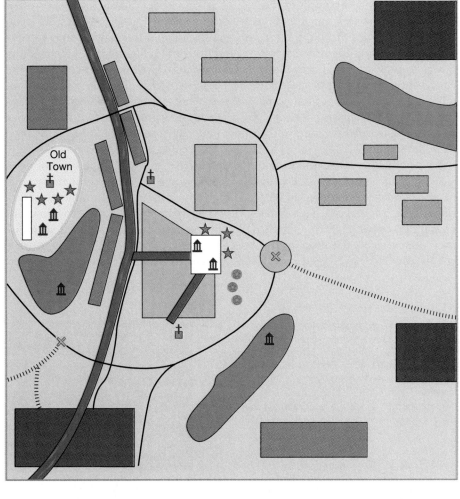

Old Town, Budapest

Housing estate, Budapest

Elite residential area, Budapest

★ Government building	◯ Shopping center	▮ Elite residential
⛪ Statue/monument	☐ Square	▨ Housing estate
⛪ Church/cathedral	▬ Pedestrian mall	☐ Retail
◉ Embassy	— Tramline	▮ Industry
✕ Railway station	� Railway	▨ Greenbelt/park

Figure 11.13 Urban structure in Eastern Europe   As indicated on this generalized diagram, cities in the formerly socialist countries of eastern Europe typically retain traces of their long history before the socialist era (1945–1989), most often in their old centers (with fortresses, cathedrals, and market squares) and elite residential areas. The socialist era left a strong imprint of mass housing, together with some monumental architecture and ceremonial spaces. Since 1989, the rapid Westernization of the economies and cultures has begun to be visible in cityscapes. (*Source:* S. Brunn and J. Williams (Eds.), *Cities of the World,* 2nd ed. New York: HarperCollins, 1993. Fig. 3.25, p. 143. © 1993. Reprinted by permission of Addison-Wesley Educational Publishers Inc. .)

Madonna graffiti on election posters, Budapest

fundamental division between three original functional components: colonial administration and commerce, military security, and indigenous commerce and residence. Usually located on a coastal site or a navigable river, the form and structure of colonial cities were dictated by European models of urban design, with a gridiron pattern of town planning and deliberate racial segregation.

The center of the military function was sometimes a walled fort but often only a camp, or cantonment. As in walled, medieval European cities, an open space was reserved around the fort or camp to allow a free field of fire. Nearby were situated the administrative offices and institutional buildings of the colonial power, around which a European-style CBD usually developed that contained the chief offices of European merchants and trading companies. The residences of most Europeans were set out some distance away, in "civil lines" (planned residential developments for officials and their families) characterized by spacious lots; adequate roads, water, and drains; and recreational facilities, clubs, and churches (Figure 11.14).

Beyond these areas, and usually separated from them by some form of open space such as a park or a golf course, a "native town" developed for indigenous peoples who were employed in servicing the administrative and commercial functions of the colonial city. This district was typically unplanned, overcrowded, and unsanitary. Urban growth after the colonial era has typically extended such districts a great deal. The more prosperous and elite groups within indigenous society, meanwhile, have taken over the residences in the civil lines and expanded

Figure 11.14 Urban structure in colonial cities The typical colonial city in South Asia was founded around a port facility, often protected by a fort, which would have had an open space around it to provide a field of fire. Colonial administrative offices were nearby, with military barracks and European civilian residences set apart from "native" residential areas. (*Source:* S. Brunn and J. Williams (Eds.), *Cities of the World,* 2nd ed. New York: HarperCollins, 1993, Fig. 9.7, p. 360. © 1993. Reprinted by permission of Addison-Wesley Educational Publishers Inc.)

Fort

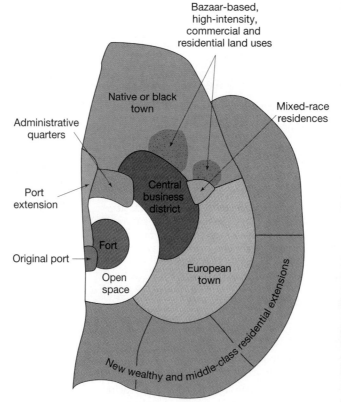

European quarter bungalow

Bazaar

into new neighborhoods around the periphery of the old former colonizers' neighborhoods.

Cities of the Periphery The cities of the world-system periphery, often still referred to as Third World cities, are numerous and varied. What they have in common is the experience of unprecedented rates of growth driven by rural "push"—overpopulation and the lack of employment opportunities in rural areas—rather than the "pull" of prospective jobs in towns and cities. Most of them have also grown a great deal in a relatively short period of time. São Paulo, Brazil, provides a good example (Figure 11.15). In 1930 the urban area of São Paulo was approximately 150 square kilometers (57 square miles), and the population was 1 million. By 1962 the city had grown to 750 square kilometers (288 square miles) with a population of 4 million. In 1980 the area was 1,400 square kilometers (538 square miles), and the population was more than 10 million. By 1995 the city covered 2,400 square kilometers (923 square miles), with a population of 15 million.

The structure of peripheral cities varies according to three factors:

- Relative levels of economic development, and the degree to which they have become industrialized and modernized
- Regional cultural values—as, for example, in the traditional layout and design of Islamic cities (see below)
- Whether society is organized more strongly on class or ethnic divisions

In many cities, however, once-distinctive patterns of spatial organization and land use have disappeared as a result of the congestion and overcrowding caused by overurbanization. To the extent that it is possible to generalize about the urban structure of the cities of the periphery (Figure 11.16, page 458), we can identify three important, common elements:

- A central concentration of modern commerce, retailing, and industry
- A distinctive zone or sector of elite residential neighborhoods
- Shanty or squatter neighborhoods that fill in every available space between and around them

Later in this chapter we will see how prominent this third element has come to be in the landscapes and the affairs of the unintended metropolises of the periphery.

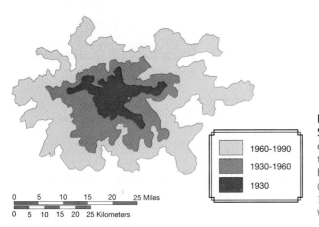

0   5   10   15   20   25 Miles

0   5   10   15   20   25 Kilometers

1960-1990

1930-1960

1930

Figure 11.15 The growth of São Paulo Since 1930, the city has grown 16-fold in extent, though the bulk of this growth has taken place since 1960. (*Source:* World Bank, *Urban Transport.* Washington, DC: The World Bank, 1986, Fig. 3, p. 5.)

Figure 11.16 Urban structure in peripheral cities   The most distinctive feature of cities in the world's periphery is their dualism, with, on the one hand, a core of modern commerce, retailing, and industry and its associated residential areas; and, on the other, an informal economy with extensive areas of makeshift, shanty housing. (*Source:* S. Lowder, *Inside Third World Cities.* London: Croom Helm, 1986, Figs. 7.1 and 7.2, pp. 211 and 214.)

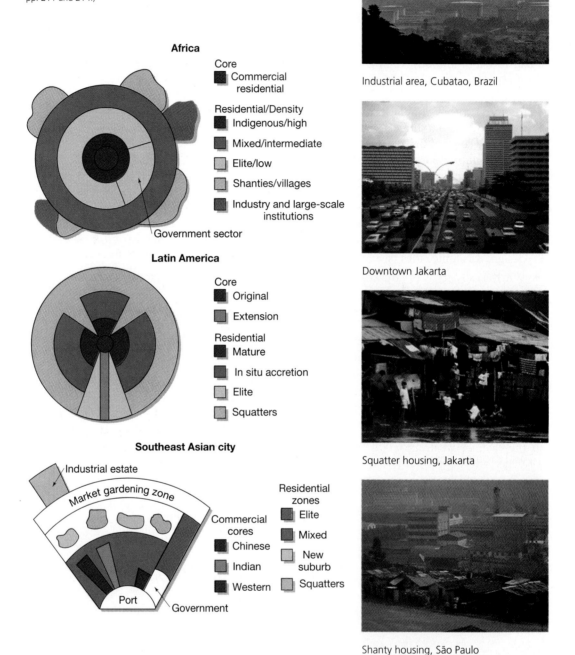

Africa

Core
■ Commercial residential

Residential/Density
■ Indigenous/high
■ Mixed/intermediate
□ Elite/low
■ Shanties/villages
■ Industry and large-scale institutions

Government sector

Latin America

Core
■ Original
■ Extension

Residential
■ Mature
■ In situ accretion
□ Elite
□ Squatters

Southeast Asian city

Industrial estate
Market gardening zone

Commercial cores
■ Chinese
■ Indian
■ Western

Port
Government

Residential zones
■ Elite
■ Mixed
□ New suburb
□ Squatters

Industrial area, Cubatao, Brazil

Downtown Jakarta

Squatter housing, Jakarta

Shanty housing, São Paulo

Urban Form and Design

Within the framework provided by the overall organization of land uses and activities, cities acquire a distinctiveness that comes from their physical characteristics: the layout of streets; the presence of monumental and symbolic structures; and the building types and architectural styles that fill out their built environment. The resulting urban landscapes are a reflection of a city's history, its physical environment, and its people's social and cultural values.

Geographers are interested in urban landscapes because, like other landscapes (Chapter 6) they can be read as multilayered texts that show how cities have developed, how they are changing, and how people's values and intentions take expression in urban form. The built environment, whether it is planned or unplanned, is what gives expression, meaning, and identity to the various forces involved in urbanization. It becomes a biography of urban change. At the same time, the built environment provides people with cues and contexts for behavior, with landmarks for orientation, and with symbols that reinforce collective values such as civic pride and a sense of identity.

Symbolic Landscapes

Some individual buildings and structures are so powerfully symbolic that they come to stand for entire cities: the Eiffel Tower or the Arc de Triomphe in Paris, the Opera House and Harbour Bridge in Sydney, the Forbidden City in Beijing, and the Houses of Parliament or Tower Bridge in London, for example (Figure 11.17, page 460). It is the generic urban landscapes of different kinds of cities that are of most interest to geographers, however. Some generic urban landscapes become symbolic of entire nations or cultures. These are ordinary cityscapes that are powerfully evocative because they are understood as being a particular *kind* of place.

Geographer D. W. Meinig identified three types of symbolic cityscapes. The stereotypical New England townscape, for example (Figure 11.18, page 461), that Meinig described as being "marked by a steeple rising gracefully above a white wooden church which faces on a village green around which are arrayed large white clapboard houses which, like the church, show a simple elegance in form and trim,"[2] is widely taken to symbolize not just a certain type of regional architecture but the best that Americans have known "of an intimate, family-centered, God-fearing, morally conscious, industrious, thrifty, democratic *community*."

Another ordinary townscape with powerful symbolic connotations is what Meinig called Main Street of Middle America (Figure 11.18). It is "middle" in several respects: between the frontier to the west and the cosmopolitan seaports to the east; between agricultural regions and industrial metropolises; between affluence and poverty. It has come to represent places with a balanced community, populated by property-minded, law-abiding citizens devoted to free enterprise and a certain kind of social morality.

A third symbolic cityscape identified by Meinig is suburbia, or, more specifically, California suburbia (Figure 11.18). The commonplace landscape of single-family dwellings standing on broad lots and fronted by open green lawns is widely attached to an image of a particular lifestyle for middle-class, nuclear families: individualistic, private, informal, and recreation- and consumption-oriented. First developed in California, this kind of suburb has diffused throughout America (and, indeed, beyond), and has come to symbolize the "American Dream" of an affluent, independent lifestyle.

The Islamic City

Islamic cities provide good examples of how social and cultural values and people's responses to their environment are translated into spatial terms through urban form and the design of the built environment. Indeed, it is because of similarities in cityscapes, layout, and design that geographers are able to talk about the Islamic city as a meaningful category. It is a category that includes thousands

[2]D. W. Meinig, "Symbolic Landscapes," in D. W. Meinig et al. (Eds.), *The Interpretation of Ordinary Landscapes.* New York: Oxford University Press, 1979, p. 165.

Venice

Tower Bridge, London

Arc de Triomphe, Paris

Opera House and Harbor Bridge, Sydney

Forbidden City, Beijing

Colosseum, Rome

Figure 11.17 Famous landmarks Some cities are immediately recognizable because of certain buildings, landmarks, or cityscapes that have come to be symbolic of them. These examples are known worldwide.

of towns and cities, not only in the Arabian Peninsula and Middle East—the heart of the Islamic Empire under the prophet Muhammad (A.D. 570–632)—but also in regions into which Islam spread later: North Africa, coastal East Africa, South-Central Asia, and Indonesia. Most of the cities in North Africa and South-Central Asia are Islamic, while many of the elements of the classic Islamic city can be found in towns and cities as far away as Seville, Granada, and Córdoba in southern Spain (the western extent of Islam), Kano in northern Nigeria and Dar es-Salaam in Tanzania (the southern extent), and Davao in the Philippines (the eastern extent).

The New England townscape symbolizes for many a particular kind of community: intimate, family-centered, industrious, thrifty, democratic, and morally aware.

Main Street of Middle America has come to be associated with the ideal of a "balanced" community in terms of income and class, with shared values of free enterprise, conservative morality, and commonsense practicality.

California suburbia is symbolic of the modern "American Dream" of a consumption- and leisure-oriented lifestyle.

Figure 11.18 Ordinary landscapes Some ordinary cityscapes are powerfully symbolic of particular kinds of places. The New England village, the Main Street of Middle America, and California suburbia are in this category, so much so that they have been taken as being symbolic of the United States itself, part of the "iconography of nationhood," the symbolic landscapes that give the country a sense of identity, both at home and abroad.

The fundamentals of the layout and design of the traditional Islamic city are so closely attached to Islamic cultural values that they are to be found in the Qur'an, the holy book of Islam. Although urban growth does not have to conform to any overall master plan or layout, certain basic regulations and principles are intended to ensure Islam's emphasis on personal privacy and virtue, on communal well-being, and on the inner essence of things rather than their outward appearance.

The single most dominant feature of the traditional Islamic city is the Friday Mosque, or *Jami*—the city's principal mosque. Located centrally, the mosque complex is not only a center of worship but also a center of education and the hub of a broad range of welfare functions. As cities grow, new, smaller mosques are built toward the edge of the city, each out of earshot from the call to prayer from the *Jami* and from one another. The traditional Islamic city was walled for defense, with several lookout towers and a Kasbah, or citadel (fortress), containing palace buildings, baths, barracks, and its own small mosque and shops (Figure 11.19, page 462).

Traditionally, gates controlled access to the city, allowing careful scrutiny of strangers and permitting the imposition of taxes on merchants. The major streets led from these gates to the main covered bazaars or street markets (*suqs*). The suqs nearest the Jami typically specialize in the cleanest and most prestigious goods, such as books, perfumes, prayer mats, and modern consumer goods.

Figure 11.19 The morphology of traditional Islamic cities The traditional Islamic city provides good examples of how social and cultural values and people's responses to their environment are translated into various aspects of urban form and architectural design.

In Islamic societies, the privacy of individual residences is paramount, and elaborate precautions are taken through architecture and urban design to ensure the privacy of women. Entrances are L-shaped and staggered across the street from one another. Walls on alleyways are often windowless, and separate rooms and courtyards are provided, when resources permit, for male guests. Architectural details also reflect climactic influences: Twisting, narrow streets help to maximize shade, and air ducts and roof funnels help create dust-free drafts.

Seen from above, the traditional Islamic city is a compact mass of residences with walled courtyards—a cellular urban structure within which it is possible to maintain a high degree of privacy. This photograph is of Yazd, Iran.

Yazd, Iran

Traditional wind towers, Yazd, Iran

The dominant feature of traditional Islamic cities is the *Jami*, or main mosque.

The *suq* is a covered bazaar or street market and is one of the most important distinguishing features of traditional Islamic cities. Typically, a suq consists of small stalls located in numerous passageways. Many important suqs are covered with vaults or domes. Within them, there is typically a marked spatial organization, with stalls that sell similar products clustered tightly together. This photograph shows the suq in Fez, Morocco.

462

Those nearer the gates typically specialize in bulkier and less valuable goods such as basic foodstuffs, building materials, textiles, leather goods, and pots and pans. Within the suqs, every profession and line of business had its own specialized alley, and the residential districts around the suqs were organized into distinctive quarters, or *ahya'*, according to occupation (or, sometimes, ethnicity, tribal affiliation, or religious sect).

Privacy is central to the construction of the Islamic city. Above all, women must be protected, according to Islamic values, from the gaze of men. Traditionally, doors must not face each other across a minor street, and windows must be small, narrow, and above normal eye level. Culs-de-sac (dead-end streets) are used where possible in order to restrict the number of persons needing to approach the home, and angled entrances are used to prevent intrusive glances. Larger homes are built around courtyards, which provide an interior and private focus for domestic life.

The rights of others are also given strong emphasis, the Qur'an specifying an obligation to neighborly cooperation and consideration—traditionally interpreted as applying to a minimum radius of 40 houses. Roofs, in traditional designs, are surrounded by parapets to preclude views of neighbors' homes, and drainage channels are steered away from neighbors' houses. Refuse and wastewater are carefully recycled. Public thoroughfares were originally designed to be wide enough to allow two fully laden camels to pass each other, and high enough to allow a camel and rider to pass through.

Because most Islamic cities are located in hot, dry climates, these basic principles of urban design have evolved in conjunction with certain practical solutions to intense heat and sunlight. Twisting streets, as narrow as permissible, help to maximize shade, as does lattice work on windows and the cellular, courtyard design of residential areas. In some regions, local architectural styles include air ducts and roof funnels with adjustable shutters that can be used to create dust-free drafts.

All these features are still characteristic of Islamic cities, though they are especially clear in their old cores, or *medinas*. Like cities everywhere, however, Islamic cities also bear the imprint of globalization. Although Islamic culture is self-consciously resistant to many aspects of globalization, it has been unable to resist altogether the penetration of the world economy and the infusion of the Western-based culture of global metropolitanism. The result can be seen in international hotels, skyscrapers and office blocks, modern factories, highways and airports, and stores (Figure 11.20).

Figure 11.20 Contemporary urban design in Abu Dhabi Like cities almost everywhere, Islamic cities reflect the imprint of globalization. Economic globalization is reflected in international hotels and the offices of transnational corporations; cultural globalization is reflected in the presence of Western consumer products and advertisements for Western popular culture.

Figure 11.21 Cairo's City of the Dead
Islamic cities have also experienced the problems of rapid population growth characteristic of peripheral countries. In Cairo, the acute housing shortage has resulted in people making their homes in the cemeteries of the "City of the Dead," a zone of elaborate cemeteries, unserviced by municipal utilities, on the eastern edge of the metropolis. These cemeteries were first occupied during the Egypt-Israeli war (1973) by refugees from the Suez Canal area. Parts of the cemeteries date from the fourteenth century. They contain elaborate funerary complexes that consist not only of tombs but also fountains, mosques, and meeting halls.

Islamic culture and urban design principles have not always been able to cope with the pressures of contemporary rates of urbanization, so that the larger Islamic cities—such as Algiers, Ankara, Cairo, Istanbul, Karachi, and Teheran, for example—now share with other peripheral cities the common denominators of unmanageable size, shanty and squatter development, and low-income mass housing (Figure 11.21).

Planned Urban Design

City planning and design have a long history. As we have seen, most Greek and Roman settlements were deliberately laid out on grid systems, within which the siting of key buildings and the relationship of neighborhoods to one another were carefully thought out. In ancient China, cities were laid out with strict regard to Taoist ideas about the natural order of the universe; with different quarters representing the four seasons of the year; and with the placement of major streets and the interior layout of buildings designed to be in harmony with cosmic energy. This kind of mystical interpretation of nature is known as *geomancy*, and its application to design is known as *feng shui*. It has always been important in Eastern cultures, and continues to influence many aspects of urban planning and design in the Far East, since many people believe that creating a positive energy flow through a home or place of business brings good luck and fortune. Feng shui was used, for example, in the design of Suntec City, a major new development in Singapore that consists of an international convention and exhibition center and a complex of office towers. In Hong Kong, the GIS database for the 1990 Land Information System included a feng shui layer that recorded areas subject to development restrictions because of feng shui principles. Meanwhile, globalization has brought feng shui to the rest of the world. East Asian immigrants have long practiced feng shui within the Chinatowns of North American cities. Recent waves of more affluent East Asian immigrants to Pacific Rim cities like Los Angeles, Vancouver, Sydney, and Auckland have required developers, architects, and interior designers to be attentive to feng shui in the construction of new suburbs. At the same time, some speculative East Asian investors in the real estate markets of world cities like New York and London have insisted on the use of feng shui in the design of skyscrapers and commercial developments.

The roots of modern Western urban planning and design can be traced to the Renaissance and baroque periods (between the fifteenth and seventeenth centuries) in Europe, when rich and powerful regimes used urban design to produce extravagant symbolizations of wealth, power, and destiny. Dramatic advances in

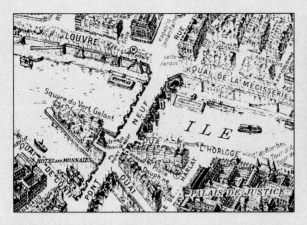

The first significant element of Renaissance urban design in Paris was the Pont Neuf (New Bridge), built across the broadest part of the River Seine and completed in 1604 under the rule of Henri IV.

The Place Royale (renamed the Place des Vosges after the Revolution) is of great significance in European history as the prototype of the residential square. It was completed in 1612.

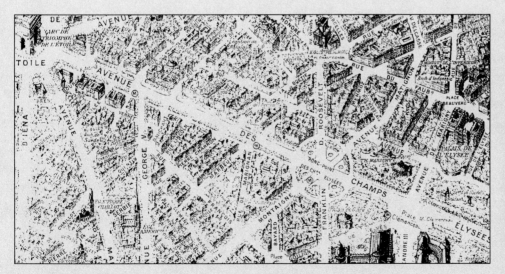

Figure 11.22 Renaissance urban design Inspired by the classical art forms of ancient Greece and Rome, Renaissance urban design sought to recast cities in a grand manner, in a deliberate attempt to show off the power and the glory of the State and the Church. Spreading slowly from its origins in Italy at the beginning of the fifteenth century, Renaissance design had diffused to most of the larger cities of Europe by the end of the eighteenth century. (*Source: Plan de Paris à Vol D'Oiseau.* Paris: Blondel la Rougery. EDIT.-imp. 1957.)

The Champs Élysées was laid out in the 1660s to provide a processional route appropriate to the king.

military ordnance (cannon and artillery) brought a surge of planned redevelopment that featured impressive fortifications, geometric shaped redoubts, or strongholds, and an extensive *glacis militaire*, or sloping clear zone-of-fire. Inside new walls, cities were recast according to a new esthetic of Grand Design—geometrical plans, streetscapes, and gardens that emphasized views of dramatic perspectives, and fancy palaces that were a deliberate attempt to show off the power and the glory of the State and the church (Figure 11.22). These developments were often of such a scale that they effectively fixed the layout of cities well into the eighteenth and even into the nineteenth century, when walls and/or glacis eventually made way for urban redevelopment in the form of parks, railway lines, or beltways.

As societies and economies became more complex with the transition to industrial capitalism, national rulers and city leaders looked to urban design to impose order, safety, and efficiency as well as to symbolize the new seats of power and authority. One of the most important early precedents was set in Paris by Napoleon III, who presided over a comprehensive program of urban redevelopment and monumental urban design. The work was carried out by Baron Georges Haussmann between 1853 and 1870. Haussmann demolished large sections of old Paris to make way for broad, new, tree-lined avenues, with numerous public open spaces and monuments. In doing so, he not only made the city more efficient (wide boulevards meant better flows of traffic) and a better place to live (parks and gardens allowed more fresh air and sunlight in a crowded city and were held to be a "civilizing" influence), but also made it safer from revolutionary politics (wide boulevards were hard to barricade; monuments and statues helped to instill a sense of pride and identity).

Beaux Arts: a style of urban design that sought to combine the best elements of all of the classic architectural styles.

The preferred architectural style for these new designs was the **Beaux Arts** style, which takes its name from L'Ecole des Beaux Arts in Paris. In this school, architects were trained to draw on Classical, Renaissance, and Baroque styles, synthesizing them in designs for new buildings for the industrial age. The idea was that the new buildings would blend artfully with the older palaces, cathedrals, and civic buildings that dominated European city centers. Haussmann's ideas were widely influential and extensively copied. One of the most famous examples is the Ringstrasse in Vienna (Figure 11.23).

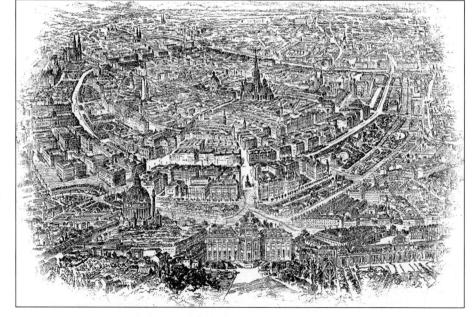

Figure 11.23 The Ringstrasse in Vienna Developed by Emperor Franz Joseph in the mid-nineteenth century on the site of derelict fortifications around the medieval Altstadt, or old town, the Ringstrasse provided a two-mile, 200-foot wide, semicircular boulevard along which Franz Joseph arrayed a series of major public buildings, including the national parliament, the city hall, the university, museums, theater, and opera house. This lithograph, viewing the city from the southwest, was published in 1873, soon after the completion of the Ringstrasse.

Figure 11.24 The Chicago Plan, 1909
Daniel Burnham's Chicago Plan of 1909 was based on esthetic means toward social objectives. By giving the city a strong visual and esthetic order, Burnham wanted to create the physical preconditions for the emergence of a harmonious social climate and strong moral order. These were popular sentiments in the Progressive Era, and much of Burnham's ambitious plan was actually carried out. (Source: R. Burnham and E. Bennett, *Plan of Chicago*. New York: Princeton Architectural Press, 1993, Plate CXXXII, p. 112.)

In the United States, the City Beautiful movement, which began in the late nineteenth century, drew heavily on Haussmann's ideas and Beaux Arts designs. The **City Beautiful movement** was a Progressive Era (c. 1890–1920) attempt to remake cities in ways that would reflect the higher values of society, using neoclassical architecture, grandiose street plans, parks, and inspirational monuments and statues. The idea, again, was deliberately to exploit urban design as an uplifting and civilizing influence, while at the same time emphasizing civic pride and power. Daniel Burnham's 1909 Plan for Chicago (Figure 11.24) is a good example. During the same period, European imperial powers imposed similar designs on their colonial capitals and administrative centers. Examples include Casablanca (Morocco), New Delhi (India), Pretoria (South Africa), Rangoon (Burma), Saigon (Vietnam), and Windhoek (Namibia).

Early in the twentieth century there emerged a different intellectual and artistic reaction to the pressures of industrialization and urbanization. This was the **Modern movement,** which was based on the idea that buildings and cities should be designed and run like machines. Equally important to the Modernists was that urban design not reflect dominant social and cultural values but, rather, help to create a new moral and social order. The movement's best-known advocate was Le Corbusier, a Paris-based Swiss who provided the inspiration for technocratic urban design. Modernist buildings sought to dramatize technology, exploit industrial production techniques, and use modern materials and unembellished, functional design. Le Corbusier's ideal city (*La Ville Radieuse*, Figure 11.25, page 468) featured linear clusters of high-density, medium-rise apartment blocks, elevated on stilts and segregated from industrial districts; high-rise tower office blocks; and transportation routes—all separated by broad expanses of public open space.

After World War II this concept of urban design became pervasive, part of what became known as the International Style: boxlike steel-frame buildings with concrete and glass facades. The International Style was avant-garde yet respectable and, above all, comparatively inexpensive to build. It has been this tradition of urban design that has, more than anything else, imposed a measure of uniformity on cities around the world. Globalization has brought the appearance of International Style buildings in big cities in every part of the world. Furthermore, the International Style has often been the preferred basis for large-scale urban design projects around the world. One of the best examples of this is Brasilia, the capital of Brazil, founded in 1956 in an attempt to shift the country's political, economic, and psychological focus away from the past; differentiate it

City Beautiful movement: attempt to remake cities in ways that would reflect the higher values of society, using neoclassical architecture, grandiose street plans, parks, and inspirational monuments and statues.

Modern movement: the idea that buildings and cities should be designed and run like machines.

Figure 11.25 La Ville Radieuse The modern era and the advent of new transportation and construction technologies encouraged the Utopian idea that cities could be built as efficient and equitable "machines" for industrial production and progressive lifestyles. One of the most famous and influential examples was *La Ville Radieuse* (1933), a visionary design by the Swiss architect Le Corbusier. His vision was for the creation of open spaces through collectivized, high-density residential areas, strictly segregated from industrial areas and highways through a geometric physical plan. (*Source:* Le Corbusier, *La Ville Radieuse.* Paris: Editions de L'Architecture D'Aujourd'hui, p. 170.)

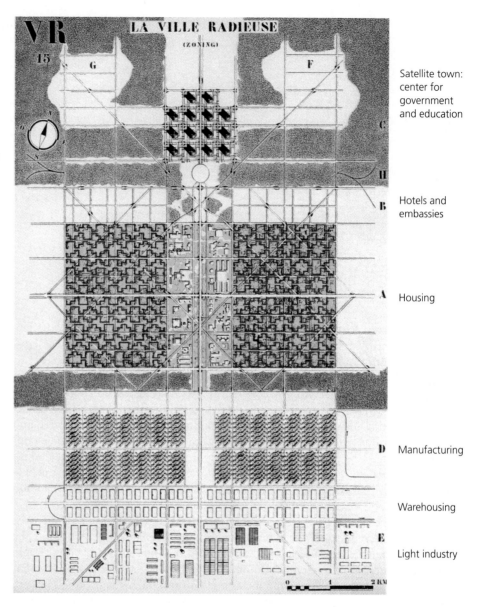

Satellite town: center for government and education

Hotels and embassies

Housing

Manufacturing

Warehousing

Light industry

from the former colonial cities on the coast; and orient the country toward the future and the interior (see Chapter 6).

Modern urban design has had many critics, mainly on the grounds that it tends to take away the natural life and vitality of cities, replacing varied and human-scale environments with monotonous and austere settings. In response to this, historic preservation has become an important element of urban planning in every city that can afford it. In addition, Postmodern urban design has brought a return to traditional and decorative motifs and introduced a variety of deliberately "playful" and "interesting" architectural styles in place of the functional designs of Modernism (Figure 11.26). **Postmodern urban design** is characterized by a diversity of architectural styles and elements, often combined in the same building or project. It makes heavy use of symbolism and of color and decoration. It is no coincidence that postmodern design has flourished in the most recent phase of globalization. Having emerged as a deliberate reaction to the perceived shortcomings of Modern design, its emphasis on decoration and self-conscious stylishness has made it a very convenient form of packaging for the new global consumer culture. It is geared to a cosmopolitan market, and it draws quite delib-

Postmodern urban design: style characterized by a diversity of architectural styles and elements, often combined in the same building or project.

Figure 11.26 Postmodern urban design Postmodern architecture and urban design is partly a reaction to the functional uniformity of Modernism. Rather than being geared to the creation of Utopian master plans, it finds its expression in artful fragments that come in many different guises, from straight copies of traditional urban form, through eclectic combinations of design styles, to new and experimental forms. This example is the Town Center at Reston (Virginia), a privately planned residential community outside Washington, DC. With intentional irony, it is in effect an outdoor shopping mall, designed to be reminiscent of a traditional, small-town Main Street.

erately on a mixture of elements from different places and times. In many ways, it has become the transnational style for the more affluent communities of the world's cities.

Urban Trends and Problems

Differences in urban growth processes between the world's core regions and the underdeveloped periphery are mirrored in patterns and processes of urban change. In the core regions, urban change is dominated by the consequences of an economic transformation to a postindustrial economy. Together with the continuing revolution in communications and information-processing technologies and the increasing dominance of transnational corporate organizations, this transformation has made for a fundamental restructuring of metropolitan areas. Traditional manufacturing and related activities have been moved out of inner-city areas, leaving decaying neighborhoods and a population of elderly and socially and economically marginalized people. New commercial activities have meanwhile begun to cluster in redeveloped CBDs and in edge cities around metropolitan fringes. The logic of agglomeration economies has created 100-mile cities—metropolitan areas that are literally 100 miles or so across—consisting of a series of cities and urban districts that are bound together through urban freeways (see Feature 11.1, "Visualizing Geography—The 100-Mile City").

These basic economic trends are accompanied by some important sociocultural trends that are shaping the details of metropolitan change. A relatively high level of affluence means that cities are continually being restructured to accommodate a variety of lifestyles—family-oriented, career-oriented, sports- and recreation-oriented, education-oriented, retirement, and so on. This affluence, however, is accompanied by a high level of social polarization, income inequality, and ethnic and racial division. The result is another trend in urban restructuring:

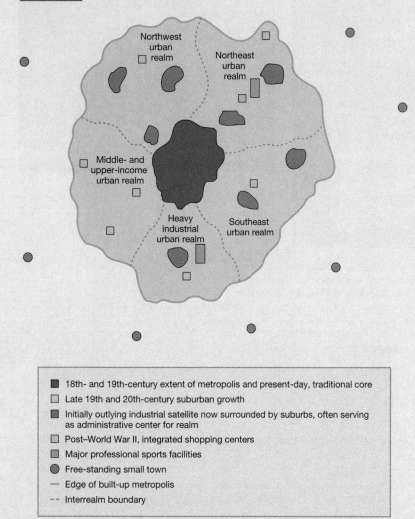

Metropolitan sprawl has left the pre-automobile city as the core of a metropolis that consists of a loose coalition of urban realms, or economic subregions. Each urban realm functions semi-independently, with a broad mix of land uses and populations of between 175,000 and 250,000. Each realm has retail, commercial, and residential subareas, as well as a commercial and retailing node that functions as a high-order central place for the majority of local residents. As a result, most residents of metropolitan areas do not make use of the entire metropolis; and many do not have much to do with the central core, except for occasional forays. (Source: J. Vance, Jr., *The Continuing City.* Baltimore: The Johns Hopkins University Press, 1990, p. 504. Copyright © 1990 The Johns Hopkins University Press.)

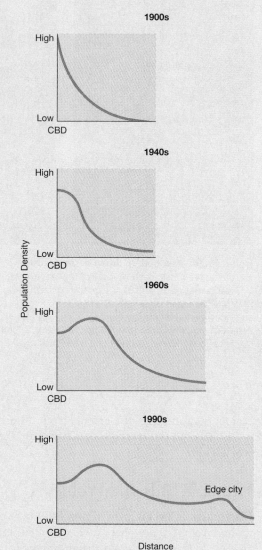

As automobile ownership has increased and highway engineering has improved, U.S. metropolitan areas have sprawled outward. Meanwhile, the density of population in central city areas has declined dramatically.

Low-density sprawl begins within a very short distance from downtown Phoenix, Arizona.

The construction of interstate highways and circumferential beltways was an important precondition for metropolitan sprawl in the United States. The 1956 Federal Aid Highway Act provided federal funding for the construction of multilane, limited-access highways between every major city, with hub-and-wheel linkages within metropolitan areas. The outer rim of the wheel was a circumferential beltway, which made outlying locations much more accessible, particularly where there were intersections with major radial spokes. Between 1956 and 1972, 41,000 miles of limited-access highway were built, at an average cost of more than $1 million per mile. The result has revolutionized metropolitan America, not only allowing metro areas to sprawl, but also allowing adjoining metro areas to coalesce, creating megalopolitan regions.

In the 1980s the extraordinary growth of suburban office and retailing centers caused them not only to overshadow their core CBD in terms of retail floorspace and office space but to compete with the CBD for highly specialized functions such as mortgage banking, corporate legal offices, accounting services, luxury hotels, and exclusive clubs. As a result, these "edge cities" developed distinctive local labor markets and strongly focused commuting patterns. Tysons Corner, an edge city located on the beltway outside Washington, DC, is an unincorporated area with 40,000 residents and over 90,000 jobs. Tysons Corner does not exist as a postal address: Residents' mail must go either to Vienna or McLean. But this anonymous city contains a huge concentration of commercial space (the eighth largest in the United States in 1995, including all downtown CBDs), including more than 20 million square feet of office space, several million square feet of retail space, nine major department stores, more than 3,000 hotel rooms, and parking for more than 90,000 cars.

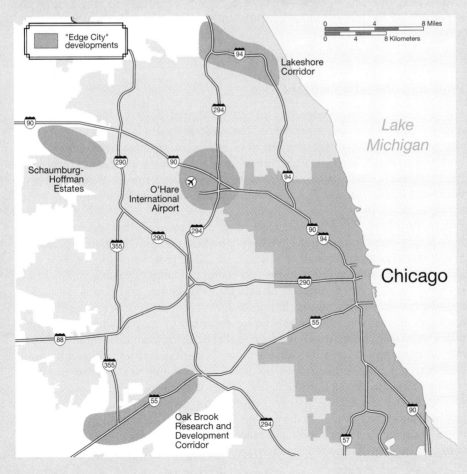

Between 1970 and 1990, the population of the Chicago metropolitan region grew by just over 4 percent. Meanwhile, the amount of land consumed by the metropolitan region increased by nearly 50 percent. Most of this sprawl was associated with the growth of edge cities on the metropolitan periphery. The importance of Schaumburg as an edge city was underlined in 1992 when Sears, Roebuck & Co. relocated 5,000 workers to Hoffman Estates from its famous skyscraper headquarters in downtown Chicago. All of the edge city environments are characterized by high levels of affluence and rapid population growth. Naperville, a community in the Research and Development Corridor to the southeast of Chicago, grew from 50,000 to over 100,000 during the 1980s. Large tracts of new middle- and upper-middle-income communities were built by developers for the thousands of office and high-tech workers whose companies had located in the corridor. Responding to its new affluence, the community reinvented itself with a new sense of place. It redeveloped its old main street downtown, created a landscaped river walk, built community recreational centers and an outdoor theater, and instituted an annual rib festival.

471

a more intensive and more fragmented mosaic of sociocultural segregation, with different groups often coalescing in separate municipalities. This fragmentation, in conjunction with the economic decline of many inner-city neighborhoods, is tied in to yet another important trend—the inability of central cities to fund basic services and maintain the essential urban infrastructure of streets, highways, schools, parks, and utilities. **Central cities** are the original, core jurisdictions of metropolitan areas.

Central cities: the original, core jurisdictions of metropolitan areas.

In contrast, the basic trend affecting the cities of the world's periphery is demographic—the phenomenal rates of natural increase and in-migration that have given rise to overurbanization. This trend is reflected in an ever-growing informal sector of the economy, in which people seek economic survival. The informal labor market is directly paralleled in informal shantytowns and squatter housing. Because so few jobs with regular wages exist, few families can afford rent or house payments for sound housing. Indirectly, it is reflected in cities' inability to provide a basic infrastructure of highways, transportation, schools, utilities, and emergency services. Because the informal sector yields no tax revenues, municipal funds are severely lacking to provide an adequate infrastructure or to maintain a safe and sanitary environment.

At the same time, however, these cities under stress represent local and regional concentrations of investment, manufacturing, modernization, and political power. As we have seen, the typical peripheral metropolis plays a key role in international economic flows, linking provincial regions with the hierarchy of world cities and, thus, to the global economy. Within peripheral metropolises, this role results in a pronounced **dualism,** or juxtaposition in geographic space of the formal and informal sectors of the economy. It is among these peripheral metropolises—Mexico City (Mexico); São Paulo (Brazil); Lagos (Nigeria); Mumbai, formerly Bombay (India); Dacca (Pakistan); Jakarta (Indonesia); Karachi (Pakistan); and Manila (the Philippines)—that we can find contenders for the title of the shock city of the late twentieth century—the city that is the embodiment of the most remarkable and disturbing changes in economic, social, and cultural life (see Feature 11.2, "Visualizing Geography—Shock City: Lagos").

Dualism: the juxtaposition in geographic space of the formal and informal sectors of the economy.

Problems of Postindustrial Cities

For all their relative prosperity, the postindustrial cities of the world's core regions have their share of problems. The most acute are localized in the central city areas that have borne the brunt of restructuring from an industrial to a postindustrial economy. In these areas there are several interrelated problems: fiscal problems, infrastructure problems, and localized spirals of neighborhood decay and cycles of poverty. A discussion of each condition follows.

Fiscal Problems Economic restructuring and metropolitan decentralization have meant that central cities have been left since the mid-1970s with a chronic problem that geographers call "fiscal squeeze." The term fiscal refers to taxes; the idea of a **fiscal squeeze** refers to the increasing difficulty of raising sufficient tax revenues to pay for the upkeep of urban infrastructure and city services. The squeeze comes from increasing limitations on revenues, and increasing demands for expenditure. The revenue-generating potential of central cities has steadily fallen as metropolitan areas have decentralized, losing both residential and commercial taxpayers to suburban jurisdictions. Growth industries, white-collar jobs, retailing, and more affluent households have moved out to suburban and exurban jurisdictions, taking their local tax dollars with them.

Fiscal squeeze: increasing limitations on city revenues, combined with increasing demands for expenditure.

At the same time, growth in property-tax revenues from older, decaying neighborhoods has slowed as the growth of property values has slowed. Yet these older, decaying neighborhoods cost more to maintain and service. Older streets, water and sewage lines, schools, and transit systems have high maintenance costs.

The residual populations of these neighborhoods, with high proportions of elderly and indigent households, are increasingly needy in terms of municipal welfare services. Large numbers of low-income migrants and immigrants also bring increased demands for municipal services. Added to all this, central city governments are still responsible for services and amenities used by the entire metropolitan population: municipal galleries and museums, sports facilities, parks, traffic police, and public transport, for example.

In some cities, fiscal squeeze has become a *fiscal crisis* as city governments have run out of money. Unable to control the costs of municipal services, and fearful that significantly reduced service levels will bring political problems or civil unrest, cities are tempted to roll over their operating deficits from one year to another, financing them the only way they can—by borrowing on high-interest, short-term bonds. This, of course, means an extra burden of debt repayment, and it almost inevitably worsens the situation to the point of crisis, leading to massive service cutbacks and layoffs of municipal workers. This is what happened in New York City in the mid-1970s and has since affected many other cities, including Baltimore, Boston, Cleveland, Detroit, Philadelphia, St. Louis and, in Europe, Birmingham, Brussels, Copenhagen, Frankfurt, Liverpool, Milan, Naples, Palermo, and Rome.

Infrastructure Problems As the rate of urban growth in core countries has slowed, public spending on the urban infrastructure of roads, bridges, parking spaces, transit systems, communications systems, power lines, gas supplies, street lighting, water mains, sewers, and drains has declined. Meanwhile, much of the original infrastructure, put in place 75 or 100 years ago with a design life of 50 or 75 years, is obsolete, worn-out, and in some cases perilously near the point of collapse.

Infrastructure problems are easily overlooked because they build up slowly. Only when a bridge collapses or a water main bursts are infrastructure problems newsworthy. Yet cities' infrastructure is crucial not only to their economic efficiency but also to public health, safety, and the quality of life. One review of infrastructure problems in American cities estimated that between half and two-thirds of the country's cities are unable to support modernized investment until major new investments can be made in their basic infrastructure. This investment, of course, is unlikely, given the chronic fiscal problems that they face.

In Boston, three-quarters of the sewer system was built a century or more ago and has now deteriorated to the point where about 15 percent of the system's overall flow is lost to leaks. Hundreds of kilometers of water mains in New York City have been identified as being in need of replacement, at an estimated cost in 1994 of over $4 billion. Almost 50 percent of all wastewater treatment systems in America are operating at 80 percent or more of their capacity, the level at which the federal government prohibits further industrial hookups in a community.

Fresh water supplies are also at risk: Old water systems are unable to cope with the leaching of pollutants into city water. Common pollutants include chlorides, oil, phosphates, and nondegradable toxic chemicals from industrial wastewater; dissolved salts and chemicals from highway de-icing; nitrates and ammonia from fertilizers and sewage; and coliform bacteria from septic tanks and sewage. Many cities still use water-cleaning technology dating to World War I. About one-third of all towns and cities in the United States have contaminated water supplies, and about 8 million people are using water that is potentially dangerous. Between 1986 and 1994, 116 outbreaks of intestinal illness occurred in U.S. cities caused by contaminated drinking water. The worst of these was in Milwaukee in April 1993, where 403,000 people, about half the city's population, suffered from an outbreak of cryptosporidium. A parasite that has been excreted by cattle or other animals and washed into lakes and reservoirs, cryptosporidium easily finds its way through obsolescent water purification systems.

A Day in the Life of Kate Adikiwe—Kate Adikiwe lives in the suburban district of Olaleye, Lagos, once a small village whose residents grew herbs, fruit, and vegetables; fished; trapped; made palm wine; and processed palm oil. The village grew rapidly when a railway line was constructed through it and as Lagos grew rapidly outward after independence. In the mid-1960s about 2,500 residents lived in Olaleye; today, about 25,000 live there. Within its small site of some 35 hectares (86 acres) is an enormous range of economic activities: a large market, beer parlors, night clubs, brothels, a makeshift cinema, tailors, shoemakers, blacksmiths, tinkers, watch-repairers, knife-sharpeners, mechanics, battery-chargers, and itinerant barbers and beauticians. Many of the women produce and sell a great variety of cooked foodstuffs, while many of the men work outside the district in factories or offices.

Kate is one of six children. Her father is a clerical worker in one of the city's department stores. Her mother is a seamstress, working from the house. The house itself contains 12 families, each with a single room and sharing the one kitchen, toilet, and bathroom in the building. One of Kate's jobs is to draw water from the nearby well each morning before school—the water is stored in plastic buckets in the living room until needed. After school, Kate has to complete her homework and help her mother prepare food for the family. Most of the cooking is done on kerosene stoves in the passageway. After the meal, Kate and her older sister help their mother with sewing—they do not expect to be able to get jobs after school and are learning to become seamstresses like their mother.

A World Bank study in 1990 found that virtually every manufacturing firm in Lagos had its own generator. The reason: The public electricity utility was too unreliable.

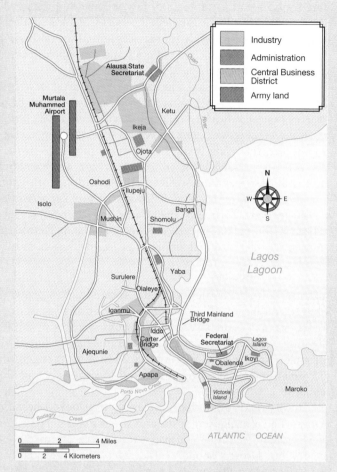

Lagos developed from an initial settlement at Iddo and on the northern shore of Lagos Island. Ikoyi, on Lagos Island, was laid out in 1918 as a Government Residential Estate to house colonial officials. Most of the city's growth, however, has been unplanned and irregular, with swamps, coves, and canals impeding efficient development. (Source: M. Peil, *Lagos*. London: Belhaven Press, 1991, p. 23.)

The cityscape on Lagos Island reflects both residential congestion and the postcolonial development of the city as a peripheral metropolis with important corporate functions.

These older neighborhoods are a legacy of the British colonial era, which ended in 1960 when Nigeria became an independent federation.

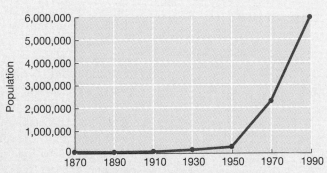

Population growth has far outstripped the city's capacity to deal with the daily movement of people, a problem that is worsened by the fact that the central city is trapped on an island site, with very limited access by road bridges.

Lagos, like most metropolises in the world's periphery, grew relatively slowly until quite recently. Triggering an explosive growth in population was the combination of the demographic transition, political independence, and an economic boom stimulated by the discovery of oil reserves in southeastern Nigeria. Because of its difficult site on sandspits and lagoons, Lagos has experienced growth characterized by irregular sprawl and, in the central area, a density of population higher than that of Manhattan, New York.

Overwhelmed by an unprecedented rate of urbanization, an economy that cannot provide regularly paid employment for a significant proportion of its residents, and a municipal government that has neither the financial resources nor the personnel to deal with the problems, Lagos has become emblematic of the problems of overurbanization. Shanty housing is a direct consequence of widespread poverty; open sewers are a consequence of limited or nonexistent municipal resources.

For many people, life in the unintended metropolis is a matter of survival. This leads to a tremendous variety of informal economic activities, from street vending to home-brewed beer, from prostitution to drug peddling. The photograph on the left shows the most common form of informal activity: street trading, which takes place on almost every unoccupied sidewalk, street, or unclaimed space. The photograph on the right shows an outdoor laundry whose operatives have temporarily invaded a lagoonshore boatyard.

In healthy humans, it causes stomach upsets and diarrhea, but for individuals with weak immune systems, it can be dangerous: Over 100 died in Milwaukee's outbreak.

Poverty and Neighborhood Decay Inner-city poverty and neighborhood decay have become increasingly pronounced since the 1960s, as manufacturing, warehousing, and retailing jobs have moved out to suburban and edge-city locations, and as many of the more prosperous households have moved out to be near these jobs. The spiral of neighborhood decay begins with substandard housing occupied by low-income households who can afford to rent only a minimal amount of space. The consequent overcrowding not only causes greater wear and tear on the housing itself but also puts pressure on the neighborhood infrastructure of streets, parks, schools, and playgrounds (Figure 11.27). The need for

Figure 11.27 Poverty areas   Poverty areas are defined by the U.S. Bureau of the Census as contiguous census tracts in which at least 20 percent of the households have an income below the official poverty level (just under $13,000 per year for a nonfarm family of four when the last census was taken in 1990). These concentrations of poverty are found not only in decaying inner-city areas, but also in newer public housing projects and in first- and second-tier suburbs that have filtered down the housing scale. In Minneapolis, Minnesota, concentrations of extreme poverty (more than 40 percent of the households in a census tract) have developed in three locations: an area of older, formerly middle-class housing in the inner South Side; in the city's major concentration of public housing in the Near North area; and around the campus of the University of Minnesota, to the east of the CBD. In St. Paul, concentrations of extreme poverty have developed in the public housing projects to the north and northeast of the CBD, and in an area of publicly assisted housing west of downtown. (*Source:* J. S. Adams, L. Lambert, and B. J. VanDrasek, "Poverty and Urban Decline." *CURA Reporter,* 25(4) [Vol. XXV, No. 4], 1995, Fig. 2, p. 4, Center for Urban and Regional Affairs, University of Minnesota.)

Sumner-Field, the first public housing project in Minneapolis, is part of a concentration of poverty on the north side of the city.

Former middle-class housing that is now part of a poverty area in the inner South Side district of Minneapolis

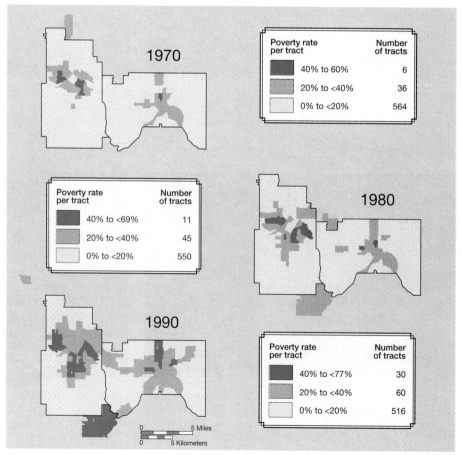

Figure 11.28 Urban decay
Landlords of older inner-city apartments have little incentive to invest in maintenance or repair, since they have a captive market of low-income households who have no other source of shelter. Equally, the low incomes of tenants mean that rents must be kept relatively low, leaving many landlords with small profit margins that leave nothing to spare for upkeep or improvement. Faced with rising property taxes and with profit margins that are unacceptably low, some landlords simply write off their property by abandoning it to long-term vacancy.

maintenance and repair increases quickly but is rarely met. Individual households cannot afford it, and landlords have no incentive to do so, since they have a captive market. Public authorities face a fiscal squeeze and are in any case often indifferent to the needs of such neighborhoods because of their relative lack of political power.

Shops and privately run services such as restaurants and hairdressers are afflicted with the same syndrome of decay. With a low-income clientele, profit margins must be kept low, leaving little to spare for upkeep or improvement. Many small businesses fail, or relocate to more favorable settings, leaving commercial property vacant for long periods. In extreme cases, such property becomes abandoned, the owners unable to find either renters or buyers. Residential buildings may also be left derelict (Figure 11.28). Faced with escalating maintenance costs and rising property taxes, and unable to increase revenues because of rent controls and the depressed state of inner-city housing markets, some landlords simply write off their property and abandon it.

There is, meanwhile, a dismal cycle of poverty that intersects with these localized spirals of decay. The **cycle of poverty** involves the transmission of poverty and deprivation from one generation to another through a combination of domestic circumstances and local, neighborhood conditions. This cycle begins with a localized absence of employment opportunities and, in turn, a concentration of low incomes, poor housing, and overcrowded conditions. Such conditions are unhealthy. Overcrowding makes people vulnerable to physical ill-health, which is compounded by poor diets. This in turn contributes to absenteeism from work, which results in decreased income. Similarly, absenteeism from school through illness contributes to the cycle of poverty by constraining educational achievement, limiting occupational skills, and so leading to low wages. Crowding also produces psychological stress, which contributes to social disorganization

Cycle of poverty: transmission of poverty and deprivation from one generation to another through a combination of domestic circumstances and local, neighborhood conditions.

and a variety of pathological behaviors, including crime and vandalism. Such conditions not only affect people's educational achievement and employment opportunities, but can also lead to the *labeling* of the neighborhood, whereby all residents may find their employment opportunities affected by the poor image of their neighborhood.

One of the most important elements in the cycle of neighborhood poverty, however, is the educational setting. Schools, obsolete and physically deteriorated like their surroundings, are unattractive to teachers—partly because of the physical environment and partly because of the social and disciplinary environment. Because of fiscal squeeze, the schools are often resource-poor, with relatively small budgets for staff, equipment, and materials. Over the long term, poor educational resources translate into poor education, however positive the values of students and their parents. Poor education limits occupational choice and, ultimately, results in lower incomes. Students, faced with the evidence all around them of unemployment or low-wage jobs at the end of school careers, find it difficult to be positive about school. The result becomes a self-fulfilling prophecy, and people become trapped in areas of concentrated poverty (Figure 11.28).

Many poverty areas are also racial ghettos although not all ghettos are poverty areas. As we have seen, ethnic and racial congregation can function to mitigate the effects of the cycle of poverty. Nevertheless, discrimination is usually the main cause of ghettoization. In the United States, discrimination in housing markets is illegal, but it nevertheless takes place in a variety of ways. One example of housing-market discrimination by banks and other lending institutions is the practice of redlining. **Redlining** involves delimiting bad-risk neighborhoods on a city map, and then using the map as the basis for determining loans. This practice results in a bias against minorities, female-headed households, and other vulnerable groups, since they tend to be localized in low-income neighborhoods. Redlining tends to become a self-fulfilling prophecy, since neighborhoods starved of property loans become progressively more run down and so increasingly unattractive to lenders. Discrimination affects education and labor markets as well as housing markets. In the case of ghetto poverty, all three types of discrimination come together, reinforcing the cycle of poverty and intensifying the disadvantages of the minority poor.

Social trends have compounded the problems of these areas in many instances. Increased divorce rates and a high incidence of teen pregnancy have led to much greater numbers of single-parent families and a feminization of poverty. These families have been portrayed as the core of a geographically, socially, and economically isolated underclass. The idea of an **underclass** refers to a class of individuals who experience a form of poverty from which it is very difficult to escape because of their isolation from mainstream values and the formal labor market. Isolated from the formal labor force and from the social values and behavioral patterns of the rest of society, the underclass has been seen as being subject to an increase in social disorganization and deviant behavior. In cities in the United States, localized inner-city poverty is now characterized by senseless and unprovoked violence; premeditated and predatory violence; domestic violence; the organized violence of street gangs; and epidemic levels of AIDS and other communicable diseases—all closely associated with drug taking and drug dealing.

One consequence of extreme poverty is *homelessness*. Chronic, long-term homelessness means not having customary and regular access to a conventional dwelling. This includes people who have to sleep in shelters, flophouse cubicles, emergency dormitories and missions, as well as those sleeping in doorways, bus stations, cars, tents, temporary shacks, cardboard boxes, park benches, and steam grates.

The number of homeless persons in the world's more affluent cities rose sharply in the mid-1970s. This was mainly a consequence of the increased

Redlining: the practice whereby lending institutions delimit "bad risk" neighborhoods on a city map and then use the map as the basis for determining loans.

Underclass: subset of the poor isolated from mainstream values and the formal labor market.

poverty and the economic and social dislocation caused by economic restructuring and the transition to a globalized, postindustrial economy. It was intensified by the fiscal squeeze confronting central cities and by the trend for governments to cut back on welfare programs of all kinds. It was also intensified by the adoption of revolving-door policies of mental health hospitals, which released large numbers of patients who had formerly been institutionalized. The homeless are now very visible throughout the major cities of the developed world (Figure 11.29): around the heating vents of shopping malls in Canadian cities, in the subterranean world of the Paris Metro, under Budapest's bridges and in Warsaw's parks, in cardboard boxes close by glass office towers in downtown Tokyo, and near abandoned factories in Sydney and Melbourne.

Estimates of the number of homeless people in U. S. cities in the mid-1990s vary a good deal. The estimate of the National Coalition for the Homeless is between 1.5 and 3 million. The U.S. Department of Housing and Urban Development estimates the number to be more like 600,000 at any one time, with about 7 million people experiencing homelessness for a time during any 5-year period. Germany, with over a million, has the most homeless in Europe, followed by Britain's 700,000 and France's 600,000. In the former Soviet Union, the "shock therapy" of free markets has brought homelessness to cities that had not experienced it for decades. In Moscow alone, almost 60,000 people became homeless between 1991 and 1995.

What makes the homelessness of the 1990s such a striking problem is not just the scale of the problem but also the nature of it. Whereas homelessness had previously involved white, adult males, relatively few of whom actually had to sleep outdoors, the new homeless are of all races and include significant numbers of women, children, and the elderly. In Canada, for example, one-quarter of the country's 250,000 homeless are children. In Europe, estimates suggest that up to 70 percent of the homeless are under 20 years old, and official statistics show that over 40 percent of the people receiving services for the homeless are women.

Figure 11.29 The new homeless In the 1980s and 1990s, changing conditions in U.S. cities created a huge increase in the number of homeless persons. The new homeless include significant numbers of women, children, and young adults, groups that had not previously been vulnerable to homelessness.

The new homeless are also much less likely to find shelter indoors, so that the homeless have become a very visible feature of the public spaces of many cities.

Problems of Unintended Metropolises

The problems of the dependent cities of the periphery stem from the way in which their demographic growth has outstripped their economic growth. High rates of long-term unemployment and underemployment mean that a significant proportion of the population of peripheral cities is forced to seek survival through the informal sector. The low and unreliable wages of informal-sector jobs lead directly to a second problem: chronic poverty and slum housing. On top of all this, the combination of rapid population growth and economic underdevelopment gives rise to transport and infrastructure problems, and to problems of environmental degradation. Together, these problems are so severe that they pose almost impossible tasks for metropolitan governance and management.

Unemployment and Underemployment   Urban unemployment rates in underdeveloped countries tend to be significantly higher than rural unemployment rates. This is a result of cities' inability to absorb the rapid population influx from the countryside. Faced with poverty in overpopulated rural areas, many people regard moving to a city much like a lottery: You buy a ticket (in other words, go to the city) in the hope of hitting the jackpot (in other words, landing a good job). As with all lotteries, most people lose. The greatest urban problem, however, is underemployment, which reflects the low productivity of the formal economic sector in underdeveloped countries. **Underemployment** occurs when people work less than full time even though they would prefer to work more hours. Underemployment is difficult to measure with any degree of accuracy, but estimates commonly range at levels between 30 and 50 percent of the employed work force in peripheral cities.

It is because of the extent of problems of unemployment and underemployment that peripheral cities have developed their characteristic informal sector of employment. People who cannot find regularly paid work must resort to various ways of gleaning a living. Some of these ways are imaginative, some desperate, some pathetic. Examples include street vending, shoe-shining, craft work, street-corner repairs, and scavenging on garbage dumps (Figure 11.30). The informal sector consists of a broad range of activities that represent an important coping mechanism. For too many, however, coping means resorting to begging, crime, or prostitution. Occupations such as selling souvenirs, driving pedicabs, making home-brewed beer, writing letters for others, and dressmaking may seem very marginal from the point of view of the global economy, but more than a billion people around the world must feed, clothe, and house themselves entirely from such occupations (Figure 11.30). In many peripheral cities, more than half of the population subsists in this way. Across Africa, the International Labor Office estimates, informal-sector employment is growing 10 times faster than formal-sector employment.

In most peripheral countries, the informal labor force includes children. In environments of extreme poverty, every family member must contribute something, and so children are expected to do their share. Industries in the formal sector often take advantage of this situation. Many firms farm out their production under subcontracting schemes that are based not in factories but in home settings that use child workers. In these settings, labor standards are nearly impossible to enforce. In the Philippines, batches of rural children are ferried by syndicates to work in garment-manufacturing sweatshops in urban areas.

Despite this side of the picture, a few positive aspects also exist to the informal sector. Pedicabs, for example, provide an affordable, nonpolluting means of transportation in crowded metropolitan settings. Garbage picking, while it may

Underemployment: when people work less than full time even though they would prefer to work more hours.

Street vending, Taegu, Korea

Street market, Surabaja, Indonesia

Garbage scavenging, Manila, Phillipines

Figure 11.30 Informal economic activities In cities where jobs are scarce, people have to cope through the informal sector of the economy, which includes a broad variety of activities, including agriculture (backyard hens, for example), manufacturing (craft work), and retailing (street vending).

Cotton spinning, Udaipur, India

seem desperate and degrading in Western eyes, provides an important means of recycling paper, steel, glass, and plastic products. One study of Mexico City estimated that as much as 25 percent of the municipal waste ends up being recycled by the 10,000 or so scavengers who work over the city's official dump sites. This positive contribution to the economy, though, scarcely balances the lives of poverty and degradation experienced by the scavengers.

Urban geographers also recognize that the informal sector represents an important resource to the formal sector of peripheral economies. The informal sector provides a vast range of cheap goods and services that reduce the cost of living for employees in the formal sector, thus enabling employers to keep wages low. Although this network does not contribute to urban economic growth or help alleviate poverty, it does keep companies competitive within the context of the global economic system. For export-oriented companies, in particular, the informal sector provides a considerable indirect subsidy to production. We must recognize, too, that this subsidy is often passed on to consumers in the core regions in the form of lower prices for goods and consumer products made in the periphery.

Consider, for example, the paper industry in Cali, Colombia. This industry is dominated by one company, Cartón de Colombia, which was established in 1944 with North American capital and subsequently acquired by the Mobil Oil Company. Most of the company's lower-quality paper products are made from recycled waste paper, and 60 percent of this waste paper is gathered by local garbage pickers. There are between 1,200 and 1,500 of these garbage pickers in Cali. Some work the city's municipal waste dump, some work the alleys and yards of shopping and industrial areas, and some work the routes of municipal garbage trucks, intercepting trash cans before the truck arrives. They are part of Cali's informal economy, for they are not employed by Cartón de Colombia, nor do they have any sort of contract with the company or its representatives. They simply show up each day to sell their pickings. This way, the company can avoid paying both wages and benefits, while dictating the price it will pay for various grades of waste paper. The company can operate profitably while keeping the price of its products down—the arrangement is a microcosm of core–periphery relationships.

Slums of Hope, Slums of Despair Unemployment, underemployment, and poverty mean overcrowding. In situations where urban growth has swamped the available stock of cheap housing and outstripped the capacity of builders to create affordable new housing, the inevitable outcome is makeshift, shanty housing that offers, at best, precarious shelter. Such housing has to be constructed on the cheapest and least desirable sites. Often, this means building on bare rock, over ravines, on derelict land, on swamps, or on steep slopes. Nearly always, it means building without any basic infrastructure of streets or utilities. Sometimes, it means adapting to the most extreme ecological niches, as in Lima, where garbage pickers actually live on the waste dumps, or in Cairo, where for generations the poor have adapted catacombs and cemeteries into living spaces (see Figure 11.21, page 464). The United Nations estimated in 1996 that one billion people worldwide live in inadequate housing.

Faced with the growth of these slums, the first response of many governments has been to eradicate them. Encouraged by Western development economists and housing experts, many cities sought to stamp out unintended urbanization through large-scale eviction and clearance programs. In Caracas (Venezuela), Lagos (Nigeria), Bangkok (Thailand), Calcutta (India), Manila (the Philippines), and scores of other cities in the periphery, hundreds of thousands of shanty dwellers were ordered out at short notice and their homes bulldozed to make way for public works, land speculation, luxury housing, urban renewal, and, on occasion, to improve the appearance of cities for special visitors. Seoul, South Korea, has probably had the most forced evictions of any city in the world. Since 1966, millions of people have been forced out of accommodations that they owned or rented, as part of a sustained government clean-up campaign. Between 1983 and 1988, in the preparation for the 1988 Olympic Games, nearly 750,000 people lost their homes in a beautification program.

Yet Seoul, more than most other cities, could afford to build new low-income housing to replace the demolished neighborhoods. Most peripheral cities cannot do so, which means that displaced slum dwellers have no option but to create new squatter and shanty settlements elsewhere in the city. Most cities, in fact, cannot evict and demolish fast enough to keep pace with the growth of slums caused by in-migration. The futility of slum clearance has led to a widespread re-evaluation of the wisdom of such policies. The thinking now is that informal-sector housing should be seen as a rational response to poverty. Shanty and squatter neighborhoods can not only provide affordable shelter but also function as important reception areas for migrants to the city, with supportive communal organizations and informal employment opportunities that help them to adjust to city life. They can, in other words, be "slums of hope." City authorities, recognizing the positive functions of informal housing and self-help improvements, are now increasingly disposed to be tolerant and even helpful toward squatters, rather than sending in police and municipal workers with bulldozers.

In fact, many informal settlements are the product of careful planning. In parts of Latin America, for example, it is common for community activists to draw up plans for invading unused land, then quickly build shanty housing before landowners can react. Part of the activists' strategy is usually to organize a critical mass of people large enough to be able to negotiate with the authorities to resist eviction. It is also common for activists to plan their invasions for public holidays, so that the risk of early detection is minimized. As the risk of eviction diminishes over time, some of the residents of informal housing are able gradually to improve their dwellings through self-help (Figure 11.31).

Nevertheless, many shanty and squatter neighborhoods exist where self-help and community organization do not emerge. Instead, grim and desperately miserable conditions prevail. These are "slums of despair," where overcrowding, lack of adequate sanitation, and lack of maintenance lead to shockingly high levels of ill health and infant mortality, and where social pathologies are at their worst. Consider, for example, the squatter settlement of Chheetpur in the city of Allahabad, India. The settlement's site is subject to flooding in the rainy season and a lack of drainage means stagnant pools for much of the year. Two standpipes (outdoor taps) serve the entire population of 500, and there is no public provision for sanitation or the removal of household wastes. In this community, most people have food intakes of less than the recommended minimum of 1,500 calories a day; 90 percent of all infants and children under four have less than the minimum calories needed for a healthy diet. More than half of the children and almost half the adults have intestinal worm infections. Infant and child mortality is high—though nobody knows just how high—with malaria, tetanus, diarrhea, dysentery, and cholera as the principal causes of death among underfives.

Figure 11.31 Self-help as a solution to housing problems Self-help is often the only solution to housing problems, because wages are so low and so scarce that builders cannot construct even the most inexpensive new housing and make a profit, and because municipalities cannot afford to build sufficient quantities of subsidized housing. One of the most successful ways of encouraging self-help housing is for municipal authorities to create the preconditions by clearing sites, putting in the footings for small dwellings, and installing a basic framework of water and sewage utilities. This "sites and services" approach has become the mainstay of urban housing policies in many peripheral countries.

Self-help housing, Ndola, Zambia

Sites-and-services housing, Lusaka, Zambia

Transport and Infrastructure Problems Even though the governments of peripheral cities typically spend nearly all of their budgets on transport and infrastructure in their race to keep up with population growth, conditions are bad and getting rapidly worse. Peripheral cities have always been congested, but in recent years the modernizing influence of formal-sector activities has turned the congestion into near gridlock. In many of the world's peripheral and semiperipheral metropolises, sharp increases have occurred in the availability and use of automobiles. One of the most dramatic examples is Taipei, Taiwan, where the number of automobiles increased from about 11,000 in 1960 to over 1 million in 1990. Not only are there more people and more traffic, but the changing spatial organization of peripheral cities has increased the need for transportation. Traditional patterns of land use have been superseded by the agglomerating tendencies inherent to modern industry and the segregating tendencies inherent to modernizing societies. The greatest single change has been the separation of home from work, however, which has meant a significant increase in commuting.

In spite of the many innovative responses to urban transportation needs (Figure 11.32), road transportation in many cities is breaking down, with poorly maintained roads, traffic jams, long delays at intersections, and frequent accidents. Many governments have invested in expensive new freeways and street-widening schemes, but because they tend to focus on city centers (which are still the settings for most jobs and most services and amenities), they ultimately fail, emptying vehicles into a congested and chaotic mixture of motorized traffic, bicycles, animal-drawn vehicles, and hand-drawn carts. Some of the worst traffic tales come from Mexico City—where traffic backups total more than 90 kilometers (60 miles) each day, on average—and Bangkok—where the 15-mile trip into

Figure 11.32 Urban transportation Creative responses to the problem of transportation come in many forms, but their success is limited by the sheer congestion of overurbanization.

town from Don Muang Airport can take three hours. The costs of these traffic backups are enormous. The annual costs of traffic delays in Jakarta, Indonesia, have been estimated at $68 million; in Bangkok, Thailand, they have been estimated at $272 million—the equivalent of more than 2 percent of Thailand's gross national product.

Water supplies and sewerage also present acute problems for many cities (Figure 11.33). Definitions of what constitutes an adequate amount of safe drinking water and sanitation vary from country to country. Although many governments classify the existence of a water tap within 100 meters (328 feet) of a house as "adequate," such a tap does not guarantee that the individual household will be able to secure enough water for good health. Communal taps often function only a few hours each day, so residents must wait in long lines to fill even one bucket. In Rajkot, India, a city of 600,000 people, piped water routinely runs for only 20 minutes each day.

The World Bank estimates that, worldwide, only about 70 percent of urban residents in less-developed countries have access to a satisfactory water source, and only about 40 percent are connected to sewers (90 percent of which discharge their waste untreated into a river, a lake, or the sea). Hundreds of millions of urban dwellers have no alternative but to use contaminated water—or at least water whose quality is not guaranteed. A small minority, usually the residents of the most affluent neighborhoods, have water piped into their homes while the majority have piped water nearby, which has to be collected. Those not served are obliged to carry water in small quantities over long distances, or to use water

Figure 11.33 Infrastructure problems   Low incomes mean that city governments are unable to raise sufficient revenues through local taxes, which in turn means that infrastructure is neglected. Putting water and sewage lines into neighborhoods that were built without these basic utilities is arduous and expensive; yet without basic utilities, public health is seriously threatened.

Installing water lines in a low-income neighborhood of Cartagena, Colombia

A shanty neighborhood in Kumasi, Ghana, where a drainage ditch and open sewer dug by residents has begun to erode, threatening the whole street

Filling containers with water from a well, Asmera, Ethiopia

Street water pump, Raipur, India

Communal water supply, Kanpur, India

Figure 11.34 Water supply problems Many peripheral cities have grown so quickly, and under such difficult conditions, that large sections of the population do not have access to supplies of clean water. Where there is a public supply—a well or an outdoor standpipe—water consumption is limited by the time and energy required to collect water and carry it home. It is not uncommon for 500 or more people to have to share a single tap. Because low-income people work extremely long hours, the time spent queuing for water and transporting buckets takes up time that could be used in earning an income. Limited quantities of water mean inadequate supplies for washing and personal hygiene, and for washing food, cooking utensils, and clothes. Where public agencies do not provide any water supply—as is common in squatter settlements—the poor often obtain water from private vendors and can pay 20 to 30 times the cost per liter paid by households with piped supplies. Water vendors probably supply about one-fourth of the population of peripheral metropolises.

Outdoor standpipe, Ankara, Turkey

from streams or other surface sources (Figure 11.34). In Colombo, Sri Lanka, about one-third of all houses have indoor piped water, and another one-fourth have piped water outside. In Dar es Salaam (Tanzania), Kinshasa (Democratic Republic of Congo), and many other peripheral cities, almost half the population has no access to piped water, either indoors or outdoors.

In many cities, including Bangkok, Bogotá, Dar es Salaam, Jakarta, Karachi, and São Paulo, only one-fourth or one-third of all garbage and solid waste is collected and removed—the rest is partially recycled informally, tipped into gullies, canals, or rivers, or simply left to rot. Sewage services are just as bad. In Bangkok, less than 5 percent of the population is connected to a sewer system;

human wastes are generally disposed of through septic tanks and cesspools with their effluents, as well as waste water from sinks, laundries, baths, and kitchens discharged into stormwater drains or canals. Jakarta has no waterborne sewage system at all. Septic tanks serve about one-quarter of the city's population; others must use pit latrines, cesspools, and ditches along the roadside. A survey of over 3,000 towns and cities in India found that only eight had full sewage treatment facilities, and another 209 had partial treatment facilities. Along one river, the Ganga, there are 114 towns and cities that dump untreated sewage into the river every day, along with waste from DDT factories, tanneries, paper and pulp mills, petrochemical and fertilizer complexes, and other industrial pollutants. Each day the Yamuna River picks up 200 million liters of untreated sewage and 20 million liters of industrial effluents as it passes through Delhi.

These problems provide opportunities for the informal sector, however. Street vendors, who get their water from private tanker and borehole operators, sell small quantities of water from 2- or 4-gallon cans. They typically charge between 5 and 10 times the local rate set by public water utilities; in some cities they charge 60 to 100 times as much. Similarly, many cities have evolved informal-sector mechanisms for sewage disposal. In many Asian cities, for example, human waste is removed overnight by handcart operators. Unfortunately, the waste is rarely disposed of properly, and often ends up polluting the rivers or lakes from which the urban poor draw their water.

Environmental Degradation With pressing problems of poverty, slum housing, and inadequate infrastructure, it is not surprising that peripheral cities are unable to devote many resources to environmental problems. Because of the speed of population growth, these problems are escalating rapidly. Industrial and human wastes pile up in lakes and lagoons, and pollute long stretches of rivers, estuaries, and coastal zones. Groundwater is polluted through the leaching of chemicals from uncontrolled dump sites; and the forests around many cities are being denuded by the demand of cities for timber and domestic fuels. This environmental degradation is, of course, directly linked to human health. People living in such environments have much higher rates of respiratory infections, tuberculosis, diarrhea, and much shorter life expectancies than people living in surrounding rural communities. Children in squatter settlements may be 50 times as likely to die before the age of five than those born in affluent core countries.

In addition, air pollution has escalated to very harmful levels in many cities (Figure 11.35). With the development of a modern industrial sector and the growth of automobile ownership, but without enforceable regulations on pollution and vehicle emissions, tons of lead, sulfur oxides, fluorides, carbon monoxide, nitrogen oxides, petrochemical oxidants, and other toxic chemicals are pumped into the atmosphere every day in large cities. The burning of charcoal, wood, and kerosene for fuel and cooking in low-income neighborhoods also contributes significantly to dirty air. In cities where sewerage systems are deficient, the problem is compounded by the presence of airborne dried fecal matter. Worldwide, according to United Nations data, more than 1.1 billion people live in urban areas where air pollution exceeds healthful levels.

A United Nations study of 20 megacities found that every one of them had at least one major pollutant at levels exceeding World Health Organization (WHO) guidelines. Fourteen of the 20 had *two* major pollutants exceeding WHO guidelines, and 7 had *three*. Such pollution is not only unpleasant but also dangerous. In Mexico City, where sulfur dioxide and lead concentrations are between 2 to 4 times higher than the WHO guidelines, and where national ozone levels are exceeded on more than half of the days throughout the year, 7 out of 10 newborns have dangerously high levels of lead in their bloodstream. In Bangkok, Thailand, where air pollution is almost as severe as in Mexico City, research has

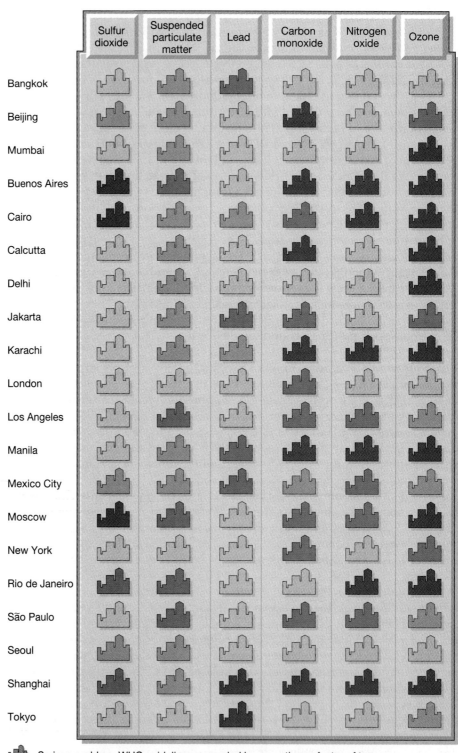

	Sulfur dioxide	Suspended particulate matter	Lead	Carbon monoxide	Nitrogen oxide	Ozone
Bangkok						
Beijing						
Mumbai						
Buenos Aires						
Cairo						
Calcutta						
Delhi						
Jakarta						
Karachi						
London						
Los Angeles						
Manila						
Mexico City						
Moscow						
New York						
Rio de Janeiro						
São Paulo						
Seoul						
Shanghai						
Tokyo						

Serious problem: WHO guidelines exceeded by more than a factor of two

Moderate to heavy pollution: WHO guidelines exceeded by up to a factor of two (short-term guidelines exceeded on a regular basis at certain locations)

Low pollution: WHO guidelines are normally met (short-term guidelines may be exceeded occasionally)

No data available or insufficient data for assessment

Figure 11.35 Air quality in megacities As this graphic shows, the inhabitants of most large cities experience a serious problem from one or more air pollutants. (*Source:* United Nations Environment Programme, *Urban Air Pollution in Megacities of the World,* United Nations Environmental Programme, 1992, Fig. 4.2.)

shown that lead-bearing air pollutants have the effect of reducing children's IQ by an average of 3.5 points per year until they are seven years old. It has also been estimated that Bangkok's pall of dust and smoke causes more than 1,400 deaths each year and $3.1 billion each year in lost productivity resulting from traffic and pollution-linked illnesses.

Amid these generally dangerous and unpleasant environments, some places stand out as particularly awful. The city of Cubatao, Brazil, for example, has acquired a reputation in the neighboring cities of São Paulo and Santos as the "Valley of Death." Cubatao contains a high concentration of heavy industry (including a Union Carbide fertilizer factory and Brazilian oil, chemical, and steel factories) that was developed in the 1960s under a military government with little or no attempt to control pollution. Most of the city's housing is of extremely poor quality, with many industrial workers living in shantytowns built on stilts above swamps. Toxic industrial wastes have been dumped in the surrounding forests and are contaminating the city's water supplies. Vegetation in and around the city has suffered substantially from air pollution; and as dying vegetation can no longer retain the soil on steeper slopes, landslides have become a danger to shanty dwellers. In 1984, hundreds of inhabitants were killed after a gasoline pipeline leaked into a swamp under one shanty and caught fire. The Cubatao River, once an important source of fish, now only supports a few eels and some hardy crabs that are too toxic for humans to eat. Local hospitals routinely accept young children to breathe medicated air, but infant mortality rates, along with the incidence of stillborn and deformed babies, and deaths from tuberculosis, pneumonia, bronchitis, emphysema, and asthma, remain high.

Proximity to industrial facilities, often the result of the need and desire of the poor to live near places of employment, poses another set of risks. In 1984, a major accident at the Union Carbide factory in Bhopal, India, caused 2,988 deaths and more than 100,000 injuries, affecting mostly residents of the shantytowns near the chemical factory.

Governance and Management The governments of towns and cities in the world's periphery are faced with tremendous problems. Just keeping up with the rate of physical and demographic growth presents an enormous challenge. Typical growth rates mean that cities' physical infrastructure of roads, bridges, and utilities need to be tripled every 10 years. Meanwhile, most city governments find it nearly impossible to take care of the daily upkeep of their existing infrastructure because of the wear and tear that is caused by overurbanization. Somehow, basic services have to be provided to populations that cannot afford to pay more than a fraction of their costs. Finally, all this has to be accomplished amid bureaucratic organizations that have little experience or expertise in the kind of capital budgeting, project programming, economic development, and spatial planning that is essential for successful metropolitan management.

The governance of most peripheral countries tends to be highly centralized, with relatively little political power allocated to city or metropolitan governments. In addition, city and metropolitan governments are typically fragmented—both geographically and functionally—as well as being understaffed and underfinanced. To make things even worse, local politics and civic administration in many countries are routinely corrupt. This means that metropolitan governance and management tends to be part of the problem rather than part of the solution.

The fundamental responsibilities of urban governance and management in the unintended metropolises of the periphery include:

- Environmental protection and public health
- The provision of a basic legislative and regulatory framework that protects citizens from exploitation by landlords and employers

- The provision of adequate access to basic services and utilities
- The efficient management of traffic systems and public transportation
- The establishment of an equitable system for financing municipal services

More often than not, municipal governments are failing badly in all these responsibilities. Worse, the actions (or inactions) and policies of many governments actually inhibit and repress the efforts of citizens to meet their basic needs through grassroots organizations and resources. Most of the shelter that people provide for themselves is deemed illegal by city governments, for example. This creates uncertainty, instability, and a disincentive for people to invest their meager resources in long-term improvements.

Although many of the individuals involved do the best they can, metropolitan governance and management seem doomed to be ineffective and inefficient until some way can be devised to improve the institutional framework (in both geographic and democratic terms) and reduce the financial constraints faced by municipal governments. This last point, regarding financial constraints, brings us back to local–global interdependencies once again, for the financial predicament of peripheral cities is ultimately tied to their dependent role in the global economic system.

Conclusion

Patterns of land use and the functional organization of economic and social subareas in cities are partly a product of the economic, political, and technological conditions at the time of the city's growth, and partly a product of regional cultural values. Geographers can draw on four particularly useful perspectives in looking at patterns of land use within North American cities: an economic perspective that emphasizes competition for space; an economic perspective that emphasizes functional linkages between land uses; a sociocultural perspective that emphasizes ethnic congregation and segregation; and an historical perspective that emphasizes the influence of transportation technology and infrastructure investment.

In many ways, the most striking contrasts in urban structure are to be found between the cities of the core regions and those of the periphery. The evolution of the unintended metropolis of the periphery has been very different from the evolution of metropolitan areas in the world's core regions. Similarly, the problems they face are very different. In the core regions, urban change is dominated by the consequences of an economic transformation to a postindustrial economy. Traditional manufacturing and related activities have been moved out of central cities, leaving decaying neighborhoods and a residual population of elderly and marginalized people. New, postindustrial activities have begun to cluster in redeveloped CBDs and in edge cities around metropolitan fringes. In a few cases, metropolitan growth has become so complex and extensive that 100-mile cities have begun to emerge, with half a dozen or more major commercial and industrial centers forming the nuclei of a series of interdependent urban realms.

In other parts of the world, patterns of land use and the functional organization of economic and social subareas are quite different, reflecting different historical legacies, and different environmental and cultural influences. The basic trend affecting the cities of the world's periphery is demographic—the phenomenal rates of natural increase and in-migration that have given rise to overurbanization. The example of Lagos provides some sobering insights into the human consequences of overurbanization. An ever-growing informal sector of the economy, in which people seek economic survival, is reflected in extensive areas of shanty housing. High rates of unemployment, underemployment, and poverty generate acute social problems which overwhelm city governments that are under-staffed and under-funded. If present trends continue, such problems are likely to characterize increasing numbers of the world's largest settlements. In the next chapter, we consider this question as part of a broader discussion of future geographies.

Main Points

- The typical U.S. city is structured around the CBD; the zone in transition; the suburbs; secondary business districts and commercial strips; industrial districts, and, in larger metropolitan areas, edge cities.

- Poorer households, unable to afford the recurrent costs of transportation, must trade off living space for accessibility to jobs, so that they end up in high densities, at expensive locations near their low-wage jobs.

- Cities experiencing high rates of in-migration tend to become structured into a series of concentric zones of neighborhoods of different ethnicity, demographic composition, and social status through processes of invasion and succession.

- In cities where growth has been less dominated by successive waves of immigrant ethnic groups, neighborhood patterns tend to be structured around the development of industrial corridors and high-class residential corridors.

- Distinctive neighborhoods emerge as similar kinds of households go through similar search patterns and make similar decisions about where to live.

- Urban structure varies a good deal from one region of the world to another because of the influence of history, culture, and the different roles that cities have played within the world-system.

- Geographers are interested in the distinctive physical features of urban landscapes because they can be read as multilayered texts that show how cities have developed, how they are changing, and how people's values and intentions take expression in urban form.

- The most acute problems of the postindustrial cities of the world's core regions are localized in the central city areas that have borne the brunt of restructuring from an industrial to a postindustrial economy. In these areas there are several interrelated problems: fiscal problems, infrastructure problems, and localized spirals of neighborhood decay and cycles of poverty.

- The problems of the cities of the periphery stem from the way in which their demographic growth has outstripped their economic growth: high rates of long-term unemployment and underemployment; low and unreliable wages of informal-sector jobs; chronic poverty; and slum housing. Their low rates of economic growth are reflective of their dependent position in the global economy.

Key Terms

Beaux Arts (p. 466)
central business district (CBD) (p. 442)
central cities (p. 472)
City Beautiful movement (p. 467)
congregation (p. 444)
cycle of poverty (p. 477)

dualism (p. 472)
edge cities (p. 443)
fiscal squeeze (p. 472)
gentrification (p. 444)
invasion and succession (p. 448)
isotropic surface (p. 446)

minority groups (p. 444)
Modern movement (p. 467)
postmodern urban design (p. 468)
redlining (p. 478)
segregation (p. 446)

underclass (p. 478)
underemployment (p. 480)
zone in transition (p. 442)

Exercises

 ### On the Internet

The Internet exercises for this chapter explore the nature of urban spaces in different parts of the world. Drawing on the many resources that the Internet offers for studying cities, these exercises will provide you with an understanding of how distinctive patterns of land use and of urban form emerge in different regions, and how the nature of urban problems varies from region to region.

Unplugged

1. On a tracing-paper overlay of a street map of your town or city, plot the distribution of churches of different denominations (you can obtain fairly accurate information on this from the yellow pages of your city's telephone book). What can you say about the spatial distribution that is revealed? How does it relate to what you know about the city's social geography?

2. Collect a week's worth of local newspapers, and review the coverage of urban problems. What kinds of problems are covered, and for what kinds of cities? Compile a list of such categories, and then carefully analyze the content of the week's coverage, calculating the number of column-inches devoted to each category of problems.

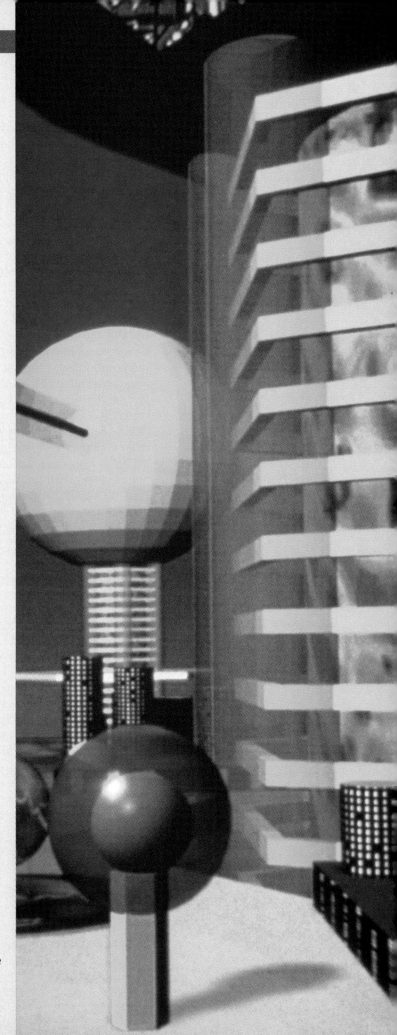

Chapter 12

Future Geographies

City of the future

In August 1996, Heather Waller, 25, of Buffalo, New York, married John Herbert, 23, of Watford, in England. Theirs was the first known marriage between people who had first met, and then fallen in love, in cyberspace. Earlier that year a New Jersey man, John Goydan, sued his wife Diane for divorce. He caught her in the act of a "virtual" affair with a cybersex partner, a man from North Carolina with the E-mail i.d. of "weasel." In Istanbul, Turkey, a few weeks later, when Sam Fahmy and Handan Saygin were married, the ceremony was carried live on the Internet; the wedding party received congratulatory E-mail instead of telegrams; and guests who could not make it to the wedding in Istanbul were able to log on to a chat area to gossip about the newlyweds or bemoan their own lack of partners. Meanwhile, Tanya Leshchenko, together with 50 other "marriage-minded" Russian women, had taken advantage of the Florida Fedorova Russian Introduction Agency's Web site, in search of a Western husband. These examples could reasonably lead us to question the future of fundamental aspects of human interaction—romance, sex, and marriage—that have until now been closely circumscribed by space and place.

Will the Internet bring about new patterns of human interaction? Will globalization bring an end to distinctive regional cultures? Will we be able to cope with the environmental stresses that increasing industrialization and rapid population growth will bring to many parts of the world? Will the United States retain its position as the world's most powerful and influential nation? Will more countries move up from peripheral status to join the semiperiphery and core of the future world-system? What kind of problems will the future bring for local, regional, and international development? What new technologies are likely to have the most impact in reshaping human geographies? These are just a few of the many questions that spring from the key themes in human geography that we have examined in Chapters 4 through 11. This chapter examines some scenarios for future geographies, drawing on the principles and concepts established in Chapters 1 and 2.

Mapping Our Futures

It is important to be able to envisage the future. We have to live in the world for a while, and we naturally want to leave it in good shape for future generations. We need, therefore, to be able to identify the changes that the future might bring, so that we can begin to work toward the most desirable outcomes. The problem is that predicting the future can be a risky business. The uncertainties of geopolitical change, the unexpectedness of technological breakthroughs, and the complexity of environmental change all compound one another to make our future seem, at first glance, highly unpredictable.

Nevertheless, there is no shortage of visionary projections (Figure 12.1, page 497). Broadly speaking, these can be divided into two sorts of scenarios: optimistic and pessimistic. Optimistic futurists stress the potential for technological innovations to discover and harness new resources; to provide faster and more effective means of transportation and communication; and to enable new ways of living. This sort of futurism is characterized by science fiction cities of mile-high skyscrapers and spaceship-style living pods; by bioecological harmony; and by unprecedented social and cultural progress through the information highways of cyberspace. It projects a world that will be stabilized and homogenized by supranational or even "world" governments. The sort of geography implied by such scenarios is rarely spelled out. Space and place, we are led to believe, will be transcended by technological fixes.

To pessimistic futurists, however, this is just "globaloney." They stress the finite limitations of Earth's resources; the fragility of its environment; and population growth rates that exceed the capacity of peripheral regions to sustain them. This sort of doomsday forecasting is characterized by scenarios that include irretrievable environmental degradation; increasing social and economic polarization; and the breakdown of law and order. The sort of geography associated with these scenarios is rarely explicit, but it usually involves the probability of a sharp polarization between the haves and have-nots at every geographical scale.

Fortunately, we don't have to choose between these two extreme scenarios. Using what we have learned from the study of human geography, we can suggest a more grounded outline of future geographies. To do so, we must first glance back at the past. Then, looking at present trends and using what we know about processes of geographic change and principles of spatial organization, we can begin to map out the kinds of geographies that the future most probably holds.

Looking back at the way that the geography of the world-system has unfolded, we can see now that a fairly coherent period of economic and geopolitical development occurred between the outbreak of World War I (in 1914) and the collapse of the Soviet Union (in 1989). Some historians refer to this period as the "short twentieth century." It was a period when the modern world-system developed its triadic core, when geopolitics was based on an East–West divide, and when geoeconomics was based on a North–South divide. This was a time when the geographies of specific places and regions within these larger frameworks were shaped by the needs and opportunities of technology systems that were based on the internal combustion engine, oil and plastics, electrical engineering, aerospace industries, and electronics. In this short century, the modern world was established, along with its now-familiar landscapes and spatial structures: from the industrial landscapes of the core to the unintended metropolises of the periphery; from the voting blocs of the West to the newly independent nation-states of the South.

Looking around now, much of the established familiarity of the modern world and its geographies seems to be disappearing, about to be overwhelmed by a series of unexpected developments—or obscured by a sequence of unsettling juxtapositions. The United States is giving economic aid to Russia; Eastern European countries want to join NATO and the European Union; Germany has unified, but Czechoslovakia and Yugoslavia have disintegrated; Israel has established a fragile peace with Egypt and Jordan; South Africa has been transformed, through an unexpectedly peaceful revolution, to black majority rule. Meanwhile, Islamic terrorists shoot up tourist buses, bomb office buildings, and sabotage aircraft; former communist Russian ultranationalists have become comradely with German neo-Nazis; Hindus, Sikhs, and Muslims are in open warfare in South Asia; and Sudanese military factions steal food from aid organizations in order to sell it to the refugees for whom it was originally intended.

The economic and cultural flux of the world provides some even more colorful examples:

> McDonald's, Pizza Hut, and American dollars are everywhere.
> Overnight jet flights and international direct dialing to North America
> afford the basic infrastructure for South American narcocapitalism. . . .
> Parts of Africa are returning to a hunting and gathering economy.
> Russia's markets are often empty and its factories are idle, but billions
> in oil, metals, lumber, and weapons are smuggled, like dope from
> Bolivia or Burma, to foreign markets through Kaliningrad. Moscow's
> GUM department store has a Benetton, while the city's nouveau riche
> mafia "entrepreneurs" ostentatiously zoom around in German BMWs
> and Mercedes Benzes, courtesy, in many cases, of lucrative car theft
> rings operating in Western Europe. Bloomingdale's sells Red Army
> watches at the costume jewelry counter. . . . Communist China's

"military" industries are making millions selling knock-off running shoes to Singapore traders and reverse-engineered strategic rockets to Saudi princes.

. . . Karaoke machines offer ancient Motown hits to American corporate managers, Hong Kong entrepreneurs, and German sex tourists in Thailand. Croatian-Canadian teenagers in Ontario hold car washes to buy weapons in South Africa for Zagreb's war efforts. . . . Global superband U2 entertains stadium audiences in Europe with channel surfing spectacles and live phone calls to the White House and trapped victims in Sarajevo. Meanwhile, Croatian, Moslem, and Serbian snipers listen to heavy metal on Sony Walkmans as they shoot up each others' families. Some of Iran's, Pakistan's, Libya's, and Mexico's largest urban populations are located in Paris, London, Milan, and Los Angeles. Disneyland now claims territory in Europe, Asia, and North America.[1]

In short, we have entered a period of transition since 1989. This is very important, because it means that we cannot simply project our future geographies from the landscapes and spatial structures of the past. Rather, we must map them out from a combination of existing structures and budding trends. We have to anticipate, in other words, how the shreds of tradition and the strands of contemporary change will be rewoven into new landscapes and new spatial structures.

While this is certainly a speculative and tricky undertaking, we can draw with a good deal of confidence on what we know about processes of geographic change and principles of spatial organization. The study of human geography has taught us to understand spatial change as a composite of local place-making processes (see Chapter 6) that are subject to certain principles of spatial organization and that operate within the dynamic framework of the world-system (Chapter 2). It has also taught us that many important dimensions exist to spatial organization and spatial change; from the demographic dimension (Chapter 3) through to the urban (Chapter 11).

As we look ahead to the future, we can appreciate that some dimensions of human geography are more certain than others. In some ways, the future is already here, embedded in the world's institutional structures and in the dynamics of its populations. We know, for example, a good deal about the demographic trends of the next quarter-century, given present populations, birth and death rates, and so on (see Figure 12.2, page 498). We also know a good deal about the distribution of environmental resources and constraints; about the characteristics of local and regional economies; and the legal and political frameworks within which geographic change will probably take place.

On the other hand, we can only guess at some aspects of the future. Two of the most speculative realms are those of politics and technology. While we can foresee some of the possibilities (maybe a spread and intensification of ethno-nationalism; perhaps a new railway era based on high-speed trains), politics and technology are both likely to spring surprises at any time. Such surprises (a political revolution in China? a breakthrough in superconductivity?) can cause geographies to be rewritten suddenly and dramatically. As we review the prospects for geographic change, therefore, we must always be mindful that our prognoses are all open to the unexpected. As we shall see, this is perhaps our biggest cause for optimism.

---

[1]G. Ó'Tuathail and T. Luke, "Present at the (Dis)integration: Deterritorialization and Reterritorialization in the New Wor(l)d Order," *Annals, Association of American Geographers,* 84, 1994, pp. 381–382.

Figure 12.1 Future Scenarios In some ways, the future is already here, embedded in the world's institutional structures and in the dynamics of its populations. On the other hand, we can only guess at some aspects of the future. Given the impossibility of knowing precisely how the future will unfold, business strategists attempt to make decisions that play out well across several possible futures. To find the most robust strategy, several different scenarios are created, each representing a plausible outcome. (*Source:* L. Beach, "How to Build Scenarios," *Wired Scenarios* 1.01, October 1995, pp. 74–81; P. Schwartz, "The New World Disorder," *Wired Scenarios* 1.01, October 1995, pp. 104–106.)

Scenario 1—The world becomes populated by consumers rather than citizens. Technology breeds unlimited, customized choices. Computers do increasing amounts of white-collar work. Real leisure increases. Governments become virtual corporations and come to rely on electronic voting. Southeast Asia and the coast of China manufacture most of the world's goods, and consume almost half themselves. Latin America is their branch office. Japan gets richer and unhappier. Russia exports trouble in the form of neoreligious cultists and mafioso. The United States and Europe become large theme parks.

Scenario 2—The world becomes dominated by a new international division of labor, based on an intensive use of networked telecommunications. Technology dominates global culture, which turns inward toward personal spaces. Old public spaces crumble, and ethnic subcultures give way to a patchwork of unbridled individual variety. Europe is wracked by civil strife as its collectively oriented civilization unravels. Russia rebounds, while Japan lags. China and the developing countries become huge flea markets where anything goes.

Scenario 3—Economic development is slowed in reaction to earlier decades of high crime and chaos. Europe experiences a second renaissance, becoming a moral beacon. Communitarian values become stronger, and governments undertake large-scale public works directed at environmental improvement. Dirty technologies are tightly regulated, and this increases the income gap between core regions and peripheral regions. Asia and Latin America become refuges for the young and restless of the core regions who find environmentalism and communitarianism too dogmatic; they settle in "free economic zones" where their education and their energy help to stimulate economic growth.

Scenario 4—The world is divided into three rigid and distinct trading blocs, but political boundaries are more fragmented than ever. The European Union, including most of Eastern Europe and Russia, has a common currency and tight border controls. The Asia-Pacific region evolves into a trading bloc in response to the European Union, but it is weakened by internal political and economic differences. Mexico collapses under civil war. Canada breaks up after Quebec's withdrawal. The third trading bloc is centered on the Indian Ocean, with India, South Africa, Saudi Arabia, and Iran as the key members. Throughout the world, political conflicts and weaknesses allow widespread terrorism, organized crime, and environmental degradation.

Scenario 5—The world settles into small, powerful city-states. Civic pride blossoms, and governments use advanced technologies to create public works of an unprecedented scale and scope. Rural areas of the world are second class but have widespread virtual hookups. Europe fractionalizes into over 50 countries; China, Russia, Brazil, and India devolve into black market ethnic states. Gangs and militia in peripheral countries and old inner-city areas transform into political law-and-order machines.

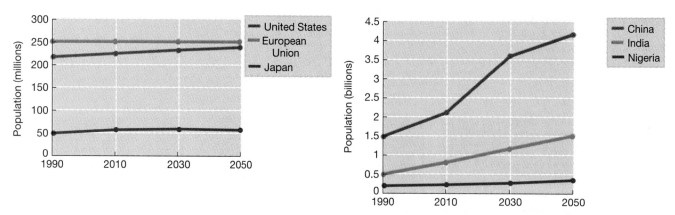

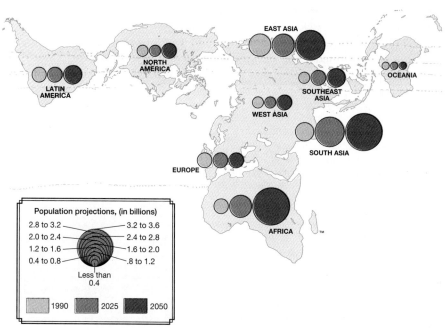

Figure 12.2 Population geography of the future Population projections show a very marked disparity between world regions, with core countries and core regions growing very little in comparison with the periphery. (*Source:* Data from United Nations Population Division, *Long-Range World Population Projections: Two Centuries of Population Growth, 1950–2150.* New York: United Nations, 1992. Map projection, Buckminster Fuller Institute and Dymaxion Map Design, Santa Barbara, CA. The word Dymaxion and the Fuller Projection Dymaxion™ Map design are trademarks of the Buckminster Fuller Institute, Santa Barbara, California, ©1938, 1967 & 1992. All rights reserved.)

Global Outlook, Local Prospects

For many years now, organizations such as the United Nations and the World Bank have prepared forecasts of the world economy. These forecasts are based on economic models that take data on macro-economic variables (for example, trends in countries' gross domestic product, their imports and exports, their economic structure, their investment and saving performance, and their demographic dynamism) and use known relationships between and among these variables to predict future outcomes. The problem is that economic projection is an inexact science. Economic models are not able to take into account the changes brought about by major technological innovations; significant geopolitical shifts; or the rather mysterious longer-term ups and downs that characterize the world economy. The present state of the world, for example, is a good deal less rosy than had been forecast in the early 1980s, just before the world economy slipped into a serious recession. Current forecasts of economic growth through 2010 are decidedly modest, with GDP per capita in core economies growing only 1.6 to 2.3 percent each year, while parts of the periphery may experience a decline (Figure 12.3).

Such forecasts, of course, are made only for the short term. In overall terms, the global economy is vastly richer, more productive, and more dynamic than it was just 15 or 20 years ago. Every prospect exists that, in the longer term, the world economy will continue to expand. Geographer Brian J. L. Berry has ana-lyzed the timing of past long-wave economic cycles and on that basis has calcu-lated that the next sustained boom might begin around 2010 (Figure 12.4, page 500). Should this in fact take place, suggests Berry, it will probably be based on the currently emerging infrastructure of information technologies, which will form the platform for a new technology system that will exploit further develop-ments in electronics (semiconductors, superconductors, artificial intelligence), bioengineering, and materials technology.

It is when we focus on the prospects of particular regions and places that things begin to look less encouraging. Future geographies seem likely to be struc-tured by an even greater gap between the haves and have-nots of the world. The Mexican Zapatist guerilla leader, Subcomandante Marcos, has described the opening up of the global economy as a death sentence for the poor. The gap between the world's core areas and the periphery has already begun to widen sig-nificantly. Even in the United States there is evidence of growing disparity. The Census Bureau, for example, reported in mid-1996 that the gap between the rich and poor in the United States was the widest it had been since World War II. The United Nations has calculated that the ratio of GDP per capita (measured at con-stant prices and exchange rates) between the developed and developing areas of

Figure 12.3 Forecasts of regional economic growth through 2010 This chart shows the actual growth of GDP per capita between 1970 and 1990, together with two World Bank estimates of growth from 1990 through 2010: one based on the continuation of past trends of slow growth and regional divergence, the other based on an assumption of the successful development of strong economic policies in all parts of the world, together with deeper international integration of trade. (*Source:* Data from "Workers in an Integrating World," *The World Bank, World Development Report 1995.* Washington, D.C., 1995.)

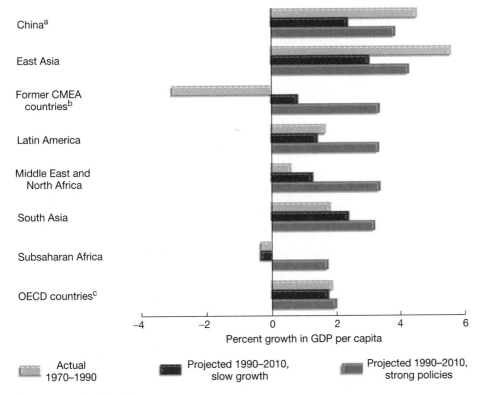

a China includes Hong Kong.
b Former CMEA countries (CMEA: Council for Mutual Economic Assistance. Member countries were Bulgaria, Cuba, Czechoslovakia, East Germany, Hungary, Mongolia, Poland, Romania, USSR, Vietnam, and Yugoslavia.)
c OECD data include Australia, Canada, European Union, Japan, New Zealand, and United States only.

Figure 12.4 The long-wave "clock" In the past, long-wave economic cycles have been fairly regular, with phases of depression occurring at 50- to 55-year intervals, interspersed with episodes of "stagflation"—a combination of economic stagnation and rapid price inflation. At that rate, the next economic boom should begin around 2010, after which we can expect a phase of inflationary speculation to lead to the next "stagflation" crisis, sometime in the 2030s. (*Source:* B. J. L. Berry, *Long-Wave Rhythms in Economic Development and Political Behavior.* Baltimore: The Johns Hopkins University Press, Fig. 70, p. 126. Copyright © 1991 The Johns Hopkins University Press.)

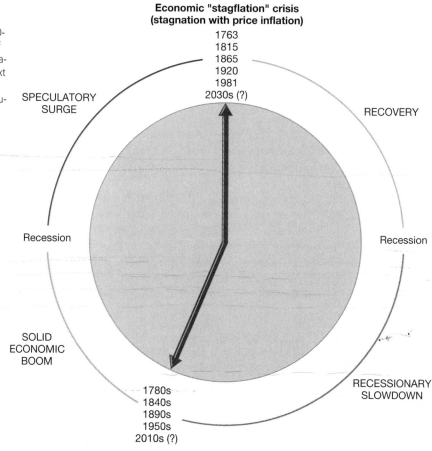

the world increased from 10:1 in 1970 to 12:1 in 1985, and predicts that by 2000 it will approach 13:1.

Little hope exists, however, that any future boom in the overall world economy will reverse this trend. In spite of the globalization of the world-system (and in many ways *because* of it), much of the world has been all but written off by the bankers and corporate executives of the core. In many peripheral countries, 20 percent or more of all export earnings are swallowed up by debt service—the annual interest on international debts (Figure 12.5). In 1995, for example, 20 peripheral countries, including Angola, Jamaica, Laos, Liberia, Mali, Nicaragua, Somalia, Syria, Tanzania, Vietnam, and Zambia, had total debts so large that they owed more than they produced. The Congo, with a debt of $5 billion, a population of only 3 million, and a total GNP of $2.3 billion, had the greatest debt burden per person of any country. Practically none of this money is ever likely to be repaid, but so long as the American, European, and Japanese banks continue to receive interest on their loans they will be satisfied. Indeed, core countries are doing extremely well from this aspect of international finance. In 1995, the world's core countries took in about $258 billion in debt servicing, while paying out a total of less than $80 billion in new loans. There is a danger to the future well-being of the better-off countries, however, in that debtor countries might act together in deliberately defaulting on their debts, which would cause a major disturbance to the global financial system. In fact, critics of globalization have charged that the mid-1990s "bailout" of the Mexican economy was really a rescue for Wall Street and the International Monetary Fund (IMF), both of which have enormous financial investments in Mexico.

Figure 12.5 The debt crisis By 1995, peripheral and semiperipheral countries owed more than $2 trillion to banks and governments in the core regions of the world-system. In some countries, the annual interest on international debts (their "debt service") accounts for more than 20 percent of the annual value of their exports of goods and services. Many countries got into debt trouble in the 1970s, when Western banks, faced with recession at home, offered low-interest loans to the governments of peripheral countries rather than being stuck with idle capital. When the world economy heated up again, interest rates rose, and many countries found themselves facing a debt crisis. The World Bank and the International Monetary Fund (IMF), working in tandem with western governments, worked to prevent a global financial crisis by organizing and guaranteeing programs that eased poor countries' debt burdens. Western banks were encouraged to swap debt for equity stakes in nationalized industries, while debtor governments were persuaded to impose austere economic policies. These policies have helped ease the debt crisis, but often at the expense of severe hardship for ordinary people. In dark humor, the IMF became known among radical development theorists as "imposing misery and famine." (*Sources:* Organization for Economic Cooperation and Development, the World Bank, and United Nations Population Division. Map projection, Buckminster Fuller Institute and Dymaxion Map Design, Santa Barbara, CA. The word Dymaxion and the Fuller Projection Dymaxion™ Map design are trademarks of the Buckminster Fuller Institute, Santa Barbara, California, ©1938, 1967 & 1992. All rights reserved.)

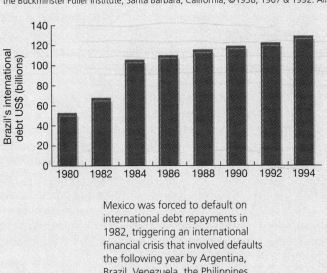

Brazil borrowed so much money in the 1970s and 1980s that it could no longer meet the interest payments. Between 1983 and 1989 the International Monetary Fund (IMF) bailed the country out but imposed austerity measures that were designed to curb imports. These included a 60 percent increase in gasoline prices and a reduction of the minimum wage to $50 a month, which gave workers half the purchasing power they had in 1940. Nevertheless, by 1994, the country's debt had reached nearly $130 billion, and annual inflation rates in the 1990s have ranged between 500 percent and 2,000 percent.

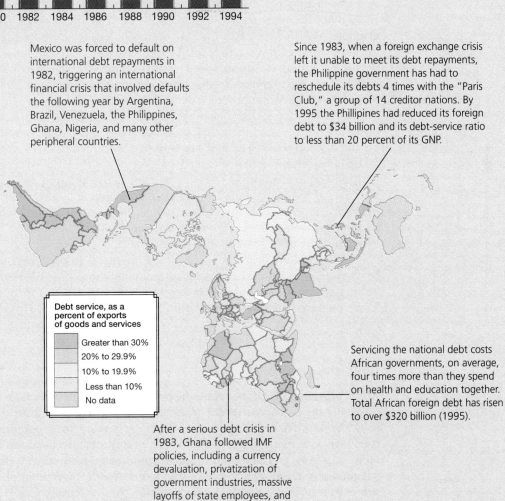

Mexico was forced to default on international debt repayments in 1982, triggering an international financial crisis that involved defaults the following year by Argentina, Brazil, Venezuela, the Philippines, Ghana, Nigeria, and many other peripheral countries.

Since 1983, when a foreign exchange crisis left it unable to meet its debt repayments, the Philippine government has had to reschedule its debts 4 times with the "Paris Club," a group of 14 creditor nations. By 1995 the Phillipines had reduced its foreign debt to $34 billion and its debt-service ratio to less than 20 percent of its GNP.

Debt service, as a percent of exports of goods and services

- Greater than 30%
- 20% to 29.9%
- 10% to 19.9%
- Less than 10%
- No data

Servicing the national debt costs African governments, on average, four times more than they spend on health and education together. Total African foreign debt has risen to over $320 billion (1995).

After a serious debt crisis in 1983, Ghana followed IMF policies, including a currency devaluation, privatization of government industries, massive layoffs of state employees, and lifting of trade restrictions.

501

Write-Offs?

As for the countries being "written-off," their future could be very bad indeed. It is not just that they have already been dismissed by investors in the core; nor that their domestic economies are simply threadbare. They face unprecedented levels of demographic, environmental, economic, and societal stress. In the worst-off regions—including much of West and Central Africa, for example—the events of the next 50 years are going to be played out from a starting point of scarce basic resources, serious environmental degradation, overpopulation, disease, unprovoked crime, refugee migrations, and criminal anarchy. African countries will be further disadvantaged as the prices of commodities produced there and in other peripheral locations have been dropping, while imported goods from the core have become more expensive. What this means is that many peripheral countries—especially in Africa—have and will probably continue to have reduced purchasing power in the global marketplace because of the decline in the value of their exports. Continuing to disable the periphery's full participation in the global economy are the combined effects of external debt crisis, dwindling amounts of foreign aid, insufficient resources to purchase technology or develop indigenous technological innovations, and the high costs of marketing and transporting commodities.

Post-independence ideals of modernization and democracy now seem remoter than ever in these regions. Corrupt dictators, epitomized by the Democratic Republic of Congo's former president, Mobutu, and Nigeria's General Abacha, have created "kleptocracies" (as in *kleptomania:* an irresistible desire to steal) in place of democracies. Of the estimated $12.42 billion in oil revenues that came to Nigeria as a result of the Persian Gulf crisis of 1990–1991, for example, $12.2 billion seems to have been clandestinely disbursed. Some governments, including Liberia and Sierra Leone, have lost control of large parts of their territories; groups of unemployed youths plunder travelers; tribal groups war with one another; refugees trudge from war zones to camps and back again; environmental degradation proceeds unchecked (Figure 12.6).

Amid this chaos, disease has prospered. Parts of Africa may be more dangerously unhealthy today than they were 100 years ago. Malaria and tuberculosis are out of control over much of sub-Saharan Africa, while AIDS is truly epidemic. In Uganda, where annual spending on health is only $3 per person (compared with debt repayments of $17 per person), one in five children die before their fifth birthday, and there is an AIDS epidemic among young adults. Of the approximately 22 million people worldwide whose blood is HIV-positive (infected with the human immunodeficiency virus that causes AIDS), 14 million are in Africa. While the rate of increase in HIV-positive cases in core countries has leveled off, it is still increasing dramatically in much of Africa. In the Congolese capital of Brazzaville, for example, 6 percent of the population was HIV-positive in 1992; in Pointe Noir, the second-largest city, it was 12 percent. Just three years later, in 1995, the rates were 15 percent and 25 percent respectively. Given all this, it is not surprising that much of Africa is unattractive to the globalizing world economy; and it should not be surprising if it remains so.

Overachievers

At the other end of the spectrum, the prospects for the core economies are bright, especially for large, core-based transnational corporations. With the end of the cold war, new markets in East-Central Europe have opened up to capitalist industry, along with more resources and a wider range of skilled and disciplined labor. New transport and communications technologies have already facilitated the beginnings of the globalization of production and the emergence of a global consumer culture. Top companies have reorganized themselves to take full advantage of this globalization. Reforms to the ground rules of international trade have

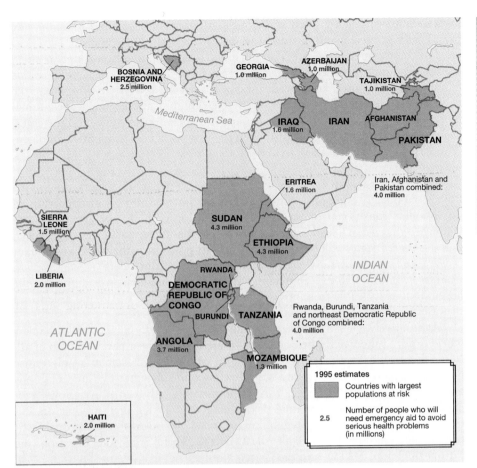

A vulture waits dispassionately as a young girl, weakened by hunger, collapses on the roadside. The photograph was taken during the civil war in Sudan in 1993. After the photographer, Kevin Carter, had shooed away the vulture, the girl stood up and walked off toward a feeding camp.

In the Sudan, 30 years of intermittent war between tribal factions from the mainly Arab north and the Christian/animist south intensified in 1983. Since that time almost 1.5 million have been killed. The United States sponsored a United Nations trusteeship of the country in 1992 but pulled its own troops out after just 2 years of low-technology guerilla warfare.

Figure 12.6 Wild Zones Parts of Africa, the Middle East, and South Asia have become "wild zones," places where national governments have lost control over economic development, ethnic conflict, and environmental degradation. In 1995, a CIA report estimated that about 40 million people were at immediate risk of malnutrition or death because of the combination of war, famine, disease, and criminal anarchy. Eighteen of the world's poorest countries are African. Caught in a downward cycle that began with the oil-price explosion of the 1970s, and accelerating with the plunge of commodities markets in the 1980s, they are getting poorer still: Per capita incomes in Africa as a whole have been decreasing at about 2 percent a year since the early 1980s.

removed many of the impediments to free-market growth; and a new, global financial system is now in place, ready to service the new global economy (see Chapter 7 for a full review of these developments). An example of this is the newly emerging World Trade Organization (WTO), an institution being created in conjunction with GATT (the General Agreement on Trade and Tariffs), which is supranational in its scope. The WTO is expected to provide a system of regulations over corporations that supercedes national-level regulations and laws. What this might mean in practice is that national restrictions over foreign corporations will be subordinated to the new rules of the WTO. It is too soon to tell the precise shape that the mulitlateral WTO will take, but its present shape suggests that local companies serving domestic markets may be significantly disadvantaged with respect to transnational corporations serving a more globally oriented market. The impacts are likely to be felt in all areas of social and economic life from health provision and the media to banking and the environment.

For the world-system core, therefore, the long-term question is not so much one of economic prosperity but of relative power and dominance. The same factors that will consolidate the advantages of the core as a whole—the end of the cold war, the availability of advanced telecommunications, the transnational reorganization of industry and finance, the liberalization of trade, and the emergence of a global culture—will also open the way for a new geopolitical and geo-economic order. This is likely to involve some new relationships between places, regions, and countries.

As we have suggested, the old order of the short twentieth century, dominated both economically and politically by the United States, is rapidly disappearing. In our present transitional phase, the new world order is up for grabs—we are coming to the end of a leadership cycle. This does not necessarily mean, however, that the United States will be unable to renew its position as the world's dominant power (see Feature 12.1, "Geography Matters—The Contenders"). As we saw in Chapter 2, Britain had two consecutive stints as the dominant world power—the hegemon that was able to impose its political view on the world and able to set the terms for a wide variety of economic and cultural practices.

Alternatively, we may not see the same kind of hegemonic power in the new world order of the twenty-first century—there may not be a new hegemon at all. Instead, the globalization of economics and culture may result in a polycentric network of nations, regions, and world cities bound together by flows of goods and capital. Order may come not from military strength rooted in national economic muscle but from a mutual dependence on *trans*national production and marketing, with stability and regulation provided by powerful international institutions (such as the World Bank, the IMF, the World Trade Organization, the European Union, NATO, and the United Nations).

Resources, Technology, and Spatial Change

Many aspects of future geographies will depend on trends in the demand for particular resources and on the exploitation of new technologies. The evolution of the world's geographies has always been shaped by the availability of key resources, and by the opportunities and constraints presented to different places and regions by successive technology systems. We should ask, therefore, what the future is likely to bring in terms of resource needs and technological shifts.

Resources and Development

The expansion of the world economy and the globalization of industry will undoubtedly boost the overall demand for raw materials of various kinds, and this will spur the development of some previously under-exploited but resource-rich regions in Africa, Eurasia, and East Asia. Raw materials, however, will be only a fraction of future resource needs. The main issue, by far, will be energy resources. World energy consumption has been increasing steadily over the recent past (Figure 12.7). As the periphery is industrialized and its population increases further, the global demand for energy will expand rapidly. Basic industrial development tends to be highly energy-intensive, however. The International Energy Agency, assuming (fairly optimistically) that energy in peripheral countries will be generated in the future as efficiently as it is today in core countries, estimates that developing-country energy consumption will more than double by 2010, lifting total world energy demand by almost 50 percent.

In 2010, peripheral and semiperipheral countries will account for more than half of world energy consumption. Much of this will be driven by industrialization geared to meet the growing worldwide market for consumer goods such as private automobiles, air conditioners, refrigerators, televisions, and household

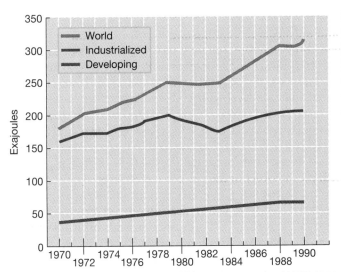

Figure 12.7 Trends in energy consumption
Global commercial energy consumption rose from less than 200 exajoules—a million million million joules, which is a unit of measurement of energy that is equivalent to 0.74 foot-pounds—in 1970 to nearly 350 exajoules in 1995. (*Source:* Data from United Nations Statistical Division, *1993 Energy Statistics Yearbook*. New York: United Nations, 1995.)

appliances. Current trends give some indication of this. For example, since 1978 the number of electric fans sold in China has increased almost twenty-fold, and the number of washing machines has increased from virtually zero to 97 million (Figure 12.8). Trends such as this caused China's oil consumption to leap by 11 percent in 1993, turning the world's sixth-largest oil producer into an oil importer.

While world reserves of oil, coal, and gas are, in fact, relatively plentiful, without higher rates of investment in exploration and extraction than at present, production will be slow to meet the escalating demand. The result might well be a temporary but significant increase in energy prices. This would have important geographical ramifications: Companies would be forced to seriously reconsider their operations and force core households into a careful reevaluation of their residential preferences and commuting behavior; and peripheral households would be forced further into poverty. If the oil-price crisis of 1973 is anything to go by (after crude oil prices had been quadrupled by the OPEC cartel), the outcome could be a significant revision of patterns of industrial location and a substantial reorganization of metropolitan form. Significantly higher energy costs may change the optimal location for many manufacturers, leading to deindustrialization in some regions and to new spirals of cumulative causation in others. Higher

Figure 12.8 The changing face of China In a dramatic break with the past, many Chinese cities now bristle with high-rise towers and construction cranes. Cognac, laser-disk players, and wide-screen television sets line department store shelves, and motorcycles jostle for space with foreign-brand automobiles.

Because of the globalization of industry, finance, and culture and the dissolution of the Soviet Union—one of the cornerstones of the old world order—the world is currently in a state of transition. Looking ahead 25 to 50 years, several contenders exist for economic and political leadership.

China

Although China is currently not even part of the core of the world-system, many observers predict a "Pacific Destiny" for the twenty-first century. In this scenario, China will be the hub of a world economy whose center of gravity is around the rim of the Pacific rather than the North Atlantic. China certainly has the potential to be a contender. It has a vast territory with a comprehensive resource base and a long history of political, cultural, and economic integration. It has the largest population of any country in the world (1.4 billion in 1995), and an economy that has been growing very rapidly. After 1978, under the leadership of Deng Xiaoping, China completely reorganized and revitalized its economy. Agriculture has been decollectivized, with communist collective farms having been modified to allow a degree of private profit taking. State-owned industries have been closed or privatized, and centralized state planning has been dismantled in order to foster private entrepreneurship. Since 1992, China has extended its "open door" policy (that is, allowing trade with the rest of the world) beyond its special economic zones and allowed foreign investment aimed at Chinese domestic markets. In the 1980s and early 1990s, when the world economy was sluggish, China's manufacturing sector grew by almost 15 percent *each year*. Almost all of the shoes once made in South Korea or Taiwan are now made in China. More than 60 percent of the toys in the world, accounting for $9 billion in trade, are made in China. Overall, China's economy is already the fourth largest in the world after the United States, Japan, and Germany (see Figure 12.1.1); on present trends, China will overtake Germany in 1998, Japan in 2003, and the United States in 2005. By the year 2020, China's GDP will be 40 percent larger than that of the United States (but, because of its huge population, its per capita GDP will still be lower than that of the United States).

Nevertheless, China's economy will remain largely agrarian until well into the twenty-first century. The task of feeding, clothing, and housing its enormous population will also constrain its ability to modernize its economy to the point where it can dominate the world economy. Meanwhile, a great deal of social and political reform remains to be accomplished before free enterprise can really flourish. Finally, China must resolve a major feature of its contemporary human geography before it can emerge as a hegemonic power. This is the dramatically uneven development that has been a consequence of its economic reforms. The positive spiral of cumulative causation has affected only the larger cities and coastal regions, while the vast interior regions of the country have become increasingly impoverished. This, ironically, is the very geographical pattern of spatial polarization that motivated Chinese communists to revolt in the 1940s. Today, it is a source of potential unrest and of political instability among the central government, the provincial governments of the interior regions, and the provincial governments of the coastal regions.

Germany and the European Union

Germany has historically been a persistent challenger for world leadership status. Its post-World War II economic renaissance, coupled with the reunification of East and West Germany in 1989, has made it a serious contender for future dominance. Germany currently has an economy that provides each household, on average, with an income of $43,418 (1994; the equivalent figure for the United States in 1994 was $39,020). It also has a very stable internal political structure and strong financial and industrial institutions that are already global in scope and orientation. Germany's greatest asset as it moves into the twenty-first century is its proximity to the labor, markets, and resources of the former satellite states of the Soviet Union. On the other hand, it will take some time to cope with the costs of reunification, modernize the infrastructure, and restructure both the economy and the social fabric of former East Germany. Another handicap to German dominance will be the deep reluctance on the part of many Germans, since World War II, to see national economic success translated into international political influence. Nevertheless, it must be recognized that with the end of the cold war and the demilitarization of German territory (both by NATO and by the Warsaw Pact), Germany is much freer to translate its economic might into political authority. This was signaled as early as 1990–1991, when Germany insisted on following an independent political course in recognizing the sovereignty of Slovenia, Croatia, and Bosnia Herzegovina (a move that, incidentally, helped set the conditions for the dreadful interethnic wars in Croatia and Bosnia).

Germany's whole future is also very much tied up in the European Union (Figure 12.1.2), and it is not stretching things too far to see a successful European Union as the main contender for world leadership. The European Union, which began in 1952 as a trading bloc (the European Economic Community, or EEC), is now an *economic union* (with *integrated* economic policies among

Figure 12.1.1 China's regional focus to economic growth

China's drive toward modernization has been based on a centrally planned liberalization of its former communist economy. Future growth will be channeled into its coastal corridor, based around the three main growth regions of Beijing, Shanghai, and Guangzhou, and a series of specially designated import-export processing zones. The most explosive growth in the immediate future will be in the region of the Pearl River Delta, around the major metropolitan areas of Hong Kong, Guangzhou, and Shenzen. In this region there are over 60 million people, and although 75 percent of them are still peasants, producing the fish, crabs, rice, bananas, and lychee fruit that overflow in city markets, the phenomenal growth of manufacturing means that the region triples its economic output every eight years. For all its poverty, the region is far more literate, educated, and healthy than most countries at its income level. This has attracted a great deal of investment, not just from China's re-emerging capitalist class, but also from Japan, South Korea, Taiwan, Hong Kong, and Thailand. The region is also the ancestral home of most of the world's 30 million overseas Chinese, who are estimated to control at least $2 trillion in liquid assets and who are now keen to invest in the region. The prospect is not so much one of the "next Taiwan" or the "next South Korea" as it is one of a core area that may well lead the entire Chinese economy into a position of global supremacy. (*Source:* After P. J. Rimmer, "Regional Economic Integration in Pacific Asia," *Environment & Planning A,* 26 (11), 1994, Figs. 4 and 5.)

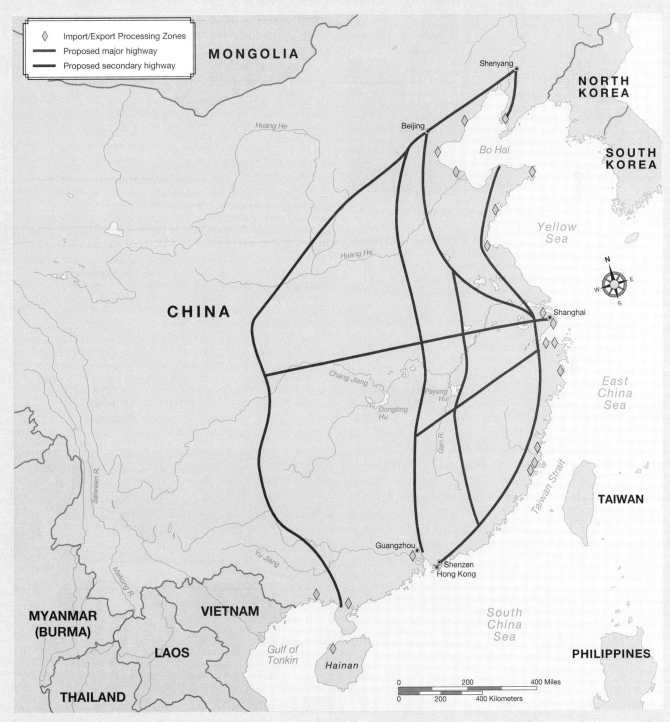

Figure 12.1.2 The headquarters of the European Union The European Union originated in 1967 as an amalgamation of three institutions—Euratom, the European Coal and Steel Community, and the European Economic Community—which had been set up in the 1950s in order to promote European economic integration. The six original members—Belgium, France, Italy, Luxembourg, the Netherlands, and West Germany—were joined in 1972 by Denmark, the Republic of Ireland, and the United Kingdom. With the addition of Greece in 1981, Portugal and Spain in 1984, and Austria, Sweden, and Finland in 1995, the European Union now has a population and an economy significantly larger than those of the United States.

member States) and has moved a long way toward its goal of becoming a *supranational political union* (with a single set of institutions and policies). With the addition of Austria, Sweden, and Finland in 1995, the European Union had a total population of 362 million (the U.S. population is just under 261 million) and an overall economy of $7.2 trillion (the GDP of the United States is just under $6 trillion). If and when it achieves full political union, its size will enable it to outvote the United States in the IMF and the World Bank. Meanwhile, however, it faces some serious difficulties, not least in achieving consensus among member States over the pathway to monetary integration.

Japan

The rise of Japan since World War II to become the second-largest national economy in the world has made it an obvious contender for the role of world leader in the first half of the twenty-first century. Despite its puny resources and the literal ruins of military defeat in 1945, it has established competitive advantages in a wide range

of industrial and financial activities that now serve global markets. The average annual income per household was $48,900 in 1994, the highest of all the major core countries. Japan is also the world's major creditor. Often attributed to a system of informal alliances among politics, business, and bureaucracy, the "Japanese miracle" includes the key elements of the Ministry of Finance, the Ministry of International Trade and Industry (MITI), and Nikkeiren, the national business association. This system, sometimes known as "Japan, Inc.," has systematically identified leading economic sectors and planned their success, first in Japan and then at a global scale. The system is still in place, and its success invites speculation that Japan will be the nation-state best equipped to cope with a globalized world economy (Figure 12.1.3).

Nevertheless, Japan faces several important handicaps as a contender. First, it is heavily dependent on external sources of raw materials. Japan is the number one importer not only of basic raw materials (petroleum,

Figure 12.1.3 The FSX fighter The FSX (Fighter Support Experimental) aircraft became symbolic of a series of larger issues surrounding the role of the United States as a world power. In 1987 the Japanese Defense Agency announced plans to develop a new aircraft to provide air cover against the possibility of an advancing Soviet fleet. A codevelopment agreement was proposed in 1989 between General Dynamics and Mitsubishi Heavy Industry, providing state-of-the-art aerospace technology to Japan in order to help it construct the FSX, an updated version of the General Dynamics F-16 fighter. Many in Congress saw the proposal as an important issue in relation to the changing world order. Should the United States help its main economic competitor to industrial secrets in one of the few areas in which the United States still held a clear advantage? And should it encourage the militarization of a country that has the economic potential to rival the United States as a world power? Under pressure from Washington, Japan agreed to allow American aircraft manufacturers to work on what was originally intended to have been an exclusively Japanese-made fighter. The first prototype was unveiled in 1996, and the Japanese Defense Agency plans to purchase 141 of the aircraft by 2007. Meanwhile, the collapse of the Soviet Union has called into question the need for the aircraft.

iron ore, and copper ore, for example) but also of agricultural goods. Second, Japan is geographically isolated from its major industrial markets and trading partners. While the Asian side of the Pacific Rim has great potential for growth, Japan's investments and partnerships are oriented overwhelmingly toward Europe and North America. Third, Japan faces a demographic trap. With an aging population, a declining birthrate, and a resolute unwillingness to allow large-scale immigration, Japan will face a serious labor shortage that can only be overcome by channeling more and more investment overseas.

Russia

If only because of its vast size and its status as the hearth of the former nuclear superpower of the Soviet Union, Russia has to be included among the contenders for economic and political leadership. Russia is the world's largest State by land area, with major reserves of important industrial raw materials and a pivotal strategic location in the center of the Eurasian landmass. Now, freed from the economic constraints of State socialism, Russia stands to benefit a great deal by establishing economic linkages with the expanding world economy (Figure 12.1.4). Similarly, the collapse of the Communist Party has removed a major barrier to domestic economic and political development. With a population of 149 million in 1995, the Russian economy also has an ample labor force and a domestic market large enough to form the basis of a formidable economy. At present, though, the Russian economy is shrinking as it withdraws from the centrally planned model. Overall, the Russian economy shrank by between 12 and 15 percent each year between 1990 and 1995. Yet, although embarrassed by the disintegration of the Soviet Union and bankrupt by the subsequent dislocation to economic development, Russia is still accorded a great deal of influence in international affairs. The embarrassment and insolvency may also prove in the long run to be the spur that pushes Russia once again to contend for great-power status.

Nevertheless, Russia has to be a long-odds contender. The latter years of the Soviet system left Russian industry with obsolescent technology and low-grade product lines, epitomized by its automobiles and civilian aircraft. Similarly, the infrastructure inherited by the Russian economy is shoddy—and often downright dangerous, as epitomized by the Chernobyl disaster of 1986, when a nuclear power plant exploded, causing a runaway nuclear reaction and widespread radiation pollution. Investment in the development of computers and new information networks was deliberately suppressed by Soviet authorities because, like photocopiers and fax machines, computers were seen as a threat to central control. As a result, the Russian economy now faces a massive task of modernization before it can approach its full potential. At the same time, the equally massive task looms of renewing civil society and the institutions of

St. Petersburg, 1993—The Lucky Strike advertisement and the retail kiosks reflect the first stage of capitalization and the emergence of a private retail sector.

Yaroslavl', 1996—This traditional-style Russian building in the city center has been converted to a boutique selling Western products. Such examples reflect dramatic changes in patterns of land use as well as the growth of the private retail sector.

St. Petersburg, 1992—During a period of food shortages, Hare Krishnas offer free food outside the Winter Palace where the Soviet revolution took place in 1917.

Yaroslavl', 1996—Teachers demonstrate outside government offices in protest against receiving no pay for three months. Public demonstrations of this kind were prohibited during the Soviet period.

Figure 12.1.4 Post–Cold War Russia In the new Russia, everyday street scenes betray the economic polarization and cultural tensions that have resulted from the "shock therapy" of competitive capitalist markets. Car dealerships selling U.S. imports, $400-a-night luxury hotels, Western-style supermarkets selling Kellogg's® corn flakes, the Russian edition of *Playboy* magazine, and French mineral water, stores selling Reebok and Benetton products, and McDonald's and Pizza Hut restaurants stand in stark contrast to dull stores, street-corner booths, and grimy restaurants. These contrasts have become symbols for Russia's polarized politics, breeding a resentment among workers and pensioners that is exploited by nationalist politicians such as Vladimir Zhirinovsky and Alexander Lebed, who want to restore Russia's *derzhava* (superpower) status through a self-reliant economy and renewed military strength.

business and democracy after 70 years of State socialism. Widespread theft of State property and the collapse of constitutional order have undermined respect for the law, and organized crime has flourished amid the factionalism and ideological confusion of the government. Meanwhile, real wages for most people have already fallen to 1950s levels.

509

The United States

The United States is more than a contender: It is the reigning hegemon. The U.S. economy is the largest in the world, with a broad resource base; a large, well-trained, and very sophisticated work force; a domestic market that has greater purchasing power than any other single country; and a high level of technological sophistication. The United States also has the most powerful and technologically sophisticated military apparatus, and it has the dominant voice and last word in international economic and political affairs. It is at least as well placed as its rivals to exploit the new technologies and new industries of a globalizing economy. It is also the only major contender for future world leadership with a global message: free markets, personal liberty, private property, electoral democracy, and mass consumption.

For the moment, however, the United States is a declining hegemon, at least in relative terms. Its economic dominance is no longer unquestioned in the way it was in the 1950s, 1960s, and 1970s. A relatively sluggish rate of growth has brought the prospect of the U.S. economy being caught and overtaken by Japan, Germany, and maybe even China. On some measures of economic development, the European Union has already overtaken the United States. More important, the globalization of the economy has severely constrained the ability of the United States to translate its economic might into the firm control of international financial markets that it used to enjoy.

Meanwhile, the end of the cold war, while a victory for the United States, has robbed it of its image as Defender of the Free World and weakened the legitimacy of its role as global policeman. The absence of a cold war enemy and the globalization of economic affairs has also made it much more difficult for the United States to identify and define the national interest. This, in turn, will foster divisiveness and volatility in domestic politics: a dynamic that will inevitably spill over into international affairs. The disarray in U.S. policies toward Somalia, Sudan, Bosnia, and Haiti in the early 1990s was symptomatic of such problems (Figure 12.1.5). In summary, while the United States must be considered the strongest contender, it is by no means a foregone conclusion that it will, in fact, become the leader for the next cycle of economic and political leadership.

Figure 12.1.5 U.S. troops leaving Somalia   In 1992, civil unrest in Somalia led to an intervention by the United Nations, with U.S. troops leading a large and heavily equipped peace enforcement task force. By March 1994, the U.S. troops had withdrawn, forced out by the low-tech guerrilla warfare of a rebel clan under the leadership of General Muhammad Farah Aideed. The withdrawal was seen around the world as a defeat for the United States, attributed to poor military planning, tactical errors, and a misunderstanding of the politics and culture of the region.

fuel costs will encourage some people to live nearer to their place of work, while others will be able to take advantage of telecommuting to reduce personal transportation costs. It is also relevant to note that almost all of the increase in oil production over the next 15 or 20 years is likely to come from outside the core economies. This means that the world economy will become increasingly dependent on OPEC governments, which control over 70 percent of all proven oil reserves, most of them in the Middle East.

On a more optimistic note, some potential efficiencies exist that might mitigate the demand for raw materials and energy in many parts of the world. Chinese industry, for example, currently uses 35 percent more energy per ton of steel than American industry, mainly because it has small, inefficient plants with outdated, energy-guzzling capital equipment. New plant and equipment would allow significant energy savings to be made. Considerable efficiencies could also be made through changes in patterns of consumption. In former socialist States, for example, years of price subsidies have made consumers (both industries and households) profligate with energy. East European countries currently consume four or five times as much energy per capita as countries of the same income levels elsewhere.

In countries that can afford the costs of research and development, new materials will reduce the growth of demand both for energy and for traditional raw materials such as aluminum, copper, and tin. Japan, for instance, may be able to reduce motor vehicle fuel consumption by 15 percent (and thereby reduce its total fuel oil consumption by 3 percent) by using ceramics for major parts of engines. It may also be possible to substitute ceramics for expensive rare metals in creating heat-resistant materials. Improved engineering and product design will also make it possible to reduce the need for the input of some resources. American cars, for example, were 15 percent lighter in 1995 than they were 20 years previously. In addition, of course, the future may well bring technological breakthroughs that dramatically improve energy efficiency or make renewable energy sources (such as wind, tidal, and solar power) commercially viable. As with earlier breakthroughs that produced steam energy, electricity, gasoline engines, and nuclear power, such breakthroughs would provide the catalyst for a major reorganization of the world's economic geographies.

New Technologies and Spatial Change

Just what new technologies are likely to have the most impact in reshaping human geographies? Given what we know about past processes of geographic change and principles of spatial organization, it is clear that changes in transportation technology are of fundamental importance. Consider, for example, the impact of oceangoing steamers and railroads on the changing geographies of the nineteenth century, and the impact of automobiles and trucks on the changing geographies of the twentieth century (Chapter 2). Among the most important of the next generation of transportation technologies that will influence future geographies are high-speed rail systems, smart roads, and smart cars. Several emerging industrial technologies also exist whose economic impact is likely to be so great that they will influence patterns of international, regional, and local development. Studies by the U.S. government, the Japanese government, the European Union, the Organization for Economic Cooperation and Development (OECD), and the United Nations have all identified biotechnology, materials technology, and information technology as the most critical areas for future economic development.

Transportation Technologies On land, the most interesting developments seem likely to center on new high-speed rail systems. Improved locomotive technologies and specially engineered tracks and rolling stock will make it possible to offer passenger rail services at speeds of 275 to 370 kilometers per hour (180 to 250 m.p.h.). With shorter check-in times and in-town rail termini, it will be quicker to travel between some cities by rail than by air. Plans have already been proposed for such systems in Florida, Texas, and California. The most advanced plans, however, are in Europe, where the European Union has drawn up a master plan for the development of 30,000 kilometers (almost 20,000 miles) of high-speed track by 2010 (Figure 12.9, page 512). The geographic implications of these systems are significant. Quite simply, once the systems become commercially viable, places that are linked to them will be well situated to grow in future rounds of economic development; places that are not will probably be left behind.

The same significance will attach to Intelligent Vehicle Highway Systems (IVHS), should they be developed from their current prototypes to become commercially viable. IVHS is a combination of so-called smart highways and smart cars. The IVHS target concept is an interactive link of vehicle electronic systems with roadside sensors, satellites, and centralized traffic management systems. This linkage allows for real-time monitoring of traffic conditions, and allows

Figure 12.9 High-speed rail in Europe Europe, with its relatively short distances between major cities, is ideally suited for rail travel, and less suited, because of population densities and traffic congestion around airports, to air traffic. Allowing for check-in times and accessibility to terminals, it is already quicker to travel between many major European cities by rail than by air. The European Union plans to coordinate and subsidize a $250 billion investment in 14,000 miles of high-speed track, to be phased in through 2012. The heart of the system will be the "PBKAL web" that will connect Paris, Brussels, Cologne (Köln), Amsterdam, and London, and which will be completed by 2003. This will be the cause of some restructuring of the geography of Europe. High-speed rail routes will have only a few timetabled stops because the time penalties that result from deceleration and acceleration undermine the advantages of high-speed travel. Places that do not have scheduled stops will be less accessible, and so less attractive for economic development.

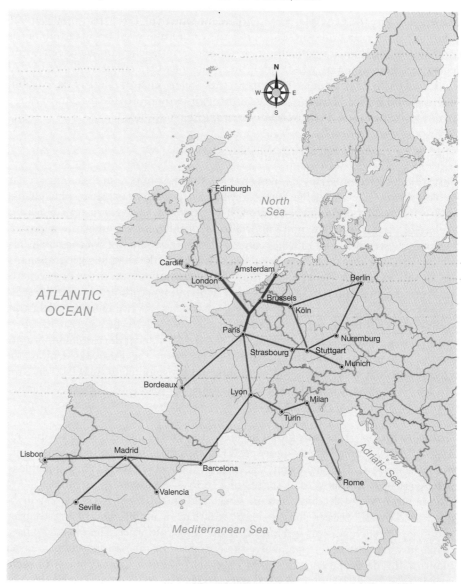

The *Eurostar* at Waterloo International Station, London.

drivers to receive alternate route information via two-way communications, on-board video screens, and mapping systems. With fewer gridlocked roads, driving would be safer, less polluting, and more efficient. Metropolitan areas and Interstate corridors that have the infrastructure of IVHS technology will be at a significant advantage in attracting new industries and their workers. The first generation of smart cars and trucks, with on-board, microprocessor-based electronics, is already on the road. More advanced prototypes are now being tested. The Rockwell Corporation is using the Department of Defense's Global Positioning System (GPS) of satellites as the foundation for an in-vehicle navigation system. An experimental stretch of "smart highway" is planned between Blacksburg, Virginia, and a nearby interstate highway (Figure 12.10). In Chicago, a fleet of vehicles is testing on-board navigation systems that route drivers around construction, accidents, and high-traffic spots.

At sea, no emerging technologies exist at present that might have an equivalent impact. In the air, however, a large untapped market encourages big investments in research and development. Plans are afoot, for example, for a new generation of wide-bodied jets that will carry close to 800 passengers (the present maximum is about 450). If we look much further ahead, a drawing-board design exists for a Japanese rocket plane that would cut the journey time from Tokyo to Los Angeles to just two hours—the current average daily commute time for business professionals in Tokyo. This conjures up a future world in which members of professional classes might commute between any pair of major world cities: Tokyo–Los Angeles, London–New York, Washington–Brussels, Beijing–San Francisco, or Zürich–Johannesburg, for example. This might not have a broad impact on human geography in the way that, say, the railroad or the automobile did, but it could have a marked impact on the geography of corporate headquarters, of supranational agencies, and on the social geography of world cities.

The projected changes in transportation technology are enormous and will have profound impacts in both the core and the periphery. One question these innovations force us to ask is: How efficient do they allow us, as a global community, to be? For instance, does it make sense to consume large amounts of fossil fuels and generate more pollution in order to transport staple food stuffs around the world when they could just as easily be grown locally? At present, in France, which possesses a long-established apple-producing industry, it is possible

Figure 12.10 Smart highway systems The idea of "smart highways" is that one or more lanes of highway be designated as "automated" lanes, in which vehicles have their speed and position controlled by computers. The result would be faster flows of higher volumes of traffic, with fewer accidents. One of the first prototype stretches of smart highway is planned outside Blacksburg, Virginia, where Virginia Tech, one of three national centers of excellence in Intelligent Vehicle/Highway Systems, is collaborating in a $150 million research consortium headed by General Motors and funded by the federal government. The initial test bed includes (1) embedded road sensors, (2) a fiberoptic backbone, (3) overhead sensors and signs, and (4) a wireless communication network.

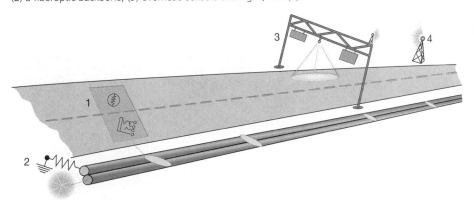

The BOSCH Travel Pilot in-vehicle, dynamically updated traveler location and direction system.

to find in the markets as many New Zealand apples as French ones. In Mongolia, where milk-producing animals are ubiquitous, the butter in the markets is imported from Germany. Does such an arrangement make economic, environmental, and social sense?

Biotechnology Although biotechnology is widely associated with both the genetic engineering of crops (through Green Revolution crops such as hybrid rice, corn, wheat, lettuce, tomatoes, sugar cane, and cotton [see Chapter 8]) and with its pharmaceutical potential (through products such as Interferon and growth hormones), biotechnology is also likely to have a profound effect on animal husbandry, industrial production, renewable energy, waste recycling, and pollution control (Figure 12.11). Genetic engineering is already being applied quite widely in animal husbandry. Examples include Japanese fish farms that use cell fusion technology to produce algae in systems that are 350 times more efficient than other methods of raising brine shrimp; and U.S. dairy farms that use bovine growth hormones to achieve a 20 percent increase in milk production at the same feed costs. Overall, the commercial output of genetically engineered products is almost certain to grow at rates of around 10 percent a year for some years hence. The U.S. Office of Technology Assessment expects that, by the year 2000, five-sixths of the annual increase in world agricultural production will result from applications of biotechnology.

Similar growth can be expected in industrial applications of biotechnology—enzymes, for example, that will be used as industrial catalysts, in the microbial recovery of metals, in waste degradation, and in biomass fuels. In geographical terms, we can expect the economic benefits of biotechnology to be greatest in the countries and regions that can afford the costs of research and development and the costs of installing and applying the new technologies. On the other hand, one of the principal advantages of many applications of biotechnology innovations is that they are economical to use on a small scale, without large infrastructure requirements. This should facilitate their use in peripheral regions. There is the prospect (but by no means the certainty) that these applications could bring not only commercial success to peripheral regions but also help solve the problem of food shortages. Similarly, it is possible that biotechnology might reverse the environmental degradation of some parts of the world because it could provide eco-

Figure 12.11   Biotechnology
Research scientists harvest kernels from genetically engineered barley at the Plant Gene Expression Center, a joint venture of the United States Department of Agriculture and the University of California, Berkeley. The research shown here is directed toward inserting new genes into barley, enabling greater resistance to plant viruses and developing a barley grain that can be used in making raised loaves of bread.

nomically viable ways of replacing chemical fertilizers and toxic sprays, recycling waste products, and cleaning up polluted water. It is also possible, unfortunately, that natural genetic diversity will be reduced as native seeds are replaced by clones, and as locally adapted forms of agriculture are replaced by industrialized ones. The result could very easily be the disappearance of thousands of plant varieties and with them the sources of natural resistance to genetically adapted pests.

Materials Technologies   Materials technologies include new metal alloys, specialty polymers, plastic-coated metals, elastothermoplastics, laminated glass, and fiber-reinforced ceramics. They are important because they can replace scarce natural resources; reduce the quantity of raw materials used in many industrial processes; reduce the weight and size of many finished products; increase the performance of many products; produce less waste; and allow for the commercial development of entirely new products. Materials design has become so sophisticated that engineers will also be able to incorporate environmental criteria into their products, rather than having to deal with environmental issues as an afterthought (though of course it may not be commercially realistic to do so). Like biotechnology, materials technology has been growing steadily for some time and is now set for a period of explosive growth in its applications, particularly in the automobile and aircraft industries. It has been estimated that specially engineered materials already account for around $1 trillion of the annual GDP of the United States.

Unlike biotechnology, applications of materials technologies will require a fairly close association with an expensive infrastructure of high-tech industry. As a result, their immediate geographical impact is likely to be much more localized within the core regions of the world-system. Peripheral regions and countries that are currently heavily dependent on the production and export of traditional raw materials—such as Guinea and Jamaica (bauxite), Zambia and the Democratic Republic of Congo (copper), Bolivia (tin), and Peru (zinc)—will probably be at the wrong end of the creative destruction prompted by these new technologies. In other words, as new materials technologies reduce the demand for traditional raw materials, production and employment in the latter will decline, and investors will probably withdraw from producer regions in order to reinvest their capital in more profitable ventures elsewhere. In contrast, some peripheral regions and countries will benefit from the increasing demand for rare Earth metals. Brazil, Nigeria, and the Democratic Republic of Congo, for example, together account for almost 90 percent of the world's production of niobium (used with titanium in making superconductive materials); Brazil, Malaysia, Thailand, Mozambique, and Nigeria together account for about 75 percent of the world's production of tantalum (used in making capacitors that store and regulate the flow of electricity in electronic components).

Information Technologies   Information technologies include all of the components of information-based, computer-driven, and communications-related activities—a wide array of technologies that includes both hardware (silicon chips, microelectronics, computers, satellites, and so on) and the software that makes it operate. In addition to telematics—the automation of telecommunications and the linkage of computers by data transmission—information technologies include developments as diverse as real-time monitoring of traffic bottlenecks, computer-controlled manufacturing, chemical and biological sensors of effluent streams, 24-hour data-retrieval systems, bar-coded retail inventory control, telemetry systems for tracking parcels and packages, and geographic information systems. Information technologies have already found widespread applications in retailing, finance, banking, business management, and public administration; yet it is estimated that, even in the more developed countries, only about one-third of the benefits to be derived from information technology–based innovations will have been realized by the year 2000.

As we have seen in earlier chapters, information technologies have already transformed certain aspects of economic geography. In terms of employment and production, an overall concentration exists in core countries, where the detailed geography of information technologies takes the form of highly localized agglomerations of activity. Examples include "Silicon Valley" (Santa Clara County) in California, the Route 128 corridor around Boston, the Research Triangle area in North Carolina, the M4 corridor in southern England, the "Silicon Glen" corridor of central Scotland, and around the edges of the Osaka and Tokyo metropolitan areas in Japan (Figure 12.12). In addition, routine production, testing, and assembly functions have been decentralized to semiperipheral countries—Hong Kong, Singapore, Taiwan, and South Korea, in particular.

Figure 12.12 Technopolis cities   Tsukuba Science City, about 40 miles northeast of Tokyo, has served as a model for the Japanese "Technopolis" program, which is a government-sponsored attempt to create a series of high-tech complexes to serve Japanese industry. Tsukuba was part of a strategic urban plan for the Tokyo metropolitan area, designed to exploit geographic principles of localization and agglomeration in order to foster Japanese high-tech industry. Although construction only began in the early 1970s, it now has a population of over 180,000, of whom more than 12,000 are scientific researchers. The Technopolis program was master-planned by the Japanese Ministry of International Trade and Industry (MITI), and its 26 high-tech complexes will be the cornerstone of future high-tech development in Japan. Each technopole, or high-tech growth-pole settlement, is designed as a garden-city-type setting for research universities, science centers, industrial research parks, joint research and development consortia, venture capital foundations, office complexes, international convention centers, and new residential towns. (Source: Map redrawn with permission from J. L. Bloom and S. Asano, "Tsukuba Science City," *Science,* 212. Copyright © 1981, American Association for the Advancement of Science.)

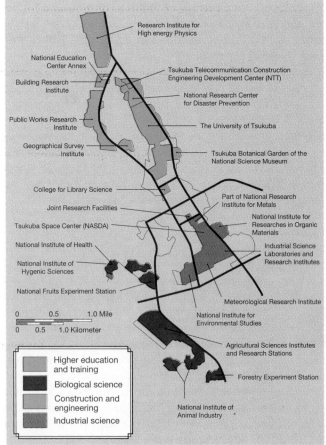

Future geographies of production and employment in information technologies will almost certainly follow the same pattern. Most research and development and high-end production will be localized within core countries, and most routine functions will be decentralized to peripheral and semiperipheral countries such as China, Indonesia, the Philippines, Sri Lanka, and Thailand. That is not to say that exceptions will not occur. India, for instance, has begun to carve a successful niche in software development, mainly because of the Indian government's provision of generous tax breaks and liberal foreign-exchange regulations for the industry.

It is the spatial effects of information technologies' *applications* that are of most interest to geographers. As we have seen in previous chapters, information technologies have already had an enormous impact in facilitating the globalization of industry, finance, and culture. At local and regional scales, while they have been instrumental in decentralizing jobs and residences, their impact has been very *uneven*. Computers, for example, are mainly used in peripheral countries for standard functions of inventory control, accounting, and payroll; and even so they remain much too expensive for widespread use in homes, small businesses, and local governments. We can expect future impacts of information technologies to exhibit the same unevenness. While the world will certainly shrink even further, a marked lag will occur in the diffusion of information technologies to many peripheral regions, thus perpetuating and even accentuating the geographical divisions between the fast world and the slow world.

It is also apparent that information technologies will increase the gap between the haves and the have-nots within the fast world. For instance, robotics and computers are replacing human labor in both industry and corporations. In the United States, for example, between 1980 and 1993, the Fortune 500 firms shed 4.4 million jobs. Most of those jobs were lost directly or indirectly through information technology innovations that replaced workers with machines. During that same period, as hundreds of thousands of workers were rendered jobless, these same corporations increased their sales 1.4 times, their assets 1.3 times, and their C.E.O.s received increased compensations of 6.1 times. Even within the fast world, disparities are increasing. In this case, the geographical impacts are greatest in the cities, whereas suburbs have been far less affected.

Adjusting to the Future

The immediate future will be characterized by a phase of geopolitical and geoeconomic transition; by the continued overall expansion of the world economy; and by the continued globalization of industry, finance, and culture. In this future, new technologies—various transport technologies, biotechnology, materials technology and, especially, information technologies—will be influential in shaping the opportunities and constraints for local, regional, and international development. The processes of change involved in shaping this kind of future will inevitably bring critical issues, conflicts, and threats. We can already identify what several of these might involve: dilemmas of scale, boundaries, and territories; fault lines of cultural dissonance; and the sustainability of development.

Scale and Territory

In a speech before the United Nations in 1993, President Bill Clinton made an observation:

> . . . Economic and technological forces all over the globe are
> compelling the world toward integration. These forces are fueling a
> welcome explosion of entrepreneurship and political liberalization, but
> they also are threatening to destroy the insularity and independence of

national economies, quickening the pace of change and making many of our people feel more insecure.

The dilemma portrayed by President Clinton is a consequence of the enormous and disruptive impact that new transport and communications technologies and new, transnational, corporate strategies have had on the political boundaries and established territorial relationships of the old world order. This order is simply not as functional as it was; and in the future it will be even less so, unless some major adjustments can be made.

The central problem is that the globalization of the economy is undermining the status of the territorial nation-state as the chief regulating mechanism of both global and local dimensions of the world-system. This is not to say that nation-states are about to become outmoded. In the future a strong logic will continue to exist for the territorial powers of nation-states, but the States themselves will become increasingly permeable to flows of capital and information. While capital, knowledge, entrepreneurship, management, and consumer tastes will continue to globalize, governments will be locked into their nineteenth-century quilt of territories and institutions.

Without strong nation-states—or some alternative—some important aspects of geographic change will escape the authority of national governments. Consider three examples:

- Commercial information, patents, stocks, bonds, electronic cash transfers, and property deeds will flow in increasing volume across national boundaries virtually unchecked and unchallenged by national governments and their agencies.

- Localities will be drawn more and more into dealing directly with overseas investors in their attempts to promote local development through their own "municipal foreign policy."

- Stealthy, temporary, "virtual states" will emerge illegally from clandestine alliances of political and military leaders and senior government officers to take advantage of the paralysis of national sovereignty.

These changes are in prospect at a time when the end of the cold war has left an enormous zone of geopolitical uncertainty. The balance of power that stabilized international politics for 40 years has gone, and its sudden disappearance has revealed the precariousness of many domestic political systems. Previously held together by a common enemy, they are now especially vulnerable to internal economic, ethnic, and cultural divisions.

Some of the consequences are fairly predictable. The nation-states of the world-system core, unable to manage national economies and protect their populations from the winds of global change, will have to cope with severe economic slumps (rather than minor recessions) and persistent problems of poverty, unemployment, and homelessness. Social-democratic parties in many countries, deprived of their main means of satisfying their liberal supporters—effective national strategies for social and economic management—will lose their appeal, leaving electorates to fragment and polarize around xenophobia, racism, family values, personal freedoms, and green politics. Because nation building had always promoted the idea of a national society and a "national culture," the permeability of nation-states also raises the prospect of national identities leaking away into local (or, via cyberspace, transnational) lifestyle communities: software clubs, multiuser dungeons, ecologists, baseball fans, fundamentalists, neo-Nazis, Corvette owners, and so on.

Some of the consequences of weak nation-states are in fact already beginning to appear. In order to stem economic leaks and gain some control over the globalizing economy, they are forming economic blocs such as North American Free Trade Association (NAFTA) and the European Union. Meanwhile, as we saw in Chapter 9, many of the same nation-states are accommodating internal cultural cleavages by decentralizing their governmental structures. In other cases, governments are dismantling expensive social welfare programs in order to open up previously publicly operated industries to the private sector—often to transnational corporations. For example, Canada's national and provincial governments are dismantling their social programs in the name of global economic competition. Once widely regarded as a public-sector success story, Canada's social welfare system is now largely being eroded as powerful corporate interests, in the form of the Business Council on National Issues, have successfully campaigned to privatize and deregulate the government sector. This deregulation has meant the loss or reduction of public control over the environment, transportation, and energy resources as well as the dismantling of universal social programs including unemployment insurance, social assistance, health care, and pensions. The result has been the emergence of a large poverty class, the shrinking of the middle class, and soaring corporate profits. With a less socially and economically secure future confronting them, Canadians have begun to organize a counterforce called the "Citizens' Agenda." The agenda is a declaration of citizen rights within a global economy. Its aim is to help citizens fight to preserve local job opportunities and conditions as well as environmental standards when they are challenged by global corporations.

If first reactions are anything to go by, the consequences of globalization will be much more dramatic in the world's peripheral regions. However strong governments may be in terms of the apparatus of domestic power, they will be next to helpless in the face of acute environmental stress, increased cultural friction, escalating poverty and disease, and growing migrations of refugees. In this situation, it is possible that people will seek liberation in violence (Figure 12.13). The question may be not so much whether war and violence will exist within the periphery, but who will fight whom, and for what purpose?

Figure 12.13 Militia factions and road warriors In some parts of Africa, civil war and the breakdown of law and order have from time to time left some areas under the control of various private armies, armed gangs, and militarized factions. Some of this conflict is a direct legacy of interethnic tensions created under nineteenth- and early twentieth-century colonialism. The photograph on the right, taken outside Monrovia, Liberia, shows the remains of civilian victims of the civil war there. The photograph below shows a reoccupied rebel position in Sierra Leone, where anarchy and armed political conflict displaced more than three-quarters of a million of the country's 3.5 million population in the early 1990s.

This is how one commentator sees the probable outcome:

> Future wars will be those of communal survival, aggravated or, in many cases, caused by environmental scarcity. These wars will be subnational, meaning that it will be hard for states and local governments to protect their own citizens physically. This is how many states will ultimately die. As state power fades—and with it the state's ability to help weaker groups within society, not to mention other states—peoples and cultures around the world will be thrown back upon their own strengths and weaknesses, with fewer equalizing mechanisms to protect them.[2]

With nation-states weakened through the transnational flow of capital, goods, and services, and the distinction between criminal violence and "legitimate" war blurred, the power of international drug cartels, local mafias, road-warrior platoons, popular militias, guerrilla factions, and private local armies will create the possibility of borderless territories that wax and wane in an ever-mutating space of chaos. Future maps of parts of the world periphery may have to be drawn without clear boundaries, just like medieval maps and the maps of European explorers. If this sounds far-fetched, it should not: in the likes of Lebanon, Liberia, Sri Lanka, Somalia, Rwanda, Peru, and Colombia, civil wars, armed insurrections, and criminal violence have made it difficult, during the 1990s, always to be certain of which group or government controls what territory. The clearly defined boundaries of international treaties and school atlases have already been rendered as fiction by *de facto* buffer zones in a few places: the Kurdish and Azeri buffer "States" between Turkey and Iran, for example, and the Turkic Uighur buffer territory between central Asia and China.

Cultural Dissonance

At one level, globalization has brought a homogenization of culture through the language of consumer goods. This is the level of "Planet Reebok," where material cultures are enmeshed by 747s, CNN, music video channels, handheld telephones, and the Internet; and swamped by Coca-Cola, Budweiser, Levis, Nikes, Walkmans, Nintendos, Hondas, and Disney franchising. Furthermore, sociologists have recognized that a distinctive culture of "global metropolitanism" is emerging among the self-consciously transnational upper-middle classes. This is simply homogenized culture at a higher plane of consumption (French wines instead of Budweiser, Hugo Boss clothes instead of Levis, BMWs instead of Hondas, and so on). The members of this new culture are people who hold international conference calls; who send and receive faxes and E-mail; who make decisions and transact investments that are transnational in scope; who edit the news; who design and market international products; and who travel the world for business and pleasure.

These trends are transcending some of the traditional cultural differences around the world. We can, perhaps, more easily identify with people who use the same products, listen to the same music, and appreciate the same sports stars that we do. In the process, however, sociocultural cleavages are opening up between the haves of the fast world and the have-nots of the slow world. By focusing people's attention on material consumption, they are also obscuring the emergence of new fault lines—between previously compatible cultural groups, and between ideologically divergent civilizations.

---

[2]R. D. Kaplan, "The Coming Anarchy," *The Atlantic Monthly*. 273, February 1994, p. 74.

Several reasons account for the appearance of these new fault lines. One is the release of pressure brought about by the end of the cold war. The evaporation of external threats has allowed people to focus on other perceived threats and intrusions. Another is the globalization of culture itself. The more people's lives are homogenized through their jobs and their material culture, the more many of them want to revive subjectivity, reconstruct we/us feelings, and re-establish a distinctive cultural identity. In the slow world of have-nots, a different set of processes is at work, however. The juxtaposition of poverty, environmental stress, and crowded living conditions alongside the materialism of "Planet Reebok" creates a fertile climate for gangsterism. The same juxtaposition also provides the ideal circumstance for the spread and intensification of religious fundamentalism. This, perhaps more than anything else, represents a source of serious potential cultural dissonance.

The overall result is that cultural fault lines are opening up at every geographical scale. This poses the prospect of some very problematic dimensions of future geographies. The prospect at the metropolitan scale is one of fragmented and polarized communities, with outright cultural conflict suppressed only through electronic surveillance and the "militarization" of urban space via security posts and the "hardened" urban design using fences and gated streets (Figure 12.14). This, of course, presupposes a certain level of affluence in order to meet the costs of keeping the peace across economic and cultural fault lines. In the unintended metropolises of the periphery, where unprecedented numbers of migrants and refugees will be thrown together, there is the genuine prospect of anarchy and intercommunal violence—unless, that is, intergroup differences can be submerged in a common cause such as religious fundamentalism.

At the regional scale the prospect is one of increasing ethnic/racial rivalry, parochialism, and insularity. Examples of these phenomena can be found throughout the world. In North America, we see increasing ethnic rivalry represented by the secessionism of the French-speaking Quebeçois, and the insistence of some Hispanic groups in the United States on the installation of Spanish as an alternative first language. In Europe, examples are the secessionism (from Spain) of the Basques, the separatist movement of the Catalans (also in Spain), the

Figure 12.14 The shape of things to come? Social, economic, and ethnic polarization in the cities of the world's core countries has led to significant changes in urban landscapes. One important aspect of these changes has been an increase in the electronic surveillance of both public and private spaces. Another has been the increased presence of private security personnel in upscale settings: Private security officers now outnumber police officers in the United States.

regional elitism of Northern Italy, and outright war among Serbs, Croats, and Bosnians. In South Asia, examples are provided by the recurring hostility between Hindus and Muslims throughout the Indian subcontinent, and between the Hindu Tamils and Buddhist Sinhalese of Sri Lanka. In Africa, a long list of ethnic rivalry, parochialism, and insularity includes the continuing and bloody conflict between the Hutus and the Tutsis in Rwanda, Burundi, northeast Democratic Republic of Congo, and western Tanzania. Where the future brings prosperity, tensions and hostilities such as these will probably be muted; where it brings economic hardship or decline, they will undoubtedly intensify.

The prospect at the global scale is of a rising consciousness of people's identities in terms of their broader historical, geographical, and racial "civilizations": Western, Latin American, Confucian, Japanese, Islamic, Hindu, and Slavic-Orthodox. According to some observers, deepening cleavages of this sort could replace the ideological differences of the cold war era as the major source of tension and potential conflict in the world. Given the steep economic differences that will almost certainly persist between the world's core regions and the periphery, this conjures up the prospect of an overt "West (and Japan) Versus the Rest" scenario for international relations.

Sustainability

The same "West Versus the Rest" theme emerges in discussions of the environmental issues that will face us in the future. As we saw in Chapter 4, the world currently faces a daunting list of environmental threats: the destruction of tropical rain forests and the consequent loss of biodiversity; widespread, health-threatening pollution; the degradation of soil, water, and marine resources essential to food production; stratospheric ozone depletion; acid rain; and so on. We also saw in Chapter 4 that most of these threats are greatest in the world's periphery, where daily environmental pollution and degradation amounts to a catastrophe that will continue to unfold, slow-motion, in the coming years.

In these peripheral regions there is, simply, less money to cope with environmental threats. In addition, the very poverty endemic to peripheral regions also adds to environmental stress. In order to survive, the rural poor are constantly impelled to degrade and destroy their immediate environment, cutting down forests for fuelwood and exhausting soils with overuse. In order to meet their debt repayments, governments feel compelled to generate export earnings by encouraging the harvesting of natural resources. In the cities of the periphery, poverty encompasses so many people in such concentrations as to generate its own vicious cycle of pollution, environmental degradation, and disease. Even climatic change, an inherently global problem, seems to pose its greatest threats to poorer, peripheral countries. The heavily populated, low-lying, delta country of Bangladesh, for example, faces extreme hazards from the rising sea levels anticipated from further global warming, while extensive regions of Africa, Asia, and Latin America are so marginal for agriculture that further drought could prove disastrous. In contrast, farmers in much of Europe and North America would welcome a local rise in mean temperatures, since it would extend their options for the kinds of crops that they could profitably raise.

Future trends will only intensify these contrasts. We know enough about the growth of population and the changing geography of economic development to be able to calculate with some confidence that the air and water pollution generated by low-income countries will more than double in the next 15 years. By 2010, China will probably account for one-fifth of global carbon dioxide emissions. We know, in short, that environmental problems will be inseparable from processes of demographic change, economic development, and human welfare. In addition, it is becoming clear that environmental problems are going to be increasingly enmeshed in matters of national security and regional conflict. The spatial *interdependence* of economic, environmental, and social problems means that some parts of the world are ecological time-bombs. The prospect of civil unrest and mass migrations resulting from the pressures of rapidly growing populations, deforestation, soil erosion, water depletion, air pollution, disease epidemics, and intractable poverty is real. These specters are alarming not only for the peoples of the affected regions but also for their neighbors.

Such images are also alarming for the peoples of rich and far-away countries, whose continued prosperity will depend on processes of globalization that are not disrupted by large-scale environmental disasters, unmanageable mass migrations, or the breakdown of stability in the world-system as a whole. The U.S. security agenda has already been cast in terms of environmental awareness. In President Clinton's address to the United Nations in 1993, for example, he noted that global environmental issues "threaten our children's health and their very security" and promised that the United States will "work far more ambitiously to fulfill our obligations as custodians of this planet not only to improve the quality of life for our citizens . . . but also because the roots of conflict are so often entangled with the roots of environmental neglect and the calamities of famine and disease."

All this has given some momentum to the notion of sustainable development. **Sustainable development** is a vision of development that seeks a balance among economic growth, environmental impacts, and social equity. In practice, sustainable development means economic growth and change should occur only when the impacts on the environment are benign or manageable and the impacts (both costs and benefits) on society are fairly distributed across classes and regions. Sustainable development is geared to meeting the needs of the present without compromising the ability of future generations to meet their needs. It envisages a future when improvements to the quality of human life are achieved within the carrying capacity of local and regional ecosystems. **Carrying capacity** is the maximum number of users that can be sustained, over the long term, by a given set of natural resources.

Sustainable development means using renewable natural resources in a manner that does not eliminate or degrade them—by making greater use, for example, of solar and geothermal energy, and by greater use of recycled materials. It means managing economic systems so that all resources—physical and human—are used optimally. It means regulating economic systems so that the benefits of development are distributed more equitably (if only to prevent poverty from causing environmental degradation). It also means organizing societies so that improved education, health care, and social welfare can contribute to environmental awareness and sensitivity and an improved quality of life.

A final and more radical aspect of sustainable development is a move away from wholesale globalization toward increased "localization": a desire to return to a more locally based economy where production, consumption, and decision making can be oriented to local needs and conditions. Thus peripheral countries, as well as workers and citizens throughout many parts of the core, are demanding a reinstatement of control over the economic events and institutions that directly shape their lives.

Put this way, sustainable development sounds eminently sensible yet impossibly Utopian. The first widespread discussion of sustainability took place in the early 1990s, and focused on the "Earth Summit" (the United Nations Conference on Environment and Development) meeting in Rio de Janeiro in 1992. Attended by 128 heads of state, it attracted intense media attention. At the conference, many examples were described of successful sustainable development programs at the local level. Some of these examples centered on the use of renewable sources of energy, as in the creation of small hydroelectric power stations to modernize Nepalese villages. Some examples centered on tourism in environmentally sensitive areas, as in the trips organized in Thailand's Phang

Sustainable development: a vision of development that seeks a balance among economic growth, environmental impacts, and social equity.

Carrying capacity: the maximum number of users that can be sustained, over the long term, by a given set of natural resources.

Nga Bay, where tourists visit spectacular hidden lagoons by sea canoe under strict environmental regulation (no drinking, littering, eating, or smoking, and limits on the number of visitors allowed each day) and with a high level of social commitment (emphasizing respect for staff, local culture and family, proper training, and good pay). Most examples, however, centered on sustainable agricultural practices for peripheral countries, including the use of intensive agricultural features such as raised fields and terraces in Peru's Titicaca Basin, techniques that had been successfully used in this difficult agricultural environment for centuries, before European colonization. After the United Nations conference, however, many observers commented bitterly on the deep conflict of interest between core countries and peripheral countries that was exposed by the summit. Without radical and widespread changes in value systems and unprecedented changes in political will, "sustainable development" will remain an embarrassing contradiction in terms.

We cannot just wait to see what the future will hold. If we are to have a better future (and if we are to *deserve* a better future) we must use our understanding of the world—and of geographical patterns and processes—to work toward more desirable outcomes. No discipline is more relevant to the ideal of sustainable development than geography. Where else, as British geographer W. M. Adams has observed, can the science of the environment (physical geography) be married with an understanding of economic, technological, social, political, and cultural change (human geography)? What other discipline offers insights into environmental change, and who but geographers can cope with the diversity of environments and the sheer range of scales at which it is necessary to manage global change?

Those of us in the richer countries of the world have a special responsibility for leadership in sustainable development because our present affluence is based on a cumulative past (and present) exploitation of the world's resources that is disproportionate to our numbers. We also happen to have the financial, technical, and human resources to enable us to take the lead in developing cleaner, less resource-intensive technologies, in transforming our economies to protect and work with natural systems, in providing more equitable access to economic opportunities and social services, and in supporting the technological and political frameworks necessary for sustainable development in poor countries. We cannot do it all at once, but we will certainly deserve the scorn and resentment of future generations if we do not try.

Conclusion

In contrast to much of the rest of the century, the end of the twentieth century and the beginning of the twenty-first is going to be a period of fluid and transitional relationships among places, regions, and nations. Nevertheless, we know enough about contemporary patterns and trends, as well as geographic processes, to be able to map out some plausible scenarios for the future. Some aspects of the future, though, we can only guess at. Two of the most speculative realms are those of politics and technology.

The future of the worst-off peripheral regions could be very bad indeed. They face unprecedented levels of demographic, environmental, and societal stress, with the events of the next 50 years being played out from a starting point of scarce basic resources, serious environmental degradation, overpopulation, disease, unprovoked crime, refugee migrations, and criminal anarchy.

For the world-system core, however, the long-term question is one of relative power and dominance. The same factors that consolidate the advantages of the core as a whole—the end of the cold war, the availability of advanced telecommunications, the transnational reorganization of industry and finance, the liberalization of trade, and the emergence of a global culture—will also open the possibility of a new geopolitical and geoeconomic order, within which the economic and political relationships among core countries might change substantially.

Many aspects of future geographies will depend on trends in demand for resources and on the exploitation of new technologies. The expansion of the world economy and the globalization of industry will undoubtedly boost the overall demand for raw materials of various kinds, and this will spur the development of previously underexploited but resource-rich regions in Africa, Eurasia, and East Asia. Yet raw materials will be only a fraction of future resource needs, however—the main issue, by far, will be energy resources.

It also appears that the present phase of globalization has the potential to create such disparities between the haves and the have-nots (as well as between the core and the periphery) that social unrest will ensue. The evidence of increasing dissatisfaction with the contemporary distribution of wealth both within and between the core and the periphery is widespread. In Russia, where public demonstrations against the government have been outlawed for decades, the new Russians are taking to the streets to protest government policies that favor transnational economic development at the expense of workers' minimum wages. In India, farmers damaged a Kentucky Fried Chicken restaurant for its role in dislocating the domestic poultry industry. In Mexico, industrial workers in transnational plants are challenging the absence of health and safety regulations that leave some exposed to harmful chemicals.

At the same time that protest against globalization and new geographies is being waged, the products of a global economy and culture are widely being embraced. The market for blue contact lenses is growing in such unlikely places as Bangkok and Nairobi; plastic surgery to reshape eyes is increasing in many Asian countries. *Baywatch* and CNN are eagerly viewed in even the most remote corners of the globe. Highway systems, airports, and container facilities are springing up throughout the periphery.

In short, future geographies are being negotiated at this very moment from the board rooms of transnational corporations to the huts of remote villagers. The outcomes of these negotiations are still in the making as we, in our daily lives, make seemingly insignificant decisions about what to wear, what to eat, where to work, how to travel, and how to entertain ourselves. These decisions help to either support or undermine the larger forces at work in the global economy, such as where to build factories, what products to make, or how to package and deliver them to the consumer. In short, future geographies can be very much shaped by us through our understanding of the relatedness of people, places, and regions in a globalized economy.

Main Points

- In some ways, the future is already here, embedded in the world's institutional structures and in the dynamics of its populations.

- New and emerging technologies that are likely to have the most impact in reshaping human geographies include advanced transportation technologies, biotechnology, materials technologies, and information technologies.

- The changes involved in shaping future geographies will inevitably bring some critical issues, conflicts, and threats, including important geographical issues that center on scale, boundaries, and territories, on cultural dissonance, and on the sustainability of development.

Key Terms

carrying capacity (p. 523)
sustainable development (p. 523)

Exercises

 On the Internet

The Internet is especially appropriate to the topic of future geographies, since it represents one of the key sets of technologies that are currently reshaping the interdependencies between people and places throughout the world. The Internet exercises for this chapter explore the relationships between resources, technology, and spatial change. After completing the exercises for this chapter you should have a better understanding of the opportunities and constraints that will shape future geographies.

 Unplugged

1. Using census data, construct a population pyramid (see Chapter 3) for any county or city with which you are familiar. What does this tell you about the future population of the locality?
2. Drawing on what you know about the geography of this locality and its regional, national, and global contexts, construct two scenarios, about 200 words each (see Figure 12.1 for brief examples), for the future, each based on different assumptions about resources and technology.

Glossary

A

Accessibility: the opportunity for contact or for interaction from a given point or location, in relation to other locations.

Acid rain: the wet deposition of acids upon Earth created by the natural cleansing properties of the atmosphere.

Age-sex pyramid: a representation of the population based on its composition according to age and sex.

Agglomeration diseconomies: the negative economic effects of urbanization and the local concentration of industry.

Agglomeration effects: cost advantages that accrue to individual firms because of their location among functionally related activities.

Agrarian: referring to the culture of agricultural communities and the type of tenure system that determines access to land and the kind of cultivation practices employed there.

Agribusiness: a set of economic and political relationships that organizes agro-food production from the development of seeds to the retailing and consumption of the agricultural product.

Agricultural density: ratio between the number of agriculturists per unit of arable land in a specific area.

Agricultural industrialization: process whereby the farm has moved from being the centerpiece of agricultural production to become one part of an integrated string of vertically organized industrial processes including production, storage, processing, distribution, marketing, and retailing.

Agriculture: a science, an art, and a business directed at the cultivation of crops and the raising of livestock for sustenance and for profit.

Ancillary activities: activities such as maintenance, repair, security, and haulage services that serve a variety of industries.

Animistic perspective on nature: the view that natural phenomena-both animate and inanimate-possess an indwelling spirit or consciousness.

B

Baby boom: population of individuals born between the years 1946 and 1964.

Backwash effects: the negative impacts on a region (or regions) of the economic growth of some other region.

Basic functions: economic activities that provide income from sales to customers beyond city limits.

Beaux Arts: a style of urban design that sought to combine the best elements of all of the classic architectural styles.

Biotechnology: technique that uses living organisms (or parts of organisms) to make or modify products, to improve plants and animals, or to develop microorganisms for specific uses.

Buddhist perspective on nature: the view that nothing exists in and of itself and everything is part of a natural complex and dynamic totality of mutuality and interdependence.

C

Carrying capacity: the maximum number of users that can be sustained, over the long term, by a given set of natural resources.

Cartography: the body of practical and theoretical knowledge about making distinctive visual representations of Earth's surface in the form of maps.

Census: count of the number of people in a country, region, or city.

Central business district (CBD): central nucleus of commercial land uses in a city.

Central cities: the original, core jurisdictions of metropolitan areas.

Centrality: the functional dominance of cities within an urban system.

Central place: a settlement in which certain products and services are available to consumers.

Central place theory: a theory that seeks to explain the relative size and spacing of towns and cities as a function of people's shopping behavior.

Centrifugal forces: forces that divide or tend to pull the State apart.

Centripetal forces: forces that strengthen and unify the State.

Chemical farming: application of inorganic fertilizers to the soil and herbicides, fungicides, and pesticides to crops in order to enhance yields.

City Beautiful movement: attempt to remake cities in ways that would reflect the higher values of society, using neoclassical architecture, grandiose street plans, parks, and inspirational monuments and statues.

Clovis point: a flaked, bifaced projectile whose length is more than twice its width.

Cognitive distance: the distance that people perceive to exist in a given situation.

Cognitive images (mental maps): psychological representations of locations that are made up from people's individual ideas and impressions of these locations.

Cognitive space: space defined and measured in terms of people's values, feelings, beliefs, and perceptions about locations, districts, and regions.

Cohort: a group of individuals who share a common temporal demographic experience.

Colonial cities: cities that were deliberately established or developed as administrative or commercial centers by colonial or imperial powers.

Colonialism: the establishment and maintenance of political and legal domination by a State over a separate and alien society.

Columbian Exchange: interaction between the Old World, originating with the voyages of Columbus, and the New World.

Commercial agriculture: farming primarily for sale, not direct consumption.

Commodity chain: network of labor and production processes beginning with the extraction or production of raw materials and ending with the delivery of a finished commodity.

Comparative advantage: principle whereby places and regions specialize in activities for which they have the greatest advantage in productivity relative to other regions-or for which they have the least disadvantage.

Confederation: a group of States united for a common purpose.

Conglomerate corporations: companies that have diversified into a variety of different economic activities, usually through a process of mergers and acquisitions.

Congregation: the territorial and residential clustering of specific groups or sub-groups of people.

Conservation: the view that natural resources should be used wisely, and that society's effects on the natural world should represent stewardship and not exploitation.

Core regions: regions that dominate trade, control the most advanced technologies, and have high levels of productivity within diversified economies.

Cosmopolitanism: an intellectual and esthetic openness toward divergent experiences, images, and products from different cultures.

Counterurbanization: net loss of population from cities to smaller towns and rural areas.

Creative destruction: the withdrawal of investments from activities (and regions) that yield low rates of profit in order to reinvest in new activities (and new places).

Crop rotation: method of maintaining soil fertility where the fields under cultivation remain the same, but the crop being planted is changed.

Crude birthrate (CBR): ratio of the number of live births in a single year for every thousand people in the population.

Crude death rate (CDR): ratio between the number of deaths in a single year for every thousand people in the population.

Crude density (arithmetic density): total number of people divided by the total land area.

Cultural adaptation: the complex strategies human groups employ to live successfully as part of a natural system.

Cultural complex: combination of traits characteristic of a particular group.

Cultural ecology: the study of the relationship between a cultural group and its natural environment.

Cultural geography: how space, place, and landscape shape culture at the same time that culture shapes space, place, and landscape.

Cultural hearths: the geographic origins or sources of innovations, ideas, or ideologies.

Cultural landscape: a characteristic and tangible outcome of the complex interactions between a human group and a natural environment.

Cultural nationalism: an effort to protect regional and national cultures from the homogenizing impacts of globalization, especially the penetrating influence of U.S. culture.

Cultural region: the area(s) within which a particular cultural system prevails.

Cultural system: a collection of interacting elements that taken together shape a group's collective identity.

Cultural trait: a single aspect of the complex of routine practices that constitute a particular cultural group.

Culture: a shared set of meanings that are lived through the material and symbolic practices of everyday life.

Cumulative causation: a spiral build-up of advantages that occurs in specific geographic settings as a result of the development of external economies, agglomeration effects, and localization economies.

Cycle of poverty: transmission of poverty and deprivation from one generation to another through a combination of domestic circumstances and local, neighborhood conditions.

D

Decolonization: the acquisition, by colonized peoples, of control over their own territory.

Deep ecology: approach to nature revolving around two key components: self-realization and biospherical egalitarianism.

Deforestation: the removal of trees from a forested area without adequate replanting.

Deindustrialization: a relative decline in industrial employment in core regions.

Democratic rule: a system in which public policies and officials are directly chosen by popular vote.

Demographic collapse: phenomenon of near genocide of native populations.

Demographic transition: replacement of high birth and death rates by low birth and death rates.

Demography: the study of the characteristics of human populations.

Dependency ratio: measure of the economic impact of the young and old on the more economically productive members of the population.

Derelict landscapes: landscapes that have experienced abandonment, misuse, disinvestment, or vandalism.

Desertification: the degradation of land cover and damage to the soil and water in grasslands and arid and semiarid lands.

Dialects: regional variations in standard languages.

Diaspora: a spatial dispersion of a previously homogeneous group.

Distance-decay function: the rate at which a particular activity or process diminishes with increasing distance.

Division of labor: the specialization of different people, regions, or countries in particular kinds of economic activities.

Domino theory: if one country in a region chose or was forced to accept a communist political and economic system, then neighboring countries would be irresistibly susceptible to falling to communism.

Double cropping: practice in the milder climates where intensive subsistence fields are planted and harvested more than once a year.

Doubling time: measure of how long it will take the population of an area to grow to twice its current size.

Dualism: the juxtaposition in geographic space of the formal and informal sectors of the economy.

E

East/West divide: communist and noncommunist countries, respectively.

Ecofeminism: the view that partiarchal ideology is at the center of our present environmental malaise.

Eco-migration: population movement caused by the degradation of land and essential natural resources.

Ecological imperialism: introduction of exotic plants and animals into new ecosystems.

Economic base: set of manufacturing, processing, trading or services activities that serve markets beyond the city.

Economies of scale: cost advantages to manufacturers that accrue from high-volume production, since the average cost of production falls with increasing output.

Ecosystem: a community of different species interacting with each other and with the larger physical environment that surrounds it.

Edge cities: nodal concentrations of shopping and office space that are situated on the outer fringes of metropolitan areas, typically near major highway intersections.

Emigration: a move from a particular location.

Environmental determinism: a doctrine that holds that human activities are controlled by the environment.

Environmental ethics: a philosophical perspective on nature that prescribes moral principles as guidance for our treatment of it.

Environmental justice: movement that reflects a growing political consciousness largely among the world's poor that their immediate environs are far more toxic than wealthier neighborhoods.

Ethnicity: a socially created system of rules about who belongs and who does not belong to a particular group based upon actual or perceived commonality.

Ethnocentrism: the attitude that one's own race and culture are superior to others'.

Ethology: the scientific study of the formation and evolution of human customs and beliefs.

Export-processing zones (EPZs): small areas within which especially favorable investment and trading conditions are created by governments in order to attract export-oriented industries.

External economies: cost savings that result from circumstances beyond a firm's own organization and methods of production.

F

Farm crisis: the financial failure and eventual foreclosure of thousands of family farms across the U.S. Midwest.

Fast world: people, places, and regions directly involved, as producers and consumers, in transnational industry, modern telecommunications, materialistic consumption, and international news and entertainment.

Federal State: a form of government in which power is allocated to units of local government within the country.

Fiscal squeeze: increasing limitations on city revenues, combined with increasing demands for expenditure.

Food manufacturing: adding value to agricultural products through a range of treatments-processing, canning, refining, packing, packaging, etc.-that occur off the farm and before they reach the market.

Food chain: five central and connected sectors (inputs, production, outputs, distribution, and consumption) with four contextual elements acting as external mediating forces (the State, international trade, the physical environment, and credit and finance).

Food regime: specific set of links that exists among food production and consumption and capital investment and accumulation opportunities.

Forced migration: movement by an individual against his or her will.

Friction of distance: the deterrent or inhibiting effect of distance on human activity.

G

Gateway city: city that serves as a link between one country or region and others because of its physical situation.

Gender: the social differences between men and women rather than the anatomical differences that are related to sex.

Genre de vie: a functionally organized way of life that is seen to be characteristic of culture groups.

Gentrification: the invasion of older, centrally located working-class neighborhoods by higher-income households seeking the character and convenience of less expensive and well-located residences.

Geodemographic analysis: practice of assessing the location and composition of particular populations.

Geographical imagination: the capacity to understand changing patterns, changing processes, and changing relationships among people, places, and regions.

Geographic information systems (GIS): integrated computer tools for the handling, processing, and analyzing of geographical data.

Geographical path dependence: the historical relationship between the present activities associated with a place and the past experiences of that place.

Geopolitics: the State's power to control space or territory and shape the foreign policy of individual States and international political relations.

Gerrymandering: the practice of redistricting for partisan purposes.

Global change: combination of political, economic, social, historical, and environmental problems at the world scale.

Globalization: the increasing interconnectedness of different parts of the world through common processes of economic, environmental, political, and cultural change.

Globalized agriculture: a system of food production increasingly dependent upon an economy and set of regulatory practices that are global in scope and organization.

Green revolution: the export of a technological package of fertilizers and high-yielding seeds, from the core to the periphery, to increase global agricultural productivity.

Gross domestic product (GDP): an estimate of the total value of all materials, foodstuffs, goods, and services that are produced by a country in a particular year.

Gross migration: the total number of migrants moving into and out of a place, region, or country.

Gross national product (GNP): similar to GDP, but in addition includes the value of income from abroad.

Growth poles: economic activities that are deliberately organized around one or more high-growth industries.

Guest workers: individuals who migrate temporarily to take up jobs in other countries.

H

Hearth areas: geographic settings where new practices have developed, and from which they have subsequently spread.

Hegemony: domination over the world economy, exercised by one national State in a particular historical epoch through a combination of economic, military, financial, and cultural means.

Hinterland: the sphere of economic influence of a town or city.

Histogram: vertical bar graph in which bar height represents frequency and in which width depicts the various frequency categories.

Historical geography: the geography of the past.

Human geography: the study of the spatial organization of human activity and of people's relationships with their environments.

Humanistic approach: places the individual—especially individual values, meaning systems, intentions, and conscious acts—at the center of analysis.

Hunting and gathering: activities whereby people feed themselves through killing wild animals and fish and gathering fruits, roots, nuts, and other edible plants to sustain themselves.

I

Immigration: a move to another location.

Imperialism: the extension of the power of a nation through direct or indirect control of the economic and political life of other territories.

Import substitution: the process by which domestic producers provide goods or services that formerly were bought from foreign producers.

Infant mortality rate: annual number of deaths of infants under one year of age compared to the total number of live births for that same year.

Informal sector: economic activities that take place beyond official record, not subject to formalized systems of regulation or remuneration.

Infrastructure (or fixed social capital): the underlying framework of services and amenities needed to facilitate productive activity.

Initial advantage: the critical importance of an early start in economic development; a special case of external economies.

Intensive subsistence agriculture: practice that involves the effective and efficient use-usually through a considerable expenditure of human labor and application of fertilizer-of a small parcel of land in order to maximize crop yield.

Internal migration: a move within a particular country or region.

International migration: a move from one country to another.

International organization: group that includes two or more States seeking political and/or economic cooperation with each other.

Intertillage: practice of mixing different seeds and seedlings in the same swidden.

Invasion and succession: a process of neighborhood change whereby one social or ethnic group succeeds another.

Islamic perspective on nature: the view that the heavens and Earth were made for human purposes.

Isotropic surface: a hypothetical, uniform plain: flat, and with no variations in its physical attributes.

J

Judeo-Christian perspective on nature: the view that nature was created by God and is subject to God in the same way that a child is subject to parents.

L

Landscape as text: the idea that landscapes can be read and written by groups and individuals.

Language: a means of communicating ideas or feelings by means of a conventionalized system of signs, gestures, marks, or articulate vocal sounds.

Language branch: a collection of languages that possess a definite common origin but have split into individual languages.

Language family: a collection of individual languages believed to be related in their prehistorical origin.

Language group: a collection of several individual languages that are part of a language branch, share a common origin, and have similar grammar and vocabulary.

Latitude: the angular distance of a point on Earth's surface, measured north or south from the equator, which is 0°.

Law of diminishing returns: the tendency for productivity to decline, after a certain point, with the continued application of capital and/or labor to a given resource base.

Leadership cycles: periods of international power established by individual States through economic, political, and military competition.

Life expectancy: average number of years an infant newborn can expect to live.

Localization economies: cost savings that accrue to particular industries as a result of clustering together at a specific location.

Longitude: the angular distance of a point on Earth's surface, measured east or west from the prime meridian (the line that passes through both poles and through Greenwich, England, and which has the value of 0°).

M

Masculinism: the assumption that the world is, and should be, shaped mainly by men, for men.

Mechanization: term applied to the replacement of human farm labor with machines.

Megacity: very large city characterized by both primacy and high centrality within its national economy.

Middle cohort: members of the population 15 to 64 years of age who are considered economically active and productive.

Migration: a move beyond the same political jurisdiction, involving a change of residence, either as emigration, a move from a particular location, or as immigration, a move to another location.

Minisystem: a society with a single cultural base and a reciprocal social economy.

Minority groups: population subgroups that are seen—or that see themselves—as somehow different from the general population.

Mobility: the ability to move, either permanently or temporarily.

Modernity: a forward-looking view of the world that emphasizes reason, scientific rationality, creativity, novelty, and progress.

Modern movement: the idea that buildings and cities should be designed and run like machines.

N

Nation: a group of people often sharing common elements of culture such as religion or language, or a history or political identity.

Nationalism: the feeling of belonging to a nation as well as the belief that a nation has a natural right to determine its own affairs.

Nation-state: an ideal form consisting of a homogeneous group of people governed by their own State.

Natural decrease: difference between CDR and CBR, which is the deficit of births relative to deaths.

Natural increase: difference between the CBR and CDR, which is the surplus of births over deaths.

Nature: a social creation as well as the physical universe that includes human beings.

Neo-colonialism: economic and political strategies by which powerful States in core economies indirectly maintain or extend their influence over other areas or people.

Net migration: the gain or loss in the total population of a particular area as a result of migration.

Nonbasic functions: economic activities that serve a city's own population.

North/South divide: the differentiation made between the colonizing States of the Northern Hemisphere and the formerly colonized States of the Southern Hemisphere.

Nutritional density: ratio between the total population and the amount of land under cultivation in a given unit of area.

O

Offshore financial centers: islands or micro-States that have become a specialized node in the geography of worldwide financial flows.

Old-age cohort: members of the population 65 years of age and older who are considered beyond their economically active and productive years.

Ordinary landscapes (vernacular landscapes): the everyday landscapes that people create in the course of their lives.

Overurbanization: condition in which cities grow more rapidly than the jobs and housing they can sustain.

P

Paleolithic period: the period when chipped stone tools first began to be used.

Pastoralism: subsistence activity that involves the breeding and herding of animals to satisfy the human needs of food, shelter, and clothing.

Peripheral regions: regions with undeveloped or narrowly specialized economies with low levels of productivity.

Plantation: a large landholding that usually specializes in the production of one particular crop for market.

Political ecology: approach to cultural geography that studies human-environment relations through the relationships of patterns of resource use to political and economic forces.

Postmodernity: a view of the world that emphasizes an openness to a range of perspectives in social inquiry, artistic expression, and political empowerment.

Postmodern urban design: style characterized by a diversity of architectural styles and elements, often combined in the same building or project.

Preservation: an approach to nature advocating that certain habitats, species, and resources should remain off-limits to human use, regardless of whether the use maintains or depletes the resource in question.

Primacy: condition in which the population of the largest city in an urban system is disproportionately large in relation to the second- and third-largest cities in that system.

Primary activities: economic activities that are concerned directly with natural resources of any kind.

Producer services: services that enhance the productivity or efficiency of other firms' activities or that enable them to maintain specialized roles.

Proxemics: the study of the social and cultural meanings that people give to personal space.

Pull factors: forces of attraction that influence migrants to move to a particular location.

Push factors: events and conditions that impel an individual to move from a location.

Q

Quaternary activities: economic activities that deal with the handling and processing of knowledge and information.

R

Race: a problematic classification of human beings based on skin color and other physical characteristics.

Range: the maximum distance that consumers will normally travel to obtain a particular product or service.

Rank-size rule: a statistical regularity in city-size distributions of cities and regions.

Reapportionment: the process of allocating electoral seats to geographical areas.

Redistricting: the defining and redefining of territorial district boundaries.

Redlining: the practice whereby lending institutions delimit "bad risk" neighborhoods on a city map and then use the map as the basis for determining loans.

Region: larger-sized territory that encompasses many places, all or most of which share similar attributes in comparison with the attributes of places elsewhere.

Regional geography: the study of the ways in which unique combinations of environmental and human factors produce territories with distinctive landscapes and cultural attributes.

Regionalism: a feeling of collective identity based on a population's politico-territorial identitification within a State or across State boundaries.

Religion: belief system and a set of practices that recognize the existence of a power higher than humans.

Remote sensing: the collection of information about parts of Earth's surface by means of aerial hotography or satellite imagery designed to record data on visible, infrared, and microwave sensor systems.

Rites of passage: the ceremonial acts, customs, practices, or procedures that recognize key transitions in human life such as birth, menstruation, and other markers of adulthood such as marriage.

Romanticism: philosophy that emphasizes interdependence and relatedness between humankind and nature.

S

Sacred space: an area recognized by individuals or groups as worthy of special attention as a site of special religious experiences or events.

Secondary activities: economic activities that process, transform, fabricate, or assemble the raw materials derived from primary activities, or that reassemble, refinish, or package manufactured goods.

Sectionalism: extreme devotion to local interests and customs.

Segregation: the spatial separation of specific population subgroups within a wider population.

Self-determination: the right of a group with a distinctive politico-territorial identity to determine its own destiny, at least in part, through the control of its own territory.

Semiotics: the practice of writing and reading signs.

Semiperipheral regions: regions that are able to exploit peripheral regions but are themselves exploited and dominated by core regions.

Sense of place: feelings evoked among people as a result of the experiences and memories that they associate with a place, and to the symbolism that they attach to it.

Sexuality: set of practices and identities that a given culture considers related to each other and to those things it considers sexual acts and desires.

Shifting cultivation: a system in which farmers aim to maintain soil fertility by rotating the fields within which cultivation occurs.

Shock city: city that is seen as the embodiment of surprising and disturbing changes in economic, social, and cultural life.

Siltation: the buildup of sand and clay in a natural or artificial waterway.

Site: the physical attributes of a location: its terrain, soil, vegetation, water sources, for example.

Situation: the location of a place relative to other places and human activities.

Slash-and-burn: system of cultivation in which plants are cropped close to the ground, left to dry for a period, and then ignited.

Slow world: people, places, and regions whose participation in transnational industry, modern telecommunications, materialistic consumption, and international news and entertainment is limited.

Society: sum of the inventions, institutions, and relationships created and reproduced by human beings across particular places and times.

Sovereignty: the exercise of State power over people and territory, recognized by other States and codified by international law.

Spatial diffusion: the way that things spread through space and over time.

Spread effects: the positive impacts on a region (or regions) of the economic growth of some other region.

Squatter settlements: residential developments that take place on land that is neither owned nor rented by its occupants.

States: independent political units with territorial boundaries that are internationally recognized by other States.

Subsistence agriculture: farming for direct consumption by the producers, not for sale.

Suburbanization: growth of population along the fringes of large metropolitan areas.

Supranational organizations: collections of individual States with a common goal that may be economic and/or political in nature and which diminish, to some extent, individual State sovereignty in favor of the group interests of the membership.

Sustainable development: a vision of development that seeks a balance among economic growth, environmental impacts, and social equity.

Swidden: land that is cleared through slash-and-burn and is ready for cultivation.

Symbolic landscapes: representations of particular values and/or aspirations that the builders and financiers of those landscapes want to impart to a larger public.

T

Taoist perspective on nature: the view that nature be valued for its own sake, not for how it might be exploited.

Technology: physical objects or artifacts, activities or processes, and knowledge or know-how.

Technology systems: clusters of interrelated energy, transportation, and production technologies that dominate economic activity for several decades at a time.

Territorial organization: a system of government formally structured by area, not by social groups.

Territoriality: the persistent attachment of individuals or peoples to a specific location or territory.

Territory: the delimited area over which a State exercises control and which is recognized by other States.

Tertiary activities: economic activities involving the sale and exchange of goods and services.

Threshold: the minimum market size required to make the sale of a particular product or service profitable.

Time-space convergence: the rate at which places move closer together in travel or communication time or costs.

Topological space: space that is defined and measured in terms of the nature and degree of connectivity between locations.

Topophilia: the emotions and meanings associated with particular places that have become significant to individuals.

Total fertility rate (TFR): average number of children a woman will have throughout the years that demographers have identified as her childbearing years, approximately ages 15 through 49.

Transcendentalism: a philosophy in which a person attempts to rise above nature and the limitations of the body to the point where the spirit dominates the flesh.

Transhumance: movement of herds according to seasonal rhythms: warmer, lowland areas in the winter; cooler, highland areas in the summer.

Transnational corporation: company with investments and activities that span international boundaries and with subsidiary companies, factories, offices, or facilities in several countries.

U

Underclass: subset of the poor isolated from mainstream values and the formal labor market.

Underemployment: when people work less than full time even though they would prefer to work more hours.

Unitary State: a form of government in which power is concentrated in the central government.

Urban ecology: the social and demographic composition of city districts and neighborhoods.

Urban form: the physical structure and organization of cities.

Urban system: an interdependent set of urban settlements within a specified region.

Urbanism: the way of life, attitudes, values, and patterns of behavior fostered by urban settings.

Urbanization economies: external economies that accrue to producers because of the package of infrastructure, ancillary activities, labor, and markets typically associated with urban settings.

Utility: the usefulness of a specific place or location to a particular person or group.

V

Virgin soil epidemics: conditions in which the population at risk has no natural immunity or previous exposure to the disease within the lifetime of the oldest member of the group.

Vital records: information about births, deaths, marriages, divorces, and the incidence of certain infectious diseases.

Voluntary migration: movement by an individual based on choice.

W

World cities: cities in which a disproportionate part of the world's most important business—economic, political, and cultural—is conducted.

World-empire: minisystems that have been absorbed into a common political system while retaining their fundamental cultural differences.

World-system: an interdependent system of countries linked by economic and political competition.

Y

Youth cohort: members of the population who are less than 15 years of age and generally considered to be too young to be fully active in the labor force.

Z

Zone in transition: area of mixed commercial and residential land uses surrounding the CBD.

Credits

Photo Credits

T = top; C = center; B = bottom; L = left; R = Right

Chapter 1

Page 1: Library of Congress.

Page 2: National Geographic Society.

Page 3: Garry Trudeau/Universal Press Syndicate.

Page 5: (TL) Rafael Macia/Photo Researchers, Inc.; (TC) Paul L. Knox; (TR) Geoff Juckes/The Stock Market; (B) Corbis Corporation.

Page 7: (T, C, B) Paul L. Knox.

Page 8: The Granger Collection.

Page 9: Library of Congress.

Page 11: (T) Library of Congress; (B) The University of Chicago Press.

Page 12: (L) Library of Congress; (C) Paul L. Knox; (R) Corbis-Bettmann.

Page 13: (L) Corbis-Bettmann; (R) Clark University.

Page 14: The Granger Collection.

Page 17: Corbis Corporation.

Page 18: (TL) Van Bucher/Photo Researchers, Inc.; (TR) Steve McCurry/Magnum Photos, Inc.; (BL) David Turnley/Black Star; (BR) Gerd Ludwig/National Geographic Society.

Page 25: (TL) Earth Observation Satellite Company (EOSAT); (TR) REGIS; (B) Tom Van Sant/The GeoSphere Project.

Page 33: Paul L. Knox.

Page 35: Paul L. Knox.

Page 39: Comstock.

Page 41: Paul L. Knox.

Page 47: (all) Paul L. Knox.

Page 48: Paul L. Knox.

Chapter 2

Page 51: RAGA/The Stock Market.

Page 56: (TR) UPI/Corbis-Bettmann; (BL) Swanstock, Inc..

Page 57: Food and Agriculture Organization of the UN.

Page 58: Macduff Everton/Swanstock, Inc..

Page 59: (L, R) Food and Agriculture Organization of the UN.

Page 60: (clockwise from TR) Paul L. Knox; Brian Brake/Photo Researchers; Larry Mulvehill/Photo Researchers, Inc.; Macduff Everton/Swanstock, Inc.

Page 61: (TL, TR, CL, CR, BL) Paul L. Knox; (BR) Macduff Everton/Swanstock, Inc.

Page 64: National Maritime Museum.

Page 66: Library of Congress.

Page 67: (T, BL) National Maritime Museum; (BR) Wesley Docxe/Photo Researchers, Inc.

Page 68: Paul L. Knox.

Page 69: (T) Topham Picture Source; (B) Joe Viesti/Viesti Associates, Inc.

Page 71: (TR, BL) Library of Congress; (BR) United States Geological Survey.

Page 73: (T, B) Paul L. Knox.

Page 75: (clockwise from CL) Marion Post Wolcott/Center for Creative Photography; Library of Congress; Library of Congress; Peter Vadnai/The Stock Market; Library of Congress.

Page 76: (CL, CR, BR) Library of Congress; (BL) The London News/ Library of Congress.

Page 77: (C) Vince Streano/Corbis Media; (BL, BR) Library of Congress.

Page 79: (R) Corbis-Bettmann; (L) Library of Congress.

Page 81: (TL, BC) Paul L. Knox; (TR) Weightman/NCGE-GPN Master Photo; (BL) Library of Congress; (BR) Gerhardt Liebmann/Photo.

Page 89: Alex S.MacLean/Peter Arnold, Inc.

Page 90: The Port Authority of New York & New Jersey.

Page 91: Philippe Brylak/Gamma-Liaison, Inc.

Chapter 3

Page 97: Paul L. Knox

Page 103: Woodruff T. Sullivan, III;

Page 105: (L) Victor Englebert/Photo Researchers, Inc.; (R) Paul L. Knox.

Page 106: Macduff Everton/Swanstock, Inc.

Page 116: Mickey Pallas/Center for Creative Photography.

Page 118: Amy Sancetta/AP/Wide World Photos.

Page 119: Paolo Koch/Photo Researchers, Inc.

Page 131: AP/Wide World Photos.

Page 132: Steve McCurry/National Geographic Society.

Page 135: FPG International.

Page 138: Thomas S.England/Photo Researchers, Inc.

Page 143: J. C. Aunos/Gamma-Liaison, Inc.

Page 144: Andrew Holbrooke/The Stock Market.

Chapter 4

Page 147: Mike Yamashita/Westlight.

Page 149: Douglas Brooker/Swanstock, Inc.

Page 150: UPI/Corbis-Bettmann.

Page 151: (L) Peter Menzel/Material World; (R) Miguel Luis Fairbanks/Material World.

Page 154: (T) New York State Library; (B) Gail Oskin/AP/Wide World Photos.

Page 158: (L) New York Public Library; (R) Dover Publications, Inc.

Page 160: Sygma.

Page 161: (T) David Ake/UPI/Corbis-Bettmann; (B) Little, Brown & Co.

Page 162: National Geographic Society.

Page 163: Sher/Anthro-Photo.

Page 164: Skeet McAuley/Swanstock, Inc.

Page 169: Emily Young.

Page 172: (T) Mark Nichter; (C) Paul L. Knox.

Page 173: Greg Gibson/AP/Wide World Photos.

Page 177: Contrast/Gamma-Liaison, Inc.

Page 179: Will & Deni McIntyre/Photo Researchers, Inc.

Page 183: Neil Smith.

Page 184: (T) Neil Smith; (B) United States Geological Survey.

Page 190: Sylvain Grandadam/Photo Researchers, Inc.

Chapter 5

Page 189: Corbis-Bettmann.

Page 192: (T) Louis Carlos Bernal/Center for Creative Photography; (L) Mimi Nichter; (R) Kaku Kurita/Gamma-Liaison, Inc.; (B) University of California, Berkeley.

Page 194: John Wiley & Sons, Inc.

Page 195: (T) John Wiley & Sons, Inc.; (B) Blaine Harrington/The Stock Market.

Page 196: Bill Hess/National Geographic Society.

Page 200: Barry Jarvinen/AP/Wide World Photos.

Page 210: Graciela Iturbide/Center for Creative Photography.

Page 212: (L, R) Michael Bonine.

Page 213: Christopher A. Klein/National Geographic Society.

Page 216: Saul Bromberger/Swanstock, Inc.

Page 217: (T) Chandrekant Trivedi; (B) Boston Pilot, March 18, 1870.

Page 218: (L) Steve Elmore/The Stock Market; (R) Association of American Geographers.

Page 219: Andrew Holbrooke/The Stock Market.

Page 222: Sallie A. Marston.

Page 223: Emily Young.

Page 226: Marvin Waterstone.

Chapter 6

Page 231: Jon-Marc Seimon/Rim Pacific.

Page 234: Cindi Katz.

Page 236: United States Department of Agriculture.

Page 237: Sherman Hines/Masterfile Corporation.

Page 238: Paul L. Knox.

Page 240: Paul L. Knox.

Page 241: Stuart L. Craig, Jr./Bruce Coleman, Inc.

Page 242: (T) Paul L. Knox; (B) MIT Press.

Page 247: (T, C, B) Cindi Katz.

Page 248: (C) Magdalene College; (B) Paul L. Knox.

Page 249: (T) National Maritime Museum; (others) Paul L. Knox.

Page 250: (T) Sallie A. Marston; (B) Paul L. Knox.

Page 251: Old Tucon Studios.

Page 252: Paul L. Knox.

Page 253: J. Holston/University of Chicago Press.

Page 255: (L) George Holton/Photo Researchers, Inc.; (R) Bernard Boutrit/Woodfin Camp & Associates.

Page 256: Carol Balacek

Page 257: M.P.L.Fogden/Bruce Coleman, Inc..

Page 260: (all) Paul L. Knox.

Page 262: (T) Mike Yamashita/Westlight; (B) Simon Jauncey/Tony Stone Images.

Page 263: Alex S. MacLean/Landslides.

Chapter 7

Page 269: Michael S. Yamashita/National.

Page 273: Sergio Dorantes/Sygma.

Page 278: Paul L. Knox.

Page 289: Laura Elliott/Comstock.

Page 296: Paul L. Knox.

Page 301: Honda of America Mfg., Inc.

Page 310: (TL) CNES/Maptec International; (TR, BL, BR) Paul L. Knox.

Page 312: McCutcheon/Visuals Unlimited.

Page 313: Jonathan Blair/Corbis Media.

Page 314: (L) EcoTraveler; (R) Brian Betts/Photographics Kaikoura.

Chapter 8

Page 317: Paulo Fridman/Sygma.

Page 320: (T, B) Food and Agriculture Organization of the UN.

Page 326: (T) Curt Carnemark/World Bank; (C) Grant Heilman/Grant Heilman Photography, Inc.; (BL) James P. Blair/National Geographic Society; (BR) Food and Agriculture Organization of the UN.

Page 327: FAO.

Page 329: FAO.

Page 331: Wolfgang Kaehler/Wolfgang Kahler Photography.

Page 332: Bernard Pierre Wolfe/Photo Researchers, Inc.

Page 334: United States Department of Agriculture.

Page 335: (L, R) United States Department of Agriculture.

Page 336: (L, R) United States Department of Agriculture.

Page 338: (L, R) Diana Liverman.

Page 342: FAO.

Page 345: Food and Agriculture Organization of the UN.

Page 347: Judith Carney, Professor.

Page 348: United States Department of Agriculture.

Page 349: Food and Agriculture Organization of the UN.

Page 351: (T) United States Department of Agriculture; (B) L. Kiff/Visuals Unlimited.

Page 352: United States Department of Agriculture.

Page 353: Vince Mannino/UPI/Corbis-Bettmann.

Page 354: United States Geological Survey.

Chapter 9

Page 357: Reuters/Corbis-Bettmann.

Page 361: (L) Sarah Leen/Matrix International; (R) U.S. Geological Survey, U.S. Department of the Interior.

Page 362: (T) Paul L. Knox; (B) Ahn Young-joon/AP/Wide World Photos.

Page 363: Paul L. Knox.

Page 374: P. Le Segretain/Sygma.

Page 379: Hulton-Deutsch Collection/Corbis Media.

Page 380: Thomas Coe/Agence France Presse/Corbis-Bettmann.

Page 398: Sallie A. Marston.

Page 400: Tony Korody/Sygma.

Page 404: Stock Montage, Inc./Historical Pictures Collection.

Chapter 10

Page 407: Owen Franken/Corbis Media.

Page 412: (T) Anthony Miles/Bruce Coleman, Inc.

Page 414: (B) Paul L. Knox.

Page 413: Georg Gerster/Comstock.

Page 414: (clockwise from TL) Paul L. Knox; Paul L. Knox; UPI/Corbis-Bettmann; Paul L. Knox; UPI/Corbis-Bettmann; E. O. Hoppe/Corbis Media; Otis Imboden/National Geographic Society; Macduff Everton/Swanstock, Inc.

Page 415: (T, BL) Paul L. Knox; (C) AP/Wide World Photos; (BR) Macduff Everton/Swanstock, Inc.

Page 417: (L) Tria Giovan/Swanstock, Inc.; (C, R) Library of Congress.

Page 419: (L, R) Paul L. Knox.

Page 420: (TL) Library of Congress; (TR) Topham Picture Source; (CR, BL, BR) Paul L. Knox.

Page 421: (TL) Stock Montage, Inc./Historical Pictures Collection; (TC, TR) Library of Congress; (C) Paul L. Knox.

Page 431: (TL) Fasol/Explorer/Photo Researchers, Inc.; (TC) Macduff Everton/Swanstock, Inc.; (BC) Sandra Baker/Swanstock, Inc.; (R) Katsumi Kasahara/AP/Wide World Photos; (BL) Joe Viesti/Viesti Associates, Inc.

Page 436: (from TL) Ed Huffman; Les Stone/Sygma; Paul L. Knox; Dan Dennehy/Swanstock, Inc.; Kjell B. Sandved/Visuals Unlimited.

Page 438: (T, C, B) John Browder.

Chapter 11

Page 441: Paul L. Knox.

Page 443: (T, B) Paul L. Knox.

Page 444: Paul L. Knox.

Page 445: (T, BL) Paul L. Knox; (BR) Stepane Compoint/Sygma.

Page 451: Roux/Explorer/Photo Researchers, Inc.

Page 452: Paul L. Knox.

Page 453: Paul L. Knox.

Page 454: Paul L. Knox.

Page 455: (T, C) Paul L. Knox; (B) Gearóid Ó Tuathail.

Page 456: (T) PNI; (C) Anthony King, Professor; (B) Jeffrey Alford/Asia Access.

Page 458: (from T) Ted Spiegel/Black Star; Paul L. Knox; World Bank; Food and Agriculture Organization of the UN.

Page 460: (TL) Andrea Braun Byrne/Swanstock; (TR, CR, BR) Paul L. Knox; (CL) Sandra Baker/Swanstock, Inc.; (BL) Dean Conger/National Geographic Society.

Page 461: (TL) Chromosohm/Sohm/Photo Researchers, Inc.; (TR) Wolfgang Kahler Photography; (B) John Humble/Swanstock, Inc.

Page 462: (T, C, BR) Michael Bonine; (BL) Stringer, Mousa Al-Shaer/Reuters/Corbis-Bettmann.

Page 463: Michael Bonine.

Page 464: (L) Paul L. Knox; (R) Michael Bonine.

Page 465: (TR, CR) Paul L. Knox; (BR) James L. Stanfield/National Geographic Society.

Page 466: Stock Montage, Inc./Historical Pictures Collection.

Page 467: D. Burnham, E. Bennett/The Art Institute of Chicago.

Page 469: Paul L. Knox.

Page 470: Francois Gohier/Photo Researchers, Inc.

Page 471: (T) Alex S. MacLean/Landslides; (C) Paul L. Knox.

Page 474: (T) Food and Agriculture Organization of the UN; (BL, BR) World Bank.

Page 475: (T, CR, BL) Efiong Etuk; (CL, BR) World Bank.

Page 476: (T, B) Paula Pentel.

Page 477: U.S. Department of Housing and Urban Development.

Page 479: Tom Arndt/Swanstock, Inc.

Page 481: (TL) Spidle/World Bank; (TR) Paul L. Knox; (BR) Ray Witlin/World Bank; (BL) Matsumoto/Sygma.

Page 483: (L) World Bank; (R) Huffman/World Bank.

Page 484: (T) Stephanie Maze/National Geographic Society; (BL) Alain Evrard/Photo Researchers, Inc.; (BR) Paul L. Knox.

Page 485: (L) Ed Huffman/World Bank; (R) Food and Agriculture Organization of the UN.

Page 486: (TR) Curt Carnemark/World Bank; (TL) Mary Heller/Swanstock, Inc.; (BL) Ray Witlin/World Bank; (BR) World Bank.

Chapter 12

Page 493: Gregory MacNicol/Photo Researchers, Inc.

Page 497: (TL) Mark Wilson/AP/Wide World Photos; (TC) Jim Witmer/Dayton Daily News/Gamma-Liaison, Inc.; (TR) Pat & Tom Leeson/Photo Researchers, Inc.; (BL) Moe Doiron/Sygma; (BR) Jim Hollander/Reuters/Corbis-Bettmann.

Page 503: (TL, BL) Food and Agriculture Organization of the UN; (C) Kevin Carter/Sygma.

Page 505: Edward Kim/National Geographic Society.

Page 508: (T) Van Parys/Sygma.

Page 509: (all) Beth Mitchneck.

Page 510: Patrick Robert/Sygma.

Page 512: Max Nash/AP/Wide World Photos.

Page 513: Aaron D. Schroeder.

Page 514: United States Department of Agriculture.

Page 516: David A. Harvey/National Geographic Society.

Page 519: (L) Mike Goldwater/Matrix International; (R) P. Robert/Sygma.

Page 521: (L) Paul L. Knox; (R) James Patelli.

Text Credit

Chapter 10, page 421: Excerpt from "Chicago" in CHICAGO POEMS by Carl Sandburg, copyright 1916 by Holt, Rinehart, and Winston, Inc. and renewed 1944 by Carl Sandburg, reprinted by permission of Harcourt Brace & Company.

Index